THE LIBRARY
ST. MARY'S COLLEGE OF MARYLAND
ST. MARY'S CITY, MARYLAND 20686

The Mathematics of Plato's Academy

Raphael's fresco *Causarum Cognitio* in the Vatican's Stanza della Segnatura represents a high-water mark of the myths surrounding Plato. The scene is a lofty and imposing building, and the composition is dominated by Plato, holding his *Timaeus* and pointing upwards with his right hand (up to the understanding of causes?), in lively and public discussion with Aristotle, who holds his *Ethics*. Geometry is clearly represented by the group at the right front; astronomy is behind them, where the king and the globe illustrate a common confusion between the astronomer Ptolemy and the dynasty of Macedonian rulers of Egypt; and music occupies the left foreground, indicated by a tablet containing a harmonic system. Arithmetic is less easily identified. Perhaps the only popular feature of the tradition that is not to be found in the fresco is an inscription 'Let no one unskilled in geometry enter'. The common name of the fresco, 'The School of Athens', has no better authority than a seventeenth-century French guidebook.

For a discussion of the fresco, see E. H. Gombrich, *Symbolic Images: Studies in the Art of the Renaissance*, 85–101. Some of our scanty evidence about Plato's Academy and the other schools of Athens will be discussed in Chapters 4 and 6 of this book.

The Mathematics of Plato's Academy

A New Reconstruction

D. H. FOWLER

*Mathematics Institute
University of Warwick
Coventry*

CLARENDON PRESS · OXFORD
1987

Oxford University Press, Walton Street, Oxford OX2 6DP
Oxford New York Toronto
Delhi Bombay Calcutta Madras Karachi
Petaling Jaya Singapore Hong Kong Tokyo
Nairobi Dar es Salaam Cape Town
Melbourne Auckland

and associated companies in
Beirut Berlin Ibadan Nicosia

Oxford is a trade mark of Oxford University Press

Published in the United States
by Oxford University Press, New York

© D. H. Fowler, 1987

All rights reserved. No part of this publication may be reproduced,
stored in a retrieval system, or transmitted, in any form or by any means,
electronic, mechanical, photocopying, recording, or otherwise, without
the prior permission of Oxford University Press.

British Library Cataloguing in Publication Data
Fowler, D. H.
The mathematics of Plato's Academy: a new
reconstruction.
1. Mathematics, Greek
I. Title
510'.938   QA22
ISBN 0-19-853912-6

Library of Congress Cataloging in Publication Data
Fowler, D. H.
The mathematics of Plato's Academy
Bibliography: p.
Includes indexes
1. Mathematics, Greek. 2. Ratio and proportion.
3. Geometry—Early works to 1800. I. Title.
II. Title: Plato's Academy.
QA22.F69 1987   509.38   86-33260
ISBN 0-19-853912-6

Typeset and printed in Northern Ireland by
The Universities Press (Belfast) Ltd

For

Annette Gruner Schlumberger
Maryse Stroh
Denise Fowler Stroh

Plato greatly advanced mathematics in general and geometry in particular because of his zeal for these studies. It is well known that his writings are thickly sprinkled with mathematical terms and that he everywhere tries to arouse admiration for mathematics among students of philosophy.

> Proclus quoting Eudemus, *Commentary on the First Book of Euclid's Elements*, tr. G. Morrow, p. 66.

In the 'Cloisters' of the Metropolitan Museum in New York there hangs a magnificent tapestry which tells the tale of the Unicorn. At the end we see the miraculous animal captured, gracefully resigned to his fate, standing in an enclosure surrounded by a neat little fence. This picture may serve as a simile for what we have attempted here. We have artfully erected from small bits of evidence the fence inside which we hope to have enclosed what may appear as a possible, living creature. Reality, however, may be vastly different from the product of our imagination; perhaps it is vain to hope for anything more than a picture which is pleasing to the constructive mind when we try to restore the past.

> O. Neugebauer, *The Exact Sciences in Antiquity*, Chapter 6.

PREFACE

Theon of Smyrna, a neo-Platonic commentator of the third century AD, wrote a book *Expositio Rerum Mathematicarum ad Legendum Platonem Utilium*; with a little less hubris than Theon, I regard this book here as 'An Account of Mathematical Topics that May be Useful for Reading Plato'. But, at the outset, I must emphasise that, like Theon, I do not deal with Plato's philosophy. My objective is to provide a background, developed around a new interpretation of aspects of early Greek mathematics, against which to set some of our evidence about the activities of Plato and his colleagues in the Academy. The exposition is divided into three parts, each with its own introduction; the following brief preface, together with these introductions, will give a summary of the aims, contents, and organisation of the book.

The subject matter is early Greek mathematics, a phrase I shall use to denote the phase of developments that culminates in the works of Euclid and Archimedes. Part One (Chapters 1–5) sets out, in some detail, an exploration of some mathematical topics suggested by ideas found in the works of Plato, Aristotle, Euclid, and Archimedes and carried out using methods derived from their procedures; and it proposes that similar investigations might, indeed, have taken place and had some influence on the development of early Greek mathematics. The main theme, though not in its treatment here, is 'well known', as mathematician's jargon expresses it: it has been intensively studied since the seventeenth century, it can be found in readily available publications, and it has, from time to time, played an important role in the development of parts of mathematics. Yet it is possible to emerge from a modern training as a fully qualified member of the mathematical community without any knowledge whatsoever of this material. Every mathematician knows the algorithm which generates the greatest common factor of two numbers: at each step, we subtract the smaller as many times as possible from the larger to leave a still smaller remainder. When the procedure terminates, the last non-zero remainder is the greatest common factor. What we shall mainly be studying, usually in the context of two line segments rather than two numbers, is the relationship between the two original quantities and the sequence of numbers that describes how many subtractions have been performed at each stage. The original pair of numbers or lines clearly determines this pattern of subtractions; but conversely, we may ask, what is it about the original pair that is determined by this pattern? Can the same pattern arise from two different pairs and, if so, how are they related?

The algorithm and the theory associated with the pattern of subtractions are now generally called the Euclidean algorithm and continued fractions. However, the word algorithm is a recent corruption of a fine old word algorism (derived from al-Khwārizmi); and I shall eventually argue that Euclid's role in the development of our understanding of the process may have been far from positive. Moreover the treatment of the material within Greek mathematics will not permit the introduction of the fractions or real numbers that are needed for the standard descriptions of continued fractions; and so on. For these and other reasons, some to be described more fully in Chapter 2, I shall refer to the cluster of ideas and procedures around this Greek algorithm (sic) by the word 'anthyphairesis', the transcription of its Greek description by the perfectly ordinary word ἀνθυφαίρεcιc, which can be roughly translated as 'reciprocal subtraction', the subtraction of one thing from another.

The main theme of Part One is the definition and use of ratio, which is described by Euclid, at *Elements* V Definition 3, as "A sort of relation in respect of size between two magnitudes of the same kind". By meditating on the questions posed at the end of the last paragraph but one, we see that anthyphairesis provides one such definition of ratio; then astronomy suggests another, and music theory another. These ideas will be developed and related to each other, and this material will be connected to the description of a mathematical curriculum that Plato sets out in *Republic* VII.

Part Two, which could equally well have been placed before Part One, discusses some aspects of our evidence about the period. Chapter 6 takes a very general point of view; it examines the evidence for one well-known and representative story about Plato's Academy, and then looks at the transmission of Greek texts in general down to our time. Then, in Chapter 7, a single topic, the treatment of numbers and fractions, is examined in fine detail, for it is out of our underlying experience with arithmetic that much of mathematics grows. In order to highlight my argument that early Greek mathematics is non-arithmetised I shall examine its obverse, arithmetised mathematics and calculation with fractions, and will argue that there is a systematic and very widespread misunderstanding about Greek arithmetical calculations. What is often taken as proven fact is, at best, an interpretation not based on any clear evidence.

Parts One and Two combine, I believe, to set out a very plausible argument that the study of various different ways of describing ratio might have played an important role in the development of mathematics within Plato's Academy; yet there seems to be no explicit mention of this subject, nor any of its techniques, discoveries, problems, and pitfalls, in any surviving Greek source of any kind. In Part Three, Chapter 8, I discuss a few of the discrepancies between what later Greek commentators tell us about Greek mathematics and what we find in our surviving early evidence. And finally, in Chapter 9, I sketch the theory and story of the development, since the seventeenth century, of the mathematics underlying my reconstruction here of anthyphairetic ratio theory, and finish by reflecting on the fact that it, too, after having intrigued and influenced Fermat (perhaps), Brouncker, Wallis, Huygens, Euler, Lambert, Lagrange, Legendre, Gauss, Galois, and many others, has subsequently been left to one side and omitted from the basic training of mathematicians, and so has been neglected by many mathematicians and historians of today.

The problems posed by trying to describe and develop mathematics that may be unfamiliar to many people, within a very novel, initially implied, historical context, have driven me, after a series of unsatisfactory rejected drafts, into adopting an unconventional expository technique. The book includes four dialogues that contain, in condensed form, the exposition of a lot of material. A substantial part of the text is then an elaboration of and commentary on material introduced in these dialogues.

The referencing system I adopt throughout does not conform to the different conventions of either classicists or mathematicians, but I hope that it can be immediately understood and used by everybody. Every book or article will normally be referred to by its author and an acronym of its title (in italic for a book, roman for an article) with, as necessary, volume, page, line, and note numbers; then these acronyms are expanded in the Bibliography. So, for example, Heath, *HGM* ii, 50–6, refers to pages 50 to 56 of volume 2 of T. L. Heath's *History of Greek Mathematics,* a discussion of Archimedes, *Measurement of a Circle.* Modern editions of ancient authors also fit into the same system, but with the refinement that the editor's name will be added, as in the example Archimedes–Heiberg, *Opera* i, 232–43, again a reference to *Measurement of a Circle.* Collected works such as this, which contain the texts in their original language, ancient or modern, with or without translations or commentaries, will be identified by a keyword: *Opera, Works, Werke,* etc. Where a particular translation of a text is referred to, the translator will be given in the same way; for example Archimedes–Heath, *WA,* 91–8, for this same work. If at any point in the text it seems that a fuller title will provide an extra grain of information, then it will be given; if an author's name requires initials for clarification, they will be given. Papyrological publications will be cited in their own standard way; for details of this, see Chapter 6, footnote 18. Euclid, Plato, and Aristotle, our major sources, need special mention. Of the nine volumes (in ten parts) of the critical edition of Euclid's surviving works, edited by J. L. Heiberg with H. Menge and M. Curtze, here called Euclid–Heiberg, *Opera,* five volumes (in six parts) have been reissued in a form described on its title page as "*Euclidis Elementa* post J. L. Heiberg edidit E. S. Stamatis"—this will be described as Euclid–Stamatis, *EE*; and there is a standard, exemplary, and readily available three-volume English translation and commentary to which I hope every half-way serious reader will have access, which will be called Heath, *TBEE* (= *The Thirteen Books of Euclid's Elements*), not Euclid–Heath, *TBEE*. 'The *Elements*' will always be Euclid's *Elements*, and most translations of passages from the *Elements* will be from Heath, *TBEE*, though some are taken from the recent comprehensive commentary, Mueller, *PMDSEE*. Similarly, Plato's and Aristotle's works will generally be referenced by title only, without editor or translator, through the translators of extended passages will be identified, often in the notes; most of these translations will be taken from Plato, *CD* and Aristotle–Ross, *WA*, now available in a convenient revised two-volume version, Aristotle–Barnes, *CW*. Three basic English reference works will be cited without author or editor: the *DSB, OCD,* and *OED* (= *The Dictionary of Scientific Biography, The Oxford Classical Dictionary,* and *The Oxford English Dictionary*). Further notes on this kind of referencing system, which I first saw and appreciated in de Ste. Croix, *CSAGW*, are given in the Bibliography. In this way, I hope that any text that is cited anywhere can be quickly identified by anybody; but whether some of these texts can then be located and consulted is quite a different matter, which may need the assistance of a very specialised library. I must, at the outset, record my gratitude to my own university library, the Ashmolean Library, and the Bodleian Library for their generous and expert help in this respect.

The book finishes with an Index of Cited Passages, an Index of Names, and a conventional General Index.

Direct quotations of texts are either displayed, or run into the text and enclosed in *double* quotation marks "..."; single quotation marks are used to set off words or phrases. I have used algebraic formulae, in contexts referring to Greek mathematics, only as a shorthand for more long-winded *geometrical* descriptions; this is explained in several places, for example Sections 1.2(e), 3.2, 4.5(b) and (e), and throughout Chapter 5. The examples just given illustrate the system of cross-referencing used in the book, and the chapters and sections are indicated throughout in the running heads.

I would like to thank the very many people, far more than I can possibly name here, who have helped me write this book. If I do mention some by name, I hope that they will not be inhibited thereby in their criticism, which has so much helped in the past, and that those who I do not name will not feel that their help is no less appreciated. My interest in Greek mathematics was kindled by reading *The Evolution of the Euclidean Elements,* and its author Wilbur Knorr has since provided further encouragement, information, and advice. Andrew Barker, Alan

Bowen, Malcolm Brown, Ivor Bulmer-Thomas, Harold Cherniss, Roger Herz-Fischler, Geoffrey Lloyd, Bernard Teissier, and Tom Whiteside have kept up a steady stream of help, encouragement, and material. It soon became apparent that I needed to enter new and vast fields of scholarship, and I wish to acknowledge warmly the help that has been given and the interest that has been shown by many classicists, and how much I have come to appreciate their information and insights. Very particular thanks are due to the papyrologists, especially Sir Eric Turner (1911–83) and his colleagues Walter Cockle, Peter Parsons, and John Tait; and I have received special assistance from Alain Blanchard, William Brashear, Myles Burnyeat, Patrice Cauderlier, Jim Coulton, D. R. Dicks, Tiziano Dorandi, Brenda Farr, Hermann Harrauer, Ludwig Koenen, Tom Pattie, Bob Sharples, David Thomas, and Nigel Wilson. The community of mathematicians and historians of mathematics has looked on, sometimes rather bemused, and offered advice and help of all kinds; notable have been E. M. Bruins, J. W. S. Cassels, Donald Coxeter, Kay Dekker, John Fauvel, Bernard Goldstein, Bill Gosper, Ivor Grattan-Guinness, Jan Hogendijk, Jan van Maanen, John Mason, Joe and Barry Mazur, Ian Mueller, David Mumford, Mike Paterson, Paddy Patterson, Mark Rafter, Rolph Schwarzenberger, Jacques Sesiano, Jeff Smith, B. L. van der Waerden, and many of my immediate colleagues of whom I should like particularly to thank, for more that I could ever hope to enumerate, Elaine Shiels and Christopher Zeeman. The staff of Oxford University Press have given me comfort, encouragement, and expert help, and have indulged my idiosyncracies. The inclusion of several items, including the frontispiece, was made possible by a grant from the Fondation Les Treilles.

The book itself has been written in three very special places, each with its own presiding genius, to whom it is jointly and affectionately dedicated.

Les Treilles, Tourtour,   D. H. F.
Les Côtes du Plan, Ollioules,   *July* 1986
and St Nicholas', Warwick

ACKNOWLEDGEMENTS

Acknowledgement is gratefully made to the following for supplying photographs and giving permission for them to be reproduced here:

The Vatican and Scala: Raphael's *Causarum Cognitio*, in the Stanza della Segnature, Vatican (Frontispiece); The Biblioteca Nazionale 'Vittorio Emanuele III', Naples: P. Herc. 1061, columns 8, 9, and 10 (in plate 1); The University of Pennsylvania Museum, Philadelphia: P. Oxy. i 29 (in Plate 2); Mount Holyoke College, Massachusetts: P. Fay. i 9 (in Plate 3); The Bodleian Library, Oxford: Ms D'Orville 301, folios 35^v and 36^r (Plate 4); The Visitors of the Ashmolean Museum, Oxford, and the Committee of the Egypt Exploration Society: O. Bodl. ii 1847 (in Plate 6); The British Library: P. Lond ii 265, columns 2 and 3 (Plate 7); The Österreichische Nationalbibliothek, Vienna: M.P.E.R., N.S. i 1, column 8 (Plate 8).

Acknowledgement is gratefully made for permission to reprint passages from the following books:

The Works of Aristotle Translated into English, ed. W. D. Ross, Oxford: Clarendon Press. Copyright © at various dates by Oxford University Press; L. Brandwood, *A Word Index to Plato*, Leeds: W. S. Maney. Copyright © 1976 by L. Brandwood; A. J. Butler, *The Arab Conquest of Egypt and the Last Thirty Years of the Roman Domination*, 2nd edition, Oxford: Clarendon Press. Copyright © 1978 by Oxford University Press; H. Cherniss, *The Riddle of the Early Academy*, Berkeley and Los Angeles: The University of California Press. Copyright © 1945 by The University of California Press; Various publications of the Egypt Exploration Society, by permission of the Committee; C. F. Gauss, *Disquisitiones Arithmeticae*, trans. A. A. Clarke, New Haven: Yale University Press. Copyright © 1965 by Yale University; H. Glucker, *Antiochus and the Late Academy*, Göttingen: Vandenhoeck & Ruprecht. Copyright © 1978 by Vandenhoeck & Ruprecht; T. L. Heath, *Mathematics in Aristotle*, Oxford: Clarendon Press. Copyright © 1949 by Oxford University Press; J. Klein, *Greek Mathematical Thought and the Origin of Algebra*, trans. E. Brann, Cambridge, Massachusetts: The MIT Press. Copyright © 1968 by The Massachusetts Institute of Technology; Various volumes of The Loeb Classical Library, Cambridge, Massachusetts: Harvard University Press, & London: William Heinemann Limited. Copyright © at various dates by the President and Fellows of Harvard College & William Heinemann Limited. In particular *Plato, The Republic*, trans. P. Shorey (Copyright © 1930 & 1935) and *Selections Illustrating the History of Greek Mathematics*, ed. & trans. I. Thomas (Copyright © 1939 & 1941); I. Mueller, *Philosophy of Mathematics and Deductive Structure in Euclid's Elements*, Cambridge, Massachusetts: The MIT Press. Copyright © 1981 by The

Massachusetts Institute of Technology; *The Commentary of Pappus on Book X of Euclid's Elements*, trans. G. Junge & W. Thomas, Cambridge, Massachusetts: Harvard University Press. Copyright © 1930 by the President and Fellows of Harvard College; R. A. Parker, *Demotic Mathematical Papyri*, Providence: Brown University Press. Copyright © 1972 by Brown University Press, reprinted here by permission of The University Press of New England; *Plato's Phaedrus*, trans. R. Hackforth, Cambridge: Cambridge University Press. Copyright © 1952 by Cambridge University Press; *Plato, Protagoras and Meno*, trans. W. K. C. Guthrie, Middlesex: Penguin Books. Copyright © 1956 by W. K. C. Guthrie, pp. 130–8 reprinted here by permission of Penguin Books Limited; *Plato's Socratic Dialogues*, trans. W. D. Woodhead, Edinburgh: Thomas Nelson & Sons. Copyright © 1952 by Thomas Nelson & Sons; *Proclus: A Commentary on the First Book of Euclid's Elements*, trans. G. R. Morrow, Princeton: Princeton University Press. Copyright © 1970 by Princeton University Press.

CONTENTS

List of plates	xxi
Note on the transcriptions of papyri	xxi

PART ONE: INTERPRETATIONS ... 1

1 The proposal — 3
  1.1 Socrates meets Meno's slaveboy — 3
  1.2 The characteristics of early Greek mathematics — 8
    (a) Arithmetised mathematics — 8
    (b) Non-arithmetised geometry — 10
    (c) Numbers and parts: the *arithmoi* and *merē* — 14
    (d) Ratio (*logos*) and proportion (*analogon*) — 16
    (e) The language of Greek mathematics — 21
  1.3 Socrates meets the slaveboy again — 25
  1.4 Notes and references — 28

2 Anthyphairetic ratio theory — 31
  2.1 Introduction — 31
  2.2 Some anthyphairetic calculations — 33
    (a) The diagonal and side — 33
    (b) The circumference and diameter — 35
    (c) The surface and section — 42
  2.3 Anthyphairetic algorisms — 42
    (a) The *Parmenides* proposition — 42
    (b) An algorism for calculating anthyphaireses — 44
    (c) An algorism for calculating convergents — 48
  2.4 Further anthyphairetic calculations — 51
    (a) Eratosthenes' ratio for the obliquity of the ecliptic — 51
    (b) The Metonic cycle — 52
    (c) Aristarchus' reduction of ratios — 53
    (d) Archimedes' calculation of circumference to diameter — 54
    (e) Pell's equation — 57
    (f) The alternative interpretation of Archimedes' *Cattle Problem* — 61
  2.5 Notes and references — 63

Contents

3 *Elements* II: The dimension of squares   **67**

  3.1 Introduction   67

  3.2 Book II of the *Elements*   68

  3.3 The hypotheses   73

  3.4 The first attempt: The method of gnomons   76

  3.5 The second attempt: Synthesising ratios   86
    (a) Introduction   86
    (b) The extreme and mean ratio   86
    (c) The nth order extreme and mean ratio   87
    (d) *Elements* XIII, 1–5   90
    (e) Further generalisations   92

  3.6 The third attempt: Generalised sides and diagonals   95
    (a) The method   95
    (b) Historical observations   100

  3.7 Summary   104

  3.8 Notes and references   105

4 Plato's mathematics curriculum in *Republic* VII   **106**

  4.1 Plato as mathematician   106

  4.2 *Arithmētikē te kai logistikē*   108

  4.3 Plane and solid geometry   117

  4.4 Academic astronomy   121
    (a) Introduction   121
    (b) The slaveboy meets Eudoxus   122
    (c) Egyptian and early Greek astronomy   126

  4.5 Academic music theory   130
    (a) Introduction   130
    (b) Archytas meets the slaveboy   131
    (c) Compounding ratios   138
    (d) The *Sectio Canonis*   143
    (e) Further problems   153

  4.6 Appendix: The words *logistikē* and *logismos* in Plato, Archytas, Aristotle, and the pre-Socratic philosophers   154
    (a) Plato   154
    (b) Archytas   156
    (c) Aristotle   156
    (d) Pre-Socratic philosophers   157

Contents xvii

5 *Elements* IV, X, and XIII: The circumdiameter and side 158
5.1 The circumdiameter and side 158
(a) The problem 158
(b) The pentagon 158
(c) The extreme and mean ratio 161
(d) Anthyphairetic considerations 162

5.2 *Elements* X: A classification of some incommensurable lines 166
(a) Introduction 166
(b) Expressible lines and areas 170
(c) Medial lines and areas 173
(d) Sums and differences 176
(e) Binomials and apotomes 182
(f) The six additive and subtractive *alogoi* lines 184
(g) The scope and motivation of Book X 190

5.3 Appendix: The words *alogos* and *(ar)rhetos* in Plato, Aristotle, and the pre-Socratic philosophers 192
(a) Plato 192
(b) Aristotle 193
(c) Pre-Socratic philosophers 194

PART TWO: EVIDENCE 195

6 The nature of our evidence 197
6.1 ΑΓΕΩΜΕΤΡΗΤΟΣ ΜΗΔΕΙΣ ΕΙΣΙΤΩ 197
6.2 Early written evidence 202
6.3 The introduction of minuscule script 216

7 Numbers and fractions 221
7.1 Introduction 221
(a) Numerals 221
(b) Simple and compound parts 226
(c) P. Hib. i 27, a parapegma 229
(d) O. Bodl. ii 1847, a land survey ostracon 230

7.2 Tables and ready reckoners 234
(a) Division tables 234
(b) Multiplication tables 238
(c) Tables of squares 239

7.3 A selection of texts 240
(a) Archimedes' *Measurement of a Circle* 240

	(b) Aristarchus' *On the Sizes and Distances of the Sun and Moon*	246
	(c) P. Lond. ii 265 (p. 257)	248
	(d) M.P.E.R., N.S. i 1	254
	(e) Demotic mathematical papyri	259
7.4	Conclusions and some consequences	263
	(a) Synthesis	263
	(b) The slaveboy meets an accountant	268
7.5	Appendix: A catalogue of published tables	270
	(a) Division tables	271
	(b) Multiplication and addition tables	277
	(c) Tables of squares	279

PART THREE: LATER DEVELOPMENTS — 281

8 Later interpretations — 283

8.1 Egyptian land measurement as the origin of Greek geometry? — 283

8.2 *Neusis*-constructions in Greek geometry — 287

8.3 The discovery and role of the phenomenon of incommensurability — 294
  (a) The story — 294
  (b) The evidence — 294
  (c) Discussion of the evidence — 302

9 Continued fractions — 309

9.1 The basic theory — 310
  (a) Continued fractions, convergents, and approximation — 310
  (b) The *Parmenides* proposition and algorithm — 320
  (c) The quadratic theory — 329
  (d) Analytic properties — 337
  (e) Lagrange and the solution of equations — 339

9.2 Gauss and continued fractions — 341
  (a) Introduction — 341
  (b) Continued fractions and the hypergeometric series — 342
  (c) Continued fractions and probability theory — 346
  (d) Gauss's number theory — 348
  (e) Gauss's legacy in number theory — 351

9.3	Two recent developments	354
	(a) Continued fraction arithmetic	354
	(b) Higher dimensional algorithms	360
9.4	Epilogue	364

Bibliography	372
Index of Cited Passages	391
Index of Names	394
General Index	399

PLATES

Plates fall between pp. 202 and 203.

1. P. Herc. 1061, columns 8, 9, and 10.
2. P. Oxy. i 29.
3. P. Fay. 9.
4. Bodleian manuscript D'Orville 301, folios 35^v and 36^r.
5. P. Hib. i 27, column 4.
6. O. Bodl. ii 1847.
7. P. Lond. ii 265 (p. 257), columns 2 and 3.
8. M.P.E.R., N.S. i 1, column 6.

NOTE ON THE TRANSCRIPTIONS OF PAPYRI

The transcriptions of papyri (except for Plate 2) are made as follows: the texts are printed in modern form, with accents and punctuation, the lectional signs occurring in the papyri being noted in the *apparatus criticus,* where also faults of orthography, etc., are corrected. Iota adscript is printed where written, otherwise iota subscript is used. Square brackets [] indicate a lacuna, round brackets () the resolution of a symbol or abbreviation, angular brackets ⟨ ⟩ a mistaken omission in the original, braces {} a superfluous letter or letters, double square brackets ⟦ ⟧ a deletion, the signs ˋ ˊ an insertion above the line. Dots printed slightly below the line within brackets represent the estimated number of letters lost or deleted, dots slightly below the line outside brackets mutilated or otherwise illegible letters, and dots under letters indicate that the reading is doubtful.

(This description has been excerpted and adapted from recent volumes of *The Oxyrhynchus Papyri.* For further discussion, see E. G. Turner, *Greek Papyri* (2nd ed.) 187 n. 22 and 203 n. 22.)

PART ONE

INTERPRETATIONS

Finding out about ratios is a secret of deep logic.
Omar Khayyam, *Discussion of Difficulties in Euclid*.

This first part will develop the main mathematical and historical structure of the interpretation. I have tried to make it as direct and explicit as possible, and have used various expository devices to get across ideas and attitudes that may be unfamiliar or unconventional. It contains no footnotes and the minimum of distractions in the main text; so even the details of the ancient texts and their translations, other references, and additional incidental material are given, to begin with, in the notes or Appendices at the end of each chapter.

Chapter 1 sets the scene by describing the context, techniques, and themes that will dominate this book. I take as model, for both the kind of mathematics and the kind of mathematical exposition, the passage from Plato's *Meno* where Socrates persuades a slaveboy to double a square. Then, after analysing the concepts and techniques that will be at our disposal, I propose a sequel to this episode in which the slaveboy, now greatly increased in confidence and knowledge, faces the question of giving a definition of ratio.

Chapter 2 starts a commentary on my dialogue, in the form of an exposition of some of the techniques it introduced. Again I adopt a Platonic rather than a Euclidean style: problems are suggested rather than formally enunciated, the motivation is treated at length while sometimes the actual evaluation is left to the reader, the treatment is very personal and discursive, and I make no attempt to set out a definitive account of the subject. I have tried to illustrate the techniques on examples that arise in our ancient evidence and my aim, here and elsewhere, is one of comprehensiveness: I wish to describe a cluster of associated methods that will provide plausible reconstructions of not one but all of these ancient examples that sometimes seem so perplexing to a modern audience.

Chapter 3 is a reinterpretation of Book II of Euclid's *Elements* and some related material. Here I have attempted to present an exhaustive account. Rather than suggesting to the reader that something can be done, I have done it and set out the details, but I encourage my readers to skip these details where this is more palatable. Here again, the aim is at comprehensiveness: a successful explanation of a coherent collection of results like Book II should account for everything that is found in the book. This chapter ends with a discussion of the texts on side and diagonals found in the Platonic commentaries by Theon of Smyrna and Proclus. This is the only substantial piece of evidence from later sources that is invoked in Part One, and my discussion indicates, incidentally, some of the difficulties with late and indirect material of this kind.

Chapter 4 presents an interpretation of Plato's curriculum in *Republic* VII. The central idea here is that Plato's *logistikē* refers to ratio theory, but that different definitions of ratio are appropriate for mathematics, astronomy, and music theory. Then the underlying unity of the curriculum comes, I suggest, from relating these different manifestations of ratio to each other. Again, the exposition of some of these ideas is done in dialogue form.

Chapter 5 gives a description of the classification of incommensurable lines found in *Elements* X and applied in *Elements* XIII. My main interest here is in the mathematical motivation of this classification, and I propose that it is a further extension of the problems suggested by the reconstructions of Chapters 3 and 4.

1
THE PROPOSAL

1.1 Socrates meets Meno's slaveboy

Plato's *Meno* is an enquiry into virtue. As part of the discussion, Plato's Socrates digresses to illustrate his opinion that we cannot teach anything, but can only prompt recollection of the opinions of a disembodied soul. But Socrates' digression takes him into the subject of our enquiry here, so let us begin with this passage, *Meno* 81e–85d. The translation is by W. K. C. Guthrie; I have suppressed all editorial additions, made one very slight alteration throughout which will be explained in Section 1.2(c), below, and restored two speeches at the end of 83b that have unaccountably been omitted.

MENO: What do you mean you say that we don't learn anything, but that what we call learning is recollection? Can you teach me that it is so?

82a SOCRATES: I have just said that you're a rascal, and now you ask me if I can teach you, when I say there is no such thing as teaching, only recollection. Evidently you want to catch me contradicting myself straight away.

MENO: No, honestly, Socrates, I wasn't thinking of that. It was just habit. If you can in any way make clear to me that what you say is true, please do.

b SOCRATES: It isn't an easy thing, but still I should like to do what I can since you ask me. I see you have a large number of retainers here. Call one of them, anyone you like, and I will use him to demonstrate it to you.

MENO: Certainly. Come here.

SOCRATES: He is a Greek and speaks our language?

MENO: Indeed yes—born and bred in the house.

SOCRATES: Listen carefully then, and see whether it seems to you that he is learning from me or simply being reminded.

MENO: I will.

SOCRATES: Now boy, you know that a square is a figure like this?

BOY: Yes.

c SOCRATES: It has all these four sides equal?

BOY: Yes.

SOCRATES: And these lines which go through the middle of it are also equal?

BOY: Yes.

SOCRATES: Such a figure could be either larger or smaller, could it not?

BOY: Yes.

SOCRATES: Now if a side is two feet long, and this side the same, how many feet will the whole be? Put it this way. If it were two feet in this direction and only one in that, must not the area be two feet taken once?

BOY: Yes.

d SOCRATES: But since it is two feet this way also, does it not become twice two feet?
BOY: Yes.
SOCRATES: And how many feet is twice two? Work it out and tell me.
BOY: Four.
SOCRATES: Now could one draw another figure double the size of this, but similar, that is, with all its sides equal like this one?
BOY: Yes.
SOCRATES: How many feet will its area be?
BOY: Eight.
SOCRATES: Now then, try to tell me how long each of its sides will be. The
e present figure has a side of two feet. What will be the side of the double-sized one?
BOY: It will be double, Socrates, obviously.
SOCRATES: You see, Meno, that I am not teaching him anything, only asking. Now he thinks he knows the length of the side of the eight-feet square.
MENO: Yes.
SOCRATES: But does he?
MENO: Certainly not.
SOCRATES: He thinks it is twice the length of the other.
MENO: Yes.
SOCRATES: Now watch how he recollects things in order—the proper way to recollect.
83a You say that the side of double length produces the double-sized figure? Like this I mean, not long this way and short that. It must be equal on all sides like the first figure, only twice its size, that is, eight feet. Think a moment whether you still expect to get it from doubling the side.
BOY: Yes I do.
SOCRATES: Well now, shall we have a line double the length of this if we add another the same length at this end?
BOY: Yes.
SOCRATES: It is on this line then, according to you, that we shall make the eight-feet square, by taking four of the same length?
b BOY: Yes.
SOCRATES: Let us draw in four equal lines, using the first as a base. Does this not give us what you call the eight-feet figure?
BOY: Certainly.
SOCRATES: But does it contain these four squares, each equal to the original four-feet one?
BOY: Yes.
SOCRATES: How big is it then? Won't it be four-times as big?
BOY: Of course.
SOCRATES: And is four-times the same as twice?
BOY: Of course not.
SOCRATES: But how much is it?
BOY: Fourfold.
c SOCRATES: So doubling the side has given us not a double but a fourfold figure?

BOY: True.
SOCRATES: And four-times four are sixteen, are they not?
BOY: Yes.
SOCRATES: Then how big is the side of the eight-feet figure? This one has given us four-times the original area, hasn't it?
BOY: Yes.
SOCRATES: And a side half the length gave us a square of four feet?
BOY: Yes.
SOCRATES: Good. And isn't a square of eight feet double this one and half that?
BOY: Yes.
SOCRATES: Will it not have a side greater than this one but less than that?
d BOY: I think it will.
SOCRATES: Right. Always answer what you think. Now tell me. Was not this side two feet long, and this one four?
BOY: Yes.
SOCRATES: Then the side of the eight-feet figure must be longer than two feet but shorter than four?
BOY: It must.
e SOCRATES: Try to say how long you think it is.
BOY: Three feet.
SOCRATES: If so, shall we add half of this bit and make it three feet? Here are two, and this is one, and on this side similarly we have two plus one, and here is the figure you want.
BOY: Yes.
SOCRATES: If it is three feet this way and three that, will the whole area be three-times three feet?
BOY: It looks like it.
SOCRATES: And that is how many?
BOY: Nine.
SOCRATES: Whereas the square double our first square had to be how many?
BOY: Eight.
SOCRATES: But we haven't yet got the square of eight feet even from a three-feet side?
BOY: No.
SOCRATES: Then what length will give it? Try to tell us exactly. If you don't
84a want to count it up, just show us on the diagram.
BOY: It's no use, Socrates, I just don't know.
SOCRATES: Observe, Meno, the stage he has reached on the path of recollection. At the beginning he did not know the side of the square of eight feet. Nor indeed does he know it now, but then he thought he knew it and answered boldly, as was appropriate—he felt no perplexity. Now
b however he does feel perplexed. Not only does he not know the answer; he doesn't even think he knows.
MENO: Quite true.
SOCRATES: Isn't he in a better position now in relation to what he didn't know?
MENO: I admit that too.

SOCRATES: So in perplexing him and numbing him like the sting-ray, have we done him any harm?

MENO: I think not.

SOCRATES: In fact we have helped him to some extent toward finding out the right answer, for now not only is he ignorant of it but he will be quite glad to look for it. Up to now, he thought he could speak well and fluently, on many occasions and before large audiences, on the subject of a square double the size of a given square, maintaining that it must have a side of double the length.

MENO: No doubt.

SOCRATES: Do you suppose then that he would have attempted to look for, or learn, what he thought he knew (though he did not) before he was thrown into perplexity, became aware of his ignorance, and felt a desire to know?

MENO: No.

SOCRATES: Then the numbing process was good for him?

MENO: I agree.

SOCRATES: Now notice what, starting from this state of perplexity, he will discover by seeking the truth in company with me, though I simply ask him questions without teaching him. Be ready to catch me if I give him any instruction or explanation instead of simply interrogating him on his own opinions.

Tell me, boy, is not this our square of four feet? You understand?

BOY: Yes.

SOCRATES: Now we can add another equal to it like this?

BOY: Yes.

SOCRATES: And a third here, equal to each of the others?

BOY: Yes.

SOCRATES: And then we can fill in this one in the corner?

BOY: Yes.

SOCRATES: Then here we have four equal squares?

BOY: Yes.

SOCRATES: And how many times the size of the first square is the whole?

BOY: Four-times.

SOCRATES: And we want one double the size. You remember?

BOY: Yes.

SOCRATES: Now does this line going from corner to corner cut each of these squares in half?

BOY: Yes.

SOCRATES: And these are four equal lines enclosing this area?

BOY: They are.

SOCRATES: Now think. How big is this area?

BOY: I don't understand.

SOCRATES: Here are four squares. Has not each line cut off the inner half of each of them?

BOY: Yes.

SOCRATES: And how many such halves are there in this figure?

BOY: Four.

	SOCRATES: And how many in this one?
	BOY: Two.
	SOCRATES: And what is the relation of four to two?
	BOY: Double.
b	SOCRATES: How big is this figure then?
	BOY: Eight feet.
	SOCRATES: On what base?
	BOY: This one.
	SOCRATES: The line which goes from corner to corner of the square of four feet?
	BOY: Yes.
	SOCRATES: The technical name for it is 'diagonal'; so if we use that name, it is your personal opinion that the square on the diagonal of the original square is double its area.
	BOY: That is so, Socrates.
	SOCRATES: What do you think, Meno? Has he answered with any opinions that were not his own?
c	MENO: No, they were all his.
	SOCRATES: Yet he did not know, as we agreed a few minutes ago.
	MENO: True.
	SOCRATES: But these opinions were somewhere in him, were they not?
	MENO: Yes.
	SOCRATES: So a man who does not know has in himself true opinions on a subject without having knowledge.
	MENO: It would appear so.
	SOCRATES: At present these opinions, being newly aroused, have a dreamlike quality. But if the same questions are put to him on many occasions and in different ways, you can see that in the end he will have a
d	knowledge on the subject as accurate as anybody's.

The passage is well known and frequently discussed, but I quote it here in full for special reasons:

(i) It is our *first* direct, explicit, extended piece of evidence about Greek mathematics; it probably dates from about 385 BC. By going forward from this time we may probe further back into the past, as in the following three typical examples: In commentaries, written in the sixth century AD, by Simplicius and Eutocius on Aristotle and Archimedes respectively, we find descriptions of work by Hippocrates (*c*.425 BC?) and Archytas (*c*.385 BC or earlier?) that come, directly or indirectly, from a now lost *History of Mathematics* written around 325 BC by Eudemus, a pupil of Aristotle. Second, Euclid's *Elements,* which is believed to have been compiled around 300 BC, gives us an edited and anonymous treatment of some of the mathematics of Euclid's predecessors, in a text that has passed through the hands of an unknown number of further scribes and editors before arriving at the versions that we now possess. Third, to find more than very short and isolated references to the mathematics of the Pythagoreans, and any information about Pythagoras himself, we have to go forward to the notoriously biased and uncritical hagiographical writings of Iamblichus in the third century AD. In this shifting and murky world of fragmentary, anonymous, secondary and tertiary accounts, this passage by Plato and other mathematical passages in his dialogues stand out as direct testimonies from someone located at the centre of that remarkable group of creative mathematicians who initiated the style and some of the techniques and preoccupations that have since dominated the subject.

(ii) It serves as a lucid and straightforward example of some of the problems, styles, and techniques of early Greek mathematics. I shall discuss this in more detail later in this chapter.

(iii) It leads directly into my exposition, at the end of this chapter, of some of the themes and preoccupations of this book.

My aim throughout will be to try to emphasise mathematical texts that survive from the fourth and third centuries BC, including another long passage in Plato's *Republic* VII which describes a curriculum for the future guardians of the state, and to attempt to fit them together into some coherent account. To begin with, I shall ignore the accounts by the commentators in late antiquity; then, when the details of the proposed reconstruction have been presented, I shall consider, in Part Three of this book, a few of the different interpretations presented by the neo-Platonic and neo-Pythagorean commentators from the third century AD onwards. I now believe that these are later misunderstandings and that they can be explained and may be almost inevitable. I introduce this argument right at the end of the book, but a proper discussion of the issues will have to be deferred to future publications.

1.2 The characteristics of early Greek mathematics

1.2(a) *Arithmetised mathematics*

There is a style of mathematics that I shall call 'arithmetised'. It is characterised by the use of some idea of number that is sufficiently general to describe some model of what is now frequently called 'the positive number line' *and its arithmetic*. The range of numbers that can be handled may be more or less comprehensive. For example, that great part of Babylonian mathematics which is restricted to the so-called regular numbers, numbers with terminating sexagesimal expansions, deals, in effect, with numbers that can be written as common fractions with denominators containing powers of two, three, and five; while, at other times, the scope of arithmetic has been extended to include informal or formal descriptions of irrational, complex, infinitesimal, and infinite quantities under this general umbrella of 'numbers'. These numbers then permeate mathematics: in geometry, lines become endowed with a 'length'; the 'area' of a rectangle is the product of the lengths of its base and height, and this basic definition is extended to the area of more and more two-dimensional figures; three-dimensional figures have a numerical 'volume'; then, more recently, higher-dimensional spaces have been created, purely numerical constructs whose properties are then described by geometrical analogies. The ratio of two quantities becomes the quotient of the two numbers that represent them; for example, the ratio of the circumference to diameter of a circle becomes a number that can be approximated by an integer 3, or by fractions such as $3\frac{1}{8}$, $\frac{22}{7}$, or $\frac{355}{113}$, or represented as a decimal number $3 \cdot 1416 \ldots$, or an infinite product

$$2\left(\frac{2}{1} \cdot \frac{2}{3} \cdot \frac{4}{3} \cdot \frac{4}{5} \cdot \frac{6}{5} \cdots\right),$$

or a continued fraction

$$4\left(\frac{1}{1+} \frac{1^2}{2+} \frac{3^2}{2+} \frac{5^2}{2+} \cdots\right).$$

Familiarity with the manipulation of numbers can then lead to the experience and confidence out of which algebra can grow; a non-mathematician, looking at a mathematics text of today, might be inclined to describe mathematics as 'all algebra', a natural reaction to the result, over the last three hundred years, of our version of this programme of arithmetisation.

Many cultures have developed their own versions of arithmetised mathematics. Our evidence about Babylonian mathematics comes from contemporary clay tablets which contain practically nothing but numbers, expressed in an efficient floating-point, decimally coded, sexagesimal notation. Egyptian mathematics is arithmetised, though its description in terms of unit fractions seems to us to be unwieldy and inconvenient. Ptolemaic astronomy and the mathematics of Heron and Diophantus seem to be a powerful and profound blend of the earlier Greek geometrical and Babylonian arithmetised methods; and these were then further developed by Arab astronomers and mathematicians. Western mathematics since the sixteenth century has, as I have already suggested, been dominated by the arithmetised point of view, and this culminated in the developments of the nineteenth and twentieth centuries when the arithmetised formulation permeated analysis; the resulting minute investigations led to the first known coherent, formal descriptions of what might be meant by 'number' and 'arithmetic'. Dedekind tells us, in the introduction of his *Stetigkeit und die irrationale Zahlen*, that he conceived his definition of 'real numbers' on Wednesday, 24 November 1858, and other versions were given by Méray, Weierstrass, and Cantor; other, different formal models for the number line have been proposed since. This recent activity shows a consciousness by mathematicians that the description of the number line and its arithmetic is at the root of our understanding and intuition about mathematics today. There may have been pockets of non-arithmetised mathematics such as, for example, nineteenth-century 'synthetic geometry', but the trend has been towards a more and more comprehensively arithmetised approach. And it seems that the power and scope of modern mathematics owes much to the completeness and effectiveness of this approach.

Greek mathematics up to the second century BC seems, to an extraordinary degree, to be different. I shall argue, throughout this book, that while the Greeks may originally have deployed techniques that could serve perfectly as labelling systems for the positive number line, they did not and could not go on to consider arithmetical operations with these labels. Thus my first characteristic of early Greek mathematics is negative: it seems to be completely non-arithmetised.

1.2(b) *Non-arithmetised geometry*

Geometry, modelled on the properties of the three-dimensional space we seem to experience about us, provides the main ingredient of Greek mathematics. And Socrates gives a good illustration of his common-sense attitude to this geometry when he says, right at the beginning of his encounter with the slaveboy:

Such a figure could be either larger or smaller, could it not?

(I shall return to this assertion much later, at the end of Chapter 4.) A typical solution of a problem or proof of a proposition will consist of a figure and a collection of statements about the figure. Thus the first two incorrect solutions to Socrates' problem of doubling a square were two figures, which the reader was left to supply, together with statements about counting equal squares; while the third and correct solution consisted of a further figure with the statements like: "these squares are cut in half"; "these lines are equal"; "this square contains two (equal) halves"; "this square contains four such halves"; and "four is twice two". Another good example of this demonstrative procedure, which also illustrates Euclid's use of the word 'equal' (*isos*), is *Elements* I 35:

Parallelograms which are on the same base and in the same parallels are equal to one another.

Here the figure is a straightforward instantiation of the enunciation of the proposition. Different configurations are possible, according to the amount of overlap and Figure 1.1 illustrates one case, slightly different from that treated by Euclid. The argument then runs: the two triangles

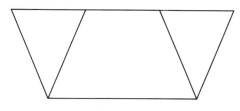

Fig. 1.1

are equal so that, when the trapezium is added to each, Common Notions 2 and 4:

If equals be added to equals, the wholes are equal.
Things which coincide with one another (*ta epharmozonta ep' allēla*) are equal to one another,

will confirm the equality of the parallelograms. We need some kind of assurance that this particular figure will ensure the universal truth of the proposition; while Aristotle worries about this kind of question, it does not seem to bother the mathematicians.

Equality, then, does not mean arithmetical equality of some numerical measure of area and it cannot mean congruence. In fact congruence does not seem to be a basic concept in the *Elements,* since Euclid has no single word for it; the idea of coincidence in Common Notion 4 might seem a candidate, but this piece of text may be a later interpolation (see Heath, *TBEE* i, 224 f.) and Euclid does not make systematic use of this terminology (it appears elsewhere only in I 4 & 8 and III 24). Another way of describing congruence is found, for the first time, in XI Definition 10:

Equal and similar (*isos kai homoios*) solid figures are those contained by similar planes equal in multitude and magnitude.

This 'definition' illustrates further how Euclid's geometry consists of assertions about space that are either taken as self-evident, as apparently this is, or that may be listed explicitly as Common Notions, such as the examples I have just quoted, or as postulates, for example Postulate 5:

That, if a straight line falling on two straight lines make the interior angles on the same side less than two right angles, the two straight lines, if produced indefinitely, meet on that side on which are the angles less than two right angles,

or asserted without comment, as in I 1:

From the point C in which the circles cut one another . . . ,

or asserted because they are held to follow from some earlier assumptions or propositions, which may be implied, or paraphrased, or quoted word-for-word, but which are almost never located by exact references, so that the two occasions when distant locations are specified (possibly in later interpolations) are very surprising:

For it was proved in the first theorem of the tenth book that, if two unequal magnitudes be set out, and if from the greater there be subtracted a magnitude greater than the half, and from that which is left a greater than the half, and if this be done continually, there will be left some magnitude which will be less than the lesser magnitude set out [XII 2],

and

For this has been proved in the last theorem but one of the eleventh book [XIII 17].

The problems of elucidating the status of Euclid's geometry have occupied and perplexed commentators since late antiquity up to the present day. Of the examples I have just cited, XI Definition 10 would now seem to us to be not a definition but a proposition, indeed a proposition that is only true if sufficiently strong additional conditions, such as convexity, are imposed, and which otherwise can now be shown, by ingenious counter-examples, to be false. Postulate 5 is now perceived not simply as the postulation of the converse of I 17:

In any triangle two angles taken together in any manner are less than two right angles,

but as a profound statement about a particular kind of geometry; and the assertions in I 1 and XII 2 are interpreted in terms of conditions that need to be imposed on our mathematical model of space. We can have no clear idea of the point of view of Euclid and his sources; we simply do not have enough reliable information. But these are not the kinds of questions I shall be considering here. I shall rather use geometry as Socrates and Euclid do, as describing some non-arithmetised idea of space through which assertions and relations are made manifest, and I will leave to one side questions of the status of the geometrical objects and procedures that participate in the figures and constructions, and refer to geometry in the way that Plato describes, at *Republic* VII, 527a:

Their [the geometers'] language is most ludicrous, though they cannot help it, for they speak as if they were doing something and as if all their words were directed towards action. For all their talk is of squaring and applying and adding and the like, whereas in fact the real object of the entire study is pure knowledge.

I shall go on to describe how geometry can be used to establish a very surprising kind of pure knowledge and will argue that this may be what Plato is referring to here.

The objects of this geometry are points, lines, and simple figures which reside in two- and three-dimensional space, and which are manipulated in various ways; and one of the main preoccupations of Euclid's geometry is the transformation, combination, and comparison of these figures—see Plato's "squaring and applying and adding and the like". The idea behind Euclid's use of 'equality' within geometry is one of size, not shape, and his concern, at first, is to see if two lines, two plane figures, or two three-dimensional regions are equal in size or, when they are not equal, which is the bigger. Let us look at these operations in a little more detail.

The comparison of two lines can be carried out using the first three propositions of Book I: these enable us to superimpose one line and its endpoint on the other, an operation that is used explicitly at the beginning of I 4. Then Book I develops some of the machinery for manipulating and comparing two rectilineal plane regions; it culminates in Proposition 44:

To a given straight line to apply, in a given rectilineal angle, a parallelogram equal to a given triangle,

and I 45 shows how to extend this to any given rectilineal figure, though Euclid here shows no concern about the one step that might today seem to require some detailed proof, the operation of triangulating the figure that is to be transformed. A crucial preliminary result is established in I 43, which asserts that the two shaded parallelograms in Fig. 1.2 are equal: the large parallelogram contains two smaller parallelograms, and each of these three is cut into equal halves by the diagonal. Hence when the two equal smaller halves are subtracted from the equal larger halves, the results will be equal. The last proposition of Book II, II 14, describes how

To construct a square equal to a given rectilineal figure,

and this theory for rectilineal plane figures is further elaborated in Book VI. Problems that arise with more general plane magnitudes and three-dimensional figures are considered in Book XII but, by that time, new ingredients have introduced new mathematical preoccupations.

Let me return, for a final illustration of non-arithmetised geometry, to Socrates and the slaveboy. At 82c–d, they have to work out the area of a square two feet by two feet. In arithmetised geometry, we would

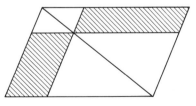

Fig. 1.2

immediately say that its square is two times two square feet. Socrates proceeds differently. He argues, with my interpolations:

If it were two feet in this direction and only one in that [it would comprise two juxtaposed one-foot squares, forming a rectangle, so that], must not the area be two feet taken once; but since it is two feet in this way also [and so comprises two such rectangles juxtaposed], does it not become twice two feet?

This illustrates the basic geometrical operation of addition, represented by juxtaposition of the corresponding figures. Similar passages occur at 83b and e.

Greek mathematics does deploy some numbers, as in the passages just considered, and these are manipulated arithmetically, even in a geometrical context, as at *Meno* 83e, again with my interpolation:

If it is three feet this way and three that, will not the whole area be [three juxtaposed rectangles each comprising three juxtaposed squares, so that it is] three-times three feet.

It is to this ingredient that I now turn.

1.2(c) *Numbers and parts: the arithmoi and merē*

Greek mathematics makes free use of the cardinal numbers; but rather than our thinking of the sequence 1, 2, 3, ..., *qua* cardinals, a much more faithful impression of the very concrete sense of the Greek *arithmoi* is given by the sequence:

duet, trio, quartet, quintet, ...

I have omitted the first term because in early Greek mathematics the unit, the *monas*, has a different status from the others, so that an argument may have to be reformulated when it applies to this case. These numbers are ordered by size ('a quartet is bigger than a trio'), and can be added by concatenation ('a trio plus a quartet makes a septet') and subtracted 'the less from the greater'.

The *arithmoi* also appear in other forms, such as the adverbial sequence:

once, twice, three-times, four-times, ...

These two particular kinds of numbers can act on each other to give multiplication, as in Socrates' "*dis duoin*" (twice two) or "*treis tris*" (three three-times). Division, or division with remainder when the less does not measure the greater, is described similarly. There are further numerical sequences, such as:

simple, double, triple, quadruple, ... ,

which Socrates also draws on, and which are used in the same way; the

1.2 Early Greek mathematics

particular choice seems to depend on grammatical or stylistic convenience. None of these other forms seems to have a common name in English, and I shall refer to them as 'repetition numbers'.

Another kind of numerical sequence is:

half, third, quarter, fifth, ... ;

see, for example, Socrates' remark at the end of *Meno* 83c. These are occasionally found in formal mathematics and are ubiquitous in certain kinds of calculations. They are frequently confused with the ordinal numbers, the different sequence:

first, second, third, fourth, fifth, ... ,

but I shall refer to them by their Greek name as the series of parts (*meros* or *morion*, plural *merē* or *morai*), and will denote them by $\acute{2}, \acute{3}, \acute{4}, \ldots$ or $2', 3', 4', \ldots$. Chapter 7 will be devoted to a discussion of these parts; meanwhile I urge that they should not be thought of as: one-half, one-third, one-quarter, one-fifth, ..., nor be written: $\frac{1}{2}, \frac{1}{3}, \frac{1}{4}, \frac{1}{5}, \ldots$.

Some examples from the *Elements* will illustrate these features. The basic definitions are VII Definitions 1 to 5:

A unit is that with respect to which each existing thing is called one; and a number is a multitude composed of units.
A number is a part of a number, the less of the greater, when it measures the greater; but parts when it does not measure it.
The greater number is a multiple of the less when it is measured by the less.

Then the answer to a question such as 'What multiple?' or 'What part?' will be a repetition number, or a phrase which describes a repetition number. See, for example, VII Definition 8:

An even-times even number (*artiakis artios arithmos*) is that which is measured by an even number according to an even number (*kata artion arithmon*).

But Euclid never tells us how to answer the question 'What parts?' Heath's hyphen, "even-times even", conveys the nuance of the adverbial form. I have added similar such hyphens to Guthrie's translation of the *Meno* passage with which this chapter began; see, for example, 83c: *tettarōn tetrakis estin hekkaideka,* four four-times are sixteen.

The multiplication that arises when a repetition number acts on a cardinal number is explicitly described in terms of addition:

A number is said to multiply a number when that which is multiplied is added to itself as many times as there are units in the other, and thus some number is produced [VII Definition 15];

and here the phrase "as many times as there are units in the other" illustrates how a cardinal number can be converted into a repetition number. We shall see, in the next chapter, how division is similarly construed in terms of subtraction.

Greek *arithmētikē* (i.e. *arithmētikē technē*) is the art of the *arithmoi*, the study of the properties of these kinds of numbers. It is frequently translated as 'arithmetic', but 'number theory' will better convey the sense of the word provided it is always interpreted within the context of these kinds of manipulations of the Greek *arithmoi*. However I shall generally leave it in its transliterated form, and refer to *arithmētikē*.

1.2(d) Ratio (*logos*) and proportion (*analogon*)

The two words, ratio and proportion, *logos* and *analogon*, refer to different kinds of mathematical entities, though they are frequently conflated and used as if they were interchangeable. A ratio is something; in modern terms, it is a function of two (or possibly more) variables, and if we want to make use of this concept, we should say what it is and how it is calculated and manipulated. I shall very often write $a:b$ as shorthand for the phrase 'the ratio of a to b'; the symbol means no more and no less than the words, and it is up to us to say what they stand for if we want to go on to make use of the idea. Euclid's definition—though description might be a better word to use—occurs at V Definition 3:

A ratio is a sort of relation in respect of size between two magnitudes of the same kind. (*Logos esti duo megethōn homogenōn he kata pēlikotēta poia schesis*).

This is elaborated in V Definition 4:

Magnitudes are said to have a ratio to one another which are capable, when multiplied, of exceeding one another.

but these do not tell us enough about ratios to enable us to prove any results about them. I shall explain this point in more detail later in this section.

On the other hand, a proportion is a condition that may or may not hold between four objects. It may appear as a question: 'Are a, b, c, and d proportional?' to which, if it is properly posed, we have to answer 'Yes' or 'No'. There are other ways of expressing the same question: 'Is a to b as c is to d?' or 'Is the ratio of a to b equal to the ratio of c to d?' (this being a question about a proportion, and not about two ratios), or the idea may be introduced as an assertion, not as a question. I shall often use '$a:b::c:d$' as shorthand for all of these forms of words. Euclid gives two different definitions of proportionality. One, at V Definition 5, is believed to be due to Eudoxus and to date from $c.350$ BC (though the evidence for this attribution and date is indirect, or late and anonymous), at the end of the period I shall mainly be considering. Since my concern here will be with the precursors of this definition, I shall not quote or try to explain it here. The second definition occurs at VII Definition 20:

Numbers are proportional when the first is the same multiple, or the same part, or the same parts, of the second that the third is of the fourth.

I have earlier suggested that the answer to 'What multiple?' or 'What part?' is an adverbial number. This fits well within the restrictions on the *arithmoi* and leads to an idea of 'same multiple' and 'same part' which makes this definition workable in these cases; but Euclid nowhere explains how to treat the case when the first number is 'parts' of the second, neither explicitly in a definition, nor implicitly in a proposition, so we do not know how to interpret 'same parts' in this definition. In fact what information he does give seems inconsistent: VII Definitions 3 and 4, already quoted earlier but repeated here for emphasis, *tell* us that

A number is part of a number, the less of the greater, when it measures the greater; but parts when it does not measure it.

so that any number is either part or parts of a greater number; but VII 4 then *proves*, at some length, that

Any number is either a part or parts of any number, the less of the greater.

Something is curiously wrong. I shall discuss some of our evidence concerning these numerical parts, *merē,* in Chapter 7.

It is often asserted that the discovery of the phenomenon of incommensurability led to a situation in which the early Greek mathematicians were unable to set the theory of ratio or proportion on firm foundations, within the means at their disposal, until the development by Eudoxus, in the middle of the fourth century BC, of the proportion theory of Book V. I wish, throughout this book, to query everything in this sentence except that which relates to the identification, dating, and attribution of Eudoxan proportion theory. As a first step, I shall address here the mathematical component of this assertion, in the following interpretation: that the ingredients of early Greek mathematics, such as I have described them so far or, alternatively, as can be inferred from what are believed to be the early books of the *Elements,* are inadequate to set up a theory of ratio or proportion sufficiently general to handle incommensurable magnitudes. At this point a long quotation from Aristotle's *Topics,* VIII 3, 158a31–159a2, is in order. The *Topics* is a manual of 'syllogistic dialectic', a kind of formal debate between a 'questioner' and an 'answerer':

There are certain hypotheses upon which it is at once difficult to bring, and easy to stand up to, an argument. Such, e.g., are those things which stand first and those which stand last in the order of nature. For the former require definition, while the latter have to be arrived at through many steps if one wishes to secure a continuous proof from first principles, or else all discussion about them wears the air of mere sophistry; for to prove anything is impossible unless one begins with the appropriate principles, and connects inference with inference till the last are reached. Now to define first principles is just what answerers do not care to do, nor do they pay any attention if the questioner makes a definition: and yet until it is clear what it is that is proposed, it is not easy to discuss it. This sort of thing happens particularly in the case of the first principles (*archē*): for while the other propositions are shown through these, these cannot be shown through anything else: we are obliged to understand every item of that sort by definition. The inferences, too, that lie too close to the first principle are hard to treat in argument.... The hardest, however, of all definitions to treat in argument are those that employ terms about which, in the first place, it is uncertain whether they are used in one sense or in several, and, further, whether they are used literally or metaphorically by the definer. For because of their obscurity, it is impossible to argue upon such terms; and because of the impossibility of saying whether this obscurity is due to their being used metaphorically, it is impossible to refute them.... It often happens that a difficulty is found in discussing or arguing a given position because the definition has not been correctly rendered.... In mathematics, too, some things would seem to be not easily proved for want of a definition, e.g. that the straight line parallel to the side, which cuts a plane figure divides similarly (*homoiōs*) both the line and the area. But once the definition is stated, the said property is immediately manifest: for the areas and the lines have the same *antanairesis* and this is the definition of the same ratio.... But if the definitions of the principles are not laid down, it is difficult, and may be quite impossible, to apply them. There is a close resemblance between dialectical and geometrical processes.

Here Aristotle (i) makes some observations about method: to avoid sophistry you must define your terms unambiguously and connect inference to inference up to your conclusion; (ii) introduces an example from mathematics. His description is loose, but it clearly means that, in Fig. 1.3, where the rectangles might equally well be replaced by parallelograms (though other such examples always seem to refer to rectangles), the ratio of A to B is equal to the ratio of a to b. Note that this can be interpreted as a statement either about ratio or about proportion; and (iii) he sketches a proof of this statement. I shall now discuss the mathematical proposition (ii), while attempting to follow the precepts in (i); and will postpone the exploration of Aristotle's proof (iii) to the next chapter.

The proposition of Fig. 1.3, which I shall hereafter refer to as 'the

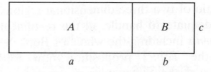

Fig. 1.3. The *Topics* proposition.

1.2 Early Greek mathematics

Topics proposition', is a version of *Elements* VI 1:

Triangles and parallelograms which are under the same height are to one another as their bases.

Euclid proves it in the usual way of Greek mathematics: he draws an appropriate figure and makes an observation about this figure which refers, implicitly, to his definition of proportionality at V Definition 5 (which I have not quoted). This makes manifest that the four magnitudes satisfy this condition and so are proportional. I shall construct an alternative definition of proportionality out of early ingredients, and then give another proof. However, I am *not* offering this as a historical reconstruction of a definition and proof that may have been used in early Greek times; on this we have no evidence. Its roles are to refute the mathematical assertion that such a proof is impossible and to introduce an important kind of manipulation. Other definitions of ratio and proportion, also constructed out of early ingredients, will be described later, and each such definition will lead to a corresponding proof of the *Topics* proposition.

We perform a manipulation that is familiar within any deductive development of mathematics: we take a proposition that is equivalent to the definition we wish to replace, and use that instead as the definition. We have to proceed case by case. For lines, we have VI 16:

If four straight lines be proportional, the rectangle contained by the extremes is equal to the rectangle contained by the means; and if the rectangle contained by the extremes be equal to the rectangle contained by the means, the four straight lines will be proportional;

the second half will now serve as our definition. For numbers, we adapt the similar proposition VII 19. For two geometrical objects a and b, and two numbers n and m, we define $a:b::n:m$ if $ma = nb$ where ma is the operation of multiplication as repeated addition, such as we have already seen in V Definition 4 and VII Definition 15. For two straight lines a and b, and two plane regions A and B, we define $a:b::A:B$ if the rectangular prism with base A and edge b is equal to the prism with base B and edge a. Dimensional restrictions prevent us from giving a similar definition for four plane regions, but we can circumvent these by proving the *Topics* proposition and then applying this to reduce one or both of the pairs of plane regions to ratios of lines. A new approach will be needed to deal with the ratio of two three-dimensional regions, but, as it stands, this definition is adequate to handle all the rectilineal plane proportion theory of the *Elements* including the whole of Book X.

The proof of the *Topics* proposition now follows immediately: $a:b::A:B$ means that the rectangular prism with base A and edge b should be equal to that with base B and edge a; but both of these are rectangular parallelepipeds with edges a, b, and c, where c is the height of the original rectangle. (The reader is highly recommended to draw a figure.)   QED

Note that this proof is expressed in terms of proportion theory; the ratios $a:b$ and $A:B$ are not here defined and they have no meaning outside a proportion. We have formal techniques, today, for passing from a definition of proportion to an idea of ratio—any reasonable definition of proportion will define an equivalence relation, and the resulting equivalence class *is* the ratio—but I believe that we cannot find anywhere, in Greek mathematics, anything that can be reasonably interpreted as implying any consideration of this kind of formal manipulation. Nor can we find any other definition of ratio, other than the statements I have quoted. Euclid's two definitions that concern ratio, V Definitions 3 and 4, do not give us enough information to construct a proof of the *Topics* proposition, which is the criterion I shall use for deciding whether a definition of ratio is mathematically adequate. Heath's comment sums up the situation:

> The true explanation of its presence [sc. V Definition 3] would appear to be substantially that given by Barrow (*Lectiones Cantabrig,* London, 1684, Lect. III of 1666), namely that Euclid inserted it for completeness' sake, more for ornament than for use, intending to give the learner a general notion of ratio by means of a metaphysical, rather than a mathematical definition; "for metaphysical it is and not, properly speaking, mathematical, since nothing depends on it or is deduced from it by mathematicians, nor, as I think, can anything be deduced." This is confirmed by the fact that there is no definition of λόγος in Book VII, and it could equally have been dispensed with here [Heath, *EE* ii, 117].

As Heath observes here, we also do not find anywhere a definition of the ratio of two *arithmoi*, $n:m$. Moreover I do not believe that we have any good evidence that anybody—mathematicians, accountants, schoolmasters, or schoolchildren—deployed anything corresponding to our conception of a common fraction n/m; all that we do have is a minute quantity of very tenuous and ambiguous early evidence on papyrus, and then abundant examples, from the ninth century onwards, in manuscripts that have been subjected to drastic modernisation of script and, possibly, numerical notation. This is a very contentious opinion, so I shall later devote much of the long Chapter 7 to this topic. Yet, notwithstanding this absence of any apparent mathematically usable definition, the word *logos* frequently appears to refer to ratio, in isolation outside a proportion: Plato and Aristotle make many such references, and ratios are 'compounded' in the *Elements,* at VI 23:

> Equiangular parallelograms have to one another the ratio compounded of the ratios of their sides,

and, in a similar proposition for the *arithmoi,* at VIII 5. The difficulties of interpretation of these last two propositions are augmented by the obscurities of the definition of compounding, itself almost certainly an interpolation, at VI Definition 5:

A ratio is said to be compounded of ratios when the sizes of the ratios, having been multiplied together, make some ratio. *(Logos ek logōn sugkeisthai legetai, hotan ai tōn logōn pēlikotētes eph' hautas pollaplasiastheisai poiōsi tina.)*

I shall return to this topic in Section 4.5(c). Further, the massive and monolithic programme of construction and classification of incommensurable lines in Book X of the *Elements* seems to be connected with ratios of pairs of lines. This programme, though not necessarily the actual text of Book X, is attributed to Theaetetus, who died in 369 BC, before Eudoxus is supposed to have introduced the general theory of proportionality of Book V, and it is based on an idea of pairs of lines being either commensurable or incommensurable 'in square'; those that are incommensurable in square are called *alogos*, without ratio. I shall discuss the content and motivation of this book later, in Chapter 5.

Our evidence seems abundantly to indicate that some idea of ratio, in isolation, not occurring within a proportion, appears to be an important ingredient of early Greek mathematics.

1.2(e) *The language of Greek mathematics*

Greek mathematicians seemed to confront directly the objects with which they were concerned: their geometry dealt with the features of geometrical thought-experiments in which figures were drawn and manipulated, and their *arithmētikē* concerned itself ultimately with the evident properties of numbered collections of objects. Unlike the mathematics of today, there was no elaborate conceptual machinery, other than natural language, interposed between the mathematician and his problem. Today we tend to turn our geometry into arithmetic, and our arithmetic into algebra so that, for example, while *Elements* I 47:

In right-angled triangles the square on the side subtending the right angle is equal to the squares on the sides containing the right angle

means literally, to Euclid, that a square can be cut into two and manipulated into two other squares (see Fig. 1.4 for the proof of I 47, or Fig. 1.5 for a proof that may have been excised from between II 8 and 9), the result is now usually interpreted as

$$p^2 + q^2 = r^2,$$

where we now must explain just what the ps, qs, and rs are and how they can be multiplied and added. To us, the literal squares have been replaced by some abstraction from an arithmetical analogy.

22 The proposal 1.2

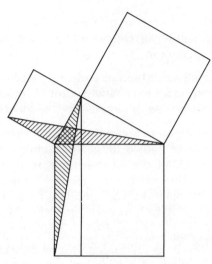

Fig. 1.4

A passage from Plato's *Gorgias,* 451a–c, illustrates the point further, and introduces new features that will come to preoccupy us:

> Suppose that somebody should ask me: ... Socrates, what is *arithmētikē*? I should reply, as you did just now [about rhetoric], that it is one of the arts which secure their effect through speech. And if he should further enquire in what field, I should reply that of the odd and the even, however great their respective numbers might be. And if he should next enquire, What art do you call *logistikē*? I should say that this art too is one of those that secure their entire effect through words. And if he should further demand in what field, I should reply, like the

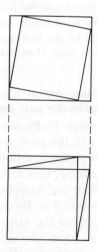

Fig. 1.5

mover of an amendment in the Assembly, that in details 'hereinbefore mentioned' *logistikē* resembles *arithmētikē*—for its field is the same, the even and the odd—but that *logistikē* differs in this respect, that it investigates how numerous are the odd and the even both relative to themselves and relative to each other.

Note how Plato emphasises speech, not writing (though this should also be seen in the context of Socrates' discussion here of the art of rhetoric); and add to this the barbed myth he tells at *Phaedrus* 274c-5b:

The story is that in the region of Naucratis in Egypt there dwelt one of the old gods of the country, the god to whom the bird called Ibis is sacred, his own name being Theuth. He it was that invented *arithmos te kai logismos,* geometry and astronomy, not to speak of draughts and dice, and above all writing. Now the king of the whole country at that time was Thamus, who dwelt in the great city of Upper Egypt which the Greeks call Egyptian Thebes, while Thamus they call Ammon. To him came Theuth, and revealed his arts, saying that they ought to be passed on to the Egyptians in general. Thamus asked what was the use of them all, and when Theuth explained, he condemned what he thought the bad points and praised what he thought the good. On each art, we are told, Thamus had plenty of views both for and against; it would take too long to give them in detail. But when it came to writing Theuth said, "Here, O king, is a branch of learning that will make the people of Egypt wiser and improve their memories; my discovery provides a recipe for memory and wisdom". But the king answered and said, "O man full of arts, to one it is given to create the things of art, and to another to judge what measure of harm and of profit they have for those that shall employ them. And so it is that you, by reason of your tender regard for the writing that is your offspring, have declared the very opposite of its true effect. If men learn this, it will implant forgetfulness in their souls; they will cease to exercise memory because they rely on that which is written, calling things to remembrance no longer from within themselves, but by means of external marks. You have found a charm not for remembering but for reminding, and you are providing your pupils with the semblance of wisdom, not the reality. For you will find that they have heard much without having been taught anything and that they will seem men of many judgements, though for the most part being without judgement and hard to live with into the bargain since they have become conceited instead of wise".

The transition in Athenian intellectual life from an oral to a written culture may not have been far in the past when Plato wrote these words, though we do not know enough to be absolutely sure about this; but Plato's sympathies lie clearly in the past, tremendous reader and writer though he was. I shall return to the question of the physical form of our evidence—just what texts we have, how reliable they might be, and how they have been edited in ancient and modern times—in Chapters 6 and 7.

To a mathematician, the point of an argument resides not in the words that make up the argument, but in the ideas that lie behind these words. In Greek mathematics, these ideas can always be expressed, and not merely described, in everyday language, although sometimes we may find this language tortured almost to extinction of its meaning. Also these words might have initially been carried within an oral tradition, and so been even more subject to the reorganisation, adaptation, adjustment, and revision that is commonplace even within the later written tradition of mathematics.

We shall therefore be concerned with a kind of mathematics that can be completely expressed in everyday language. Whenever I resort to any symbolic notation in the context of Greek mathematics, the reader should confirm that it is being used only as a convenient shorthand and that the symbols themselves have neither more nor less meaning than the words that they replace. For example, I shall introduce the notation $\sqrt{n}$ to stand for the side of a square equal to n times the square on a given line; and if a and b are two lines, then $a.b$ will denote the rectangle with sides a and b, the construction introduced in the first proof of the *Topics* proposition given above. These ideas and notations are prompted by *Elements* II Definition 1 and Proposition 14:

Any rectangular parallelogram is said to be contained by the two straight lines containing the right angle

and

To construct a square equal to a given rectilineal area.

So, to write $\sqrt{2} \cdot \sqrt{3}$ is to employ a convenient shorthand to describe a particular rectangle. But, within this interpretation, to write $\sqrt{2} \cdot \sqrt{3} = \sqrt{6}$ is to make a meaningless assertion, while if we render it meaningful by writing, for instance, $\sqrt{2} \cdot \sqrt{3} = \sqrt{6} \cdot 1$, then it becomes a non-trivial geometrical statement about the equality of two rectangles which must be established by some geometrical argument based, for example, on the configuration of I 43 (see Fig. 1.2, above, and the end of Sections 4.5(b) and (e)). If, however, we are tempted to interpret the statement arithmetically, then we should heed Dedekind's words, in his *Stetigkeit und die Irrationale Zahlen*. After having defined addition, he goes on:

Just as addition is defined, so can the other operations of the so-called elementary arithmetic be defined, viz., the formation of differences, products, quotients, powers, roots, logarithms, and in this way we arrive at real proofs of theorems (as, e.g., $\sqrt{2} \cdot \sqrt{3} = \sqrt{6}$) which, to the best of my knowledge have never been established before [p. 22].

Our familiarity, since the sixteenth century, with the convenient notation for decimal fractions and, since the nineteenth century, with well-founded arithmetised mathematics, has tended to dull our awareness of the machinery that must lie behind a formally correct account of the arithmetical manipulations of real numbers. These problems will not be our concern here, but we must be ever vigilant that any symbolism that is

1.3 Socrates meets the slaveboy again

Early Greek mathematics, I have suggested, is not arithmetised and did not even have at its disposal our notation or conception of common fractions. Yet, on the other hand, we find that mathematicians of the highest calibre, who elsewhere were exercised to attempt precise and explicit explorations of their basic concepts, seemed to find no difficulty in using freely the idea of the ratio of two numbers or of two geometrical magnitudes, and nobody at that time seemed to have expressed any concern about its meaning, not even in the face of knowledge of the phenomenon of incommensurability.

Can this picture possibly be consistent?

I shall now take seriously Socrates' final remark about *Meno's* slaveboy's potential as a mathematician and will make him the mouthpiece for some of my proposals. I will try to imagine how he could have answered if Socrates had asked him for the meaning of the ratio of two numbers. I shall suppose that Socrates' previous question in the *Meno* some time earlier has led the boy to do a substantial amount of private study and this has developed a latent ability in mathematics. And, to keep the exposition moving briskly, I shall make him a little more forthcoming than he was when Socrates first met him. So picture a talented youth, who advances towards learning and investigation smoothly and surely and successfully, with perfect gentleness, like a stream of oil that flows without a sound, so that one marvels how he accomplishes all this at his age (Plato's description of another mathematician, borrowed from *Theaetetus* 144b). I imagine the conversation set around 370 BC; at this time, Plato is involved with his highly philosophical and mathematical middle dialogues (*Republic, Theaetetus, Parmenides, Sophist, Statesman,* and *Timaeus*), Archytas and Theaetetus have made considerable contributions to mathematics, and Eudoxus' influence is beginning to be felt, though his proportion and exhaustion theories of *Elements* V and XII lie in the future; but the dialogue contains one blatantly anachronistic fantasy in Socrates' reference to 'my young friend Euclid'. Suppose, then, that Socrates meets the slaveboy again in Meno's house, and is tempted to resume his earlier conversation. (Each speech is numbered for later reference.)

SOCRATES$_1$: Tell me, boy, what is the relationship of size between this heap of sixty stones, and that heap of twenty-six stones?
BOY$_2$: Do you mean the number of times the smaller goes into the larger?
SOCRATES$_3$: Try it.
BOY$_4$: It goes more than twice, but less than three-times.

SOCRATES$_5$: Can you be more precise?
BOY$_6$: It goes twice with eight stones left over.
SOCRATES$_7$: Those 'eight stones' aren't related to anything now.
BOY$_8$: I just omitted to say that they're still in relation to the other heap of twenty-six stones.
SOCRATES$_9$: Go on.
BOY$_{10}$: Give me time. I can now describe that relationship in the same way. So let's say: twice at the first step, and now I'll tell you about the relationship between eight stones and twenty-six stones.
SOCRATES$_{11}$: Go on.
BOY$_{12}$: Again that will be the number of times the smaller goes into the larger: that's three-times and then the relation between two stones and eight stones; and that's four-times-exactly.
SOCRATES$_{13}$: So you've extracted a relationship expressed by: first step, twice; second step, three-times; third step, four-times-exactly. The technical name for that is a ratio.

$$(60:26 = [2, 3, 4])$$

BOY$_{14}$: So that's what a ratio is! I've always wondered what those mathematicians mean when they use the word *logos*.
SOCRATES$_{15}$: It's just one of the different possible definitions. Can you deduce anything from that last step?
BOY$_{16}$: Yes, that we could have manipulated throughout in twos. So heaps of thirty and thirteen stones will give the same ratio.
SOCRATES$_{17}$: Now let's compare this ratio with some other ratios. Is it bigger than the ratio: *three*-times, three-times, four-times exactly?
BOY$_{18}$: You haven't said that ratios can always be compared or told me what it means; but I'd say that, however it's defined, it ought to imply that anything that starts 'first step, three-times' will be bigger than anything that starts 'first step twice'.

$$([2, \ldots] < [3, \ldots])$$

SOCRATES$_{19}$: Good. It's sometimes better in mathematics to leave those formal points of definition and verification until later, and first find out what kind of thing you want to do. Try another comparison, this time with the ratio twice, *four*-times, four-times-exactly. Think carefully.
BOY$_{20}$: That's strange. Since the second remainder comes from the first, originally bigger, pile, we can see that anything that starts 'twice, four-times' will be *less* than anything that starts 'twice, three-times'.

$$([2, 4, \ldots] < [2, 3, \ldots])$$

That kind of thing seems to happen generally, since it depends on which of the two original piles the remainder comes from.
SOCRATES$_{21}$: Yes. To see which of two ratios is the greater, you have to note whether their terms differ first at the first, third, fifth, ... step, or the second, fourth, sixth, ... step since, in the latter case, the reverse of the relationship of less and greater between the *terms* holds between the *ratios*; this will be the first way we shall find that the greater and less is connected with the odd and

even. So do you see how we can sandwich our ratio: twice is less than twice, four-times, which is less than our ratio twice, three-times, four-times, which is less than twice, three-times, which is less than three-times.

$$([2] < [2, 4] < [2, 3, 4] < [2, 3] < [3])$$

BOY$_{22}$: Yes, and I can see how it will work in general. With each step you get closer and closer, alternating above and below.

SOCRATES$_{23}$: Will you always get there?

BOY$_{24}$: I suppose so. The heaps are always getting smaller and smaller, so the process can't go on for ever; I wonder how long it takes. Can I ask you a question?

SOCRATES$_{25}$: Is that your question? Ignore that remark; we're doing mathematics, not logic! Go ahead.

BOY$_{26}$: What two heaps will have the ratio twice, four-times, three-times-exactly?

SOCRATES$_{27}$: It's easy to work that out backwards: the final step will be three stones to one stone (or six to two, and so on). So the penultimate step will be four-times three plus one (that's thirteen) stones to three. So the answer will be twenty-nine stones to thirteen stones, or fifty-eight to twenty-six, and so on. But if you now want to work out the ratio twice, four-times, three-times, followed by anything else, you'll have to start all over again. Now there's a very convenient and easy forward method, but my young friend Euclid is baffled by the problem of explaining it formally.

BOY$_{28}$: So we can now translate that chain of inequalities you just told me back into heaps of stones: the ratio of two stones to one stone is less than the ratio of nine stones to four stones, which is less than our ratio of sixty stones to twenty-six stones, which is less than the ratio of seven stones to three stones, which is less than the ratio of three stones to one stone.

$$([2] < [2, 4] < [2, 3, 4] < [2, 3] < [3] \text{ becomes}$$
$$2:1 < 9:4 < 60:26 < 7:3 < 3:1)$$

SOCRATES$_{29}$: Yes, and if you do a few more conversions that stretch out a bit further than three steps, and explore the idea a bit, you'll soon find out the forward method for yourself. Sometimes I think in terms of heaps of stones, and sometimes I convert them into ratios, since I find that gives me a lot more information.

BOY$_{30}$: What do you mean?

SOCRATES$_{31}$: Not too fast! I can't keep up with your questions.

BOY$_{32}$: Well here's another question: Can you add or multiply these ratios?

SOCRATES$_{33}$: What nonsense! Who ever thought of adding the ratio of twenty-nine to thirteen stones to the ratio of sixty to twenty-six stones. It doesn't make sense!

BOY$_{34}$: But suppose we have three heaps of stones; how can we calculate the ratio of the first to the third from the ratio of the first to the second and the second to the third? Or a similar question about adding: how can we get the ratio of the first and second combined to the third out of the ratios of the first to third and second to third? Surely these are sensible questions.

SOCRATES$_{35}$: You've got me there! This definition of ratio has the peculiar feature that these kinds of calculations seem curiously intractable; even Theaetetus can't do them directly and if he can't, I'm certainly not going to try. But if you think about ratios in quite a different way, the way that Archytas and the music theorists have been doing, then these kinds of manipulations make good sense in general and become quite easy; but their approach doesn't allow them to describe all ratios that can occur in geometry. Eudoxus and the theoretical astronomers have just started to think about yet another description of ratio, but they haven't been able to do much with their definition so far. Then linking these different but equivalent ideas poses further problems. There are even technical difficulties with the procedures used by scribes and taxmen when they work out their accounts, but there's no interest in cleaning up that mess! So you can see that there are lots of problems still to be solved with your 'adding and multiplying ratios'. Let's move on to a different topic: what about the ratio of two lines?

BOY$_{36}$: Surely I can just use the same process.

SOCRATES$_{37}$: And of two plane regions or two solids?

BOY$_{38}$: I can use geometrical constructions to convert two rectilineal regions into two adjacent rectangles with equal vertical sides; then the ratio of the rectangles is the ratio of the bases, so we're back to the ratio of lines. But I foresee problems in doing these operations with curvilinear regions in general and most kinds of solids.

SOCRATES$_{39}$: Yes. Eudoxus is working on just these problems now. What about the ratio of the diagonal to a side?

BOY$_{40}$: Of what?

SOCRATES$_{41}$: A square, a pentagon, anything you like. Or the circumference to a diameter? Or the surface to a section? Or the circumdiameter to a side, a diagonal, or an edge? Or the dimensions of squares and cubes?

BOY$_{42}$: Not so fast! All of this is very intriguing. How can I find out more?

SOCRATES$_{43}$: Go and see Plato and his friends in the garden of Academia out in the north-western suburbs; they don't seem to talk about anything else now. But be warned: studies that demand more toil in the learning and practice than this you will not discover easily, nor find many of them; it looks like a long time before this idea of ratio will be understood.

This last sentence begins with another quotation from Plato (*Republic* VII, 526c); and my slaveboy should perhaps also be warned about the reception he is likely to get at the garden of Academia if the description at *Republic* VI, 495c–6a, a passage that I am not going to quote here, is a reliable indication of Plato's aristocratic attitude to other social classes and ungentlemanly activities. This is an important aspect of any evaluation of the beliefs and activities of Plato and his associates, but it will only make a very bizarre and fleeting appearance in my story here, in Chapter 6, note 11.

1.4 Notes and references

For a description of the system of referencing books and articles, see the introduction. There is a detailed bibliography, arranged alphabetically, at the end of the book.

1.4 Notes and references

1.1 The translation by Guthrie of the *Meno* is taken from Plato, *CD*; this translation is reprinted, with an interesting collection of articles, in Brown, *PM*. For a critical edition of the Greek text with full commentary and discussion of the date of the dialogue, see Plato–Bluck, *PM*.

1.2(a) Dedekind's *Stetigkeit* is translated in his *ETN*. This essay remains the most lucid description of the 'real numbers' that I know. Dedekind's opinions on Eudoxus and his assertion about arithmetic, quoted in Section 1.2(e), have often been challenged, but I believe they stand, correct.

One by-product of the development of computers and the consequent study of algorithms has been to make evident some of the subtleties and problems that underlie arithmetised mathematics. An excellent general reference is Knuth, *ACP* ii; and, for an informal description of some of the difficulties underlying arithmetic with decimal fractions, see my *FHYDF*. I shall return to this topic in Sections 4.5(b) and (e), and elsewhere.

1.2(b) Heath's discussions of equality in the notes in *TBEE* i are unsatisfactory: up to I 35 he treats the word as if it meant congruence, and then he writes (*TBEE* i, 327): "Now, without any explicit reference to any change in the meaning of the term, figures are inferred to be *equal* which are equal in *area* or in *content* but need not be of the same *form*". Surely 'equal' throughout means 'equal in magnitude'. However his remarks on the following page on 'content' are very valuable; they enable us to say, in modern terms, that Euclid is developing the results needed for a non-arithmetised theory of ordered content, for line magnitudes in Books I 1–3, rectilineal plane areas in I, II, and VI, and simple solids in XI. Eudoxus' theory of exhaustion enables this theory to be extended to circles, pyramids, cones, and spheres in Book XII; see my dialogue, Socrates$_{39}$, in reply to Boy$_{38}$. Using these theories, we can say when two geometrical magnitudes are equal or, if they are unequal, which is the greater, which is exactly what is needed for the ratio theories I shall be describing.

There are four places in the *Elements* where a local reference such as "as was proved in the theorem preceding this" is given: VIII 19 and 21, IX 12, and X 30. There are other places where general references to earlier results are found: "It is then manifest from what was proved before that ... (X 55)"; "For let the same construction be made as before ..., then, in manner similar to the foregoing" (X 56; also see 57–59, 61–65, and 91); and the general and vague "for we have learnt how to do this" (X 10). These are the *only* specific references in the *Elements*. (I am grateful to Malcolm Brown for this information, which does not seem to have been pointed out or discussed in any of the literature on the *Elements*.) The various references added, in editorial square brackets, to Heath's translation *TBEE* are far from complete; a more detailed analysis of the logical structure of the *Elements* can be found in Mueller, *PMDSEE* which incorporates the earlier analyses of Neuenschwander, *EVBEE* and *SBEE*. Some such back-references are found in marginal annotations (scholia) to our earliest complete manuscript of the *Elements*; for an illustration, see the frontispiece of Heath, *TBEE* i with its brief description on p. xi.

For instructions on how to make a model of a non-convex counter-example to *Elements* XI Definition 10, see the article by its discoverer, Connelly, FS.

1.2(c) See Klein, *GMTOA,* for a thorough discussion of the *arithmoi, arithmetikē*, and related topics. I know of no place where the role of repetition numbers is considered, but have not made an exhaustive search; I was tempted to use the language of the schoolroom and call them 'gozinter numbers', as in 'two gozinter four twice', but have resisted. Greek numerical notation, and its extensions to fractional quantities, will be considered in Chapter 7.

1.2(d) The translation of the *Topics*, quotation is based on Aristotle–Ross, tr. Pickard–Cambridge, *WA* i, Heath, *MA*, 80, and Knorr, *EEE*, 257. The role of the *Topics* proposition in Book X is analysed in detail in Knorr, *EEE*, Chapter 8, and there is an excellent discussion of Book V proportion theory in Knorr, APEPT. The way in which the Greek word *parallēlogrammon* often refers to a rectangle is well illustrated in *Elements* X; this book will be described in detail in Chapter 5. *Elements* VI, Definition 5, quoted at the end of this section, is not listed by Heath with the other definitions on *TBEE* ii, 188, but it can be found on p. 189.

1.2(e) Klein, *GMTOA*, discusses the conceptualisation and language of Greek mathematics; see especially p. 63: "Greek scientific arithmetic and logistic are founded on a 'natural' attitude to everything countable as we meet it in daily life. This closeness to its 'natural' basis is *never* betrayed in ancient science."

The translations of the *Gorgias* and *Phaedrus* quotations are adapted from those, by Woodhead and Hackforth respectively, in Plato, *CD*. I have corrected the last line of the first, and taken the last two pungent sentences of the second from Cherniss, *REA*, 3. The status of books, writing, and education in early Greek life is controversial; see Goody, *DSM* and Havelock, *LRG* for different points of view to that expressed here. I have followed Turner, *Athenian Books in the Fifth and Fourth Centuries B.C.*, and the final sentences of this brilliant analysis will serve as an envoi to my first chapter:

> Plato ... devotes some pages of the *Phaedrus* to an analysis of the deficiencies of books. Even if it be granted that the writer is disinterested and knows his theme inside out, a book suffers from many deficiencies. It inclines its reader to rely on someone else's bottled-up memory, and turns him into a pseudo-philosopher, a *doxosophos*. Its message, once frozen into writing, is rigid—a book cannot answer questions, if defective, or defend itself, if attacked. The author of a book is like a reckless gardener, writing in water or sowing in ink. It is impossible not to feel that Plato, tremendous reader though he was, is fighting a rearguard action against the written word's inhibiting effect on thought; that Plato realises the day is past when he could undo the harm done by a book by publicly convicting the author of ignorance. By the first thirty years of the fourth century books have established themselves and their tyranny lies ahead.

2
ANTHYPHAIRETIC RATIO THEORY

2.1 Introduction

In response to the opening question of my dialogue, which was couched in the terms of *Elements* V Definition 3, my slaveboy develops a process of repeated reciprocal subtraction which is then used to generate a definition of ratio as a sequence of repetition numbers (see S_{13} and B_{14}, i.e. the 13th speech, by Socrates, and the 14th, by the boy). As my Socrates says (S_{15}), and as I have already illustrated in Section 1.2(d), this is not the only possible way of defining ratio or proportion, even within the context of early Greek mathematics; so, to distinguish it, I shall often refer to this as 'the anthyphairetic ratio', and call the underlying subtraction process 'anthyphairesis'. These words are derived from the Greek verb *anthuphairein* used, at *Elements* VII 1 and 2, and X 2 and 3, to describe this operation of reciprocal subtraction. Euclid's only explicit reasons for introducing the process are to find the greatest common measure (B_{16}) and, in X 2, as a criterion for incommensurability which he then never uses explicitly; but I shall be arguing that substantial portions of the *Elements,* Plato's mathematical references, and other fourth- and third-century testimonies can be interpreted very plausibly in anthyphairetic terms. The sequence of repetition numbers that arises from anthyphairesis represents the result of repeated subtractions (B_2), not divisions, and this interpretation is corroborated by the Greek name, derived from *anti-hypo-hairesis,* 'reciprocal sub-traction'. I wish to avoid the modern names, the 'Euclidean algorithm' and 'continued fractions', by which the process and associated mathematical objects are now known, for the following reasons, among others that will underlie my reconstruction: the Euclidean algorithm is now generally construed as a division process, whereas anthyphairesis is based on repeated subtraction; continued fractions are now handled using the real numbers and a sophisticated generalisation of fractions, whereas I shall propose a different approach in which a heuristic stage, based on the properties of what are now called Farey series and making use only of manipulations of *arithmoi,* is followed by geometrically formulated proofs in the style of, but much more elaborate than, those we have seen in *Meno* 81–85; and Euclid's *Elements* may even have contributed to the decline of interest in and knowledge of the process.

It seems natural to expect that ratios should themselves be comparable in size though, as the boy points out (B_{18}), it is not obvious that the comparison of cardinal (B_2) and repetition (B_4) numbers should automatically imply a comparison of anthyphairetic ratios, sequences of repetition numbers. Euclid makes a similar kind of silent assumption in his proportion theory at *Elements* V 10: he has defined proportion and disproportion (a condition that *a* to *b* should be less than or, implicitly, greater than *c* to *d*) in V Definitions 5 and 7, but in V 10 he assumes a trichotomy law, that any four magnitudes must either be proportional or disproportional. The comparison of anthyphairetic ratios reveals the curious feature that the lexicographic ordering of ratios is reversed at the second, fourth, sixth, etc. steps (B_{20} & S_{21}). This is only the first of the instances we shall find in which an even–odd parity plays a fundamental role in anthyphairetic ratios, and, I shall argue, it is only one of the many examples of behaviour to which Plato might be referring when he talks of the 'mutual relationships' of 'the even and the odd' and, sometimes, the 'greater and the less'; one such passage, *Gorgias* 451a–c, was quoted in Section 1.2(e), and others will be considered in Section 4.2.

As my slaveboy observes (B_{36}), the same process can be applied to two lines and, more generally, to any two geometrical magnitudes which allow of the necessary operations of comparison and subtraction. The *Topics* proposition provides a very convenient configuration for evaluating the ratio of two plane regions (B_{38}), and, as Aristotle says at *Topics* 158b29 ff. (quoted in full in Section 1.2(d)),

The areas and the lines have the same *antanairesis* and this is the definition of the same ratio (*ton auton logon*).

The word used here is derived from the compound *anti-ana-hairesis*, 'reciprocal re-traction'. Some have argued that the two words have different meanings, but I shall treat them as synonymous, as did Alexander of Aphrodisias, in his commentary on this passage (Alexander–Wallies, *IT*, 545):

The said [fact] expressed in such-like terms is not familiar, but it becomes familiar once the definition of 'in proportion' (*analogon*) is stated, that the line and the space are divided in proportion by the drawn parallel. Now this is the definition of proportionals which the ancients used: those magnitudes are in proportion to each other (and *homiōs* to each other) of which the *anthuphairesis* is the same. But he [sc. Aristotle] has called *anthuphairesis antanairesis*.

These words also seem to be used interchangeably in commercial documents on papyrus, to describe a deduction from a financial account. (There is here no suggestion of a process, the *repeated* subtraction, that is conveyed by the adverb *aei* in Euclid's usage; see B_{10}.) Of the many examples, here are two from the same collection of documents, the Zenon Archive: *anthuphairein* in P. Lond. vii 1994, lines 164, 176, 233, and 321, and vii 1995, line 333 (both dated Autumn 251 BC); and *antanairein* in P. Cair. Zen. iii 59355, lines 95 and 150 (dated Summer

243 BC). A third word, *antaphairein*, is found in Nicomachus, *Introduction to Arithmetic* I 13.11, where it is used with exactly the same sense.

It is a natural step within *arithmetised* geometry to assign a unit length to some important or basic line within a figure, and then calculate the length of the other lines. For example, we would say that the length of the diagonal of a square with unit side is $\sqrt{2}$ (a real number); the longer diagonal of a hexagon with unit side is 2, the shorter is $\sqrt{3}$; and so on. The equivalent step within *anthyphairetic* geometry is to evaluate the ratio of two lines, areas, or volumes that are related by some construction. My Socrates suggests several such problems (S_{39} & S_{41}) and he also hints that this approach might reveal useful or interesting behaviour (S_{29}). The rest of this chapter will embark on the study of some of these examples. Since the mathematical approach may be novel, even to many mathematicians, I shall emphasise here this aspect of the subject, but will illustrate the procedures, wherever possible, with ancient examples. Until these mathematical techniques become more familiar, the historical questions of whether or not some of them might be plausible reconstructions of ancient practice cannot be evaluated.

My concern throughout this book will be to explore the properties of anthyphairetic and other kinds of ratios, and not to develop abstract theories of ratio or proportion, akin to those we find in *Elements* V and VII. Investigations of the formal aspects are very important, but a full treatment of the topic can be found in Knorr, *EEE*, Chapter 8 and Appendix B. I see no need to duplicate that work here.

2.2 Some anthyphairetic calculations

2.2(a) *The diagonal and side*

Consider first the problem of finding the ratio of a diagonal and side, proposed at S_{39} and S_{41}. The general problem is as follows: given a line, construct a regular polygon with this line as side, take a diagonal (there will be a choice of different diagonals when there are six or more sides) and calculate the ratio of this diagonal to the side. In each case, we need some additional geometrical construction which will embody the information we seek, just as the first step of the solution of Socrates' problem in the *Meno* was to draw a figure which contained the obliquely placed square. (Note in passing that the word *diagramma* seems, in Plato and Aristotle, to refer ambiguously to either a geometrical figure or a proof.) The constructions for the square, pentagon, and hexagon, the three examples I shall consider here, are given in Fig. 2.1; these having been drawn, it remains to say the appropriate words about them.

First, the square. Figure 2.1(a) shows how, starting from the small

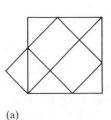 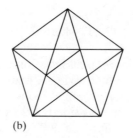 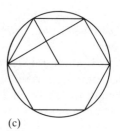

Fig. 2.1

square placed obliquely in the corner, we can construct a large square with side equal to the small side plus small diagonal; when the figure is completed, as shown, it becomes clear that the large diagonal is twice the small side plus the small diagonal. Observe that this figure is not unlike the successful third *Meno* construction, and it also resembles the proof of Pythagoras' theorem that may have been excised from between *Elements* II 8 and 9 (see Fig. 1.5): all consist of a square or rectangle placed obliquely inside a larger square. We now calculate the ratio of the large diagonal plus large side to the large side: this clearly is twice, followed by the ratio of the large diagonal minus side to the large side. Again, from the figure, this latter ratio, taken the greater to the less (B_2), will be the ratio of the small diagonal plus side (equal to the large side) to the small side (equal to the large diagonal minus side). Since the absolute size of the figure does not affect these ratios (cf. *Meno* 82c), we have proved that the ratio of a diagonal plus side to side of any square is equal to twice, followed by the same ratio; and so is equal to twice, twice, followed by the same ratio; and so is equal to twice, twice, twice, twice, followed by the same ratio; and so is equal to an unending sequence twice, twice, twice, etc. Now we can modify the first term to get the result that the ratio of the diagonal to side of a square is equal to once, twice, twice, twice, etc. (Any readers who find such a complete but informal description unsatisfactory or unconvincing should, here and elsewhere, rework the material to their own satisfaction, within the stylistic restrictions imposed by the historical context.) I shall discuss some evidence that may relate to Fig. 2.1(a) in Section 3.6, below.

This proof depends on the result that my slaveboy appreciated right from the start (B_8 & B_{10}), that some terms of a given ratio, of numbers or magnitudes, can be evaluated and then followed by the ratio of the remaining numbers or magnitudes. Here this gave rise to the recursive behaviour in which a ratio was equal to twice, followed by the same ratio, and so the original ratio, by eating its own tail, is completely evaluated. This remarkable kind of phenomenon will itself recur time and time again. The proof started from the ratio of diagonal plus side to side purely for reasons of brevity and convenience; the evaluation of the ratio of diagonal to side works equally well, as the reader is encouraged to verify; but is one step longer, since the recursive property does not then manifest itself until the second subtraction step has been performed.

Next the pentagon. It is easy to prove that the diagonals of a pentagon will cut each other so as to generate a smaller pentagon, in a configuration known as a pentagram; see Fig. 2.1(b). The many isosceles triangles in the figure enable us to see that the side of the large pentagon is equal to the side plus diagonal of the small, while the large diagonal is the small side plus twice the small diagonal. The evaluation of the ratio of the large diagonal to the large side then leads, after two steps, to the same kind of recursive behaviour; and this shows that their ratio is equal to the unending sequence once, once, once, etc. The reader is again strongly encouraged to work out the details.

There are many propositions on or involving the pentagon in the *Elements*. Of these, the closest in spirit to this figure seems to be XIII 8:

If in an equilateral and equiangular pentagon straight lines subtend two angles taken in order, they cut one another in extreme and mean ratio, and their greater segments are equal to the side of the pentagon,

which can be elucidated, pruned, and extended as follows:

If in an equilateral and equiangular pentagon [two adjacent diagonals] cut one another,... their greater segments are equal to the side of the pentagon [and their lesser segments are equal to the diagonals of the smaller pentagon formed by the larger diagonals].

Next the hexagon; see Fig. 2.1(c). Here there are two different diagonals: the larger is the diameter of the circumscribing circle and so is twice the side of a hexagon (this, of course, is a description of the ratio in any sense of the word), while the shorter is such that the square on it is three times the square on the side (this result is contained, explicitly, in the proof of XIII 12). The shorter diagonal is an example of a problem my Socrates called 'the dimensions of squares' (S_{41}): given a line ρ and two numbers n and m, construct squares equal to n times and m times the square on ρ; what then can we say about the ratio of their sides? I shall return to this example in Section 2.3(b) below, and discuss the solution of the problem in detail in the next chapter.

Other ratios associated with regular polygons and polyhedra will be considered in Chapter 5.

2.2(b) *The circumference and diameter*

One of the simplest of geometric figures is a circle. Clearly there are problems in treating the circumference of the circle as a geometrical magnitude, to be added to or compared with other line magnitudes; but Archimedes, in his *Sphere and Cylinder* I, extends the scope of his geometry to include well-behaved curved one- and two-dimensional geometrical objects. The book is prefaced by some Axioms (*Axiōmata*), which start:

There are in a plane certain finite bent lines which either lie wholly on the same side of the straight lines joining their extremities or have no part on the other side. I call concave in the same direction a line such that, if any two points whatsoever are taken on it, either all of the straight lines joining the points fall on the same side of the line, or some fall on one and the same side while others fall on the line itself, but none falls on the other side;

and some Postulates (*Lambanomena*), which start:

Of all lines which have the same extremities, the straight line is the least. Of other lines lying in a plane and having the same extremities, [any two] such are unequal when both are concave in the same direction and one is either wholly included between the other and the straight line having the same extremities with it, or is partly included by and partly common with the other; and the included line is the lesser.

As a result of these principles, Archimedes is able to say, at the end of his postulates, that

With these premises, if a polygon be inscribed in a circle, it is clear that the perimeter (*perimetros*) of the inscribed polygon is less than the circumference (*peripherias*) of the circle; for each of the sides of the polygon is less than the arc of the circle cut off by it,

and, in Proposition 1, he then proves that the perimeter of a circumscribed polygon is greater than the circumference. The first six propositions then establish theoretical results about approximating the circumference by inscribed and circumscribed polygons; but the actual calculation of the magnitudes of their perimeters is carried out elsewhere, in his short work *Measurement of a Circle*. Proposition 3 of this treatise evaluates that:

The perimeter of every circle is three-times the diameter and further it exceeds by [a line] less than a seventh part of the diameter but greater than ten seventy-first [parts of the diameter].

I shall examine some textual details of Archimedes' argument in Section 7.3(a), where I shall describe some of the corruptions, additions, and emendations to our surviving manuscripts. I will illustrate there how Greek procedures may be based on manipulations that are very different in kind from, though equivalent to, our fractional manipulations; so that the last words here, "ten seventy-first [parts]" (*deka hebdomēkostomonois*), are problematic and do not carry the same connotations as our $\frac{10}{71}$. From the anthyphairetic point of view, the calculation shows explicitly that the ratio of circumference to diameter is less than three-times, seven-times, and greater than three-times followed by the ratio of seventy-one to ten (taken, as always, the greater to the less; see B_2), i.e. greater than three-times, seven-times, ten-times:

$$[3, 7, 10] < c:d < [3, 7].$$

Other parts of the calculation also fit very well into an anthyphairetic context, as I shall illustrate in Section 2.4(d), below. We also have a short report of another closely related calculation by Archimedes, at Heron, *Metrica* I 25:

Archimedes proves in his work on plinthides and cylinders that of every circle the perimeter has to the diameter a greater ratio than 211 875 to 67 441, but a lesser ratio than 197 888 to 62 351. But since these numbers are not well-suited for practical measurements, they are brought down to very small numbers, such as 22 to 7.

Unfortunately Heron gives no further details, the work by Archimedes to which he refers is now lost, the alleged lower bound to the ratio is in fact a close upper bound, and the upper bound is a poorer estimate than 22:7.

Now to the illustration of some further properties of ratios, using this ratio $c:d$ as an example, as a first elucidation of my Socrates' remark (S_{29}) that these anthyphairetic ratios contain a lot of information. I encourage any reader who is daunted by the material in this or any subsequent section of this chapter and the next to skip-read or cut to the beginning of the following section, and return later, when the material becomes necessary, or more accessible, or more appealing. Each section starts from a modest level; some rise to encounter quite subtle phenomena.

The expansion of this ratio of circumference to diameter actually starts

$$c:d = [3, 7, 15, 1, 292, 1, 1, 1, \ldots];$$

I shall extract some of the information that is coded in these numbers. Let us label the lines involved as follows:

$$c - 3d = r$$
$$d - 7r = s$$
$$r - 15s = t$$
$$s - t = u$$
$$t - 292u = v$$
$$u - v = w$$

etc.

```
                                           First c=3d+r
c └────────────────┴──────────────────────┘
                                           Third r=15s+t

      Second d=7r+s
d └──┴──┴──┴─┴─┴─┘
      Fourth s=t+u
```

Fig. 2.2 Four steps in the anthyphairesis of $c:d$.

(This is not algebra, i.e. symbolic manipulation, but mere shorthand; and a similar comment applies to the later manipulations in which we count intervals.) The lines r, t, and v first arise as subintervals of c; while s, u, and w are subintervals of d; see Fig. 2.2. Now consider the question of estimating the sizes of r, s, t, etc. We can make a series of increasingly precise statements:

(i) If we ignore the remainder at any step, and so treat the subtraction process as if it terminated at that step, we will get an approximation to the ratio. The remainders seem to be getting very small very quickly, so this should yield increasingly good approximations. The remainders at the first, third, fifth, etc. steps all arise as subintervals of the larger original magnitude c, so ignoring them should give underestimates since it will reduce the size of c; while ignoring the second, fourth, etc. remainders should similarly give overestimates (B_{20} & S_{21}). Hence (S_{21} & B_{22}):

$$[3] < [3, 7, 15] < [3, 7, 15, 1, 292] < \ldots < c:d < \ldots < [3, 7, 15, 1] < [3, 7];$$

then (B_{28} & S_{29}) we can convert this into a string of ratios of numbers:

$$3:1 < 333:106 < 103\,993:33\,102 < \ldots < c:d < \ldots < 355:113 < 22:7.$$

(ii) The instruction to ignore a remainder at a certain step is rather vague, since this would seem to alter all of the preceding steps of the process. What we can do, instead, is to alter the *original* larger or smaller interval so that the subtraction process performed with this modified interval will terminate at the required step. We can see by how much to decrease or increase by counting intervals, as in the following analysis of Fig. 2.2. Consider the first two steps:

$$c = 3d + r = 3(7r + s) + r$$
$$= 22r + 3s$$
$$d = 7r + s.$$

Look at $7c$ and $22d$:

$$7c = 7 \times 22r + 21s$$
$$22d = 22 \times 7r + 22s;$$

2.2 Anthyphairetic calculations 39

they are very close to each other. If we increase $7c$ by s, or decrease $22d$ by s, they will be equal:

$$(c + s/7) : d = c : (d - s/22) = 22 : 7.$$

A similar calculation will hold at the third step, where the decomposition $r = 15s + t$ is introduced:

$$c = 22r + 3s = 22(15s + t) + 3s = 333s + 22t$$
$$d = 7r + s = 7(15s + t) + s = 106s + 7t.$$

Now look at $106c$ and $333d$. Since $106 \times 22 = 333 \times 7 + 1$, this time we have

$$(c - t/106) : d = c : (d + t/333) = 333 : 106;$$

and so on. The process seems to be perfectly general and seems to hold for any ratio; if the nth step of $c : d$ introduces a new interval x as remainder, then the associated ratio $p : q$ which terminates on the nth step is got by adding (n even) or subtracting (n odd) the qth part of x to the antecedent c, or subtracting (n even) or adding (n odd) the pth part of x to the consequent d:

$$(c \pm x/q) : d = c : (d \mp x/p) = p : q,$$

where the signs alternate as n increases. In Section 2.3(c) I shall address the further problem of calculating these ratios $p : q$.

Again, let me emphasise that the symbolism introduced here is merely a convenient abbreviation for longwinded descriptions of how the intervals can be decomposed into more and more, smaller and smaller intervals.

(iii) We can now estimate the error in these approximations. A first approach yields the following: we want to evaluate $c : d$, and have shown that $(c + s/7) : d = 22 : 7$, also, since $d = 7r + s$, we know that $s < d/7$. Hence $22 : 7$ is greater than $c : d$ by an amount that can be described by increasing the antecedent c by less than $d/7^2$. Alternatively, we write $c : (d - s/22) = 22 : 7$, so that we can decrease the consequent d by an interval less than $d/7 \times 22$.

At the next step, we have $(c - t/106) : d = 333 : 106$, and t is less than s (since it is a remainder at the third step when s is subtracted), while $s < d/106$ (from the decomposition $d = 106s + 7t$). Hence $333 : 106$ is greater than $c : d$ by an amount that can be described by decreasing the antecedent c by less than $d/106^2$. Alternatively we can increase the consequent; this gives a slightly less simple and convenient result.

This result also appears to be perfectly general: at the kth stage, if we estimate the remainder x in this way we will get

$$(c \pm x/q) : d = c : (d \mp x/p) = p : q,$$

where x is less than d/q; so the error corresponds to decreasing or increasing the consequent c by an amount less than d/q^2.

(iv) We get a more refined error analysis by exploiting the following heuristic principle: If, at any stage of the evaluation of an anthyphairetic ratio, the lesser term goes a large number of times into the greater, then it must be very much smaller than the greater and the remainder it will leave will be even smaller. So if we neglect this remainder by modifying the antecedent or consequent as in (i) or (ii), and so truncate the anthyphairesis before this large term, we should expect to get a particularly good approximation. Consider, for example, the third step of this ratio $c:d$,

$$r - 15s = t.$$

This shows that $s < r/15$, which combines with the description

$$d = 7r + s$$

to yield that $s < d/15 \times 7$. Hence the first estimate of (iii) can be improved to the conclusion that $22:7 = [3, 7]$ is greater than $c:d$ by an amount that can be described by increasing the antecedent c by less than $d/15 \times 7^2$.

Again, this result is perfectly general: if the $(k+1)$st stage of the process gives

$$w - nx = y,$$

(where the case of large n is of particular interest), then

$$(c \pm x/q) : d = p : q \quad \text{where} \quad x < d/nq,$$

so the error corresponds to decreasing or increasing the consequent c by an amount less than d/nq^2. For example, the ratio $355:113 = [3, 7, 15, 1]$ is in excess by an amount that corresponds to increasing the circumference less than $1/292 \times 113^2$ of the diameter.

It is possible to extend this method further to give a *lower* bound for the error, since if

$$w - nx = y \quad \text{with} \quad y < x$$

we know that x will not go $(n+1)$-times into w. Hence

$$w/(n+1) < x < w/n,$$

and this leads to the estimate

$$d/(n+2)q < x < d/nq.$$

(v) Two historical remarks are in order concerning the calculations I have just evaluated. First, the only fractional quantities involved are the unit fractions or 'parts', $2', 3', 4', \ldots$, described briefly in Section 1.2(c) and to be analysed in Chapter 7. Second, we today would tend to describe the difference between a smaller and a larger ratio, $p:q$ and $r:s$, by giving some arithmetical expression for their difference $r:s - p:q$, either exactly or by specifying some bound such as 'less than one part in a thousand' or the like. Greek practice is different. To my knowledge, ratios, even ratios of numbers, are never subtracted like this; instead a typical manipulation would be to use a fourth proportional, write $r:s = x:q$, and then compare x and p, just as we have been doing.

(vi) The argument of (iv) developed the heuristic principle that a large term in the anthyphairetic expansion indicates the existence of an unusually accurate approximation to the given ratio by the ratio of the relatively small numbers got by truncating the anthyphairesis just before the appearance of this large term. Conversely, if we are calculating, somehow, term by term, the anthyphairesis of a given ratio, then the amount of work needed to calculate a given term cannot be predicted in advance, since it will depend on the size of that term. Suppose, then, that someone, possibly Archimedes, set out to calculate the anthyphairesis of $c:d$; he would already find difficulty in calculating the third term, 15, of the expansion, and would have enormous problems in evaluating the fifth term, 292.

(vii) Consider, further, what will happen if we try to calculate the anthyphairesis of some commensurable ratio by finding closer and closer approximations to it; an illustration, to fix the idea, would be to calculate the ratio of the surface of a sphere to an equatorial section by approximating these by figures that are more amenable to manipulation. As the approximations get better and better, so the remainder corresponding to the last step of the exact calculation will get smaller and smaller, and so the corresponding term of the anthyphairesis will grow bigger and bigger. (In fact, a terminating anthyphairesis should often be considered as finishing with an additional infinite term; see Section 2.3(b) below.) So, for example, let us try to imagine the reaction of anyone (again possibly Archimedes) who set out to calculate the anthyphairesis of circumference to diameter of a circle with a view to formulating a hypothesis, for example on whether it is commensurable or not: as the fifth term appears to grow in size, they might be tempted to conclude that the ratio was indeed commensurable.

If we, today, are tempted to feel patronising about such ignorance, then we should reflect that π was only proved to be irrational (that being our arithmetised formulation of this problem) by Lambert, in 1768, and to be transcendental, by Lindemann, in 1882; that numbers of the form α^β, such as $2^{\sqrt{3}}$, which look as if they ought to be transcendental, were only proved to be so by Gelfond in 1934, and this important result was generalised to $\alpha_1^{\beta_1} + \alpha_2^{\beta_2} + \ldots + \alpha_k^{\beta_k}$ by Baker only in 1966; that, while

Euler investigated the zeta function

$$\zeta(s) = \frac{1}{1^s} + \frac{1}{2^s} + \frac{1}{3^s} + \ldots$$

and found that $\zeta(2) = \pi^2/6$, $\zeta(4) = \pi^4/90$ and, generally, that $\zeta(2n) = \pi^{2n}/k$, (a result described in Weil, *NT*, 184 and 256–76, as "one of Euler's most sensational early discoveries, perhaps the one which established his growing reputation most firmly.... Characteristically, before solving it, Euler had engaged in extensive numerical calculations"), it was not until 1978 that it was proved, by Apéry, that $\zeta(3)$ is irrational, and it is still not known if it is transcendental; and that it is not yet known whether Euler's constant

$$\gamma = \lim \left[1 + \frac{1}{2} + \frac{1}{3} + \ldots + \frac{1}{n} - \log n \right]$$

is rational or irrational. Some questions like this are extraordinarily difficult to answer.

2.2(c) *The surface and section*

The obvious two-dimensional generalisation of the problem of finding the ratio of circumference to diameter is to find the ratio of the surface of a sphere to its equatorial section. At first sight, this ratio of two curved regions would seem to be of even greater difficulty (see Section 2.2(b)(vii), above); in fact the answer is as simple as it could be. Archimedes proves, at *Sphere and Cylinder* I, Proposition 33:

The surface of any sphere is four-times the greatest of the circles in it.

Many other results by Archimedes yield similar examples of ratios which are commensurable, and have anthyphairetic expansions containing only one or two terms.

2.3 Anthyphairetic algorisms

2.3(a) *The Parmenides proposition*

Plato gives, at *Parmenides* 154b–d, a special case of a proposition that we shall use:

If one thing actually is older than another, it cannot be becoming older still, nor the younger younger still, by any more than their original difference in age, for if equals be added to unequals, the difference that results, in time or any other magnitude, will always be the same as the original difference.... [But] if an equal time is added to a greater time and to a less, the greater will exceed the less by a smaller part (*morion*).

2.3 Anthyphairetic algorisms

To paraphrase and simplify slightly Plato's homely paradox: as two people grow older together, the difference between their ages will remain constant, but the ratio of their ages will get closer to 1:1. Let r and s denote the different ages, and p the equal time added to both, then

if $p:p < r:s$, (i.e. $r > s$)   then $p:p < (p+r):(p+s) < r:s$.

A slightly more general form of this result will be basic to my reconstruction:

if $p:q < r:s$ then $p:q < (p+r):(q+s) < r:s$.

I shall call this 'the *Parmenides* proposition'.

An illustration of this *Parmenides* proposition, in modern usage, would be that the average speed for a day's journey must lie between the average speeds for the morning and afternoon. Note, however, that this illustration does not apply to early Greek mathematics, and indeed such an inhomogeneous ratio, of distance to time, is prohibited by *Elements* V Definition 3. We can see a possible context for this prohibition by observing how this ratio, as now conceived, depends on an arithmetised foundation of mathematics, in which it is interpreted as something like (distance in miles) ÷ (time in hours). Also observe that we cannot evaluate the anthyphairesis of such a ratio of incomparable magnitudes without first converting it into some homogeneous ratio, so that we can apply the operation of B_2; typically we now arithmetise, though we could equally well geometrise and represent both distance and time by line segments. Similar comments will apply to other definitions of ratio that I shall give in Chapter 4 (cf. S_{35}).

We shall use the proposition in the case where p, q, r, and s are numbers, *arithmoi*. Euclid's theory of proportion for numbers in *Elements* VII contains no definitions or propositions about disproportion, 'p to q is greater or less than r to s', but to explore the *Parmenides* proposition in this context, we can appeal to the device I described in Section 1.2(d), and extend a proposition into a new definition. We take *Elements* VII 19, which asserts that

if $p:q :: r:s$ then $ps = qs$, and conversely,

and make some extension of the theory to enable us to assert, either as a definition or as a proposition, that

if $p:q < r:s$ then $ps < qr$, and conversely.

The *Parmenides* proposition now follows easily: we wish to prove that

$$p:q < (p+r):(q+s),$$

which is equivalent to

$$p(q+s) < q(p+r), \quad \text{i.e.} \quad ps < qr,$$

and this now is equivalent to our assumption that $p:q < r:s$. The second inequality follows by a similar argument. The result can be conceived as belonging either within a theory of ratio, or within a theory of proportion that has been extended to include disproportion.

It is worth looking at this result in the case when p, q, r, and s are magnitudes. The proof I have just given cannot then apply, since the product of two magnitudes cannot be defined in many concrete cases, such as for two plane regions, two solid regions, two intervals of time, or even two non-rectilineal lines. But all the ingredients for a different proof can be found in Book V, and while Euclid does not actually formulate the proposition for magnitudes, he does prove an important related result on equality, at V 12:

$$\text{if } p:q::r:s, \quad \text{and} \quad p:q::t:u, \text{ etc.,}$$
$$\text{then } p:q::(p+r+t+\text{etc.}):(q+s+u+\text{etc.}).$$

We shall use this in our proof. We proceed as follows: if $p:q < r:s$ then, by a step that Euclid assumes in V 18, the so-called existence of a fourth proportional, there is a magnitude x such that

$$p:q::r:x.$$

(Note that this step is not true for numbers; for example we cannot find a number x such that $3:4::5:x$.) By V 8, we know that $x > s$. An application of V 12 gives

$$p:q::(p+r):(q+x),$$

and another application of V 8 yields the result that

$$p:q < (p+r):(q+s).$$

The other inequality is proved similarly.

The *Parmenides* proposition, together with the proof I have just given, is found in Pappus, *Collection* VII 8. It was rediscovered by Chuquet, in his *Triparty en la Science des Nombres* of 1484, where it is called "la règle des nombres moyens". It is a basic property of Farey series. And—the end to which I shall put it here—it can be used to explore many of the properties of anthyphairesis.

2.3(b) *An algorism for calculating anthyphaireses*

Two basic problems within anthyphairetic ratio theory are as follows. First, given some ratio θ, find a procedure for calculating the anthyphairesis $[n_0, n_1, n_2, \ldots]$ of θ; and, conversely, given such an anthyphairetic ratio, find some convenient procedure for calculating the sequence of approximations, $[n_0, n_1, \ldots, n_k] = p_k:q_k$, called the *convergents* of the ratio (B_{26} to S_{29}). I shall now describe an algorism to solve the first of these problems, and illustrate it with the example that arose as the ratio of the shorter diagonal of the hexagon to the side, a typical example of the following general class of the ratio of sides of squares: given a line ρ and numbers n and m, construct squares equal to n and m times the square on ρ and call their sides $\sqrt{n}$ and $\sqrt{m}$ respectively (so this notation stands for the lines, the geometrical objects, and not their arithmetical interpretation as lengths, real numbers); now evaluate $\sqrt{n}:\sqrt{m}$. The example considered here will be the ratio $\theta = \sqrt{3}:1$.

The procedure must be formulated in terms of, and only depend on, the kind of interpretations and manipulations of the *arithmoi* that were described in 1.2(c). For example, it is not appropriate to arithmetise in terms of decimal numbers and proceed as we usually do today:

$$\sqrt{3} - 1 \times 1 = 1 \cdot 732 \ldots - 1 \times 1 = 0 \cdot 732 \ldots$$
$$1 - 1 \times 0 \cdot 732 \ldots = 0 \cdot 268 \ldots$$
$$0 \cdot 732 \ldots - 2 \times 0 \cdot 268 \ldots = 0 \cdot 196 \ldots$$
$$0 \cdot 268 \ldots - 1 \times 0 \cdot 196 \ldots = 0 \cdot 072 \ldots$$

etc.

so that $\sqrt{3}:1 = [1, 1, 2, 1, \ldots]$; or to use common fractions and write $[1, 1, 2, 1]$ as

$$1 + \cfrac{1}{1 + \cfrac{1}{2 + \cfrac{1}{1}}} = 1 + \cfrac{1}{1 + \cfrac{1}{3}} = 1 + \frac{3}{4} = \frac{7}{4}.$$

These kinds of manipulations occur in the modern theory of continued fractions, and they can be adapted into convenient ways of proceeding with numerical calculations on a pocket calculating machine, but they do not correspond to anything that we know of the style and techniques of early Greek mathematics.

Instead we use the *Parmenides* proposition. To operate it, we shall need two ingredients: a test for whether $p:q$ is less than, equal to, or greater than the given ratio $\sqrt{3}:1$, and the initial under- and overestimates. The test will be

$p:q$ is $<$, $=$, or $> \sqrt{3}:1$ according as p^2 is $<$, $=$, or $> 3q^2$.

I shall not stop here to consider the formal status of this proposition within ratio theory, but will rather adopt the attitude of my Socrates, at S_{19}. The initial estimates will be the most natural ones, like those that my slaveboy produced at B_4:

$$1:1 < \sqrt{3}:1 < 2:1$$

or, in general,

$$n:1 < \theta < (n+1):1,$$

which I shall refer to as the 'standard' under- and overestimates. Also I have taken the given ratio θ as 'the greater to the less'. There is never any difficulty in either setting up any problem so as to consider this case, or in handling the reciprocal ratio, 'the less to the greater'; I shall give an illustration of this in Section 2.4(a), below.

Now proceed as follows. By the *Parmenides* proposition, $1:1<3:2<2:1$ and, since $3^2<3.2^2$, $3:2$ is an underestimate. Hence $3:2<\sqrt{3}:1<2:1$. A second application of the proposition leads to the underestimate $5:3$, hence $5:3<\sqrt{3}:1<2:1$. This then generates an overestimate $7:4$; and so on. The operation is conveniently set out as in Table 2.1.

The process will generate runs of improving under- and overestimates. If, for convenience and simplicity, we retain only the estimate at the end of each run (they are underlined in the table), we get

$$1:1<5:3<19:11<\sqrt{3}:1<26:15<7:4<2:1$$

which we can now convert into anthyphairetic ratios,

$$[1]<[1,1,2]<[1,1,2,1,2]<\sqrt{3}:1<[1,1,2,1,2,1]<[1,2,1]<[2].$$

Anybody who performs these calculations—and everyone who would understand should—will notice an ambiguity of a terminating expansion that arises by holding over the very last subtraction to a new step:

$$[n_0, n_1, \ldots, n_k + 1] = [n_0, n_1, \ldots, n_k, 1].$$

This is the only possible ambiguity of anthyphairetic expansions. Also there is a subtlety of inequalities between terminating ratios, beyond that

TABLE 2.1. Approximations to $\sqrt{3}:1$.

Under-estimate	Over-estimate	New estimate $p:q$	p^2	$3q^2$	Under/equal/over	Run length
1:1	2:1	3:2	9	12	3 under	
3:2	2:1	5:3	25	27	2 under	2
5:3	2:1	7:4	49	48	1 over	1
5:3	7:4	12:7	144	147	3 under	
12:7	7:4	19:11	361	363	2 under	2
19:11	7:4	26:15	676	675	1 over	1
19:11	26:15	45:26	2025	2028	3 under	
etc.						

described in S_{21}. Consider the two relationships

$$[2, 1] = [3] < [3, 2],$$

neither of which seems to conform to S_{21}. The equality has just been explained; the inequality arises because any terminating expansion should be conceived as finishing with a large term, bigger than any number that will occur elsewhere; let us write this as 'm' for 'many' or 'millions' or 'myriads'. Then

$$[2, m] < [2, 1, m]$$

now does satisfy Socrates' description. So S_{21} should be modified to:

S'_{21}: To see which of the two ratios is the greater, proceed as follows. Express any terminating ratio in the standard form, in which the last term is not 'once', and add an extra notional very large step of 'many-times'. Now note whether their terms differ first at the first, third, fifth, ... step, or the second, fourth, sixth, ... step, since, in the latter case, the reverse of the relationship of less and greater between the *terms* holds between the *ratios*. ...

Now consider the string of inequalities that we have just calculated. It is reminiscent of those that appeared in my dialogue, at S_{21} and B_{28}, and this suggests that it may be related to the anthyphairesis. In fact, the connection is very close indeed, since we can deduce immediately from our innermost inequalities,

$$[1, 1, 2, 1, 2] < \sqrt{3}:1 < [1, 1, 2, 1, 2, 1],$$

that

$$\sqrt{3}:1 = [1, 1, 2, 1, 2, 1 \text{ or more}, \dots].$$

Alternatively, we can extract this information directly from Table 2.1, by reading off the lengths of the runs of under- and overestimates, except for some slight difficulty over the first two terms of the expansion. This difficulty, and its resolution, can best be explained to a modern audience by starting off the algorism with the universal, but un-Greek, approximations

$$0:1 < \theta < 1:0,$$

as in Table 2.2 where the expansion of a ratio $\theta = [n_0, n_1, \dots]$ is illustrated. The first term n_0 accounts for the first n_0 lines of Table 2.2; the next line generates the overestimate $(n_0 + 1):1$; and the $(n_0 + 2)$nd line, marked by an asterisk, corresponds to the first line of Table 2.1. Hence n_1 is one plus the number of overestimates at the beginning of Table 2.2. In the case of $\sqrt{3}:1$, there are no initial *over*estimates, so this yields $\sqrt{3}:1 = [1, 1, \dots]$. Thereafter, the lengths of the runs of under- and overestimates of Table 2.1 give the successive terms of the anthyphairesis.

TABLE 2.2. Expansion of $\theta = [n_0, n_1, \ldots]$ starting from $0:1 < \theta < 1:0$. The schema of Table 2.1 starts at the line marked *.

Under-estimate	Over-estimate	New estimate	Type	Run length
0:1	1:0	1:1	under	
1:1	1:0	2:1	under	
...				
$(n_0-1):1$	1:0	$n_0:1$	under	n_0
$n_0:1$	1:0	$(n_0+1):1$	over	
* $n_0:1$	$(n_0+1):1$	$(2n_0+1):2$	over	*
...				
$n_0:1$	$((n_1-1)n_0+1):(n_0-1)$	$n_1n_0+1:n_1$	over	n_1
$n_0:1$	$(n_1n_0+1):n_0$	$((n_1+1)n_0+1):(n_1+1)$	under	
etc.				

I recommend anybody who is unfamiliar with any of these manipulations to calculate a few terms of the anthyphairesis of $\sqrt{n}:1$ for $2 \leq n \leq 10$.

2.3(c) *An algorism for calculating convergents*

As my Socrates explains at S_{27}, there is no problem in evaluating individual convergents: $[n_0] = n_0:1$, $[n_0, n_1] = (n_0n_1+1):n_1$, $[n_0, n_1, n_2] = (n_0n_1n_2 + n_0 + n_2):(n_1n_2+1)$, and so on. What we now develop is a much more convenient forward method. I shall describe and illustrate it on the examples of convergents of the two ratios $[2, 3, 4, 5, \ldots]$ and $[5, 5, 10, 5, 10, \ldots]$.

One problem with $[2, 3, 4, 5, \ldots]$ is that we do not know if any ratio θ exists that has this as anthyphairesis. This, I feel, is not a problem that would have had any meaning for the Greeks, even within the most extreme developments of my reconstruction, so I shall ignore it here; its solution depends either on the sort of eighteenth-century manipulations at which Euler excelled, or on nineteenth-century ideas of the foundations of mathematics. I choose this example since the numbers involved illustrate clearly and unambiguously the manipulations to be performed. This problem does not arise with $[5, 5, 10, 5, 10, \ldots]$, since this is the anthyphairesis of some explicit ratio that I shall identify later in this section, and establish by rigorous geometrical proof in the next chapter.

So suppose that $\theta = [2, 3, 4, 5, \ldots]$. This means that, if we expand θ using the algorism of the previous section, then the pattern of runs of under- and overestimates will be determined by this expansion. For convenience, I shall adopt, in Table 2.3, the format of Table 2.2, and

2.3 Anthyphairetic algorisms

TABLE 2.3. *The approximation procedure for the ratio* $\theta = [2, 3, 4, \ldots]$

Under-estimate	Over-estimate	New estimate $p:q$	Type	Run length
0:1	1:0		under	
		☐	under	2
			over	
* 2:1	3:1		over	*
		☐	over	3
			under	
			under	
			under	
		☐	under	4
etc.				

start with the initial estimates

$$0:1 < \theta < 1:0;$$

the more strictly historical algorism will start instead (at the line in Table 2.3 marked with an asterisk) with the initial estimates

$$2:1 < \theta < 3:1.$$

With this information, we can set out the skeleton format of Table 2.3, which contains enough information for us now to insert the remaining information by an automatic procedure. In particular we are most interested in the derivation of the run-end estimates, the ratios that go into the boxes in column 3; these, it can be seen, are the successive convergents. Clearly the first box contains $(2 \times 1 + 0) : (2 \times 0 + 1) = 2:1$. Then the next line starts with the estimates $2:1 < \theta < 1:0$ and so, after three steps of overestimates, the next box will contain $(3 \times 2 + 1) : (3 \times 1 + 0) = 7:3$. Then the next line starts with the estimates $2:1 < \theta < 7:3$ and so, after four steps of underestimates, the next box will contain $(4 \times 7 + 2) : (4 \times 3 + 1) = 30:13$; and so on. We can abstract this information in Table 2.4 where, for convenience, it is written out in a compact horizontal format.

Once two columns of Table 2.4 are known, the rest of the table can be calculated easily and quickly. The table can be started either from the first two columns, which correspond to the estimates

$$0:1 < \theta < 1:0,$$

or the next two columns can be written down, since $[2] = 2:1$ and $[2, 3] = 7:3$, and the algorism applied from this point.

TABLE 2.4. The convergents of $[2, 3, 4, 5, 6, \ldots]$

n_k			2	3	4	5	6
p_k	0	1	2	7	30	157	$(6 \times 157 + 30) = 972$
q_k	1	0	1	3	13	68	$(6 \times 68 + 13) = 421$
type	under	over	under	over	under	over	under

As a second example, consider $\varphi = [5, 5, 10, 5, 10, \ldots]$; the reader is encouraged to compute the first five convergents of this ratio and confirm immediately that

$$5:1 < 265:51 < 13775:2651 < \varphi < 1351:260 < 26:5.$$

The point of this example is that early results in what I have called the problem of dimensions of squares (see S_{41}, and Sections 2.2(a) and 2.3(b) above) will be that

$$\sqrt{(n^2 + 1)}:1 = [n, 2n, 2n, 2n, \ldots]$$
$$\text{and } \sqrt{(n^2 + 2)}:1 = [n, n, 2n, n, 2n, n, 2n, \ldots],$$

where the expansions are periodic, repeating indefinitely—this will be proved in the next chapter. Hence $\sqrt{27}:1 = [5, 5, 10, 5, 10, \ldots]$, and the calculation of the convergents then tells us that

$$265:51 < \sqrt{27}:1 < 1351:260,$$

in which the error estimates of Section 2.2(b) show that

if $\sqrt{(27 + x)}:1 = 265:51$, then $x < 1/(5 \times 51^2)$, and
if $\sqrt{(27 - y)}:1 = 1351:260$, then $y < 1/(10 \times 260^2)$.

But since $\sqrt{27} = 3\sqrt{3}$ (this is a geometrical statement, about fitting together nine squares each equal to three times a given square, something in the style of Plato's slaveboy's first two attempts at doubling a square), this inequality implies that

$$265:153 < \sqrt{3}:1 < 1351:780,$$

with the corresponding error estimates also being divided by 3. I shall return to this statement in Section 2.4(d), below.

There is one place in the *Elements* where one looks, in vain, for details of this kind of calculation of convergents: At the beginning of Book X we find anthyphairesis introduced as a criterion of incommensurability:

If, when the less of two unequal magnitudes is continually subtracted in turn (*anthuphairein aei*) from the greater, that which is left never measures the one before it, the magnitudes will be incommensurable [X 2],

2.4   Further calculations   51

and as a construction to solve the problems

Given two (or three) commensurable magnitudes, to find their greatest common measure [X 3 (or 4)].

These are then followed by a block of four propositions that are variations on X 5:

Commensurable magnitudes have to one another the ratio which a number has to a number.

At this point, it would surely be an interesting and obvious step to relate the pattern of the anthyphairesis to the numbers involved in this ratio, but on this the text is silent. There would be no difficulty in describing the backwards procedure (Socrates' explanation at the beginning of S_{27} can easily be translated into Euclidean idiom), but it is not easy to see how to express the much more convenient forward procedure within the stylistic restrictions of Euclid's *Elements,* where the enunciations of propositions cannot contain any reference to particular figures or numbers, and where no symbolic manipulation is permitted (see the end of S_{27}). Nevertheless, at no point does either algorism transcend any of the restrictions on *arithmētikē*. The absence of any such exploration from the *Elements* has meant that very few of those who have learned their mathematics from Euclid, either directly or indirectly, have any intimate experience of the operation of anthyphairesis, in any guise.

2.4   Further anthyphairetic calculations

2.4(a)   *Eratosthenes' ratio for the obliquity of the ecliptic*

Ptolemy explains, at *Syntaxis* I 12, how to measure the obliquity of the ecliptic, and then concludes:

We found that the arc between the northernmost and southernmost points, which is the arc between the solsticial points, is always greater than $47\frac{2}{3}°$ and less than $47\frac{3}{4}°$. From this we derive very much the same ratio as Eratosthenes, which Hipparchus also used. For [according to this] the arc between the solstices is approximately 11 parts where the meridian is 83 [Ptolemy–Toomer, *PA*, 63].

This asserts that $11:83$ lies between $47\frac{2}{3}:360$ and $47\frac{3}{4}:360$; and this curious ratio $11:83$ has provoked occasional perplexed comment. Let us apply the approximating algorism. We start with the standard under- and overestimates to the interval,

$$1:8 < 47\tfrac{2}{3}:360 < 47\tfrac{3}{4}:360 < 1:7,$$

and apply the *Parmenides* proposition until the first intermediate ratio is generated. The calculation is set out in Table 2.5; the result, $11:83$, arises after four lines of calculation.

TABLE 2.5. Eratosthenes' ratio for the obliquity of the ecliptic.

Under-estimate	Over-estimate	New estimate $p:q$	$360p$	$47\frac{2}{3}q$	$47\frac{3}{4}q$	Type
1:8	1:7	2:15	720	715	$716\frac{1}{4}$	over both
1:8	2:15	3:23	1080	$1096\frac{1}{3}$	$1098\frac{1}{3}$	under both
3:23	2:15	5:38	1800	$1811\frac{1}{3}$	$1814\frac{1}{4}$	under both
5:38	2:15	7:53	2520	$2526\frac{1}{3}$	$2530\frac{1}{4}$	under both
7:53	2:15	9:68	3240	$3241\frac{1}{3}$	3247	under both
9:68	2:15	11:83	3960	$3956\frac{1}{3}$	$3963\frac{1}{4}$	between

By-products of this calculation are the evaluations:
$$47\tfrac{2}{3}:360 = [0, 7, 1, 1, 1, 4, \ldots] \text{ and}$$
$$47\tfrac{3}{4}:360 = [0, 7, 1, 1, 5 \text{ or more}, \ldots].$$
In fact,
$$47\tfrac{2}{3}:360 = [0, 7, 1, 1, 4, 3, 1, 3,],$$
$$11:83 = [0, 7, 1, 1, 5], \text{ and}$$
$$47\tfrac{3}{4}:360 = [0, 7, 1, 1, 5, 1, 6, 2].$$

This example is of a ratio taken 'the less to the greater' which has the effect of introducing an initial step of zero which reverses antecedent and consequent; then the ratio 'the greater to the less' follows. We can handle the situation in words, by saying something like: 'the antecedent of the given ratio is less than the consequent, and the anthyphairesis of the ratio is seven-times, once, ...'. So the use of a zero here carries no implications about the introduction of a zero into the Greek number system.

Reciprocation seems to be a process that is handled fluently and with confidence in Greek mathematics.

2.4(b) *The Metonic cycle*

Astronomers in Babylonia from the time of Darius (*c.*490 BC), and in Greece after the mid-fifth century BC, used a cyclical calendar in which 235 months, some (called 'hollow) of 29 days, the rest of 30 days ('full'), are made equal to 19 years. In Greece, this astronomical (not civil) lunar-solar calendar is attributed to Euctemon and Meton, and is said to have started on the summer solstice, 432 BC. It was then modified by Callippus, in the fourth century, into a cycle of $76 = 4 \times 19$ years.

Although this calendar may seem to embody some very precise astronomical parameters, it is shown in Goldstein, NMC, how it can in

2.4 Further calculations

fact be deduced from the following relatively crude data:

(i) The year is a little more than 365 days long,

(ii) The length of 12 synodic months (i.e. new moon to new moon) is a little more than 354 days (note that $\frac{354}{12} = 29\frac{1}{2}$), and

(iii) The average length of a synodic month (the 'mean' synodic month) is a little more than $29\frac{1}{2}$ days.

We now look for a cycle of p years in which the $365-354 = 11$ days discrepancy, called the 'epact', approximates an integral number q of months; so we have to find p and q such that

$$(29\tfrac{1}{2})q < 11p < 30q,$$

i.e.

$$29\tfrac{1}{2}:11 < p:q < 30:11.$$

This can easily be done by trial and error; or it can be used as another illustration of the algorism. We take the standard under- and over-estimates to the interval:

$$2:1 < 29\tfrac{1}{2}:11 < 30:11 < 3:1$$

and apply the algorism to generate a ratio which lies within the interval, exactly as in the previous example. This gives

$$29\tfrac{1}{2}:11 < 19:7 < 30:11,$$

the cycle in which 19 years are approximately equal to $19 \times 12 + 7 = 235$ months and 6940 days.

The Metonic year is $365\tfrac{1}{4} + \tfrac{1}{19}$ days long. Callippus' calendar incorporates the more accurate parameter that a year is approximately $365\tfrac{1}{4}$ days long, so that 76 years are approximately 27 759 days; and so it relates the three principal astronomical periodic rotations: the earth around its axis, the moon around the earth, and the earth and moon around the sun. Another similar, slightly less accurate but more convenient cyclical calendar in which 25 years, 9125 days, are made equal to 309 lunar months, will be described in Section 4.4.

2.4(c) *Aristarchus' reduction of ratios*

Aristarchus observes, within the proof of Proposition 13 of his *On the Sizes and Distances of the Sun and Moon*,

But 7921 has to 4050 a ratio greater than that which 88 has to 45,

and in Proposition 15,

But 71 755 875 has to 61 735 500 a ratio greater than that which 43 has to 37.

We could generate these results by applying the *Parmenides* proposition directly, as we did with Eratosthenes' ratio. But since we do not know, in advance, how long this will take, apart from the very rough bound that, starting with $1:1 < 7921:4050 < 2:1$, we must arrive at a ratio of $88:45$ at the latest within about forty steps (also see my slaveboy's unanswered reflection in B_{24}), it will generally be quicker, whenever possible, to calculate the anthyphairesis and use this to compute the convergents. An easy evaluation gives

$$7921:4050 = [1, 1, 21, 1, 1, 1, 2, 22],$$

from which we can calculate:

n	1	1	21	1	1	1	2	22
p	1	2	43	45	88	etc.		
q	1	1	22	23	45			
type	under	over	under	over	under			

and so the first of Aristarchus' results. (At this point we can see that the *Parmenides* proposition would have required 23 lines and 46 multiplications to arrive at this answer.)

The second inequality follows similarly. In this particular case, it might be sensible to remove the obvious factors $3^3 \times 5^3$; but, in general, it is a notorious fact that factorisation can be an enormously long process, while finding the common measure of two numbers using anthyphairesis, as in *Elements* VII 2, is extremely rapid. This means that the very process for removing common factors efficiently is precisely the same process which generates the anthyphairetic ratio and so, with very little extra effort, the approximations we are seeking.

I remark, in passing, that Aristarchus' treatise is an excellent illustration of what I have called 'non-arithmetised' mathematics. Relationships are expressed as ratios of *arithmoi,* and the only fractional quantities are simple parts. I shall go into this question in more detail in Section 7.3(b), where I shall describe the two mild exceptions to the statement I have just made.

2.4(d) *Archimedes' calculation of circumference to diameter*

Proposition 3 of Archimedes' *Measurement of a Circle* is a mine of calculations, many of which can be interpreted anthyphairetically. I have already discussed its enunciation in Section 2.2(b). The first half of the calculation, which evaluates an upper bound for $c:d$, starts from an underestimate of the ratio of the diameter of the circle to the side of a circumscribing hexagon, $\sqrt{3}:1 > 265:153$. I showed, in Section 2.3(c), how this can be easily and quickly evaluated from the result that

$$\sqrt{27}:1 = [5, 5, 10 \ldots].$$

2.4 Further calculations

The second half, the calculation of a lower bound, starts from an overestimate of the ratio of the shorter diagonal to side of an inscribed hexagon; this (see Section 2.2(a)) is again $\sqrt{3}:1$, and Archimedes' overestimate of $1351:780$ is again a third of the next convergent of $\sqrt{27}:1$, also calculated in Section 2.3(c). Both of these approximations are also convergents of $\sqrt{3}:1$, the ninth and twelfth respectively, but the fact that Archimedes did not take the eighth and ninth or the ninth and tenth convergents to $\sqrt{3}:1$, viz.:

$$265:153 < \sqrt{3}:1 < 97:56 < 362:209 < 1351:780,$$
$$\text{9th} \qquad\qquad \text{8th} \quad\ \text{10th} \qquad\quad \text{12th}$$

any of which would provide sufficient accuracy for his subsequent calculation, has caused some perplexity. A derivation that starts from the expansion of $\sqrt{27}:1$ accounts for this anomaly. Also, the terms of $\sqrt{27}:1$ are relatively large, and this will guarantee the very rapidly improving accuracy of the convergents; see Section 2.2(b)(iv).

Archimedes performs each half of the calculation by successively doubling the number of sides of each hexagon four times, and each time estimating the ratio of the side of the new polygon to the diameter of the circle, always rounding any subsequent approximation upwards for the circumscribed polygon and downwards for the inscribed polygon. Each doubling of the polygons (except the final step of the circumscribing polygon) involves estimating a square root. For example, the first step which replaces the circumscribing hexagon by a dodecagon, requires an underestimate for $\sqrt{349450}:153$. Since

$$\sqrt{349450}:1 = [591, 6, 1, 196, \ldots],$$

this has the extremely accurate underestimate $591\frac{1}{7}:153$. But Archimedes uses instead the approximation $591\frac{1}{8}:153$, either for reasons of computational convenience, or as a result of his computational techniques, or because he can estimate that his initial approximations to $\sqrt{3}:1$ allow him some latitude, or for other reasons; or, as I shall illustrate later, in Section 7.3(a), it is not impossible that this particular fractional part might be a later editorial addition. We shall see there how the text has been re-worked and altered so much that we cannot, any longer, be sure of some of the details of the original calculations.

The calculation of the overestimate ends with the following passage in which I have adapted the translation so as to imitate, in the English text, the different ways in which the numbers are expressed.

The ratio of $A\Gamma$ [the diameter] to the perimeter of the 96-sided polygon is greater than the ratio of 4673 2' to 14 688, which [now taken 'the greater to the less'] is greater than threefold and there are remaining 667 2', which [number] is less than the seventh part of 4673 2'; so that the circumscribed polygon is threefold the diameter and bigger by less than the seventh part; so therefore the perimeter of the circle is less by more than threefold and more by a seventh part.

The description here is reminiscent of the opening lines of my dialogue, but so also would be a conventional arithmetical description of a ratio as an integer part plus a unit-fractional part; any difference would become evident if the ratio needed a more accurate or less simple description. The corresponding passage at the end of the second part of the calculation, the evaluation of the underestimate, reads:

Conversely then the perimeter of the polygon bears to the diameter a ratio greater than 6336 has to 2017 4', which is bigger than 2017 4' or threefold and ten 71st [parts]. And so the perimeter of the inscribed 96-sided polygon is threefold the diameter and greater than the 10 71st [parts], so that also the circle is still more than threefold and greater than the 10 71st [parts].

Unfortunately this passage which, if more explicit, might have shed some light on the way Archimedes conceived the ratios involved, is tantalisingly brief; nor is it made any more informative by being repeated three times. What is more, each occurrence of the expression '10 71st [parts]' shows a wide range of variation between apparently incorrect or meaningless alternatives in the surviving manuscripts. Indeed, in many cases throughout the text the 'correct' numbers cannot be traced back any earlier than early sixteenth-century annotations to the manuscript; for example, the best manuscripts do not contain here the ratio $6336:2017\frac{1}{4}$, but $6301\frac{1}{6}:7017\frac{1}{4}$. More details will be given in Section 7.3(a).

Both of these calculations illustrate again the process that Heron called, in the passage from *Metrica* I 25 quoted in Section 2.2(b), "bringing down to small numbers". For example we can evaluate that

$$14688:4673\tfrac{1}{2} = [3, 7, 667, 2]$$

and

$$6336:2017\tfrac{1}{4} = [3, 7, 10, 2, 1, 36],$$

and hence the overestimate $[3, 7] = 22:7$ and underestimate $[3, 7, 10] = 223:71$. However, it is here more illuminating to describe the same calculation in terms of the *Parmenides* proposition. The starting estimates

$$3:1 < c:d < 4:1$$

follow from the figure of a hexagon inscribed in, and a square circumscribed about a circle. The initial run of overestimates is (read from the right):

$$3:1 < c:d < 22:7 < 19:6 < 16:5 < 13:4 < 10:3 < 7:2 < 4:1;$$

all can be verified from the inequalities

$$c:d < 14688:4673\tfrac{1}{2} < 22:7$$

since
$$14688 \times 7 = 102816 < 4673\tfrac{1}{2} \times 22 = 102817.$$

The next run of underestimates is:
$$3:1 < 25:8 < 47:15 < 69:22 < 91:29$$
$$< 113:36 < 135:43 < 157:50 < 179:57$$
$$< 201:64 < 223:71 < 6336:2017\tfrac{1}{4} < c:d < 22.7.$$

However, since the next estimate satisfies
$$6336:2917\tfrac{1}{4} < 245:768 < 14688:4673\tfrac{1}{2},$$

there is not enough information to decide whether 245:78 is less than, equal to, or greater than $c:d$. In this sense, the ratios given in the enunciation extract all the anthyphairetic information from Archimedes' surviving calculation.

2.4(e) *Pell's equation*

The problem is best explained in the following two texts. First, a challenge to European mathematicians proposed by Fermat in 1657. The translation is from Diophantus–Heath, *DA*, 285–6:

There is hardly anyone who propounds arithmetical questions, hardly anyone who understands them. Is this due to the fact that up to now arithmetic has been treated geometrically rather than arithmetically? This has indeed generally been the case both in ancient and modern works; even Diophantus is an instance. For, although he has freed himself from geometry a little more than others in that he confines his analysis to the consideration of rational numbers yet even there geometry is not entirely absent, as is sufficiently proved by the *Zetetica* of Viete, where the method of Diophantus is extended to continuous magnitude and therefore to geometry.

Now arithmetic has, so to speak, a special domain of its own, the theory of integral numbers. This was only lightly touched upon by Euclid in his *Elements*, and was not sufficiently studied by those who followed him (unless, perchance, it is contained in those Books of Diophantus of which the ravages of time have robbed us); arithmeticians have therefore now to develop it or restore it.

To arithmeticians therefore, by way of lighting up the road to be followed, I propose the following theorem to be proved or problem to be solved. If they succeed in finding the proof or solution, they will admit that questions of this kind are not inferior to the more celebrated questions in geometry in respect of beauty, difficulty, or method of proof.

Given any number whatever which is not a square, there are also given an infinite number of squares such that, if the square is multiplied into the given number and unity is added to the product, the result is a square.

Example. Let 3, which is not a square, be the given number; when it is multiplied into the square 1, and 1 is added to the product, the result is 4, being a square.

The same 3 multiplied by the square 16 gives a product which, if increased by 1, becomes 49, a square.

And an infinite number of squares besides 1 and 16 can be found which have the same property.

But I ask for a general rule of solution when any number not a square is given.

E.g. let it be required to find a square such that, if the product of the square and the number 149, or 109, or 433 etc. be increased by 1 the result is a square.

Second, a passage from Theon of Smyrna–Hiller, *ERMLPU*, 42–5, adapted from the translation of Thomas, *SIHGM* i, 132–7, to bring out the evocation on the word *logos* at the beginning:

Just as numbers potentially contain triangular, square, and pentagonal ratios (*logoi*), and ones corresponding to the remaining figures, so also we can find side and diagonal ratios (*logoi*) appearing in numbers in accordance with the generative principles (*logoi*); for it is from these that the figures acquire balance. Therefore since the unit, according to the supreme generative principle (*logos*), is the starting-point of all the figures, so also in the unit will be found the ratio (*logos*) of the diagonal to the side. For instance, two units are set out, of which we set one to be a diagonal and the other a side, since the unit, as the beginning of all things, must have it in its capacity to be both side and diagonal. Now there are added to the side a diagonal and to the diagonal two sides, for as great as is the square on the side, taken twice, [so great is] the square on the diagonal taken once. The diagonal therefore became the greater and the side became the less. Now in the case of the first side and diagonal the square on the unit diagonal will be less by a unit than twice the square on the unit side; for units are equal, and 1 is less by a unit than twice 1. Let us add to the side a diagonal, that is, to the unit let us add a unit; therefore the [second] side will be two units. To the diagonal let us now add two sides, that is, to the unit let us add two units; the [second] diagonal will therefore be three units. Now the square on the side of two units will be 4, while the square on the diagonal of three units will be 9; and 9 is greater by a unit than twice the square on the side 2.

Again, let us add to the side 2 the diagonal of three units; the [third] side will be 5. To the diagonal of three units let us add two sides, that is, twice 2; there will be 7. Now the square from the side 5 will be 25, while that from the diagonal 7 will be 49; and 49 is less by a unit than twice 25. Again, if you add to the side 5 the diagonal 7, there will be 12. And if to the diagonal 7 you add twice the side 5, there will be 17. And the square of 17 is greater by a unit than twice the square of 12. When the addition goes on in the same way in sequence, the proportion will alternate; the square on the diagonal will be now greater by a unit, now less by a unit, than twice the square on the side; and such sides and diagonals are both expressible (*rhētos*).

The squares on the diagonals, alternating one by one, are now greater by a unit than double the squares on the sides, now less than double by a unit, and the alternation is regular. All the squares on the diagonals will therefore become double the squares on the sides, equality being produced by the alternation of excess and deficiency by the same unit, regularly distributed among them; with the result that in their totality they do not fall short of nor exceed the double. For what falls short in the square on the preceding diagonal exceeds in the next one.

2.4 Further calculations

There is also a similar passage in Proclus' commentary on Plato's *Republic*, which contains an additional description of 'side and diagonal lines'; this will be quoted in Section 3.6(b), where the passages will be discussed in more detail.

Both of these texts show an interest in finding, for a given number n (Theon and Proclus deal only with the case $n = 2$), those numbers q for which $nq^2 + 1$ is a square number p^2. Theon also refers to the further cases where $nq^2 - 1$ is to be a square and, while Fermat does not explicitly mention this case, the choice of the three examples he proposes shows without doubt that he was also interested in this related problem. Euler named the equation

$$x^2 - ny^2 = 1,$$

to be solved in integers, after John Pell (1611–85), a minor British mathematician who had nothing to do with the problem, and the name has stuck. More details of the recent history of the topic will be given in Chapter 9; my concern here is to point out that our basic approximating algorism of Section 2.3(b) gives a practical method for solving the general problem that would be compatible with our knowledge of early Greek mathematics, and to introduce some further Greek evidence. Again I shall concentrate mainly on the mathematical issues; textual and historical problems will be considered in Section 3.6(b), and developments since the seventeenth century will be surveyed in Section 9.1(c).

Suppose we apply the basic algorism for calculating convergents to $\sqrt{n} : 1$. At each step of the process, each application of the *Parmenides* proposition, we need to compare the new approximation $p : q$ with $\sqrt{n} : 1$, and this is done in columns 4 and 5 of Table 2.1, by comparing p^2 and nq^2. In the example of $\sqrt{3}$, calculated in Table 2.1, we see that this gives a pattern of 3 under, 2 under, 1 over, 3 under, 2 under, 1 over, and this pattern will appear to continue indefinitely; it is a further example of the periodic behaviour exhibited by the process which I hope has been obvious to everybody, and which will soon come to dominate our discussion. This same kind of periodic behaviour will seem to appear whatever value of n is chosen, and moreover the value '1 over' will always seem to occur, sometimes alternately with the value '1 under'. In other words, the algorism set out in Table 2.1 will always seem to solve the problem. What is more, the solutions will always seem to be the penultimate convergents before the end of the period of $\sqrt{n} : 1$; if the period contains an even number of terms, they all will be solutions of $x^2 - ny^2 = 1$; if an odd number, the solutions of $x^2 - ny^2 = \mp 1$ will alternate. Hence, if we already know in advance the anthyphairesis of $\sqrt{n} : 1$, we can proceed directly to the calculation of the convergents and hence the solution. For example, the convergents of $\sqrt{27} : 1$, calculated in

Section 2.3(c), suggests that the solutions of $x^2 - 27y^2 = 1$ may be $x = 26$; $y = 5$; $x = 1351$, $y = 260$; and so on. We can then check that this is so.

The heuristic arguments of Section 2.2(b)(iv) again hint at what may be happening: a solution of $x^2 - ny^2 = \pm 1$ will correspond to a very good approximation, $x:y$, to $\sqrt{n}:1$. Now the run-end approximations that occur just before the appearance of a long run of under- or overestimates, being the convergents that precede a large term in the anthyphairesis, will be particularly good approximations. Moreover the largest term in the expansion of $\sqrt{n}:1$ (but not of $\sqrt{n}:\sqrt{m}$) always seems to be the term '$2n_0$' that marks the end of the period. Hence we might expect that the penultimate convergent of each period would give rise to a small value of $x^2 - ny^2$; in fact it turns out to generate the smallest possible value—it cannot be zero, but it is ± 1. If the period contains an even number of terms, the first penultimate convergent and all subsequent penultimate convergents will be overestimates, and hence the positive signs; if odd, they will alternate between under- and overestimates, and so the signs will also alternate.

The other piece of Greek evidence that gives rise to further speculation concerning Pell's equation in antiquity is Archimedes' remarkable *Cattle Problem*, an epigrammatic puzzle involving the sizes of four herds of cattle. The problem has two possible interpretations, which then give rise to the following two mathematical problems: either solve a collection of linear indeterminate equations

$$nx - my = 1,$$

where n and m satisfy

$$nm = 23014894 = 2 \cdot 7 \cdot 353 \cdot 4657;$$

or solve the Pell's equation

$$x^2 - 410286423278424y^2 = 1$$

(in which the coefficient of y^2 factorises to $2^3 \cdot 3 \cdot 7 \cdot 11 \cdot 29 \cdot 353 \cdot 4657^2$), an equation whose smallest solution is the stupendous

$$x = 7 \cdot 760 \ldots \times 10^{206544}$$
$$y = 3 \cdot 831 \ldots \times 10^{206537}.$$

The existence of this problem may imply that either the interpretation which leads to the Pell's equation, or the interpretation as a linear problem, or both interpretations, could be solved, at least 'in principle', by the setter of the problem. So let us therefore look at the mathematics that lies behind the alternative interpretation.

2.4 Further calculations

2.4(f) *The alternative interpretation of Archimedes' Cattle Problem*

Consider the linear indeterminate problem expressed by the equation

$$nx - my = 1,$$

where, as always, n and m are integers, and the equation is to be solved in integers. Since my will then be less than but close to nx, this means that $y:x$ will have to be less than but close to $n:m$; and this again suggests that we look at approximations to $n:m$. Again we can proceed in two similar ways, either by working with the approximating algorism of Section 2.3(b), or by proceeding directly to the evaluation of the anthyphairesis and convergents of 2.3(c). In addition, each approach splits into two distinct cases.

First, look again at a typical sequence of approximation $p:q$, generated by the basic algorism of Section 2.3(b); take, for example, those that arise in column 3 of Table 2.1, and compare successive pairs. We see that

$$3 \cdot 3 - 2 \cdot 5 = -1$$
$$5 \cdot 4 - 3 \cdot 7 = -1$$
$$7 \cdot 7 - 4 \cdot 12 = +1$$
$$12 \cdot 11 - 7 \cdot 19 = -1$$

etc.

and, in general, if $p:q$ and $p':q'$ are consecutive ratios, they seem to satisfy

$$pq' - p'q = \pm 1,$$

the negative sign occurring when $p:q$ increases to $p':q'$, the positive sign when it decreases; column 6 tells us which of these cases applies.

So suppose we wish to solve

$$26x - 15y = +1.$$

We execute the approximation algorism on $26:15$, which will then generate all but the last line of Table 2.1 (at which point it will terminate with an equality). The penultimate estimate $19:11$ will be less than $26:15$. Hence, we guess (and can verify) that

$$26 \cdot 11 - 15 \cdot 19 = 1,$$

and the equation is solved. Other solutions (in fact, every other solution) then arise from the identity

$$26(11 + 15k) - 15(19 + 26k) = 1.$$

Suppose, however, that we wish to solve the dual equation

$$26x - 15y = -1.$$

We now want an approximation that will be *greater* than $26:15$; so we look back in Table 2.1 to the previous overestimate $7:4$ and check that it is a solution:

$$26 \cdot 4 - 15 \cdot 7 = -1.$$

Then, as before, the general solution is given by

$$26(4 + 15k) - 15(7 + 26k) = -1.$$

This second case has led us to the previous convergent, and this suggests that we look at the behaviour of the convergents and ignore the rest of the intermediate approximations. So we can repeat this kind of exploration on Table 2.4, where we see that the cross products of adjacent convergents again satisfy the same kind of relationship. This leads to the second technique for solving these equations, in which we proceed directly to the anthyphairesis and convergents, without passing via the calculations of Table 2.1. Hence, for example, to solve

$$972x - 421y = -1,$$

we evaluate that $972:421 = [2, 3, 4, 5, 6]$; work out the convergents (exactly as in Table 2.4, except that the table will terminate with an equality); and so read off and check that the penultimate convergent, which is an overestimate, satisfies

$$972 \cdot 68 - 421 \cdot 157 = -1,$$

from which we can write down the general solution, as before. If, however, we want to solve the dual equation

$$972x - 421y = +1,$$

we adapt this method by exploiting the ambiguity of the terminating expansion (see Section 2.2(b)): We now write $972:421 = [2, 3, 4, 5, 5, 1]$, and take the penultimate convergent of this modified expansion. This gives the solution

$$972 \cdot 353 - 421 \cdot 815 = +1.$$

These results appear to be perfectly general, and appear to solve every example of such problems. They have been expressed, purely for convenience of exposition, in the form of equations, but they are, in fact, straightforward statements about differences of products of integers that can easily be expressed and manipulated in everday language.

2.5 Notes and references

2.1 Aspects of the modern theory of continued fractions, together with a sketch of the history of its development, will be given in Chapter 9. But for someone who is merely interested in understanding my proposals here, I do not think it is necessary and it may even be distracting to know this modern theory in advance. (In common with perhaps nine-tenths of mathematicians of today, I had never met the topic in my undergraduate or graduate mathematical training, and was only vaguely aware of a revival of interest in it among certain mathematicians when I accidentally stumbled on the path which eventually led to this anthyphairetic interpretation of early Greek mathematics.)

The translation of the passage of Alexander of Aphrodisias, *IT*, is taken from Knorr, *EEE*, 258. This book contains an excellent summary of previous proposals concerning anthyphairesis, in particular, the work of Allman, Becker, von Fritz, Heller, van der Waerden, and Zeuthen, together with Knorr's own interpretations. (I have also taken the brief description of the etymology of *anthuphairesis* and *antanairesis* from it; see p. 290, n. 26.) My reconstruction will be different. In brief, all preceding proposals have reconstructed an anthyphairetic theory of *proportion,* but I shall argue that there may have been a highly developed exploration of some remarkable properties of anthyphairetic *ratios*; see Section 1.2(d), above. Also, previous anthyphairetic interpretations have invoked the modern arithmetised theory of continued fractions, while the version developed here is based on the non-arithmetised *Parmenides* algorism; see Section 9.1(b) and note 13.

Knorr, *EEE,* Chapter 8 and Appendix B, gives a very useful discussion of the formal development of the proportion theories of the *Elements*; but his interpretation makes crucial use of the surprising difficulties in an anthyphairetic proof of *Elements* V 9: if $A:C::B:C$ then $A=B$. Surely a practising mathematician, faced with the difficulties that Knorr has uncovered, would proceed indirectly, via the *alternando* property that he has just proved. For then $A:C::B:C$ is equivalent to $A:B::C:C$. Now the anthyphairesis of $C:C$ is [1]; so also must be that of $B:C$; so $B=C$.

For an example of an interpretation in which *anthuphairesis* and *antanairesis* do not mean the same thing, see Heron–Bruins, *CC* iii, 60 ff.

For details of abbreviations used to refer to papyri, see Liddell, Scott, & Jones, *GEL,* pp. xl ff; or Turner, *GP.* Several aspects of the Zenon archive are discussed by Turner; see the index, s.v. Zenon. Some topics in papyrology will be discussed more systematically in Chapters 6 and 7 below.

2.2(a) The evaluation of the ratio of diagonal to side of a square given here is based on an interpretation of the description of side and diagonal lines found in Proclus' commentary on Plato's *Republic,* to be described in detail in Sections 2.4(e) and 3.6(b). It is the one important place where my reconstruction draws in a crucial way on mathematical evidence that is only found in uncorroborated reports in later commentators—here, in Theon of Smyrna and Proclus.

The anthyphairesis of the side and diagonal of a pentagon is made the basis of a reconstruction of early Greek mathematics in von Fritz, DIHM; for a discussion, see Knorr, *EEE,* 29–32.

The usual Greek word for both diagonal and diameter is *diametros,* 'the through measure'. It can refer to the diagonal of a parallelogram in contexts where there is no corresponding diameter of a circumcircle; see for example, *Elements* I 34: "The opposite sides and angles of parallelogrammic areas (*parallēlogramma chōria*) are equal to one another and the diagonal (*diametros*) bisects the areas"; here *diametros* cannot mean the diameter of a circle, in any interpretation. There is a very rare word *diagōnios* which is only used twice in the *Elements*, at XI 28 and 38, and in very few places elsewhere (I am grateful to Malcolm Brown for this information.) I shall adapt translations to read 'diagonal' or 'diameter' as appropriate. For example, the usual translation as 'side and *diameter* lines (or numbers)' is, I believe inappropriate.

2.2(b) The translations of Archimedes' *Sphere and Cylinder* are taken from Thomas, *SIHGM* ii, 40ff. There is a photographic reproduction of our sole surviving manuscript of Heron's *Metrica* in Heron–Bruins, *CC* i (with transcription in vol. ii and annotated translation in vol. iii); the translation of the passage quoted here comes from Knorr, *AMC*. Also see the notes to Sections 2.3(c) and 2.4(d), below.

An anecdote from Reid, *H*, 164, illustrates the difficulty and unpredictability of problems associated with irrationality and transcendence: "[Siegel] was always to remember a lecture on number theory which he heard [as a student in 1919] from Hilbert.... Hilbert wanted to give his listeners examples of the characteristic problems of the theory of numbers which seem at first glance so very simple but turn out to be incredibly difficult to solve. He mentioned Riemann's hypothesis [concerning the zeros of $\zeta(s)$], Fermat's theory [the hypothesis concerning the insolubility of $x^n + y^n = z^n$ in integers if $n \geq 3$], and the transcendence of $2^{\sqrt{2}}$ (which he had listed as his seventh problem at Paris) as examples of this type of problem. Then he went on to say that there had recently been much progress on Riemann's hypothesis and he was very hopeful that he himself would live to see it proved. Fermat's problem had been around for a long time and apparently needed entirely new methods for its solution—perhaps the youngest members of his audience would live to see it solved. But as for establishing the transcendence of $2^{\sqrt{2}}$, no one present in the lecture hall would live to see that! The first two problems which Hilbert mentioned are still unsolved. But less than ten years later, a young Russian named Gelfond established the transcendence of $2^{\sqrt{-2}}$. Utilising this work, Siegel himself was shortly able to establish the desired transcendence of $2^{\sqrt{2}}$."

2.3 (a), (b) & (c) The '*Parmenides* proposition' is described in Steele, *MRCP*, on p. 179: "The *Parmenides* (154b-d) notes that the ratio $(a + x):(b + x)$ lies nearer to 1:1 than does $a:b$ (all terms being positive) and enunciates a strangely explicit principle of continuity." There is also a useful discussion in Plato–Allen, *PP*, 258–9 which concludes: "It has been said that Plato was not an original mathematician. But it is worth remarking that the theorems he here uses are not attested before Hellenistic times."

My original treatment, in *REGM*, of the approximation algorism and the evaluation of convergents was more diffident than the account given here; see, e.g., p. 823: "the algorithm described here is not proposed as a reconstruction of an original procedure" (in italics there!). At that stage I had not appreciated the implications of the *Parmenides* passage, I was unaware of the existence of the proposition in Pappus, and my treatment of the question of the starting estimates was offhand and unsatisfactory. I now believe that it is not unreasonable to propose some appropriate formulation of this algorism as a reconstruction of an early Greek procedure. Also see, for example, Knorr, *AMC*, 137 ff., where a much more sophisticated variant of the algorism, there expressed in terms of the manipulation of common fractions, is proposed as an explanation of Archimedes' approximation to $\sqrt{3}:1$.

2.5   Notes and references   65

The *Parmenides* proposition is an example of a mean: given a smaller and larger ratio, it will generate an intermediate ratio. We have many reports of early interest in different kinds of means among Greek mathematicians and music theorists (see e.g. Heath, *HGM* i, 85–9); but in all these cited examples, the mean generates an intermediate between two magnitudes or, in special cases, between two numbers, while the *Parmenides* proposition generates an intermediate between two ratios. There is a connection between the arithmetic mean and the *Parmenides* proposition: if $c = \frac{1}{2}(a+b)$ is the arithmetic mean between two magnitudes a and b with $a < b$, the *Parmenides* proposition, applied to $a:1$ and $b:1$ (with $a:1 < b:1$) will generate the ratio $a+b:2 = c:1$, and this kind of description can be extended to the general case of weighted arithmetic means. I shall return to this question in Section 4.5(b); see especially B_{74} there.

In contexts in which ancient mathematics is being discussed, I use the word 'algorism' in preference to the now more common 'algorithm', castigated by the *Oxford English Dictionary* (1884), s.v. algorism: "English also shows two forms, the popular *augrime*, ending in *agrim*, *agrum*, and the learned *algorism* which passed through many pseudo-etymological perversions, including a recent *algorithm* in which it is learnedly confused with Greek αριθμός, number." The *Supplement* to the *OED*, Vol. 1 (1972), admits the word and cites, as its first example, the first edition of Hardy & Wright, *ITN* (1935): "The system of equations... is known as Euclid's algorithm" (p. 135); but in the second edition this is changed to "The system of equations... is known as the *continued fraction algorithm*" (p. 134)! (This suggests the proportion, (continued fractions):(algorithms)::(anthyphairesis): (algorisms)!) Of course, the word 'algorithm' has a much longer and well-established history in Latin; for an example, see Euler's article 'De usu novi algorithmi in problemate Pelliano solvendo' (= UNAPPS).

2.4(a) For discussions of Eratosthenes' ratio, see Ptolemy–Toomer, *PA*, 63 n. 75; Dicks, *GFH*, 167 ff. & *DSB* s.v. Eratosthenes; Neugebauer, *HAMA* ii, 734 (esp. n. 15); Rawlins, EGU with my subsequent note EROE; Goldstein, OEAGE; and Taisbak, EET. This translation uses common fractions, but the Greek text follows the practice described briefly in Section 1.2(c) and to be examined in detail in Chapter 7: Ptolemy describes Eratosthenes' ratio as μζ *kai meizonos men ē dimoirou tmēmatos, elassonos de hēmisous tetartou*, "47 and more than the two parts of a division, but less than a half fourth".

2.4(b) There is a full discussion of calendric cycles in Neugebauer, *HAMA* ii, 615–29. Of the precise data that can be extracted from some of them, he observes drily (p. 616): "We may, of course, say that [Callippus' calendar] implies that 1 month = 29; 31, 53, 3, ... days and be pleased by the accuracy of this value but one may well ask whether Callippus was ever interested (or able) to carry out the division..." (this sexagesimal value of the mean synodic month is equal to the decimal 29·53085... days; the actual value is 29·53059 days), and (p. 623): "Neither the Metonic nor the Callippic cycle have anything to do with an accurate determination of the length of the mean synodic month".

2.4(c) There is a critical edition of *On the Sizes and Distances of the Sun and Moon* in Aristarchus–Heath, *AS*. I shall describe the style of Aristarchus' calculations in Section 7.3(b).

2.4(d) Also see 2.2(b), above. There is a complete text of Archimedes' *Measurement of a Circle* with translation, filled out with extracts from Eutocius' commentary, in Thomas, *SIHGM* i, 316–33, but this omits the apparatus to be found in the critical edition, Archimedes–Heiberg, *Opera* i, 232–43; and Heiberg's apparatus needs to be corrected by and collated with the Latin translation by William of Moerbeke in Clagett, *AMA* ii. See Section 7.3(a) for more details, and a discussion of the treatment of the fractions in the original.

On the excessive accuracy of Archimedes' approximations of $\sqrt{3}:1$, see Knorr, AMC, 137 f. This article contains the best overall analysis of Archimedes' general strategy that I know; my only reservation concerns the arithmetised context that is assumed for the calculation. I shall discuss the text in more detail in Section 7.3(a).

2.4(e) Fermat's challenge, together with an exchange of letters between Frenicle de Bessy, Brouncker, and Wallis, is published in Wallis, *CE*; there is a discussion of the different methods of solving the problem in Edwards, *FLT*, 25 ff. The translation of Theon of Smyrna, on side and diagonal numbers, the last paragraph excepted, is taken from Thomas, *SIHGM* i, 132–37, except that I have again replaced 'diameter' by 'diagonal' throughout.

2.4(f) I know of no discussion of the implications of this alternative interpretation of Archimedes' *Cattle Problem*. My own exploration of these mathematical aspects of the problem, ACPPM, has never been published.

3
ELEMENTS II:
THE DIMENSION OF SQUARES

3.1 Introduction

This chapter will be entirely devoted to the problem of the dimension of squares (see S_{41} and Section 2.2(a)): given a line ρ, and two numbers n and m, what can we say about the anthyphairesis of $\sqrt{n}$ to $\sqrt{m}$, the ratio of the sides of the two squares equal to n and m times the square on ρ? We shall only consider the case where $\sqrt{n}$ and $\sqrt{m}$ are incommensurable, and so the anthyphairesis does not terminate.

The algorisms of the last chapter have begun to reveal some remarkable properties of the anthyphairetic ratio $\sqrt{3}:1$, and a similar arithmetical exploration, leading to the formation of a hypothesis, will be the first step of the investigation of any ratio $\sqrt{n}:\sqrt{m}$. This hypothesis will then determine the second step, the construction of a geometric figure. Although some of these figures are very elaborate, they are built up out of simple ingredients such as squares, rectangles, and parallelograms; their geometrical properties are, on the whole, straightforward and will be taken for granted, just as Socrates, in the *Meno*, omitted to verify the details of the various geometrical figures he described. Finally, in each case, we shall see that the hypothesis with which we started will correspond to some property exhibited by the figure.

These detailed mathematical arguments underly a historical argument: I propose that this whole programme may indeed correspond to an early Greek investigation of which only the basic geometrical figures survive. Thus the situation is complementary to that which we find in the *Meno*: our surviving manuscripts of Plato's dialogues do not—and perhaps none of them ever did—contain any of the figures on which the mathematical arguments depend. (In Socrates' conversation with the slaveboy, the information in the text is so detailed that we have no difficulty in reconstructing the figures with confidence but, at *Meno* 86e–87b, there is another mathematical passage that has provoked an enormous variety of different geometrical reconstructions.) For the problem of the dimension of squares, I shall propose that the basic figures needed in this investigation of the dimension of squares survive in the form of a series of simple though convoluted geometrical exercises: the substantial, coherent, and idiosyncratic material comprising Book II of the *Elements* and the block of cognate propositions at XIII 1–5. It is now a much harder and more hazardous process to proceed in the opposite direction, to reconstruct the mathematical investigation which gave rise to these geometrical figures. One important methodological principle can be stated at the outset, and will be observed hereafter throughout this book: that the evidence must be taken as a whole. Since the geometrical techniques set out in the propositions of Book II can be adapted to other purposes in other contexts, various applications of some of these propositions can be found elsewhere. Because of this versatility, and because of the coherence of this material within the *Elements,* if we wish to try to explain its role within the development of pre-Euclidean mathematics, it is not sufficient to provide some retrospective justification in which *some* of these propositions are related to topics studied *somewhere* throughout the long history of the development of mathematics, from ancient Babylonian arithmetical techniques or before up to modern preoccupations with algebraic structures; we must rather try to provide some acceptable reconstruction which leads forward, up to *all* of them. And since this material also seems to relate to the construction and classification of incommensurable lines in Book X, the interpretation should connect with the developments of this book also.

The arguments of this chapter will be arduous. They will involve novel mathematical procedures, elaborate numerical explorations, and detailed textual examinations. If it is any consolation to the non-mathematician, I think that a conventional mathematical training may generate difficulties in following the line of reasoning, for much of the work fits uneasily into our arithmetised viewpoint of today; to the non-specialist in ancient mathematics, a similar consolation applies, since my explanation requires that a great deal of the received interpretation be put aside. It is not important to follow the mathematical argument in detail to understand the proposals that follow in later chapters, and I again encourage my reader to skip many of the technical details that arise. However, historical considerations now start moving into the foreground, and I believe it is necessary to appreciate the kinds of mathematical techniques I shall be describing here in order to evaluate the plausibility of some of my later historical proposals.

3.2 Book II of the *Elements*

Book II comprises two definitions:

Any rectangular parallelogram is said to be contained by the two straight lines containing the right angles. And in any parallelogramic area let any one whatever of the parallelograms about its diameter with the two complements be called a gnomon,

and fourteen propositions about squares, rectangles, and gnomons. The subject of each proposition is best conveyed by its figure (and it is these figures, not what is made of them in their enunciations or proofs, that will enter my proposed reconstruction); these are given here as Figs. 3.1–14, each corresponding to Book II Propositions 1–14. (The additional Figs. 3.9(a) and 3.10(a) will be explained presently.) The caption of each figure gives an abbreviated statement of the enunciation. For example the enunciation of II 4:

If a straight line [AB] be cut at random [at C], the square on the whole [AB²] is equal to the squares on the segments [AC² + CB²] and twice the rectangle contained by the segments [2AC . CB]

has become the caption of Fig. 3.4:

$$AB^2 = AC^2 + CB^2 + 2AC \cdot CB.$$

It is important to realise that all these operations are geometrical; neither Euclid's propositions, nor my notation, makes or needs any reference to the manipulation of any kind of number other than the *arithmoi*, as with the 'twice' that appears in this enunciation.

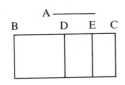

Fig. 3.1
A . BC = A . BD + A . DE + A . ED

Fig. 3.2
AB . BC + BA . AC = AB²

Fig. 3.3 AB . BC = AC . CB + BC²

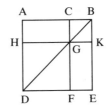

Fig. 3.4
AB² = AC² + CB² + 2AC . CB

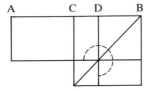

Fig. 3.5 (AC = CB)
AD . DB + CD² = CB²

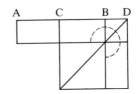

Fig. 3.6 (AC = CB)
AD . DB + CB² = CD²

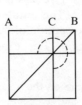

Fig. 3.7
$AB^2 + BC^2 = 2AB \cdot BC + CA^2$

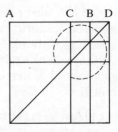

Fig. 3.8 (BC = BD)
$4AB \cdot BC + AC^2 = (AB + BC)^2 = AD^2$

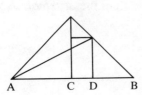

Fig. 3.9 (AC = CB)
$AD^2 + DB^2 = 2AC^2 + 2DC^2$

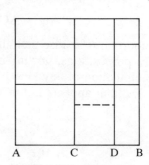

Fig. 3.9(a) (AC = CB)

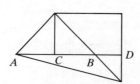

Fig. 3.10 (AC = CB)
$AD^2 + DB^2 = 2AC^2 + 2CD^2$

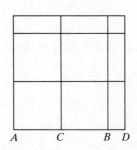

Fig. 3.10(a) (AC = CB)

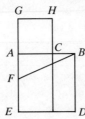

Fig. 3.11
$AB \cdot BC = AC^2$

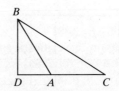

Fig. 3.12
$BC^2 = BA^2 + AC^2 + 2CA \cdot AD$

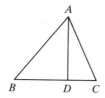

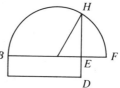

Fig. 3.13
$AC^2 + 2CB \cdot BD = CB^2 + BA^2$

Fig. 3.14 (EF = ED)
$A = BE \cdot EF = EH^2$

The demonstration of Proposition 4 is typical of Propositions 1–8. The setting-out is brief (lines 4 to 7 in Heath's translation); it identifies the various components of the figure: "For let the straight line AB be cut at random at C," then "The square on the whole" is "the square on AB", and so on. The construction follows (lines 8 to 14): the square ADEB, the diagonal BD, CF parallel to AD or EB, and HK through G parallel to AB or DE. The proof itself is a long and meticulous verification of the consequences of the construction: that BC equals CG (lines 15 to 22), that CGKB is a square, the square on CB (lines 23 to 35), and that HDFG is the square on AC (lines 36 to 38, repeated on line 39). Then it is asserted that AG (i.e. the rectangle AHGC) equals GE (line 40), and that AG and GE are equal to the rectangle contained by AC and CB (lines 41 to 42). The results so far are summarised (lines 43 to 45), and this summary is repeated (lines 46 to 48):

Therefore the four areas HF, CK, AG, GE are equal to the squares on AC, CB, and twice the rectangle contained by AC, CB,

and the fact that this gives a decomposition of the square, the ostensible point of the proposition, is merely stated (lines 49 to 50):

But HF, CK, AG, GE are the whole ADEB, which is the square on AB.

This is then followed by a repetition of the setting-out (lines 51 to 52) and the enunciation (abbreviated, as usual, to "Therefore etc." in Heath's translation). The proposition has an alternative proof and is followed by a porism:

From the demonstration it is manifest that in square areas the parallelograms about the diameter are squares,

but both are believed to be interpolations by Theon of Alexandria; see Heath, *TBEE* i, 381.

The elaborate verifications in this proof, for example that CBGK is a square, are not repeated in the later propositions, so that the proofs of Propositions 5, 6, and 7 are shorter. Proposition 8 again has a long proof, since the verification that the double gnomon gives rise to equal squares in the corner and equal rectangles along the arms is carried out in detail, again with repetitions. In all of these propositions, almost all of the text of the demonstrations concerns the construction of the figures, while the substantive content of each enunciation is merely read off from the constructed figures, at the end of the proofs, as in lines 49 to 50 of II 4, the proposition just considered.

Proposition 9 starts in the same style; its enunciation follows the well-established pattern:

If a straight line be cut into equal and unequal segments, the squares on the unequal segments of the whole are double of the square on the half and of the square on the straight line between the points of section;

it has the same kind of setting-out, but the associated figure and proof differs from the previous propositions. Euclid's proof proceeds by drawing two isosceles right-angled triangles which fit together into a larger isosceles right-angled triangle (Fig. 3.9), and applying I 47, the so-called Pythagoras' theorem; alternative proofs are easily constructed which avoid the use of I 47 and continue to apply the techniques of dissecting squares. One such proof is given in Heath's note to II 9, and an even simpler version of it follows directly from Fig. 3.9a. A similar situation occurs with Proposition 10: Euclid's proof is based on Fig. 3.10, and an alternative proof can be read off from Fig. 3.10a. Proposition 11 solves a specific problem:

To cut a straight line so that the rectangle contained by the whole and one of the segments is equal to the square on the remaining segments;

Propositions 12 and 13 give versions of a generalisation of I 47 which correspond to what we now call the cosine law for obtuse- and acute-angled triangles respectively; and Proposition 14, which concludes the book, shows how

To construct a square equal to a given rectilineal figure.

This construction starts with a typical Euclidean back-reference, to I 45:

For let there be constructed the rectangular parallelogram BD equal to the rectilineal figure A,

and the new and important contribution of the rest of the proposition is to show how to convert this rectangle into an equal square.

With the exception of implied uses of I 47 and 45, Book II is virtually self-contained in the sense that it only uses straightforward manipulations of lines and squares of the kind assumed without comment by Socrates in the *Meno*. Moreover the only reference to I 45, just quoted above, occurs in the last proposition, II 14, where it is tacked on to contribute extraneous generality, out of keeping with the style of the rest of the book. Also, it can be argued that a proof of 'Pythagoras' theorem', referred to in Section 1.2(e), has been excised from between Propositions 8 and 9. This proof, which I shall call Proposition 8a, exploits the manipulation of gnomons, the basic technique of Book II, and, with its obliquely placed square, is reminiscent of the successful third *Meno* figure; so the proof also conforms in style with the testified ingredients of early Greek mathematics. Moreover the stylistic variations in the proof of II 9 and 10 could then arise from the wish of the book's author, compiler, or editor to integrate this new result further into the logical development before it is required in II 11 and generalised in II 12 and 13; then a later editor might have excised the proposition to remove the duplication of I 47 and II 8a. The argument is plausible, and further considerations in favour of it can be found in Knorr, *EEE*, 174–79. I shall return, in Section 3.6, to further discussion of Propositions 12 and 13.

3.3 The hypotheses

The algorisms of Chapter 2 enable us to evaluate as many terms as we want of the anthyphairesis of any ratio $\sqrt{n}:\sqrt{m}$, the only restriction being our ability to express or evaluate the squares of the integers involved. We also evaluate, at the same time, the convergent of these ratios. For example:

$$\sqrt{2}:1 = [1, 2, 2, 2, 2, \ldots],$$

with convergents $1:1, 3:2, 7:5, 17:12, 41:29, \ldots$;

$$\sqrt{3}:1 = [1, 1, 2, 1, 2, 1, 2, \ldots],$$

with convergents $1:1, 2:1, 5:3, 7:4, 19:11, 26:15, 71:41, \ldots$;

$$\sqrt{3}:\sqrt{2} = [1, 4, 2, 4, 2, 4, 2, \ldots],$$

with convergents $1:1, 5:4, 11:9, 49:40, 109:89, 485:396, 1079:881, \ldots$;

$$\sqrt{4}:1 = 2:1 = [2];$$
$$\sqrt{4}:\sqrt{2} = \sqrt{2}:1, \quad \text{given above;}$$
$$\sqrt{4}:\sqrt{3} = [1, 6, 2, 6, 2, 6, 2, \ldots],$$

with convergents $1:1, 7:6, 97:84, 209:181, 1351:1170, 2911:2521, \ldots$; etc.

The obvious hypothesis to draw from these and further examples is that the expansion of $\sqrt{n}:\sqrt{m}$ is periodic, with a period that starts with the second term; and we have already verified this for the example $\sqrt{2}:1$, the ratio of the diagonal to side of a square. Table 3.1 lists the expansions, but not the convergents, of $\sqrt{n}:1$ and $\sqrt{(2n+1)}:\sqrt{2}$ for $1 \leq n \leq 50$. (When n is a square, the ratio $\sqrt{n}:1$ is commensurable and the entry for the anthyphairetic ratio contains only one term; otherwise the entry is terminated by a comma after one apparent period. Hereafter

74　　The dimension of squares　　3.3

TABLE 3.1

n	$\sqrt{n}:1$	$\sqrt{2n}+1:\sqrt{2}$
1	1	1, 4, 2,
2	1, 2,	1, 1, 1, 2,
3	1, 1, 2,	1, 1, 6, 1, 2,
4	2	2, 8, 4
5	2, 4,	2, 2, 1, 8, 1, 2, 4,
6	2, 2, 4,	2, 1, 1, 4,
7	2, 1, 1, 1, 4,	2, 1, 2, 1, 4,
8	2, 1, 4,	2, 1, 10, 1, 4,
9	3	3, 12, 6,
10	3, 6,	3, 4, 6,
11	3, 3, 6,	3, 2, 1, 1, 3, 1, 12, 1, 3, 1, 1, 2, 6,
12	3, 2, 6,	3, 1, 1, 6,
13	3, 1, 1, 1, 1, 6,	3, 1, 2, 14, 2, 1, 6,
14	3, 1, 2, 1, 6,	3, 1, 4, 4, 1, 6,
15	3, 1, 6,	3, 1, 14, 1, 6,
16	4	4, 16, 8,
17	4, 8,	4, 5, 2, 5, 8,
18	4, 4, 8,	4, 3, 3, 8,
19	4, 2, 1, 3, 1, 2, 8,	4, 2, 2, 2, 8,
20	4, 2, 8,	4, 1, 1, 8,
21	4, 1, 1, 2, 1, 1, 8,	4, 1, 1, 1, 3, 18, 3, 1, 1, 1, 8,
22	4, 1, 2, 4, 2, 1, 8,	4, 1, 2, 1, 8,
23	4, 1, 3, 1, 8,	4, 1, 5, 1, 1, 3, 2, 1, 18, 1, 2, 3, 1, 1, 5, 1, 8,
24	4, 1, 8,	4, 1, 18, 1, 8,
25	5	5, 20, 10,
26	5, 10,	5, 6, 1, 3, 3, 1, 6, 10,
27	5, 5, 10,	5, 4, 10,
28	5, 3, 2, 3, 10,	5, 2, 1, 20, 1, 2, 10,
29	5, 2, 1, 1, 2, 10,	5, 2, 3, 6, 1, 20, 1, 6, 3, 2, 10,
30	5, 2, 10,	5, 1, 1, 10,
31	5, 1, 1, 3, 5, 3, 1, 1, 10,	5, 1, 1, 1, 1, 2, 1, 1, 1, 1, 10,
32	5, 1, 1, 1, 10,	5, 1, 2, 2, 1, 10, 1, 2, 2, 1, 10,
33	5, 1, 2, 1, 10,	5, 1, 3, 1, 2, 1, 1, 22, 1, 1, 2, 1, 3, 1, 10,
34	5, 1, 4, 1, 10,	5, 1, 6, 1, 10,
35	5, 1, 10,	5, 1, 22, 1, 10,
36	6	6, 24, 12,
37	6, 12,	6, 8, 12,
38	6, 6, 12,	6, 4, 1, 7, 2, 7, 1, 4, 12,
39	6, 4, 12,	6, 3, 1, 1, 24, 1, 1, 3, 12,
40	6, 3, 12,	6, 2, 1, 2, 1, 24, 1, 2, 1, 2, 12,
41	6, 2, 2, 12,	6, 2, 3, 1, 4, 2, 1, 1, 1, 7, 1, 24, 1, 7, 1, 1, 1, 2, 4, 1, 3, 2, 12,
42	6, 2, 12,	6, 1, 1, 12,
43	6, 1, 1, 3, 1, 5, 1, 3, 1, 1, 12,	6, 1, 1, 2, 8, 2, 1, 1, 12,
44	6, 1, 1, 1, 2, 1, 1, 1, 12,	6, 1, 2, 26, 2, 1, 12,
45	6, 1, 2, 2, 2, 1, 12,	6, 1, 2, 1, 12,
46	6, 1, 3, 1, 1, 2, 6, 2, 1, 1, 3, 1, 12,	6, 1, 4, 1, 1, 8, 1, 1, 4, 1, 12,
47	6, 1, 5, 1, 12,	6, 1, 8, 3, 1, 4, 1, 3, 8, 1, 12,
48	6, 1, 12,	6, 1, 26, 1, 12,
49	7	7, 28, 14,
50	7, 14,	7, 9, 2, 2, 9, 14,

3.3 The hypotheses

I shall always silently assume that n and m are such that $\sqrt{n}$ and $\sqrt{m}$ are incommensurable, so that all examples such as $\sqrt{8}:\sqrt{2}$ have been suppressed, as in the second column of this table.)

The examples in the table lead us to formulate a slightly more refined hypothesis, that

$$\sqrt{n}:\sqrt{m} = [n_0, \overline{n_1, n_2, \ldots, n_k, 2n_0}].$$

where the period, identified by the superior bar, terminates with a term equal to twice the first term of the expansion. It also seems that we can guess the general behaviour of some examples that lie around the commensurable ratio $\sqrt{n^2}:1$, for example:

$$\sqrt{(n^2 - 1)}:1 = [(n-1), \overline{1, 2(n-1)}],$$
$$\sqrt{(n^2 + 1)}:1 = [n, \overline{2n}], \text{ and}$$
$$\sqrt{(n^2 + 2)}:1 = [n, \overline{n, 2n}].$$

This chapter will be dedicated to the geometrical verification of statements of this sort.

While the computation of the whole of Table 3.1 using the algorisms of Chapter 2 might be straightforward, it would be very time-consuming and tedious, as can be seen by calculating the example of $\sqrt{19}:1$, the first example of a ratio $\sqrt{n}:1$ which involves a long period containing several terms greater than one in its period. (However, a skilled and resourceful calculator would notice further patterns of behaviour that could be exploited to reduce greatly the amount of routine calculation.) In fact I calculated the results of this table using a pocket calculating machine and a simple arithmetical algorithm:

$$x = x_0; \quad x_k = n_k + \varphi_k \quad \text{with} \quad 0 \leq \varphi_k < 1; \quad \text{if} \quad \varphi_k \neq 0 \quad \text{then} \quad \varphi_k^{-1} = x_{k+1};$$
$$\text{then} \quad x:1 = [n_0, n_1, n_2, \ldots],$$

and I verified the one doubtful example, $\sqrt{83}:\sqrt{2}$, which lay beyond the limits of the accuracy of the machine, using the algorithm to be described in Chapter 9. So I do not wish to propose that such a systematic and large-scale set of calcualtions as is embodied in Table 3.1 might ever have been performed in antiquity. It would seem more reasonable to assume that only some very restricted corpus of examples was ever explored, at most, for example, the ratios $\sqrt{n}:1$ for $2 \leq n \leq 20$, and a few other isolated examples of $\sqrt{n}:\sqrt{m}$. Therefore the quite startling general hypothesis, that the period comprises a *palindromic* block terminated by twice the first term, would not present itself so forcefully from this limited exploration. (In the case of $\sqrt{(n^2 + 1)}:1$, this palindromic block is void, while in some other cases it is reduced to only one term.) I know of no evidence, of any kind whatsoever, that leads me to suggest this hypothesis might have any part in a reconstruction of Greek mathematics; all of the abundant Greek evidence about palindromes seems to be later in time and connected with charms, magic, and gratuitous word-play.

An example of a false hypothesis which results from an inference made from an insufficiently universal collection of examples is that the period has been evaluated when the first occurrence of a term equal to $2n_0$ is reached. A more extensive search would reveal examples such as

$$\sqrt{15}:\sqrt{8} = [1, 2, 1, 2, 2, \ldots].$$

However, it is true that in any ratio of the form $\sqrt{n}:1$, the first occurrence of a term $2n_0$ will mark the end of the period, though I shall not prove this here in this book and I know of only one obscure place in the mathematical literature (Muir, *EQSCF*) where this type of result is investigated.

It seems curiously difficult to prove any of these arithmetical conjectures directly from the properties of the *Parmenides* proposition, from the operation of the algorism that generated them, the more so if the proof is restricted to the ingredients of Greek *arithmētikē* such as I have described them in Section 1.2(c). But we can reformulate the problem in various ways and seek indirect proofs; for example, in modern style, we can arithmetise these conjectures by converting them into statements about the manipulation of real numbers, and then explore them algebraically. I shall proceed differently, and geometrise. For us here, the hypotheses that $\sqrt{2}:1 = [1, \bar{2}]$ will be a statement about the anthyphairesis of the diagonal and side of a square, and one version of such an interpretation was proved in 2.2(a). So we now turn to the problem of constructing further figures which may help in establishing and explaining some of the hypotheses that arise from our heuristic investigations of these ratios $\sqrt{n}:\sqrt{m}$.

3.4 The first attempt: The method of gnomons

The first method is a direct confrontation of the problem: we build up squares of the appropriate sizes by adding gnomons to the unit square (the square on some fixed line ρ, chosen as unit), introduce some additional construction lines, and then convert the statements about the anthyphairesis into statements about these figures which we then verify.

PROPOSITION $\sqrt{2}:1 = [1, \bar{2}]$.

PROOF We prove the equivalent result that $(1 + \sqrt{2}):1 = [\bar{2}]$. Start with the unit square P; then the square of size 2P with the same bottom left-hand corner differs from P by a gnomon Q + R + S, equal to P (Fig. 3.15). This figure now contains the sides AB, equal to 1, and AC, equal to $\sqrt{2}$. We can check that BC, the width of the gnomon, is less than AB,

3.4 The method of gnomons 77

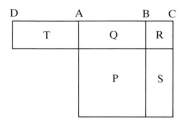

Fig. 3.15

since a gnomon of width AB is equal to 3P, which is too big. Add a line AD, equal to AB; then DC = 1 + √2, and all the lines needed in the enunciation of the proposition have been constructed. The first step of the anthyphairesis of DC: AB = (1 + √2): 1 is

$$DC - 2AB = BC \quad \text{with} \quad BC < AB.$$

Hence

$$DC:AB = [2, AB:BC].$$

We now prove that AB:BC = DC:AB. This is equivalent to the statement that the square on AB is equal to the rectangle contained by DC and BC; hence we must verify that

$$P = T + Q + R$$

where T is the appended rectangle contained by AD and BC. But these two plane regions are obviously equal, since T = Q = S, and Q + R + S = P. Hence, by the same recursive property that was described and exploited in Section 2.2(a), we see that

$$DC:AB = [2, DC:AB] = [\bar{2}]. \qquad \text{QED}$$

The same technique, starting from a square of side $n\rho$ and size $n^2 P$, will prove the generalisation:

PROPOSITION   For any number n, $\sqrt{(n^2 + 1)} = [n, \overline{2n}]$.

PROPOSITION $\sqrt{3}:1 = [1, \overline{1, 2}]$.

PROOF   We prove the equivalent result that $(1 + \sqrt{3}):1 = [\overline{2, 1}]$. Start with the unit square P, a gnomon Q + R + S of size 2P, and the square with side AC = √3, size 3P (Fig. 3.16). We can check that ½AB < BC < AB, since a gnomon with width ½AB is too small, while a gnomon of width AB is too big. Add a new gnomon T + U + V + W + X to make the total width of added gnomons, BD, equal to AB; and append a rectangle Y with sides EA = AB and CD. We can see that Q = S, T = X = Y, and

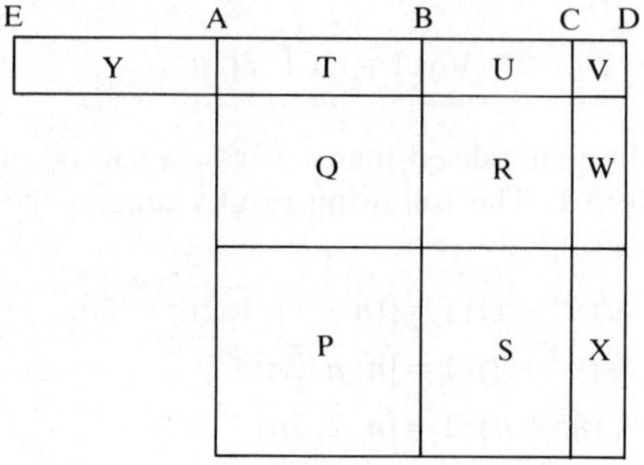

The dimension of squares

Fig. 3.16

$U = W$. We now proceed to the anthyphairesis of $EC:AB = (1 + \sqrt{3}):1$.

$$EC - 2AB = BC \quad \text{with} \quad BC < AB,$$

and, since $AB = BD$,

$$BD - 1BC = CD \quad \text{with} \quad CD < BC.$$

Hence

$$EC:AB = [2, 1, BC:CD].$$

We now prove that $BC:CD = EC:AB$. This is equivalent to the statement that the rectangle with sides AB and BC is equal to the rectangle with sides EC and CD; hence we must verify that

$$R + U = Y + T + U,$$

that is,

$$R = 2T.$$

But

$$2Q + R = 2P,$$

hence

$$2Q + 2T + R = 2P + 2T,$$

and

$$Q + T = P.$$

Thus

$$R = 2T,$$

and the equality is established.

Therefore

$$(1 + \sqrt{3}):1 = EC:AB = [2, 1, BC:CD]$$
$$= [2, 1, EC:AB]$$
$$= [\overline{2, 1}],$$

3.4   The method of gnomons   79

and so
$$\sqrt{3}:1 = [1, \overline{1, 2}]. \qquad \text{QED}$$

This result can be generalised many ways, as can be guessed from an inspection of Table 3.1. The following results suggest themselves almost immediately:

$$\sqrt{(n^2 - 1)}:1 = [(n-1), \overline{1, 2(n-1)}],$$
$$\sqrt{(n^2 + 2)}:1 = [n, \overline{n, 2n}],$$
$$\sqrt{(n^2 + n)}:1 = [n, \overline{2, 2n}],$$
$$\sqrt{(2n^2 + n)}:\sqrt{2} = [n, \overline{4, 2n}].$$

The evaluation of further expansions suggests yet more results; for example, a table of $\sqrt{n}:\sqrt{3}$ will yield

$$\sqrt{(3n^2 + 2n)}:\sqrt{3} = [n, \overline{3, 2n}], \quad \text{and}$$
$$\sqrt{(3n^2 + n)}:\sqrt{3} = [n, \overline{6, 2n}].$$

These expansions are particular cases of the result that

$$\sqrt{(p^2 q + 2p)}:\sqrt{q} = [p, \overline{q, 2p}],$$

the general form of the expansion which has two terms in its period; I shall give a geometrical derivation of this formula in the next section. All of these results can be verified with elaborations of the procedure used to evaluate $\sqrt{3}:1$; I shall go on to consider a few examples, continuing as long as the procedures exhibit new features, and will discuss only the new features that enter each calculation. But first, let us relate what we have done so far to the propositions of Book II.

The verification of $\sqrt{2}:1 = [1, \bar{2}]$ started with the construction of a particular case of a figure of the type of II 4 or 7, to which a rectangle was appended to make a figure of type II 5 and 6. The lines $\sqrt{n}$ needed in this and other examples can all be constructed using II 14. The statement, in Fig. 3.15, that the square on AC is equal to 2P is a particular example of II 4, and that the rectangle contained by DC and BC can be decomposed into T + Q + R is an example of II 1. Figure 3.16 introduces the configuration of double gnomons of II 8 (except that, in this proposition, the gnomons are of equal width; we shall soon encounter examples of multiple equal gnomons), and the decomposition there of P into Q + T is an example of II 2, while the decomposition of the rectangle with sides AB and BC into R + U is an example of II 3.

Propositions II 1 to 10 refer to properties of the general configurations of Figs. 3.1–10, each of which is based on "a line cut at random" (*hos etuchen*), while the configurations of Figs. 3.15 and 3.16, which arise from particular anthyphairetic ratios, are precisely specified. (In fact, while Figs. 3.15 and 3.16 are drawn approximately to scale, in the following Figs. 3.17–19 the correctly proportioned gnomons will be too narrow to be labelled conveniently.) My proposed interpretation is that the author of the archetype of Book II is introducing the type of figure and kind of argument that are needed for anthyphairetic calculations, but adapted into self-contained propositions or exercises that do not refer to anthyphairesis. Therefore general statements have been chosen to illustrate the particular kinds of manipulations that arise in anthyphairesis. So, for example, the very particular assertion, in Fig. 3.15, that

$$AC^2 = P + Q + R + S = P + 2P = 3P$$

has become, in II 4, the general statement that

$$AC^2 = P + Q + R + S,$$

and a variation, which introduces overlapping regions such as arise later in these anthyphairetic examples, occurs in II 7

$$AC^2 + R = 2(Q + R) + P.$$

(This proposition reads:

If a straight line be cut at random, the square on the whole and that on one of the segments both together are equal to twice the rectangle contained by the whole and the said segment and the square on the remaining segment.)

Simililarly, the configuration of Fig. 3.16 gives rise to the two variations of II 5 and 6, according as whether the gnomon is added to or subtracted from the initial square. The correspondence between these anthyphairetic arguments and Book II will be of this nature, and my arguments can only be sustained if every proposition of Book II can eventually be accounted for in terms of some reasonable anthyphairetic procedure. This first method of gnomons will, in this sense, account for II 1–10 and 14.

Let us return to further examples of this method of gnomons.

PROPOSITION $\sqrt{(n^2 - 1)} : 1 = [(n - 1), \overline{1, 2(n - 1)}]$.

PROOF The Euclidean technique for handling a general proposition such as this, which I shall follow in this proof, is to consider the result for a small fixed typical value of n, e.g. $n = 4$. We construct Fig. 3.17 as follows: P_1 is the unit square, P_9 a square of size $(n - 1)^2 P_1 = 9P_1$, $Q_1 + Q_2 + Q_3 + R + S$ a gnomon of size $2(n - 1)P_1 = 6P_1$, so that $AC = \sqrt{15}$, the side of a square of size $(n - 1)^2 P_1 + 2(n - 1)P_1 = (n^2 - 1)P_1 = 15P_1$; $BD = AE = 1$; and $FA = AB = (n - 1) = 3$. As before, we verify that $\frac{1}{2} AE < BC < AE$.

The first two steps of the anthyphairesis of $FC : AE = (3 + \sqrt{15}) : 1$ are:

$$FC - 6AE = BC$$

3.4 The method of gnomons 81

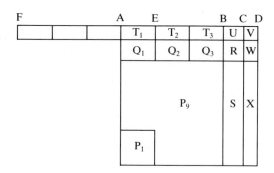

Fig. 3.17

and, since AE = BD,

$$BD - 1BC = CD.$$

Hence

$$FC:AE = [6, 1, BC:CD].$$

We now prove that BC:CD = FC:AE, i.e. that BC . AE = FC . CD, i.e. that

$$R + U = 6T + U$$

where T denotes the equal rectangles T_1, T_2, and T_3. But

$$6Q + R = 6P_1.$$

So, adding 6T and observing that $Q + T = P_1$, we get the result. QED

PROPOSITION $\sqrt{(n^2 + 2)}:1 = [n, \overline{n, 2n}]$.

REMARK Instead of following the Euclidean technique of choosing a fixed n, e.g. $n = 3$, I shall here illustrate the point that a general proof involves little more than the device of writing Q_1, Q_2, Q_3 as Q_1, $Q_2, \ldots, Q_n$ and recognising where to put the dots and the ns; this kind of notational trick is a modern way of automatically indicating generality of an argument. Euclid's use of everyday language, which does not permit the symbolic manipulation involved, does not allow him the possibility of that kind of expression of generality, but he seems aware of the underlying technique. For a clear and celebrated example, see his proof of IX 20.

PROOF Construct Fig. 3.18 as follows: Start with a square P_{n^2} of area n^2P_1; add a gnomon $Q_1 + Q_2 + \ldots + Q_n + R + S$ of size $2P_1$; add $(n-1)$ further gnomons of the same width; the total width of the gnomons will then be less than the unit (see below); so add a final gnomon

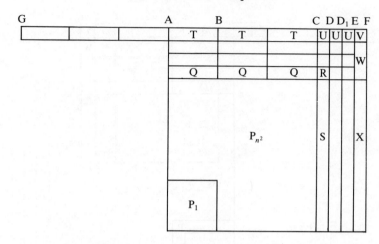

Fig. 3.18

$T_1 + T_2 + \ldots + T_n + U_1 + U_2 + \ldots + U_n + V + W + X$ to bring up the total width of added gnomons to the unit AB. Finally adjoin GA = AC.

Thus the following lengths have been constructed: AB, the unit; AC = n; AD = $\sqrt{(n^2 + 2)}$; GD = $n + \sqrt{(n^2 + 2)}$; CD = $DD_1 = D_1D_2 = \ldots = D_{n-2}E$; and CF = AB.

We need to check that $(1/(n+1))AB < CD < (1/n)AB$; this follows by the usual kind of argument, since $(2n/(n+1))AB^2 + (1/(n+1)^2)AB^2 < 2P_1$ but $(2n/n)AB^2 + (1/n^2)AB^2 > 2P_1$.

The first two steps of the anthyphairesis are:

$$GD - 2nAB = CD, \quad \text{where} \quad CD < (1/n)AB,$$

and, since AB = CF,

$$CF - nCD = EF, \quad \text{where} \quad EF = AB - nCD < CD.$$

Hence

$$GD:AB = [2n, n, CD:EF].$$

Finally, we prove that CD:EF = GD:AB, i.e. since AB = CF, CD . CF = GD . EF. As before, this is equivalent to

$$nR + U = 2nT + U, \quad \text{i.e.} \quad 2T = R.$$

But, by the original construction,

$$2nQ + R = 2P_1,$$

and

$$nQ + T = P_1,$$

from which the conclusion follows. QED

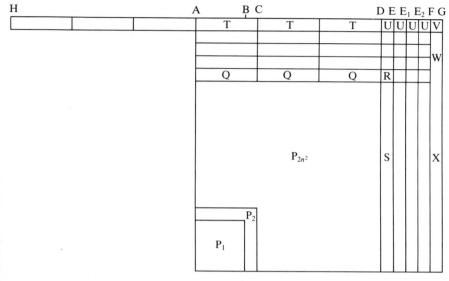

Fig. 3.19

PROPOSITION $\sqrt{(2n^2+2)} : \sqrt{2} = [n, \overline{4, 2n}]$.

When n is even, this case reduces to the result

$$\sqrt{(4n^2+n)} : 1 = [2n, \overline{4, 4n}]$$

which can be evaluated using the techniques already described. Therefore we need only consider the case of odd n, and Fig. 3.19 illustrates the case of $n = 3$, around which the reader is invited to construct a proof.

The only new idea of the construction of the figure for this proposition is to start with a square P_2 other than the unit square P_1; in fact, ratios of the form $\sqrt{n} : \sqrt{m}$ are no more difficult to handle than ratios $\sqrt{n} : 1$. So the reader is invited to investigate, in either the Euclidean or a more general geometric interpretation, the result:

PROPOSITION $\sqrt{(mn^2+2n)} : \sqrt{m} = [m, \overline{m, 2n}]$.

This shows that this method of gnomons can verify every expansion of the form $\sqrt{n} : \sqrt{m}$ which has one or two terms in its period. We now move on to the case of periods of length three, which must therefore have the form $[n, \overline{m, m, 2n}]$ since the first two terms of the period are palindromic. The first such example of a ratio $\sqrt{n} : 1$ occurs with $n = 41$:

$$\sqrt{41} : 1 = [6, \overline{2, 2, 12}].$$

Since it is no more difficult to handle ratios of the form $\sqrt{n} : \sqrt{m}$, I shall illustrate the technique on a ratio involving smaller numbers.

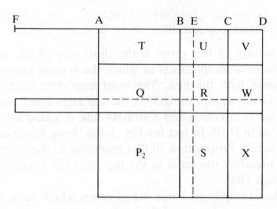

Fig. 3.20

PROPOSITION $\sqrt{5}:\sqrt{2} = [1, \overline{1, 1, 2}]$.

PROOF See Fig. 3.20: Start with a square $P_2 = 2P_1$; add a gnomon $Q + R + S$ of size $3P_1$. By the usual argument, this gnomon will have a width less than $\sqrt{2}$, so we can add a further gnomon $T + U + V + W + X$ to bring the total width of added gnomons up to $\sqrt{2}$. Adjoin the line $FA = AB$. Hence $AB = \sqrt{2}$; $AC = \sqrt{5}$; $FC = \sqrt{2} + \sqrt{5}$; and $AB = BD$. We can check, as usual, that $\frac{1}{2}AB < BC < AB$.

The first two steps of the anthyphairesis of $FC:AB = (\sqrt{2} + \sqrt{5}):\sqrt{2}$ are

$$FC - 2AB = BC, \quad \text{where} \quad BC < AB,$$

and, since $AB = BD$,

$$BD - 1BC = CD \quad \text{where} \quad \tfrac{1}{2}BC < CD < BC.$$

Let $CE = CD$; the third step is then

$$BC - 1CE = BE.$$

Hence

$$FC:AB = (\sqrt{2} + \sqrt{5}):\sqrt{2} = [2, 1, 1, CE:BE],$$

and, as usual, we must now prove that

$$FC:AB = CE:BE, \text{ i.e. } FC \cdot BE = AB \cdot CE$$

which is equivalent to

$$2(Q - T) + (R - U) = U + V,$$

i.e.

$$2Q + R = 2T + 2U + V.$$

But $2Q + R$ is the first gnomon, of size $3P_1$, and $2T + 2U + V$ is the second gnomon, also of size $3P_1$. QED

The new feature of this proof is the third step of the anthyphairesis, which introduces a complication in which the second gnomon folds back in a comparison with the first. The most important details of this new figure occur in the square on BD, in the top right-hand corner, in which "a straight line is bisected, and a straight line is added to it in a straight line" exactly as in II 10. In fact my Fig. 3.10a, from which can be read off a proof of Euclid's Proposition 10 that conforms to the style of the proofs of II 4–8, is precisely the same as the top right-hand corner of Fig. 3.20, rotated through 180°.

Most of the examples in Table 3.1 of ratios which have three terms in their period are special cases of the general result:

PROPOSITION $\sqrt{(2n^2 + 2n + 1)} : \sqrt{2} = [n, \overline{1, 1, 2n}]$,

that can be established by the same technique. The remaining cases are

$$\sqrt{41} : 1 = [6, \overline{2, 2, 12}], \quad \text{and}$$
$$\sqrt{37} : \sqrt{2} = [4, \overline{3, 3, 8}],$$

which can also be verified in the same way. However it is not easy, from an arithmetical search, as from an extension of Table 3.1, to find general formulae to describe these cases. In fact, with a bit of algebra (which I do not propose was available in any form to the Greeks) we can now easily derive the following general result:

PROPOSITION $\sqrt{(n^2m^2 + n^2 + 2nm + 1)} : \sqrt{(m^2 + 1)} = [n, \overline{m, m, 2n}]$.

It is possible to verify this for small values of n and m by the same method, and even to see that this verification is perfectly general.

The procedure can still be used for examples with four terms within the period, provided they involve only small numbers. For example, the table yields

$$\sqrt{7} : 1 = [2, \overline{1, 1, 1, 4}]$$

and a short further search reveals

$$\sqrt{8} : \sqrt{3} = [1, \overline{1, 1, 1, 2}],$$

both examples of the general result:

PROPOSITION $\sqrt{(3n^2 + 4n + 1)} : \sqrt{3} = [n, \overline{1, 1, 1, 2n}]$.

It is even just possible to verify the

PROPOSITION $\sqrt{13} : 1 = [3, \overline{1, 1, 1, 1, 6}]$,

which has five terms in its period, but anything that involves a longer period, or larger numbers within the period, becomes unfeasible. For example, the conjecture:

HYPOTHESIS $\quad \sqrt{19}:1 = [4, \overline{2, 1, 3, 1, 2, 8}]$

generates a figure that seems to me to defy analysis by this method.

3.5 The second attempt: Synthesising ratios

3.5(a) Introduction

Since the method of gnomons has run into difficulties, we now try a different kind of attack on the problem: we change direction, and approach a solution from the other side. Hitherto we had started in each case from a hypothesis that

$$\sqrt{n}:\sqrt{m} = [n_0, \overline{n_1, n_2, \ldots, n_k, 2n_0}]$$

which was then actually handled in the equivalent form

$$(\sqrt{n} + n_0\sqrt{m}):\sqrt{m} = [\overline{2n_0, n_1, \ldots, n_k}],$$

and we exploited in an essential way the feature that this anthyphairesis was now purely periodic. Let us now instead try to explore the questions: can we construct the ratio whose anthyphairesis is some purely periodic expansion $[\overline{n_0, n_1, \ldots, n_k}]$, and can we then show how, or explain why, if n_0 is even and $n_1, n_2, \ldots, n_k$ is palindromic, then this ratio will be of the form $(\sqrt{n} + \frac{1}{2}n_0\sqrt{m}):\sqrt{m}$? The arguments that can be plausibly reconstructed within the Greek context will only give a very partial answer to these general problems; indeed the ultimate resolution of this question, to be described in Section 9.1(c), will require contributions from some of the most celebrated mathematicians of the seventeenth, eighteenth, and nineteenth centuries AD.

3.5(b) The extreme and mean ratio

These questions about purely periodic expansions have been phrased very generally in order to describe, at the outset, the approach and objective. Much more reasonable, as a reconstruction of ancient explorations, would be to study the particular ratios $[\overline{1}], [\overline{2}], \ldots, [\overline{1, 2}], [\overline{1, 3}], \ldots, [\overline{2, 1}], [\overline{2, 3}], \ldots, [\overline{1, 2, 3}], \ldots$. The explicit surviving Greek evidence then relates only to the simplest case of $[\overline{1}]$: if a line AB is to be cut at the point C such that

$$AB:AC = [\overline{1}],$$

then

$$AB - 1AC = CB \quad \text{with} \quad CB < AC$$

3.5 Synthesising ratios

and
$$AB:AC = AC:CB.$$

Hence, by the usual transformation of this proportion, the problem of finding C is precisely that solved by the construction of II 11:

To cut a straight line [AB] so that the rectangle contained by the whole and one of the segments [CB] is equal to the square on the remaining segment [AC].

Euclid's construction is illustrated in Fig. 3.11: ABDE is the square on AB, F bisects AE, FG = FB, and AGHC is the square on AG. The proof is a straightforward verification, using II 6 applied to GAFE, and Pythagoras' theorem applied to the triangle ABF.

This construction is then invoked in IV 10:

To construct an isosceles triangle having each of the angles at the base double of the remaining one. Let any straight line AB be set out, and let it be cut at the point C so that the rectangle contained by AB, BC is equal to the square on CA; .... Therefore the isosceles triangle ABD [with AB = AD, and BD = AC] has been constructed having each of the angles at the base DB double of the remaining one,

and this particular triangle is then used in IV 11:

In a given circle to inscribe an equiangular and equilateral pentagon. ... Let the isosceles triangle FGH be set out having each of the angles at G, H double the angle at F; ....

Then the same construction, now described using proportion theory, is named at VI Definition 3:

A straight line is said to have been cut in extreme and mean ratio (*akron kai meson logon tetmēsthai*) when, as the whole line is to the greater segment, so is the greater to the less,

and its metrical properties are explored in a series of propositions at the beginning of Book XIII. Finally, the properties of the extreme and mean ratio and the pentagon are used in XIII 8, 9, 11, 16, and 17. The only other surviving references to extreme and mean ratio are found in Ptolemy's *Syntaxis*, Pappus' *Collection*, Hypsicles' 'Book XIV' of the *Elements*, and an anonymous scholium on Book II of the *Elements*; no other Greek text of any kind makes any explicit allusion to the construction.

3.5(c) *The nth order extreme and mean ratio*

Consider now the case of $[\bar{2}]$. The ratio $\sqrt{2}:1$, the diagonal to side of a square, is $[1, \bar{2}]$; hence, if ABDE is the square on AB, then $AB:(EB - AB) = [\bar{2}]$. We can construct the point C_1 on AB with $AC_1 = EB - AB$,

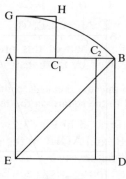

Fig. 3.21

as in Fig. 3.21, where EG = EB. If $C_1C_2 = AC_1$, then

$$AB:AC_1 = [2, AC_1:C_2B];$$

so the rectangle contained by AB and C_2B is equal to the square on AC_1.

This construction is reminiscent of II 11, and it prompts us to explore the following generalisations:

DEFINITION The point C_1 is said to divide AB in the noem ratio (read: nth order extreme and mean ratio) if, taking n points $C_1, C_2, \ldots, C_{n-1}, C_n$ on AB such that $AC_1 = C_1C_2 = \ldots = C_{n-1}C_n$, then C_n lies between A and B, and $AB \cdot C_nB = AC_1^2$. (See Fig. 3.22.)

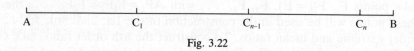

Fig. 3.22

PROPOSITION If C_1 cuts AB in the noem ratio, then $AB:AC_1 = [\bar{n}]$.

PROOF $AB:AC_1 = [n, AC_1:C_nB]$, and $AB:AC_1 = AC_1:C_nB$.   QED

The definition of the noem ratio implies that C_nB is less than AC_1; it will be called the 'lesser segment' of the noem ratio. The 'greater segment' of the extreme and mean ratio generalises in two ways: to AC_1, which will be called the 'initial segment' of the noem ratio; and to AC_n, which will again be called the 'greater segment'. Care must sometimes be exercised in making the appropriate choice. I shall describe two typical results about the 3rd order ratio, but will indicate the generality of these kinds of procedures by labelling the division points C_1, C_{n-1}, and C_n, thus following a compromise path between the different Euclidean and modern styles of handling general statements.

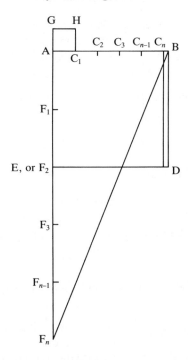

Fig. 3.23

CONSTRUCTION To divide a line AB at C_1 in the noem ratio, construct the square ABDE (see Fig. 3.23) and on AE, produced as necessary, take points F_1, $F_2(=E)$, F_3, F_4, ..., with $AF_1 = F_1F_2 = F_2F_3, \ldots$; then these points will be used in the construction of the 1st, 2nd, 3rd, 4th, ... order extreme and mean ratios. To construct the nth order ratio, take G on EA produced with $F_nG = F_nB$, and C_1 on AB with $AC_1 = AG$. Let $AC_1 = C_1C_2 = \ldots = C_{n-1}C_n$; we must prove that C_n lies between A and B and that $AB \cdot C_nB = AC_1^2$. Now

$$F_nG^2 = (AF_n + AG)^2 = (AF_n + AC_1)^2$$
$$= AF_n^2 + AC_1^2 + 2AF_n \cdot AC_1, \quad \text{(by II 4)}$$

and
$$2AF_n \cdot AC_1 = nAE \cdot AC_1 = AB \cdot AC_n.$$

But
$$F_nG^2 = F_nB^2 = AF_n^2 + AB^2. \quad \text{(by I 48 or II 8a)}$$

Hence
$$AC_1^2 + AB \cdot AC_n = AB^2.$$

Therefore $AC_n < AB$ and, subtracting $AB \cdot AC_n$ from both sides, we see that $AC_1^2 = AB \cdot C_nB$. QED

The case where n is even, $n = 2m$, is connected with the dimension of squares: here $AF_n = mAB$, an integral multiple of AB, so $F_nB^2 = (m^2 + 1)AB$, and $AC_1 = \sqrt{(m^2+1)}AB - mAB$ If AB is the assigned unit length, then

$$AB : AC_1 = 1 : (\sqrt{(m^2+1)} - m) = [\overline{2m}].$$

Thus we have again arrived, by a circuitous route, at the result that

$$\sqrt{(m^2+1)} : 1 = [m, 1 : (\sqrt{(m^2+1)} - m)] = [m, \overline{2m}].$$

3.5(d) *Elements* XIII, 1–5

The metrical properties of the extreme and mean ratio are considered in XIII 1–5, a series of propositions reminiscent of the style and techniques of Book II. All of these propositions can be easily generalised to the noem ratio; here, as an illustration, is XIII 1 followed by its generalisation:

If a straight line be cut in extreme and mean ratio, the square on the greater segment added to half of the whole is five-times the square on the half.

Euclid's proof is based on Fig. 3.24; the lettering of the vertices is his but the labelling of the regions has been added, for our convenience. He proceeds as follows: If AB is cut in extreme and mean ratio at C, and $DA = \tfrac{1}{2}AB$, we must prove $DC^2 = 5AD^2$. Draw the squares on DC and AB and complete the figure as shown. We then know that

$$AB \cdot CB = AC^2, \quad \text{i.e.} \quad P = Q,$$

and that

$$AB \cdot AC = 2AD \cdot AC, \quad \text{i.e.} \quad R = S_1 + S_2.$$

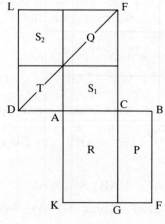

Fig. 3.24

3.5 Synthesising ratios

Hence
$$P + R = Q + S_1 + S_2.$$
Add $AD^2 = T$, and assemble the result into squares:
$$DC^2 = AB^2 + AD^2.$$
But $AB^2 = 4AD^2$, so $DC^2 = 5AD^2$. QED

We tend, today, to refer to the extreme and mean ratio by its modern name, the 'golden section', and to arithmetise and identify it with the 'golden number', $\frac{1}{2}(\sqrt{5}+1)$. The proposition expresses this result in the equivalent Greek idiom: if $AB = 2$, twice some assigned line, then $CD = \sqrt{5}$, the side of a five-fold square, so we have proved that
$$2:(\sqrt{5}-1) = [\bar{1}].$$
Note that this is not, nor can it be converted into, a ratio of two sides of integral squares.

The generalisation of this proposition explains the role of the number 5:

PROPOSITION If a straight line be cut in the noem ratio, the square on the initial segment added to n times the half of the whole is $(n^2 + 4)$ times the square on the half.

PROOF If AB is cut in noem ratio at C_1, and $D_n A = n/2\, AB$, we must prove that $D_n C_1^2 = (n^2 + 4) AD_1^2$ (see Fig. 3.25). Draw the squares on

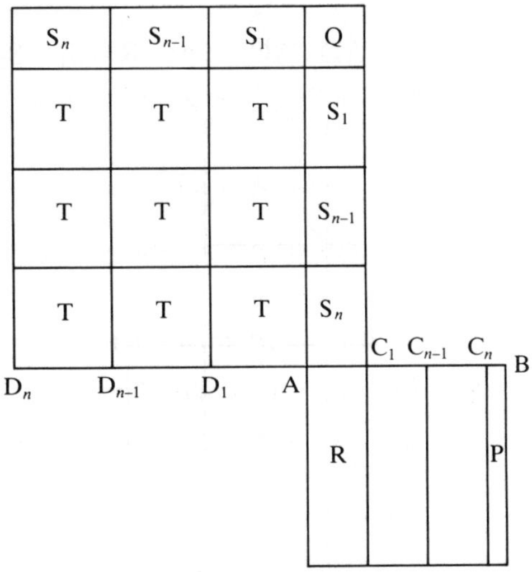

Fig. 3.25

$D_n C_1$ and AB, and complete the figure as shown. We know that
$$AB \cdot C_n B = AC_1^2, \quad \text{i.e.} \quad P = Q$$
and that
$$AB \cdot AC_1 = 2AD_1 \cdot AC_1, \quad \text{i.e.} \quad R = 2S.$$
Hence
$$P + nR = Q + 2nS.$$
Add $AD_n^2 = n^2 T$, and assemble the result into squares:
$$\begin{aligned} D_n C_1^2 &= AB^2 + n^2 T \\ &= (n^2 + 4)T \quad \text{(since } AB^2 = 4T\text{)} \\ &= (n^2 + 4)AD_1^2. \end{aligned} \qquad \text{QED}$$

Propositions 2 to 5 of Book XIII can easily be generalised in this way; complete details are given in my article GGS.

3.5(e) *Further generalisations*

Now consider the problem of synthesising a ratio containing two terms in its period. In advance, I must say that I feel that this will be implausible as a reconstruction of a Greek procedure.

PROPOSITION If
$$a_0 - na_1 = a_2$$
and
$$a_1 - ma_2 = a_3$$
are the first two steps of the anthyphairesis of the ratio $a_0 : a_1$ of two lines, and $a_0 : a_1 = [n, m]$, then
$$ma_0 \cdot a_2 = na_1^2.$$

PROOF Since $[n, m]$ is an anthyphairetic ratio, $n \geq 1$, $m \geq 1$, and $a_0 > a_1$. We have
$$a_0 : a_1 = [n, m, a_2 : a_3]$$
and hence, since the anthyphairesis is periodic,
$$a_2 : a_3 = a_0 : a_1 \quad \text{so} \quad a_0 \cdot a_3 = a_1 \cdot a_2.$$
Hence
$$\begin{aligned} a_0 \cdot a_1 &= a_0 \cdot (ma_2 + a_3) \\ &= ma_0 \cdot a_2 + a_0 \cdot a_3 \\ &= ma_0 \cdot a_2 + a_1 \cdot a_2, \end{aligned}$$

3.5 Synthesising ratios

and so

$$ma_0 \cdot a_2 = a_0 \cdot a_1 - a_1 \cdot a_2$$
$$= a_1 \cdot (a_0 - a_2)$$
$$= na_1^2. \qquad \text{QED}$$

Note that, although this proposition is expressed symbolically, everything can be interpreted as a shorthand for the geometrical manipulations of rectangles and squares. However, the algebraic content and the amount of autonomous identity that is attributed to the symbols and their manipulations are much greater here than anything we have so far encountered.

We are really more concerned with the converse result:

PROPOSITION If $a_0 - na_1 = a_2$, where $a_0 > a_1 > a_2$ and $ma_0 \cdot a_2 = na_1^2$, then $a_0 : a_1 = [n, m]$.

PROOF Write $a_1 - ma_2 = a_3$; we want to prove that $a_3 < a_2$ and that $a_2 : a_3 = a_0 : a_1$, i.e. that $a_0 \cdot a_3 = a_1 \cdot a_2$. Now

$$a_0 \cdot a_3 = a_0 \cdot (a_1 - ma_2)$$
$$= a_0 \cdot a_1 - ma_0 \cdot a_2$$
$$= a_0 \cdot a_1 - na_1^2$$
$$= a_1 \cdot (a_0 - na_1) = a_1 \cdot a_2.$$

Also $a_0 \cdot a_3 = a_1 \cdot a_2 < a_0 \cdot a_2$, and so $a_3 < a_2$. QED

CONSTRUCTION To divide AB at C_1 so that $AB : AC_1 = [n, m]$ (see Fig. 3.26, which illustrates the case of $[4, 3]$), first carry out a preliminary construction to locate B′ on AB with $nAB'^2 = mAB^2$; this can be done as follows, using II 14. Draw the square ABDE and, on EA produced, take J with $nAJ = mAB$. On EJ as diameter, draw a semicircle meeting AB at B′; then $AJ \cdot AE = AB'^2$, so $nAB'^2 = nAJ \cdot AE = mAB^2$. Next, on AE, produced if necessary, take the point F_m with $AF_m = \frac{m}{2} AB$; on EA produced, take G with $F_m G = F_m B'$; and C_1 on AB with $AC_1 = AG$. Take $C_2, \ldots, C_n$ with $AC_1 = C_1 C_2 = \ldots = C_{n-1} C_n$; we shall prove that $mAB \cdot C_n B = nAC_1^2$. This verification follows the same pattern as the earlier constructions:

$$F_m G^2 = (AF_m + AG)^2 = (AF_m + AC_1)^2$$
$$= AF_m^2 + AC_1^2 + 2AF_m \cdot AC_1$$
$$= AF_m^2 + AC_1^2 + mAB \cdot AC_1.$$

But

$$F_m G^2 = F_m B'^2 = AF_m^2 + AB'^2;$$

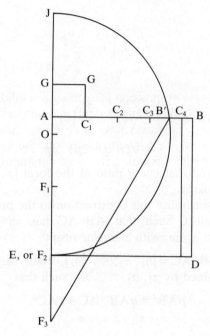

Fig. 3.26

hence
$$AC_1^2 + mAB \cdot AC_1 = AB'^2.$$

Now multiply by n, and write $nAB'^2 = mAB^2$ and $nAC_1 = AC_n$; then
$$nAC_1^2 + mAB \cdot AC_n = mAB^2.$$

Hence
$$nAC_1^2 = mAB^2 - mnAB \cdot AC_1$$
$$= mAB \cdot (AB - nAC_1)$$
$$= mAB \cdot C_nB,$$

and so $AB : AC_1 = [\overline{n, m}]$. QED

If we take AB as the unit, we can evaluate AC_1 as follows:
$$AC_1 = GF_m - AF_m$$
$$= \sqrt{(AF_m^2 + AB'^2)} - AF_m$$
$$= \sqrt{\left(\frac{m^2}{4} + \frac{m}{n}\right)} - \frac{m}{2}$$

With this evaluation, we can now synthesise the general ratio having the

3.6   Generalised sides and diagonals   95

form $[p, \overline{q, 2p}]$. We start by writing q for n and $2p$ for m, to get

$$[\overline{q, 2p}] = 1 : \left(\sqrt{p - \frac{2p}{q}} - p\right),$$

and hence

$$[p, \overline{q, 2p}] = \sqrt{\left(p^2 - \frac{2p}{q}\right)} : 1$$
$$= \sqrt{(p^2 q - 2p)} : \sqrt{p}.$$

Thus we have proved that every ratio of the form $[p, \overline{q, 2p}]$ is indeed a ratio of sides of squares.

It is possible to generalise this construction to the problem of cutting a line AB at a point C such that AB:AC has an arbitrary periodic expansion. We start again, with a similar result:

PROPOSITION If $AB:AC = \overline{[n_1, n_2, \ldots, n_k]}$, then there exist numbers p, q, and r, determined by $n_1, n_2, \ldots, n_k$, such that

$$p AB^2 = q AB \cdot AC + r AC^2,$$

and conversely.

Then a construction similar to that of Fig. 3.26, in which $r AB'^2 = p AB^2$ and $AF_3 = p/r AB$, will yield the required division point. But the evaluation of p and q requires so much detailed symbolic manipulation that it is quite unthinkable as a reconstruction of an ancient procedure. More details of this construction, which is due to Christopher Zeeman, are given in my BTEE.

3.6 The third attempt: Generalised sides and diagonals

3.6(a) *The method*

We now take the procedure of sides and diagonals that was used in Section 2.2(a) to compute the ratio $\sqrt{2}:1$ and generalise it so that it will apply to a ratio $\sqrt{n}:\sqrt{m}$ of any sides of squares. I shall discuss in more detail the historical evidence relating to the procedure in the next section; first, I describe its operation.

Behind the construction of Fig. 2.1(a), repeated here as Fig. 3.27 (with a slight addition that makes explicit how the side of the larger square is constructed out of the diagonal plus side of the smaller), there lay a relation

$$S = s + d$$
$$D = 2s + d.$$

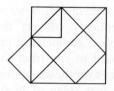

Fig. 3.27

There is no problem in drawing a square corresponding to a prescription $S = as + bd$, for any values of a and b, and deriving the corresponding relation $D = 2bs + ad$; but only very special values of a and b will lead to the recursive behaviour which will confirm the periodicity of $D:S = \sqrt{2}:1 = [1, \bar{2}]$ (or, equivalently and more conveniently, of $(S + D):S = (1 + \sqrt{2}):1 = [\bar{2}]$); the reader is encouraged to draw the figure for $a = 2$, $b = 3$, and then try to use this to evaluate $D:S$, to see how the periodic behaviour fails to manifest itself, in contrast with the case of $a = 3$, $b = 2$. The historical evidence, plus some experience with anthyphairetic ratios, suggests that we look at the behaviour of the convergents to identify what is so special about the successful side and diagonal relations.

The anthyphairesis and convergents can always be calculated using the algorisms of Section 2.3, though I shall now, for convenience, transpose the second and third lines of the standard format of Table 2.4 and will refer to the convergents as $d_k:s_k$ (for diagonal and side, of course) rather than $p_k:q_k$. For the example of $\sqrt{2}:1$, we have:

$$\sqrt{2} = [1 \quad 2 \quad 2 \quad 2 \quad 2 \quad \ldots]$$
$$s_k \quad\quad 1 \quad 2 \quad 5 \quad 12 \quad 29$$
$$d_k \quad\quad 1 \quad 3 \quad 7 \quad 17 \quad 41$$

The algorism of 2.3(c) described how to generate the successive convergents by the relations

$$s_{k+1} = n_{k+1}s_k + s_{k-1}.$$
$$d_{k+1} = n_{k+1}d_k + d_{k-1}.$$

The side and diagonal *numbers* described by Theon (see Section 2.4(e)) and Proclus (see below) appear to be the different relations that, for the case of $\sqrt{2}:\sqrt{1}$,

$$s_{k+1} = s_k + d_k$$
$$d_{k+1} = 2s_k + d_k$$

with

$$s_0 = d_0 = 1;$$

these also seem to generate the convergents. It is a plausible assertion

3.6 Generalised sides and diagonals

that these two relations are equivalent; however we do not attempt to prove this here, but immediately convert the arithmetical side and diagonal relation for $\sqrt{2}:1$ into a geometrical prescription for manipulating squares, which is then used to construct the particular figure which embodies the proof of our hypothesis about the periodicity of the anthyphairesis of $D:S$. The first stage of heuristic investigation of the convergents is then again left behind, incomplete but superseded.

I encourage the reader to explore further how this new kind of relation is apparently satisfied by the convergents of any ratio, by looking at the sequence of convergents to ratios whose anthyphaireses have successively longer and longer periods which (for convenience only) contain only small terms: $\sqrt{5}:1$, $\sqrt{3}:1$, $\sqrt{5}:\sqrt{2}$, ... Here I shall present some of the conclusions of such an exploration, illustrated in the case of $\sqrt{7}:1$.

The algorism of 2.3(b) applied to $\sqrt{7}:1$ gives, after fourteen lines of calculation:

$$\sqrt{7} = [2 \quad 1 \quad 1 \quad 1 \quad 4 \quad 1 \quad 1 \quad 1 \quad 4 \ldots]$$

| s_k | 1 | 1 | 2 | 3 | 14 | 17 | 31 | 48 | 372 |
| d_k | 2 | 3 | 5 | 8 | 37 | 45 | 82 | 127 | 717 |

and so the main hypothesis we wish to prove is that $\sqrt{7}:1 = [2, \overline{1, 1, 1, 4}]$. Since the period is four terms long, we are led to seek a side and diagonal relation of the form

$$s_{k+4} = as_k + bd_k$$
$$d_{k+4} = 7bs_k + ad_k,$$

and further exploration suggests that the appropriate values of a and b always seem to be given by the convergent just before the occurrence of the end-of-period term '$2n_0$'; here the end-of-period term appears to be $n_4 = 4 = 2n_0$, which suggests that $a = 8$ and $b = 3$. Hence we guess that

$$s_{k+4} = 8s_k + 3d_k$$
$$d_{k+4} = 21s_k + 8d_k.$$

(This 'penultimate' convergent 8:3 has a very special relation to the ratio $\sqrt{7}:1$, some aspects of which have already been described in Section 2.4(e). The side and diagonal relation will thus always be built around solutions of Pell's equation; here $8^2 - 7.3^2 = 1$.)

We now convert this into a prescription for a geometrical construction:

$$S = 8s + 3d$$
$$D = 21s + 8d,$$

which we apply to an appropriate figure built out of the lines 1 and $\sqrt{7}$. A convenient choice, which can be generalised to any ratio of sides of squares $\sqrt{n}:\sqrt{m}$, is a small parallelogram with one side s equal to the assigned unit (or, in general, to $\sqrt{m}$), one diagonal d equal to $\sqrt{7}$ (or, in general, to $\sqrt{n}$), and the other side an integral multiple ps of s. The basic inequalities for the sides of the triangle (*Elements* I 20) tell us that p must satisfy $d - s \leq ps \leq d + s$, and hence, when $d = \sqrt{7}$ and $s = 1$, $p = 2$ or 3. Let us choose $p = 2$. So we take the parallelogram with sides $s = 1$ and $2s = 2$, and (longer) diagonal $d = \sqrt{7}$, and apply the prescription to construct a larger similar parallelogram with shorter side $S = 8s + 3d$, and longer side $2S = 16s + 6d$. Fig. 3.28 illustrates this case of $n = 7$, $m = 1$, $p = 2$, $a = 8$, and $b = 3$, and once its complexities have been explained it can be seen to be a straightforward generalisation of Fig. 3.27 with $S = as + bd$.

We now evaluate the length of the longer diagonal D of this larger parallelogram and verify that $D = nbs + ad$. But, from the figure, we see that

$$AB = bs, \qquad DE = ad, \qquad EF = p^2bs,$$

so it remains only to calculate BD. We invoke *Elements* II 12:

In obtuse-angled triangles the square on the side subtending the obtuse angle is greater than the squares on the sides containing the obtuse angle by twice the rectangle contained by one of the sides about the obtuse angle, namely that on which the perpendicular falls, and the straight line cut off outside by the perpendicular towards the obtuse angle,

and apply it to ABG. Note that this triangle is obtuse-angled at B (since d is the longer diagonal of the parallelogram) and BD is equal to "twice [BC], ..., the straight line cut off outside [the triangle ABG] by the perpendicular [GC] towards the obtuse angle [at B]". Hence

$$b^2d^2 = b^2s^2 + p^2b^2s^2 + bs \cdot BD,$$

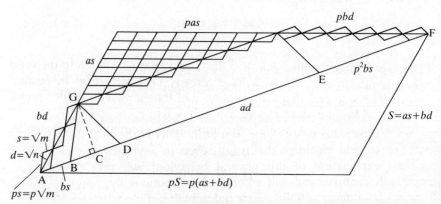

Fig. 3.28

and so, since $d^2 = ns^2$, we get

$$BD = (n - p^2 - 1)bs.$$

(The derivation of this last statement should be interpreted in terms of the converse of II 1.) We now add AB, BD, DE, and EF, which gives

$$D = nbs + ad,$$

and the geometric prescription is confirmed. A different kind of figure, which generates an acute-angled triangle and which the reader is encouraged to explore, will occur when d is the shorter diagonal of the parallelogram; in this case the verification requires *Elements* II 13.

Note that this construction can be carried out for any values of a and b. However, the next and final step depends in a subtle way on the underlying relation $a^2 - nb^2 = \pm 1$, but the general elucidation of this kind of result will not be achieved before the eighteenth and nineteenth centuries. So here we must leave the general case $\sqrt{n}:\sqrt{m}$ and revert to the verification of the particular case of $\sqrt{7}:1$ for which $S = 8s + 3d$ and $D = 21s + 8d$. We evaluate $(2S + D):S$, as follows:

$$\begin{aligned}
(2S + D):S &= [4,\ S:(D - 2S)] \\
&= [4,\ (8s + 3d):(5s + 2d)] \\
&= [4,\ 1,\ (5s + 2d):(3s + d)] \\
&= [4,\ 1,\ 1,\ (3s + d):(2s + d)] \\
&= [4,\ 1,\ 1,\ 1,\ (2s + d):s] \\
&= [4,\ 1,\ 1,\ 1,\ (2S + D):S] \\
&= [\overline{4,\ 1,\ 1,\ 1}].
\end{aligned}$$

Hence $\sqrt{7}:1 = [2, \overline{1, 1, 1, 4}]$. QED

This method will appear to verify any example of a ratio of sides of squares, no matter how long its period may be and, in this sense, the problem is solved. However, I cannot conceive of a general proof which could be formulated within the techniques of Greek mathematics that this procedure will always work, no matter what values of n and m are chosen. Nor, as I observed earlier, do I know of any evidence that Greek mathematicians observed or explored the palindromic behaviour of these anthyphairetic ratios $\sqrt{n}:\sqrt{m}$, nor can I conceive of how Greek techniques could elucidate this result, even in any particular case. The eventual explanation of this general behaviour will take us into the nineteenth century, and will combine contributions by, among others, Fermat, Brouncker, Wallis, Euler, Lagrange, Legendre, and Galois. Details and references will be given in Section 9.1(c).

3.6(b) *Historical observations*

It is, at the very least, a noteworthy coincidence that the programme of the verification of anthyphairesis of sides of squares can be completed by a procedure whose only significant mathematical prerequisites, other than some very basic results of Book I of the *Elements,* are those two slightly incongruous Propositions 12 and 13 of Book II that were left over from our earlier partially successful assaults on the same problem. This correlation between mathematical problems and ancient evidence could immediately be carried further if I were to proceed, as I now could, to consideration of Book X, because just as the material of Book II seems to be connected with and preparatory for Book X (though this in a way that is difficult to accommodate in detail within many interpretations of Book II) so does this anthyphairetic interpretation of Book II lead into an anthyphairetic motivation of Book X. But to embark straightaway on this arduous topic would be to tax cruelly the stamina of even my most sympathetic reader, so I shall leave this next step until later (see Chapter 5) and will, in the next chapter, turn to a discussion of Plato's mathematics within this interpretation.

Consider, now, the historical evidence concerning this procedure of 'sides and diagonals'. In the previous chapter (at 2.4(e)), I quoted, in its entirety, what Theon of Smyrna writes about side and diagonal numbers. Our other source of information on the topic is Proclus' commentary on Plato's *Republic* (Proclus–Kroll, *IR* ii, 24–9), written in the fifth century AD, in which Proclus makes a wide variety of observations concerning Plato's mysterious nuptial number, *Republic* VIII, 546b–d. Proclus' discussion of side and diagonal numbers starts at ii 24.16:

> The Pythagoreans demonstrate by numbers (*dia tōn arithmōn*) that the expressible (*rhētos*) [squares] constructed on the inexpressible (*arrhētos*) diagonals are greater or less than double by a unit. For since the unit is, generatively, everything, it is clear, they say, that it is both side and diagonal. So let there be two units, the one as side, the other as diagonal, ...

and the text continues in a very similar vein to Theon's description, with the addition only of a customary reference to Plato:

> Whence Plato said that the number forty-eight is the square of the expressible diagonal of five minus one, and of the inexpressible minus two, since the square of the diagonal is double that of the side [i.e. $48 = (\sqrt{(50-1)})^2 - 1 = (\sqrt{50})^2 - 2$ where $(\sqrt{50})^2 = 2 \times 5^2$].

It seems apparent that both passages derive ultimately from a common archetype. I leave a detailed analysis of this textual question to others more competent than I (I know of no published discussion of this issue), and start with the following three observations:

(i) The information that this construction derives from the Pythagore-

3.6 Generalised sides and diagonals 101

ans is found only in Proclus, not in Theon, though Theon does elsewhere frequently mention the Pythagoreans. Therefore, since both authors seem to be drawing on a common source, it is not improbable that this is an addition, either by Proclus himself or by some intermediate source of his text.

(ii) Neither Theon nor Proclus actually demonstrates the arithmetical result that $d_k - 2s_k^2 = \pm 1$ in any sense beyond verifying that it seems to hold for the first few cases.

(iii) The mathematical implications of the words (*ar*)*rhētos* are not entirely clear. (I shall return to this question in Section 5.2(a).)

All of these points are further illustrated by the sentences with which this first passage from Proclus ends (ii 25.9–13):

And if we take all the [squares] on the diagonals of this kind, of which each is greater or less than double by a unit, they will be really double: for instance nine and 49 [is double] of 25 and 4. This is why the Pythagoreans too had confidence in this method.

After this inconsequential description of side and diagonal numbers, and after some obscure and sometimes corrupt passages in which Proclus discusses various apparently unrelated views of Dercyllides, Nicolas, Magnus, and Pythagoras, Proclus returns to considerations of sides and diagonals (*IR* ii, 27.1–28.12):

Since it is impossible that the diagonal be expressible when the side is expressible (since there does not exist a square number double of a square number, from which it is also clear that there are incommensurable (*asummetros*) magnitudes, and that Epicurus was wrong to make the atom a measure of every body, and Xenocrates was wrong to make the indivisible line the measure of lines) the Pythagoreans and Plato thought to say that, the side being expressible, the diagonal is not absolutely (*haplōs*) expressible, but, in the squares whose sides they are, [the square of the diagonal] is either less by a unit or more by a unit than the double ratio which the diagonal ought to make: more, as for instance is 9 than 4, less as is for instance 49 than 25. The Pythagoreans put forward the following kind of elegant theorem of this, about the diagonals and sides, that when the diagonal receives the side of which it is diagonal it becomes a side, while the side, added to itself and receiving in addition its own diagonal, becomes a diagonal. And this is demonstrated by lines (*grammikōs*) through the things in the second [book] of *Elements* by him. If a straight line be bisected and a straight line be added to it, the square on the whole line with the added straight line and the square on the latter by itself are together double the square on the half and of the square on the straight line made up of the half and the added straight line. Let AB be a side and let $B\Gamma$ be equal to it, and let $\Gamma\Delta$ be the diagonal of AB, having a square double that of it [i.e. AB]; by the theorem, the square on $A\Delta$ with that on $\Delta\Gamma$, will be double that on AB and on $B\Delta$. Of these, the square on $\Delta\Gamma$ is double that on AB; and it remains that the square on $A\Delta$ is double that on $B\Delta$, for if as whole is to whole, so is what is taken away to what is taken away, the remainder, also, will be to the remainder as the whole is to the whole. Then the diagonal ΓΔ, receiving in addition the side BΓ, is a side; and AB, taking in addition itself, [i.e.] the [side] BΓ, and its own diagonal ΓΔ, is a diagonal; for it has a square double that of the side [ΔB]. These things, then, this way. Let us now prove by numbers (*arithmētikōs*) the result on the expressible diagonals which are, as we said earlier, by a unit greater or less. Let there be a unit and let there be another unit around it; . . . ,

and with this Proclus reverts to another inconsequential description of side and diagonal numbers not dissimilar to Theon's and his own earlier description at ii 24.11–25.6, but carried now up to a side of 12 and diagonal of 17. Let me continue the list of observations on these passages:

(iv) Proclus, only, gives a description of an "elegant theorem" about side and diagonal *lines*.

(v) This elegant theorem is said to be "demonstrated by lines through the things in the second book of *Elements*"—surely Euclid's *Elements*. (Alternatively, the end of this phrase can be construed to mean "through the elements in the second book".)

(vi) There is no proof of, or reference to, this elegant theorem in *Elements* II as we have it.

(vii) Proclus cites the enunciation of *Elements* II 10 with slight, but surely unimportant, variations in the wording, and he uses this to verify the result.

(viii) The manuscript of Proclus' commentary contains no figure but it seems certain that the configuration to which his text refers is that provided by the editors Kroll in his discussion of the passage (at ii 393–400), and given here as Fig. 3.29, though Kroll and most modern commentators continue the diagram further to the right.

(ix) Proclus makes no further use of this elegant geometrical theorem but reverts abruptly, at the end, to a further inconsequential discussion of side and diagonal numbers.

(x) Anybody seeking a connection between the elegant theorem and *Elements* II, as we now have it, would be led to Propositions 6, 8

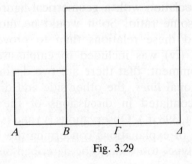

Fig. 3.29

(perhaps), and 10, since all of these propositions deal, more or less explicitly, with one line added to another (like $s + d$), then with the line added to itself and added to the other (like $2s + d$); and, of these, Propositions 6 and 8 involve rectangles, while 10 involves only squares. Thus the elegant theorem seems, in the allusive style in which such references are made in Greek mathematics, to be referring to II 10, and this proposition does indeed provide a quick proof; but it leads no further.

(xi) Proclus' completion of the proof by invoking a general result on proportions, in fact *Elements* V 19 or VII 11, is unnecessarily and irrelevantly heavy-handed; all that is needed is Common Notion 3.

To these, let me add the subjective opinion:

(xii) Proclus' proof of the elegant theorem is pedestrian, unilluminating, and inconsequential, in contrast to the perspicuous proof based on Fig. 3.27.

Now let me fit these observations together. Proclus and Theon are drawing on some earlier source, now lost to us. Either Theon's source is already less complete than Proclus', or Theon only abstracts a portion of it; in either case, having made observation (i) above, we can set Theon's testimony aside, as being contained within Proclus' text. Next, from (ii), we can infer that this earlier source did not contain a complete proof of the arithmetical results, though such is easy to provide since it follows from any acceptable description of the manipulation

$$d_{k+1}^2 - 2s_{k+1}^2 = (2s_k + d_k)^2 - 2(s_k + d_k)^2 = -(d_k^2 - 2s_k^2)$$
$$= \pm(d_1^2 - 2s_1^2) = \pm 1.$$

Hence, we infer, this source may not have been primarily concerned with proving this relation, but with illustrating or investigating the connection between its essential ingredients: both side and diagonal equal to one as starting values; the recurrence relation for s_k and d_k; and the relation $d_k^2 - 2s_k^2 = \pm 1$. This opinion is reinforced by the way the same discussion is repeated, each time at length. (Note also how, in my proposed reconstruction, it is not the proofs of these kinds of relations that is pertinent, but their association with a geometrical figure, and thence to the anthyphairesis of some ratio. So it would be more important to explore and understand these relations than to prove one particular example.) Observation (iv) was included to emphasise a feature that often passes without comment: that there are two distinct relations, one involving side and diagonal *lines*, the other side and diagonal *numbers*, and these are often conflated in discussions of the topic (see, for example, Heath, *TBEE* i, 398 ff.). Observations (v) and (vi) are compatible with a very wide range of explanations concerning the form and role of *Elements* II or its archetype, too wide to be described and analysed here;

but observations (vii) to (xi) seem to indicate that the citation of II 10, along with the figure (see (viii)), and the details of the proof, was a later addition to the original text, possibly by Proclus or his source.

This leaves (xii). Underlying this observation is an inconsistency within Proclus' testimony, that the theorem as he describes and demonstrates it is *not* elegant. This is my subjective judgement but, as an illustration of how it does seem to be more widely held, see how the alternative proof by Fig. 3.27 is described by Heath, *TBEE* i, 400–1, as an "an ingenious suggestion", and by Burkert, *LSAP*, 430, n. 16 (in a footnote which contains the only geometrical figure in this long book) as "still more perspicuous". Just as Meno's slaveboy must have marvelled at Socrates' ability to draw the figure which led to the solution of the problem of doubling a square (*Meno* 84d–e), so we can still marvel at the construction of Fig. 3.27, and the way this diagram makes evident the anthyphairesis of $\sqrt{2}:1$. Moreover, Fig. 3.27 fits well in the style of attested fourth-century Greek mathematics, since it bears a striking resemblance to the successful *Meno* figure, as well as to the alternative reconstructed proof of Pythagoras' theorem II 8a (see Fig. 1.5 and its associated text, and Section 3.2, above); all consist of obliquely placed squares or rectangles within larger squares. And further, once this figure has been drawn, it is not too difficult to see some of the remarkable ways in which this simple construction can be exploited and generalised within the context of anthyphairetic mathematics. These qualities of simplicity, surprise, power, and generality are, today, some of the features that lead mathematicians to use words like 'elegant', and I can conceive of it being so also in antiquity. But, removed from its fertile context, for example as we find it in Proclus, the theorem scarcely seems to merit the description.

I suggest therefore that Proclus' description of a figure in which "when the diagonal receives the side of which it is diagonal it becomes a side" may indeed originally have been a reference to the construction of Fig. 3.27, and that this construction was investigated in the context of its generalisation to diagrams like Fig. 3.28 with a view to proving the observed periodicity in the problem of the dimension of squares. Then, as the original anthyphairetic content was excised from the material to be included in Euclid's *Elements,* all that remained of this procedure were *Elements* II, Propositions 12 and 13, and a haunting echo of an elegant theorem about diagonals and sides.

3.7 Summary

My proposal is that Book II of the *Elements* may be a collection of figures and results whose connecting feature is that they arose in three different attempts to verify and explore the periodicity of the anthyphairetic ratios of incommensurable sides of squares, $\sqrt{n}:\sqrt{m}$. Ultimately, the third and

3.8 Notes and references

3.1 Much of the material of this chapter has already been published in my three articles BTEE, BTEE2, and GGS.

3.2 The most common interpretation of Book II is in terms of what is now called 'geometrical algebra', a name that was apparently coined by Tannery and Zeuthen for a much older point of view. This kind of interpretation has recently been the focus of vigorous argument; for a general discussion see Mueller, *PMDSEE*, 41–53 and 301–3, and Berggren, *HGM*, 397–8. My proposals here provide an entirely different reconstruction of the antecedents and original context of the book.

3.3 The hypotheses about the periodic palindromic behaviour of $\sqrt{n}:\sqrt{m}$ will be proved in Section 9.1(c). Lack of space has precluded any discussion of the Greek evidence on palindromes; I recommend the curious reader to start with Horsfall, SB, and the references contained therein. I cannot see any evident connection between this material and mathematics. I do not know of any palindromes in Plato's writings or other fourth-century texts beyond isolated words like αρα, σοφος, etc.

3.4 There is a discussion that relates incommensurability, anthyphairesis, and Book II-style arguments in Zeuthen, CLAEE, but this only deals with a few simple cases; it does not involve the figures of II 5 and 6; and the arguments seem to be set in an arithmetised context.

3.5(b) The extreme and mean ratio has been known under a variety of names. There is a brief history of these in the introduction to my GGS; in particular the most common name, the golden section, was coined by Martin Ohm, the younger brother of the physicist Georg Simon Ohm, in 1834 or 1835. The dramatically simple construction of the extreme and mean ratio of Fig. 3.30, based on the mid-points of two sides of an equilateral triangle inscribed in a circle, was published, perhaps for the first time, in 1984; see Odom, EP. I am indebted to Roger Herz–Fischler for much detailed information about the extreme and mean ratio; see Curchin & Fischler, HANTDEMR, DQDPR, and his forthcoming monograph *MHDEMR*.

Fig. 3.30

3.6 I know of nowhere in the mathematical literature where the underlying property of the generalised method of sides and diagonals (that the convergents of a ratio $\sqrt{m}:\sqrt{n}$ whose period contains p terms satisfies a recurrence relation of the form $s_{k+p} = nas_k + bd_k$, $d_{k+p} = mbs_k + ad_k$, where $a:b$ is the convergent that occurs just before the end-of-period term $2n_0$) is proved. I am grateful to J. W. S. Cassels for sketching a proof of this result.

On the contrast between arguments that proceed *dia tōn arithmōn* or *dia tōn grammōn*, see Neugebauer, *HAMA* ii, 771 n. 1.

4
PLATO'S MATHEMATICS CURRICULUM IN *REPUBLIC* VII

4.1 Plato as mathematician

My attitude to Plato and the Academy has been greatly influenced by the sceptical and rigorous analysis described in Cherniss, *The Riddle of the Early Academy*, especially Lectures 1 and 3 therein. And, to a historian of ancient mathematics, an additional riddle is raised by Cherniss's scrupulous exploration of the fragmentary evidence, in the tension between his conclusion, on the one hand, that:

All the evidence points unmistakably to the same conclusion: the Academy was not a school in which an orthodox metaphysical doctrine was taught, or an association the members of which were expected to subscribe to the theory of ideas.... The metaphysical theories of the director were not in any way 'official' and the formal instruction in the Academy was restricted to mathematics [pp. 81 f.],

and

Philodemus says that mathematics made great progress under the direction of Plato, who formulated problems which the mathematicians zealously investigated. Proclus, too, in his famous summary, which appears to derive ultimately from the *History of Mathematics* written by Eudemus, credits Plato's concern for mathematics with the great progress of these studies and particularly of geometry. Besides Theaetetus, Leodamas, and Phillip of Opus, he names six specialists in mathematics who, he says, passed their time together in the Academy and pursued their investigations in common. [The next sequence will be quoted below.] [Plato] is said to have induced Phillip to turn his attention to the subject, to have originated the theorems about the section, the number of which Eudoxus increased, and to have communicated to Leodamas the method of analysis [p. 65],

while, on the other hand, Cherniss is forced to add, in the sentence omitted from the previous quotation, that:

It cannot be imagined that Plato *taught* any of these men mathematics, though he is said ... [p. 65, original italics].

This opinion is then amplified:

Plato's influence on these men, then, was that of an intelligent critic of method, not that of a technical mathematician with the skill to make great discoveries of his own; and it was by his criticism of method, by his formulation of the broader problems to which the mathematician should address himself, and, as the summary of Proclus says, by arousing in those who took up philosophy an interest in mathematics that he gave such a great impulse to the development of the science [p. 66].

(I have omitted the detailed references to ancient literature and modern scholarship that fill out these quotations; the reader is strongly encouraged to consult the original book and the cognate review article, Cherniss, PM.) I believe that, within my new interpretation, we may argue that while Plato's principal interest was in dialectic, for which he regarded mathematics only as a preliminary, he does none the less show detailed knowledge of important characteristics and problems of technical mathematics, and there is no indication that he could not communicate on equal terms with the mathematicians who seem to have dominated, if not comprised, the group of friends and associates that assembled round him. This thesis I shall now attempt to illustrate by sketching an interpretation of his mathematics curriculum in *Republic* VII.

This curriculum (*Republic* VII 521c–531c) is proposed as part of the education of future guardians of the state; it will occupy men from age twenty (537b–c) to thirty (537d). It will be preceded by early training in childhood, imparted through play (536d–e), by sightseeing trips to the battlefield (537a), and by a two- or three-year break for gymnastics (537b), and only those who show promise will go on to study mathematics. At age thirty, and after a second selection, the students will pass on to a training in dialectic (537d–e) for five years (539e). They will then be compelled "to hold commands in war and other offices suitable to youth" for a further fifteen years, up to the age of fifty (539e–40a); and those who come through this programme of preparation will, from time to time for the rest of their lives, be called on to rule the state (540a–b). In this way, the objective will be achieved:

What we require is that those who take office should not be lovers of rule. Otherwise there will be a contest with rival lovers. What others, then, will you compel to undertake the guardianship of the city than those who have most intelligence of the principles that are the means of good government and who possess distinctions of another kind and a life that is preferable to the political life? [521b]

This life is that of the true philosopher as conceived by Plato. A related programme of training for citzens of the state, at a lower technical level and excluding the training in dialectic, is described in *Laws* VII, especially at 817e–822c.

If this state is not to remain a day-dream, then the first generation of true philosophers must be trained. Plato finished the book with a chilling description of the speediest and easiest way of doing this (540e–541a).

The curriculum is in five parts: *arithmētikē te kai logistikē* (524d–526c);

plane geometry (526c–527c); three-dimensional geometry (528a–d); astronomy, which was prematurely introduced as the third subject (527d–528a), then replaced in this position by solid geometry and reintroduced as the fourth topic (528d–530c); and, finally, music theory (530d–531c). Plato emphasises, at the end of his description, that this is not a disparate collection of topics (531c–d):

> I take it that if the investigation of all these [mathematical] studies goes far enough to bring out their community and kinship with one another, and to infer their affinities, then to busy ourselves with them contributes to our desired end and the labour taken is not lost; but otherwise is in vain, ... , but it is a huge task. ... Do we not know that all this is but the preamble of the law itself, the prelude of the strain we have to apprehend?

It is such a unity that I shall emphasise, by illustrating how each topic can contribute to our understanding of ratio. (All quotations from the *Republic* will be given in the translation of P. Shorey in the Loeb Classical Library and reprinted in Plato, *CD*.)

4.2 Arithmētikē te kai logistikē

From the aphorisms of Heraclitus, for example Fragment 1:

> *Tou de logou toud' eontos aei axunetoi ginontai anthrōpoi kai prosthen ē akousai kai akousantes to prōton.* ... (Of the *logos* which is as I describe it men always prove to be uncomprehending, both before they have heard it and when once they have heard it. ...),

and Fragment 50:

> *Ouk emou alla tou logou akousantas homologein sophon estin hen panta einai* (Listening not to me but to the *logos* it is wise to agree that all things are one),

up to the opening sentence of St John's gospel:

> *En archē ēn ho logos, kai ho logos ēn pros ton theon, kai theos ho logos,* (In the beginning was the word, and the word was with God, and God was the word (*sic*)),

and beyond, the word *logos* has carried an enormous range of connotations. It occupies five dense columns of text in the *Greek–English Lexicon* by Liddell, Scott, & Jones, with a substantial entry in the *Supplement*; this is one of the longest entries in the whole lexicon. I shall here be concerned only with its very restricted range of mathematical meanings and uses.

An unambiguous ingredient of our surviving contemporary references to Greek mathematical and scientific thought in the first half of the fourth century, especially in Plato's and Aristotle's writings, is the frequent appeal to the idea of *logos* as ratio, and the use of the derived words that may even have been coined by Plato and his associates: *logistikē*, the art (i.e. *technē*, understood) of *logos*, usually translated as calculation, reckoning, computation, etc., and *logismos*, also used for calculation, reckoning, computation, etc. or, more abstractly, for judgement and rationality; an index to the occurrences in Plato, Archytas, Aristotle, and the pre-Socratic philosophers is given in an appendix to this chapter. While these words are used in a range of contexts and with a range of meanings that may be irrelevant to my mathematical enquiry here, a very substantial number of explicitly technical occurrences remain. And while, to many readers with other interests in mind, it often does not matter whether Plato is making a precise allusion to mathematics or not, the identification and interpretation of these references is crucial to a specialised investigation of early Greek mathematics. So, to take one minute example, that the sentence:

Logistikē pou tis hēmin hēn technē [Statesman 259e]

can be translated (by Skemp in Plato, *CD*) as

There exists, we agree, an art of counting,

or (by H. N. Fowler, in the Loeb Classical Library edition) as

We recognised a sort of art of calculation

may be a matter of little general concern; but we must here make a more serious attempt to identify just what Plato meant by *logistikē* in this mathematical reference. This issue is complicated by three factors:

(i) Plato frequently refers to both *logistikē* and *arithmētikē* together, sometimes treating them as if they are overlapping parts of the same mathematical study; see, for example, the first part of the curriculum, which we shall be discussing in more detail below. Then, at other times, he is prepared to draw a fine distinction between them; see, in particular, *Gorgias* 451a–c, quoted in 1.2(e), and the related passage, *Charmides* 165e–166a, which does not even mention *arithmētikē*. In other contexts, for example *Protagoras* 356a–357b, he talks of *metrētikē*, the art of measurement, in similar terms:

If the saving of our life depended on the choice of odd and even, and on knowing when to make the right choice of the greater and when of the less—taking each by itself or comparing it with the other, and whether near or distant—what would save our life? Would it not be knowledge; and knowledge of *metrētikē*, since the art here is concerned with excess and defect, and of *arithmētikē* as it has to do with the odd and even?

(ii) *Logistikē* has an everyday meaning of practical calculation and financial accounts, and Plato himself uses it with this sense, for example at *Philebus* 56d–57a, where he again talks of *arithmētikē*, *logistikē*, and *metrētikē*:

> Socrates: Are there not two kinds of *arithmētikē*, that of the people and that of philosophers?... And how about *logistikē* and *metrētikē* as used in building and trade when compared with the geometry and *logismoi* studied (or practised) in philosophy (*hē kata philosophian geōmetria te kai logismoi katameletōmenoi*)— shall we speak of each of them as one or two?
> Protarchus: I should say that each of them was two.

Similar references to everyday *arithmētikē* and *logistikē* can be found in the first part of the curriculum. I shall distinguish the two different types by talking of 'practical' or 'theoretical' *arithmētikē* and *logistikē*.

(iii) By the first century BC, at the latest, theoretical *logistikē* had come to refer to that kind of artificial calculation that mathematicians tend to devise about numbered collections such as 'pebbles' or 'cattle on the isle of Sicily'; and neo-Platonic commentators, when discussing *logistikē*, seem to dwell exclusively on this interpretation. See, for example, Proclus–Morrow, *CFBEE*, 38–40:

> Others, like Geminus, think that mathematics should be divided differently; they think of one part as concerned with intelligibles only and of another as working with perceptibles and in contact with them.... Of the mathematics that deals with intelligibles they posit *arithmētikē* and geometry as the two primary and most authentic parts, while the mathematics that attends to sensibles contains six sciences: mechanics, astronomy, optics, geodesy, canonics, and *logistikē*.... Geodesy and *logistikē* are analogous to these sciences [sc. geometry and *arithmētikē*], since they discourse not about intelligible but about sensible numbers and figures.... Nor does the student of *logistikē* consider the properties of number as such, but of numbers as present in sensible objects; and hence he gives them names from the things being numbered, calling them sheep numbers or cup numbers.

Another such description, credited to Anatolius, is cited in the Heronian *Definitions* (Heron–Heiberg, *Opera* iv, 164); and an anonymous scholium to the passage at *Charmides* 165e–166a gives a similar description of *logistikē* as the science of numbered objects. (Both of these passages are translated in Thomas, *SIHGM* i, 16–19.) But in his discussion of the curriculum in *Republic* VII, Plato is emphatic that this theoretical *arithmētikē te kai logistikē*, his astronomy, and his music theory are theoretical sciences which do *not* concern themselves with sensibles. There is a great gulf between what Plato is describing and what the neo-Platonic commentators understand by these terms. So, for the moment, I shall continue to ignore the later evidence.

A comprehensive discussion of *logistikē* can be found in J. Klein, *Greek Mathematical Thought and the Origin of Algebra*. Since I agree in large measure with Klein's conclusion about the meaning of Plato's *logistikē*, but wish to propose a different reconstruction of the mathematics that lies behind it, I would like to proceed by making explicit some of these agreements and divergences, and recommend this book for an immensely detailed enumeration and evaluation of the evidence; and also the summary, with comprehensive references, in Burkert, *LSAP*, 446 n. 119. Klein concludes, and I concur, that

> The ['theory of ratios and proportions'] does seem identical with the 'theoretical logistic' postulated by Plato [p. 6],

though I disagree with an implication of the word 'postulated', that while the possibility of this theory was assumed, little of it had actually been achieved. The next passages indicate Klein's conception of this theoretical logistic. (I have occasionally silently arranged the syntax of these quotations slightly to fit the context here, without, I hope, disturbing the sense of the originals. The italics and quotation marks are all Klein's. The first passage is followed by my annotations.)

> In the face of definite multitudes of things we habitually *determine their exact number*—we 'number', i.e., count, the things.... In order to be able to count we must know and distinguish the single numbers, we must "distinguish the one and the two and the three" (*to hen te kai ta duo kai ta tria diagignōskein—Republic* VII, 522c). Plato calls the totality of this science of all possible numbers the 'art of number'—'arithmetic'.[a] But we are also in the habit of *multiplying or dividing these multitudes*. This means that we are no longer satisfied with the number by which we have enumerated the things in question,[b] but that we bring to bear on this number new 'numbers', whether we wish to separate off a 'third' part of the respective quantity or wish to produce a multitude which amounts to 'four' times the given one.[c] In such multiplications and divisions, or, more generally, in all *calculations* which we impose on multitudes, we must *know beforehand* how the different numbers *are related to one another* and how they are constituted *in themselves*.[d] This whole science... is called the 'art of calculation'—'logistic' [pp. 18–19].

Notes:

[a] In fact, at *Republic* VII, 522c, Plato actually writes: "I mean, in sum, *arithmos te kai logismos*". See (i), above.

[b] I take this and the following sentence to be a reference to the unit in which we are counting, which is then cut up or multiplied to make a new unit. But Plato insists, explicitly and emphatically, that this is not permitted in theoretical *logistikē*; see *Republic* VII, 525e–526a. Klein is well aware of this, as the quotations below will indicate.

[c] Klein rightly insists throughout on the concreteness of the Greek *arithmoi* as numbered collections of objects, as cardinal numbers; but he does not analyse how repetition numbers describe relations between these cardinal numbers. So here he must put 'third' and 'four times' in quotation marks, presumably because they are not cardinal numbers, and so do not fit within the scope of his *arithmoi* without some further explanation. See Section 1.2(c).

[d] This is, without doubt, a reference to the explanations of *logistikē* at *Gorgias* 451a–c (quoted in 1.2(e)) and *Charmides* 165e–6a (quoted above).

The reason why Klein needs such an elaborate description of multiplication and division is, I believe, partly because he has not identified the role of repetition numbers in Greek mathematics, but mainly because he cannot get away from the common conception that a ratio of two numbers *must* ultimately be manipulated and conceived mathematically as a fraction, or some more acceptable formulation of our idea of a fraction, and this contravenes the Platonic prohibition of "cutting up the unit" at *Republic* VII, 525e. This is made quite explicit in his Chapter 5, 'Theoretical Logistic and the Problem of Fractions', which opens:

The question before us is: What prevents later writers from interpreting the arithmetical theory of relations, i.e., proportions, as theoretical logistic? [p. 37],

and his answer follows:

The crucial obstacle to theoretical logistic—keeping in mind its connection with calculation—arises from *fractions,* or more exactly, from the fractionalisation of the unit of calculation [p. 39],

and

A calculation which is intended to be as exact as possible simply cannot be effected within the realm of these pure numbers [whose noetic character is expressed precisely in the indivisibility of the units]. The immediate consequence of this insight, at least within the Platonic tradition, is the exclusion of all computational problems from the realm of the 'pure' sciences [p. 43].

A further difficulty, suggested in the introduction, is not developed in detail:

An additional reason was the elaboration of the theory of ratios into a *general* theory of proportion, which depended on the discovery of incommensurable magnitudes and which led altogether beyond the realm of counted collections [p. 7].

I shall postpone discussion of the topic of this last quotation until much later (see Section 8.3) and here only comment that I find the choice of 'depended' (*beruhen*, in the original German) puzzling.

It is, I hope, now clear that it *is* possible to handle exact calculations, ratios of incommensurable magnitudes, approximations, and other such manipulations without transcending any of the restrictions on the *arithmoi* given in 1.2(c); many anthyphairetic illustrations of such manipulations were given in Chapters 2 and 3, and further procedures based on other definitions will be described below, in Sections 4.4 and 4.5. The essential liberating step comes from rejecting the techniques and preoccupations of our own present-day brand of arithmetised mathematics, which *is* founded on the arithmetical manipulations of fractions, with its further extensions to real numbers and beyond, and developing in its place techniques more suited to the Greek context.

I propose, then, that we should conceive of *logistikē* (*technē*) and *logismos* as 'ratio theory'. There are different varieties of ratio theory: I have, so far, developed only some of the theory of anthyphairetic *logistikē*, but I shall also illustrate astronomical and music-theoretical *logistikē* below and will introduce an accountant's ratio theory, with several variations, in Section 7.4(b). The unity of Plato's curriculum will then depend on being able to pass freely between these different ideas of ratio; we shall see how this will pose tricky problems for the mathematicians and will direct their attention away from the particular contexts of astronomy and music, towards abstract problems. Let us start by looking at Plato's description of *arithmētikē te kai logistikē* (*Republic* VII, 522c–526c), to see how well it corresponds to this point of view.

The study of ratio is intimately bound up with the use and manipulations of the *arithmoi*; in fact ratios are characterised by various kinds of patterns that can then be described by sequences of numbers. In this manner *arithmētikē* and *logistikē* enter in a fundamental way. As Plato says:

Is it not true of them that every art and science must necessarily partake of them? [sc. *arithmos te kai logismos*] [522c]

Plato describes how this study leads naturally to the awakening of thought by illustrating how the comparison of three lines (three fingers is the example he chooses in 523c–524c) leads to a true comparison, not of how they seem to appear—which may depend on how near or far each is—or of their irrelevant characteristics of colour or touch or weight, but purely in relation to the "great and small". Now the theories of geometrical magnitudes, abstract magnitudes, and number that we find in the *Elements* also develop and exploit this kind of comparison—"the less of the greater" (*ho elassōn tou meizonos*) is a ubiquitous phrase in *Elements* V and VII—and this comparison of *arithmoi* and magnitudes is precisely the ingredient needed to define any kind of ratio (see, for example, my dialogues, B_2–S_{13} & B_{36-8} and E_{67} & B_{85}–A_{88}, below). We shall see, indeed, that:

It is one of those studies which we are seeking that naturally conduce to the awakening of thought [523a],

and we will go on to see just how valid is the assertion (S_{43}) that

Studies that demand more toil in the learning and practice than this we shall not discover easily nor find many of them [526b],

though the difficulty of the subject is offset by its extraordinary attractiveness and charm (cf. 528c–d).

One feature of these ratios that has provoked a lot of subsequent toil is very strikingly evocative of the assertion that

> It occurs to me, now that the study of *logismos* has been mentioned, that there is something fine in it ... that it strongly directs the soul upwards and compels it to discourse about pure numbers, never acquiescing if anybody proffers to it in the discussion numbers attached to visible and tangible bodies [525c–d].

Practical *logistikē*—the activities of merchants and hucksters (525c), *arithmētikē, logistikē,* and *metrētikē* as used in building and trade (*Philebus,* 56e), and the calculation of financial accounts and taxes (see Chapter 7)— is mainly concerned with arithmetical calculations. But let us look more closely at what is implied by this arithmetic. The ontological status of arithmetic with *ratios* is far from evident, and my slaveboy's question at B_{32} is a blatant anachronism: however fluent we might now be in performing these kinds of operations, and however basic they might now be to the development of our kind of mathematics, they have presented enormous obstacles in the understanding of mathematics, both phylogenically and ontogenically. (It is not too difficult to explain to a beginner what we might want a fraction like 'five-eighths' to mean; but to explain, from basic principles, what a calculation like 'five-eighths divided by three-quarters' might reasonably signify is of a different order of difficulty, and I suspect that many who can perform such manipulations would have serious problems in elucidating what they are doing.) Moreover, we find little or no explicit trace of such fractional arithmetic in our evidence on Greek mathematics. As an illustration of this, observe how reactions to the text of the 'arithmetical' Book VII of the *Elements* can vary from a recognition that this book does not seem to deal with the arithmetic of fractions in any form (see, for example, Mueller, *PMDSEE,* Chapters 2 and 3) to what I believe is a wholesale rewriting of this book in terms of fractional calculations (see, for example, van der Waerden, *SA,* 48–9 & 110–16, quoted in Section 4.5(c), below), which then sometimes involves a comprehensive apparatus of equivalence relations, purely formal tricks of the last hundred years (see, for example, Taisbak, *DL*). I believe that my Socrates' reply S_{33} is also anachronistic: these topics are not found in the *Elements,* or elsewhere in early Greek mathematics, because they are not a part of the non-arithmetised mathematics which is being developed there, and neither my boy's question B_{32}, nor the reply S_{33}, fits naturally within this non-arithmetised approach. But if we excise these two speeches, and modify the dialogue at this point to:

BOY$'_{32-4}$: Well here's another question. Suppose we have three heaps of stones; how can we calculate the ratio of the first to the third from the ratio of the first to the second and the second to the third? Or a similar question about adding ... , we now have something that is historically acceptable and mathematically interesting; my slaveboy is now asking about the operation known to Euclid as 'compounding' ratios. Not only is Euclid's text corrupt and confused (see the brief comments at the end of 1.2(d), and the more detailed analysis to follow in Section 4.5(c), below) but also the underlying mathematical problem is curiously difficult. It is a remarkable feature of anthyphairetic and astronomical ratios that their arithmetic is intractable; it is difficult to pass directly from the information that $\sqrt{2}:1 = [1, 2, 2, 2, \ldots]$ and $\sqrt{3}:1 = [1, 1, 2, 1, 2, 1, 2, \ldots]$ to the results that $\sqrt{6}:1 = [2, 2, 4, 2, 4, 2, 4, \ldots]$ and that $(\sqrt{2} + \sqrt{3}):1 = [3, 6, 1, 5, 7, 1, 1, \ldots]$, where this latter ratio seems to exhibit no evident pattern and its evaluation, using the algorisms of Chapter 2, is a great deal more difficult than all of the ratios we have met so far except for the circumference and diameter of a circle. (This 'binomial' line $\sqrt{2} + \sqrt{3}$ is one of the *alogoi* of the classification of *Elements* X; see Chapter 5, below.) Even the relation between the ratios $b:c$ and $2b:c$ is far from obvious. As a modern textbook puts it:

There is, however, another and yet more significant practical demand that the apparatus of continued fractions [i.e. anthyphairetic ratios, modern style] does not satisfy at all. Knowing the representations of several numbers we would like to be able, with relative ease, to find the representations of the simpler functions of these numbers (especially, their sum and product). In brief, for an apparatus to be suitable from a practical standpoint, it must admit sufficiently simple rules for arithmetical operations; otherwise it cannot serve as a tool for calculation. We know how convenient systematic fractions are in this respect. On the other hand, for continued fractions there are no practically applicable rules for arithmetical operations; even the problem of finding the continued fraction for a sum from the continued fraction representing the addends is exceedingly complicated, and unworkable in computational practice [Khinchin, *CF,* p. 20].

The situation is quite extraordinary: on the one hand, in the world of the sensibles, these everyday operations of arithmetic with general kinds of numbers are commonplace. We have many, many examples of financial accounts on papyrus from Graeco-Roman Egypt which contain explicit examples of such arithmetic, expressed throughout in unit fractions. This is practical *logistikē*; there seems to be no insuperable difficulty in doing it, but it is very difficult to describe in any correct and complete way how it is done, or, at a more subtle level, to say just what is being done. (See the last part of S_{35}; more details of our evidence about these calculations will be given in Chapter 7.) In some kinds of theoretical *logistikē*, on the other hand, we can define our terms precisely, but we cannot now perform some of these everyday manipulations. (Now see the beginning of S_{35}.) The reaction of a practical calculator to a description of the insights and power of this theoretical *logistikē* is likely, even today, to be not dissimilar to the reception that Plato describes, at the beginning of

Republic VII, 514a–518b, that will be accorded to someone who returns to the cave of shadows after seeing the reality of the world of sunlight outside! The final ironic twist in this story of the search for algorisms for arithmetic—that anthyphairetic arithmetic is actually not very difficult while decimal arithmetic is much more subtle than is generally thought—will be described in Section 9.3(a).

So, on the one hand, theoretical *logistikē* seems, in the ancient evidence, to show little preoccupation with the arithmetic of ratios and, even today, the corresponding mathematical problem of arithmetic with continued fractions is still little discussed because of its supposed difficulty. On the other hand, the cluster of mathematical ideas round the subject as described by Plato seems to involve the ingredients of 'the odd and the even', 'the greater and the less' (or 'the excess and the defect') and 'their mutual relationships'; does this resonate with anything of significance in connection with ratio theories? In fact, the associations, in particular with anthyphairetic ratios, are now so rich and diverse that any attempt to enumerate them completely would spread through many pages, but here, in summary, are a few instances:

(i) At the ground level of the definition and evaluation of any kind of ratio of two numbers or two magnitudes we shall need to manipulate 'the greater and the less' in some kind of comparison process (see S_1 to S_{13} and the further definitions to follow); and this comparison tends to implicate *arithmētikē* to describe the pattern of the process. In anthyphairesis, the even-numbered terms, $n_0, n_2, \ldots$, describe subtractions from the initially larger pile, the odd, $n_1, n_3, \ldots$, from the initially smaller.

(ii) Now ascend to the level of understanding anthyphairetic ratios, so calculated. Here the order relationship ('the less and greater') between ratios involves taking account, in a surprising and subtle way, of 'the less and the greater' numbers that arise in 'the odd and the even' steps; see S_{17} to B_{22}, and S'_{21} in Section 2.3(b).

(iii) Look now at the algorisms for calculating anthyphairetic ratios, described in Section 2.3. These are built around the alternating manipulations of under- and overestimates, 'the mutual relationship of the greater and the less and the odd and the even'.

(iv) Pass to the properties of anthyphairetic ratios as explored in Section 2.4. Again we find an increasingly subtle exploration of the greater and the less; and the odd-even parity enters in the behaviour of Pell's equation (see the generalised side and diameter numbers and Archimedes' *Cattle Problem*) in a way that still has not received a satisfactory characterisation.

(v) There is an odd-even parity underlying the astronomical ratios to be described in Section 4.4, below, but it plays a less prominent role there.

When I try to identify the mathematical features which underlie and permeate the definition, development, and application of anthyphairetic ratio theory, I find it hard to better these allusive descriptions that can be borrowed from Plato.

4.3 Plane and solid geometry

Plato does not tell us what his plane geometry is, but what it does (*Republic* VII 526d–527c):

[SOCRATES:] What we have to consider is whether the greater and more advanced part of it tends to facilitate the apprehension of the idea of good. ... If it compels the soul to contemplate essence, it is suitable; if genesis, it is not. ... [It] will not be disputed by those who have even a slight acquaintance with geometry, that this science is in direct contradiction with the language employed in it by its adepts. ... Their language is most ludicrous, though they cannot help it, for they speak as if they were doing something and as if all their words were directed towards action. For all their talk is of squaring and applying and adding and the like, whereas in fact the real object of the entire study is pure knowledge ... the knowledge of that which always is, and not of a something which at some time comes into being and passes away.
[GLAUCON:] ... Geometry is the knowledge of the eternally existent ....
[SOCRATES:] It would tend to draw the soul to truth, and would be productive of a philosophical attitude of mind, directing upward the faculties that now wrongly are turned earthward. ... We must require that the men of your Fair City shall never neglect geometry, for even the by-products of such study are not slight.

We can only infer from this that geometry is thought by Plato to be very important, but it is difficult to tell, from this report alone, just what kind of geometry is being referred to.

About solid geometry, Plato is much more informative (528a–d):

[SOCRATES:] After plane surfaces, we went on to solids in revolution [sc. astronomy] before studying them in themselves. The right way is next in order after the second dimension to take the third. This, I suppose, is the dimension of cubes (*auxē kubōn*) and of everything that has depth.
[GLAUCON:] ... But this subject does not appear to have been investigated yet.
[SOCRATES:] There are two causes of that: first, inasmuch as no city holds them in honour, these enquiries are languidly pursued owing to their difficulty. And secondly, the investigators need a director, who is indispensable for success and who, to begin with, is not easy to find, and then, if he could be found, as things are now, seekers in this field would be too arrogant to submit to his guidance. But if the state as a whole should join in superintending these studies and honour them, these specialists would accept advice, and continuous and strenuous investigation would bring out the truth. Since even now, lightly esteemed as they are by the multitude and hampered by the ignorance of their students as to the true reasons for pursuing them, they nevertheless in the face

of all of these obstacles force their way by their inherent charm and it would not surprise us if the truth about them were made apparent.
[GLAUCON:] It is true that they do possess an extraordinary attractiveness and charm.

Note particularly how Plato insists that his solid geometry has "not been investigated", is "languidly pursued", is "lightly esteemed", is "hampered by ignorance", and is "absurdly neglected".

The standard interpretations of this passage are that Plato is referring either to the construction of the regular solids, or to solid geometry in general, or to the duplication of the cube. However, this does not seem compatible with our evidence about Hippocrates, Archytas, and Theaetetus, which indicates that these subjects *do* "appear to have been investigated" by the specialists and are *not* "lightly esteemed by the multitude". Rather than develop these objections in detail here, I shall proceed directly to my alternative interpretation.

The first stage of the problem of duplicating the cube is to find some geometrical configuration from which we can construct the double cube. This generalises the successful duplication of a square in the third section of Socrates' encounter with the slaveboy in the *Meno,* and we have evidence that this aspect of the problem for cubes had been resolved by the contributions of Hippocrates and Archytas, and perhaps even of Eudoxus and Plato, at the time of the composition of the *Republic.* We now try to pass to the next stage of the problem of duplicating the cube, and try to 'count up' the solution. This will correspond to the further stages in the problem of doubling (or tripling, etc.) the square, in which we evaluate the ratio $\sqrt{2}:1$ (or, more generally, the ratio $\sqrt{n}:\sqrt{m}$)—what I have called 'the dimension of squares'—and it will be my proposed interpretation of Plato's problem of the "dimension of cubes and everything that has depth" (*Esti de pou touto peri tēn tōn kubōn auxēn kai to bathous metechon,* 528b). Since I have so heavily loaded the terminology by referring, in my S_{41} and throughout Chapter 3, to the earlier problem as the 'dimension of squares', an explanation of the Greek text is perhaps in order at the outset. The word used here is *auxē,* as in *deutera auxē,* 'second dimension' or *auxē kubōn,* 'dimension of cubes'. This carries connotations of growth and increase, as in the example at *Republic* VI 509b: *tēn genesin kai auxēn kai trophēn,* "generation and growth and nurture"; also see the other examples in Liddell, Scott, & Jones, *GEL*. The first step in the plane problem is the generation (*genesis*) of the square (see *Meno* 82b–c or *Elements* I 46). Then we pass on to the 'growth' of larger squares; for example, two juxtaposed squares can be converted into a double square (see *Meno* 82b–5c or *Elements* II 14) whose side is then 'counted up' (*arithmeō, Meno* 84a) in relation to the original square; three juxtaposed squares are

4.3   Plane and solid geometry   119

converted into a triple square, whose side is 'counted up' against the original and the double square, etc. In such a way, all squares can be systematically 'grown' and 'counted up' against each other in a programme that uncovers some remarkable mathematical phenomena, some of which were described in the previous chapter. (The interesting proposal that Plato is not completely satisfied with the conclusion of Socrates' encounter with the slaveboy in the *Meno,* since they do not attempt to 'count up' the side of the double square, is developed in Brown, PDSA.)

Let us try a similar procedure for cubes, and so pass to this next step of trying to 'count up' the double cube. As with the dimension of squares (see Section 3.3), we start with a heuristic exploration. The generalised *Parmenides* proposition and a table of cubes up to 99 enable us to apply the algorism of 2.3(c) to yield the expansion

$$\sqrt[3]{2} : 1 = [1, 3, 1, 5, 1, 1, \ldots].$$

No pattern is yet evident. A table of cubes up to 999 enables us to extend this calculation to

$$\sqrt[3]{2} : 1 = [1, 3, 1, 5, 1, 1, 4, 1, 1, \ldots].$$

Still no pattern manifests itself. (The calculation so far is set out in Table 4.1. Note that it has already exceeded the standard Greek numeral system, to be described in Chapter 8.) The evaluation can be continued as long as the arithmetical operation of cubing the numbers is practically feasible, or we can go on to develop new and more elaborate procedures for extending the calculation further. The first 75 terms of the ratio are:

$$\sqrt[3]{2} : 1 = [1, 3, 1, 5, 1, 1, 4, 1, 1, 8, 1, 14, 1, 10, 2, 1, 4, 12, 2, 3,$$
$$2, 1, 3, 4, 1, 1, 2, 14, 3, 12, 1, 15, 3, 1, 4, 534, 1, 1, 5, 1,$$
$$1, 121, 1, 2, 2, 4, 10, 3, 2, 2, 41, 1, 1, 1, 3, 7, 2, 2, 9, 4,$$
$$1, 3, 7, 6, 1, 1, 2, 2, 9, 3, 1, 1, 69, 4, 4, \ldots]$$

Still no hypothesis manifests itself; there is neither any apparent periodicity, nor any associated apparent systematic behaviour of $p^3 - 2q^3$ that might lead to a generalisation of Pell's equation. The only regularity that persists, since it is a fundamental property of the *Parmenides'* proposition and not of the particular example to which it is applied, is the behaviour of the cross-product of consecutive approximations in column 3 of Table 4.1:

$$3^2 - 2 . 4 = 1$$
$$4^2 - 3 . 5 = 1$$
$$5 . 7 - 4 . 9 = -1$$
$$\text{etc.}$$

TABLE 4.1. Evaluation of $\sqrt[3]{2}:1 = [1, 3, 1, 5, 1, 1, 4, 1, 1, \ldots]$

Under-estimate	Over-estimate	New estimate $p:q$	p^3	$2q^3$	Under/over	anthy-phairesis
0:1	1:0	1:1	1	2	under	1
1:1	1:0	2:1	8	2	over	
1:1	2:1	3:2	27	16	over	
1:1	3:2	4:3	64	54	over	3
1:1	4:3	5:4	125	128	under	1
5:4	4:3	9:7	729	686	over	
5:4	9:7	14:11	2744	2662	over	
5:4	14:11	19:15	6859	6750	over	
5:4	19:15	24:19	13824	13718	over	
5:4	24:19	29:23	24389	24334	over	5
5:4	29:23	34:27	39304	39366	under	1
34:27	29:23	63:50	250047	250000	over	1
34:27	63:50	97:77	912673	913066	under	
97:77	63:50	160:127	4096000	4096766	under	
160:127	63:50	223:177	11089567	11090466	under	
223:177	63:50	286:227	23393656	23394167	under	4
286:227	63:50	349:277	42508549	42507866	over	1
286:227	349:277	635:504	256047875	256048128	under	1
635:504	349:277	984:781	952763904	952759082	over	
etc.						

The difficulty with the problem of the dimension of cubes is that we cannot even begin, at this stage, to formulate any hypothesis that we could attempt to prove. The hunt for results cannot start until we have identified the quarry.

This problem is still only partially understood. It was not until Gauss's work on the continued fraction expansion of a randomly chosen real number, a problem that he abandoned incomplete because of its difficulty, that any reasonable general hypothesis could be formulated, and only very partial results have so far been established, notwithstanding more and more extended explorations made possible by the development of electronic computers; see, for example, von Neumann & Tuckerman, CFE, and Churchhouse & Muir, CFANMI. Alternatively we might ask the simpler question: are the terms bounded or, if not, can we obtain some idea of their rate of growth? Again, only partial results have been found, very weak bounds like $n_k \leq a^{b^k}$ where a and $b > 1$. More explanations and details will be given in Section 9.2(c).

Within this interpretation, Plato's words about the problem of the dimension of cubes can apply, in their entirety, to this problem today;

and we still have no informed idea of its possible ramifications, though the charm of the problem still attracts a few mathematicians. But I have tried, in my description, to show how this problem also fits naturally within an anthyphairetic formulation of early Greek mathematics, and Plato's words can fit no less appropriately there also, without any anachronism. Moreover we can now go back to Plato's plane geometry. We read later that

> It is by means of problems, then, as in the study of geometry, that we will pursue astronomy... [*Republic* VII, 530b].

Plato does not appeal to the deductive method, the proving of theorems, or the development of formal theories, but to "problems". I propose that one of the things that has so impressed Plato about geometry is the initial success there of the anthyphairetic programme: the fruitful pursuit of problems such as my Socrates outlined at S_{41}. I shall describe, in Chapter 5, how a clever kind of argument enables this programme to be extended further to yield some information about the circumdiameter and side of the pentagon and the circumdiameter and edge of all the regular solids, and so provide an interpretation of, and motivation for, Book X of the *Elements* and bring out its relationship to Books II, IV, and XIII. In this way Books X and XIII, which surely deserve the description as "the greater and more advanced parts" of geometry (*Republic* VII, 526e), will be seen to fit together within the anthyphairetic programme. But I know of no technique, even today, that will yield any anthyphairetic result about any other problem in solid geometry.

4.4 Academic astronomy

4.4(a) *Introduction*

Plato's views on theoretical astronomy have earned him the contempt of observational and computational astronomers. In answer to Glaucon's query about

> How... ought astronomy to be taught contrary to the present fashion if it is to be learned in a way to conduce to our purpose? [529c],

Socrates explains (529c–530c):

> These sparks that paint the sky, since they are decorations on a visible surface, we must regard, to be sure, as the fairest and most exact of material things; but we must recognise that they fall far short of the truth, the movements, namely, of real speed and real slowness in true number and in all true figures both in relation to one another and as vehicles of the things they carry and contain. These can be apprehended only by reason (*logos*) and thought, but not by sight.... We must use the blazonry of the heavens as patterns to aid in the study of those realities,

just as one would do who chanced upon diagrams drawn with special care and elaboration by Daedalus or some other craftsman or painter. For anyone acquainted with geometry who saw such designs would admit the beauty of the workmanship, but would think it absurd to examine them seriously in the expectation of finding in them the absolute truth with regard to equals or doubles or any other ratio (*isōn ē diplasiōn ē allēs tinos summetrias*).... Do you not think that one who was an astronomer in very truth would feel in the same way when he turned his eyes upon the movements of the stars? He will be willing to concede that the artisan of heaven fashioned it and all that it contains in the best possible manner for such a fabric; but when it comes to the proportions of day and night, and of their relation to the month, and that of the month to the year, and of the other stars to these and one another, do you not suppose that he will regard as a very strange fellow the man who believes that these things go on forever without change or the least deviation.... It is by means of problems, then, as in the study of geometry, that we will pursue astronomy too, and we will let be the things in the heavens, if we are to have a part in the true science of astronomy. [Glaucon:] You enjoin a task that will multiply the labour of our present study of astronomy many times.

There is a related passage at *Timaeus,* 37c–39e, where Plato describes how the universe was made out of uniformly rotating motions which generate time and the pattern of eternal nature. See, for example, 39b–c:

In order that there might be a clear measure of the relative speeds, slow and quick, ... God kindled a light which we now call the Sun, to the end that it might shine, so far as possible, throughout the whole Heaven, and that all the living creatures entitled thereto might participate in number, learning it from the revolution of the Same and Similar. In this wise and for these reasons were generated Night and Day, which are the revolution of the one and most intelligent circuit; and Month, every time that the Moon having completed her own orbit overtakes the Sun; and Year, as often as the Sun has completed his own orbit.... [Translation by R. G. Bury, in the Loeb Classical Library.]

I shall again describe my interpretation by a dialogue in which my slaveboy shows further evidence of his considerable mathematical abilities. The subject will again be "the absolute truth in regard to equals or doubles or any other ratio", but now in the context of "the relation [of the day] to the month, and that of the month to the year".

4.4(b) *The slaveboy meets Eudoxus*

BOY$_{44}$: Excuse me, Eudoxus, Socrates told me [see S$_{35}$] that you astronomers have your own way of handling ratios. Would you explain it to me?

EUDOXUS$_{45}$: How do you think the relationship of day to month can best be described?

BOY$_{46}$: Well, I could do the reciprocal subtraction process, but I can't now point to heaps of stones or manipulations with geometric figures while I'm doing it.

EUDOXUS$_{47}$: Yes, that technique is well adapted to *arithmētikē* and geometry, but it doesn't seem to be the right thing for astronomy. Tell me: what would happen if the lunar month was always thirty days long?

BOY$_{48}$: I suppose that the next new moon would always appear sometime on the thirtyfirst day after the previous one.

EUDOXUS$_{49}$: Anytime?

BOY$_{50}$: That's possible. I've heard talk of astronomical anomalies, variations in the regularity of the heavens, so that the month could sometimes be a bit shorter, sometimes a bit longer, provided the new moon always appeared sometime on the thirtyfirst day after the previous new moon.

EUDOXUS$_{51}$: Let's do some theoretical astronomy. Suppose that the motions are uniform, and go on forever without change or the least deviation, and that the month is exactly thirty days long.

BOY$_{52}$: Then the moment of new moon must always take place at the same time on the thirtyfirst day after the previous new moon.

EUDOXUS$_{53}$: Why?

BOY$_{54}$: If it didn't, it would slowly creep backwards or forwards through the day. Now the first time the new moon can actually be observed is at dusk just after sunset. So, if the new month starts at the sunset of this observation, and if the moment of new moon is actually creeping forwards in time, then sometimes one month will start with a thin observable crescent and the next month will start with a thick crescent, thirty-one days later. Or, if the month were a bit shorter than thirty days, it would sometimes give rise to a twenty-nine day month, in the same kind of way.

EUDOXUS$_{55}$: Let's draw this out schematically on a line which represents time flowing by uniformly. We'll mark each sunset, the beginning of each day, to give a scale of equal time intervals:

D_1 D_2 D_3 D_4 D_5 D_6 D_7 D_8 D_9 D_{10} D_{11} D_{12} D_{13} D_{14} D_{15} D_{16} D_{17} D_{18} D_{19} D_{20}

Then, on the same line, we mark off the conjunction of sun and moon, the theoretical moment of new moon. For convenience, I'll suppose the 'month' is very short, just one or two days long, or the line will have to stretch out a very long way for us to see what's happening:

D_1 D_2 D_3 D_4 D_5 D_6 D_7 D_8 D_9 D_{10} D_{11} D_{12} D_{13} D_{14} D_{15} D_{16} D_{17} D_{18} D_{19} D_{20}
M_1 M_2 M_3 M_4 M_5 M_6 M_7 M_8 M_9 M_{10} M_{11} M_{12} M_{13} M_{14}

BOY$_{56}$: It's beginning to look like geometry now.

EUDOXUS$_{57}$: Yes, you can do the same thing with two lines or two numbers. But don't be misled by first appearances; I haven't been able to find any problems of interest in geometry with this yet—just a few formal definitions and propositions about ratio and proportion.

BOY$_{58}$: I see now how you're using the day like the dripping of a water-clock, to mark the rhythm of the passage of time.

EUDOXUS$_{59}$: Then describe the rhythm of the months.

BOY$_{60}$: You mean the pattern of Ds and Ms? I can simply say:

D_0 & M_0 D_1 M_1 D_2 M_2 D_3 D_4 M_3 D_5 M_4 D_6 D_7 M_5 D_8, etc., where D&M denotes a conjunction at sunset that I can suppose starts the sequence; you did say that we were doing theoretical astronomy! Or I can say: a one-day month, a one-day month, a two-day month, a one-day month, a two-day month, etc., and then I can code it as { ..., †1, 1, 2, 1, 2, ... } where that obelus again indicates the coincidence. I'm also supposing here that the new month always starts at the sunset coinciding with or following conjunction, and that's another highly theoretical supposition.

EUDOXUS$_{61}$: These sequences are the astronomical ratios, of course. Can any sequence of numbers arise in this way?

BOY$_{62}$: Clearly not. You couldn't have { ..., 1, 2, 3, ... } or, for subtler reasons, { ..., 1, 1, 2, 2, ... }; I wonder how you can distinguish a permitted sequence? I suppose, just like anthyphairetic ratios, that there must be lots of information coded in those numbers.

EUDOXUS$_{63}$: Quite so. Here's an example of a longer sequence; see if you can decode it:

{ ..., †1, 1, 2, 1, 2, 1, 2, 1, 1, 2, 1, 2, 1, 2, 1, 1, 2, 1, 2, 1, 2, 1, 1, 2, 1, 2,

1, 2, 1, 1†, ... }

BOY$_{64}$: Since it contains two obeli it must be a commensurable ratio and then—because I can count the numbers and the commas and do a bit of fiddling at the ends—I can see that twenty-nine commas represent thirty months, while the ones and twos, added together, make forty-two which represents forty-three days. So it's the ratio 43:30.

Incidentally I can represent that 43:30 anthyphairetically as [1, 2, 3, 4]; and that seems to me to be shorter, neater, and clearer.

EUDOXUS$_{65}$: Socrates told me that you were quick! But I did help you a lot by that gift of two coincidences; what would you have done without them? And you are right: these astronomical ratios are very dilute, small beer compared with the pure spirit of anthyphairetic ratios.

BOY$_{66}$: I can't yet see what to do if there aren't any concidences. But I can see how that first anthyphairetic term $n_0 = 1$ comes from the way that the astronomical sequence contains only 1s and 2s, since that means that the month is more than one but less than two days long. So the interval between D_1 and M_1 will give the difference between a day and a month, here less than a day; then the interval between D_2 and M_2 is twice that difference, still less than a day; but the interval between D_3 and M_3, three-times that difference, is bigger than a day, since the third month contains an extra day. So $n_1 = 2$. I suppose that now that new remainder, the interval between M_2 and D_3, between two months and three days—that's a convergent!—can be measured against the previous remainder, the difference between D_1 and M_1, in one of the larger patterns that seem to occur in the sequence.

EUDOXUS$_{67}$: Not bad! Since you're obviously so keen on anthyphairetic ratios, I'll show you how to generate the astronomical ratio which corresponds to the anthyphairetic ratio [1, 2, 3, 4, ...]. I'll describe longer and longer sequences which start from an initial coincidence M_0&D_0. I, too, can only see how to do this starting from a concidence, though I can also see that lots of these astronomical ratios won't ever have a coincidence in them, however far you go backwards or forwards through them.

4.4 Academic astronomy 125

Since $n_0 = 1$, we must start M&D D M, ... Let's call that initial pattern S_0—that doesn't include the coincidence M&D; so $S_0 = $ D M. In fact, it's better, for the moment, to forget that initial coincidence.

Since $n_1 = 2$, we now repeat this pattern S_0 twice, and then we'll have to slip in an extra day: <u>DM</u> <u>DM</u> D. Let's call that S_1 and write $S_1 = S_0{}^{n_1}D$.

Now we look at the interval from M_2 to D_3, and compare that with D_1 to M_1. Since $n_2 = 3$, when M_2 to D_3 has been repeated three-times it will still be less than D_1 to M_1 but, on the next time, we'll have to slip in an extra D_1 to M_1. So we get $S_2 = $ <u>DMDMD</u> <u>DMDMD</u> <u>DMDMD</u> <u>DM</u> $= S_1{}^{n_2}S_0$. At the end of that, we have the interval from D_{10} to M_7—another convergent!

Next, since $n_3 = 4$, we repeat S_2 four-times and then, on the next time round, we have to slip in an extra S_1 sequence; so $S_3 = S_2{}^{n_3}S_1$. Work it out, and you'll see we get the sequence I've been talking about [in E_{55} and E_{63}, above].

If the anthyphairetic ratio terminated there, [1, 2, 3, 4], this would end with a coincidence. Now you know that terminating anthyphairetic sequences behave as if they finish with very, very big numbers [see S'_{21} in Section 2.3(b)]. So if it wasn't exactly terminating, but actually was [1, 2, 3, 4, 1 0000 0000, ...]—that myriad–myriad is an everyday approximation to infinity—then, as this cycle S_3 was repeated, so the gap between M_{30} and D_{43} would be doubled to M_{60} to D_{86}, tripled to M_{90} to D_{129}, and so on, and would eventually manifest itself. (If, on the other hand, D_{43} came before M_{30}, then we'd be dealing with a ratio like [1, 2, 3, 3, 1, 1 0000 0000, ...], which is close to but bigger than [1, 2, 3, 4,]—this time the month is slightly longer.) This is another illustration of how these large terms in anthyphairetic expansions code large amounts of information; an astronomical ratio would have to go on for scores of myriads of myriads of terms to encode this same information. But theoretical astronomomy doesn't face any problems with detecting exact coincidences; it exploits the repetition of precisely uniform events and the 'before, coincident, or after' in exactly the same way that geometry exploits addition and the 'less, equal, or greater'. Complaining that the heavens aren't like that, or saying that these kinds of measurements aren't possible, is like objecting that a geometer's figure of a circle isn't exactly round.

Do you follow?

BOY$_{68}$: I'm sorry, Eudoxus, I wasn't listening! Take that long sequence you gave me and count the number of one-day months lying between consecutive two-day months and add one to each. (That's probably just the usual trouble with counting points and intervals, that there's always one more point than intervals.) This gives a sequence:

$$\{\ldots, \dagger 3, 2, 2, 3, 2, 2, 3, 2, 2, 3, 2, 2, 3\dagger, \ldots\}.$$

Note that this sequence contains just twos and threes. Now count the number of twos between the threes and add one to each: $\{\ldots, \dagger 3, 3, 3, 3\dagger, \ldots\}$. I think this tells you that the anthyphairetic ratio is $[1, 2, 3, \ldots]$ and, since this contains four terms—those coincidences introduce complications, and I think incommensurable ratios without coincidences may be easier to handle—I think I can deduce from this that $n_4 = 4$. And I think that the sequences that arise as astronomical ratios are precisely those that, when you go on transforming them in this way, will always generate new sequences only containing ns and $(n + 1)$s, where that n will be the next term of the corresponding anthyphairetic ratio.

EUDOXUS$_{69}$: That's brilliant! Let me think.
[And they both walk off, thinking.]

4.4(c) *Egyptian and early Greek astronomy*

One feature of the early phase of astronomy in Egypt and Greece, parallel to and distinct from the cosmological speculations of pre-Socratic philosophers, is the construction of cyclical calendars. The earliest Greek examples, known as *parapegmata,* are associated with the names of Meton, Euctemon, Democritus, Eudoxus, and Callippus, and several examples survive, either directly on papyrus or stone, or indirectly, as described by later authors. There is an enormous literature on these parapegmata which I shall not attempt to describe here; it is summarised and reviewed in Neugebauer, *History of Ancient Mathematical Astronomy* ii, 615–29. Lunar calendars and the phenomena of first and last visibility of the moon (see B_{54}) are very clearly described and analysed in Parker, *The Calendars of Ancient Egypt.* Also see Bulmer-Thomas, PA; Goldstein & Bowen, NVEGA; Riddell, EMES; and van der Waerden, GAC.

Any discussion of calendars is complicated by the distinction that must be made between solar and lunar calendars, a distinction that frequently corresponded, in ancient astronomy, to the difference between civil and astronomical calendars. The solar calendar is based on the movement of the sun through the zodiac: the year therefore has 12 zodiacal months, each 30 or 31 days long. In a lunar calendar, the first day of the month is tied to some lunar phenomenon, for example the appearance of the new crescent after sunset, the last appearance or first disappearance of the old crescent before sunrise, or the observation or estimation of the moment of full moon; the lunar month is 29 or 30 days long; and the year will contain either 12 or 13 lunar months.

Another complication is caused by astronomical anomaly. Since the motions of the heavenly bodies vary in speed, the intervals they determine will vary in duration. Here are two examples. First, while the *mean* synodic month (conjunction to conjunction, averaged) is 29·53059 days, the synodic months themselves may vary between about $29\frac{1}{4}$ and $29\frac{4}{5}$ days long, and the observations of crescent visibility in Egypt can give up to three 29-day or five 30-day lunar months in a row. Second, because the sun's motion is not uniform, the intervals between the solstices and equinoxes will vary during the year. These periods, starting with the summer solstice, are 92, 89, 90, and 94 days long, and they are so given in Callippus' parapegma as reported in the 'Eudoxus papyrus',

P. Par. 1. (Geminus' later description of Callippus' parapegma is slightly different; see Neugebauer, *HAMA* ii, 627 n. 9.) On the other hand, Eudoxus' parapegma, as described in this same text, apparently makes these intervals into [91], 92, 91, and 91 days (the slight doubt is caused by restoration of damaged text), in clear defiance of the other recorded calendric evidence and astronomical observations. In my interpretation of Plato's astronomy, I consider a kind of theoretical astronomy in which anomaly is ignored (see E_{47-51}), and I suggest that Eudoxus may indeed have been involved in this kind of study.

Here, briefly, are some examples of ancient cyclical calendars. First a Greek text which contains an early lunar calendar: P. Ryl. [inv.] 666 = P. Ryl. iv 589, dated 180 BC, first published in Turner & Neugebauer, GDNM. This starts with a list of financial accounts, perhaps owed by the members of a gymnasium, on which interest is charged at the implied rate of two per cent per month; then the second half of the text is a lunar calendar. This calendar omits the meteorological, astronomical, and religious data that is a feature of most of the surviving Greek parapegmata and, in the preserved section at least, and despite what it says in its introduction, it simply lists lunar months. The text reads:

Year 1 of Queen Cleopatra and King Ptolemy the son, gods Epiphaneis. Parapegma of lunar new moons, showing how they are related to the days of the Egyptian twelvemonth. The period of the table is twenty-five years, three hundred and nine months [including intercalary months], nine thousand one hundred and twenty-five days. It indicates the lunar months and which of them are full [i.e. 30 days long], which hollow [i.e. 29 days long], which intercalated [i.e. the 13th month]; and in what sign of the Zodiac the sun will be during each month. When the sun has traversed the twenty-five years it will return to the same starting point and revolve in the same manner. . . . The lunar new moons in the first year are

Thoth 20	[29 days]
Phaophi 19	[30 days]
Hathyr 19	[30 days]
Choiak 19	[29 days]
. . . .	

The text containing the month lengths, given at the end in square brackets, is lost at this point through damage to the manuscript, but a later similar fragment allows it to be reconstructed with confidence. If the calendar had continued for twenty-five years, it would have stretched over some two metres of roll; only about half a metre survives. Another, more extensive, example of this same twenty-five year lunar calendar is found in the demotic papyrus P. Carlsberg 9 (see Neugebauer & Parker, *EAT* iii, 220–5, Neugebauer, *HAMA* i, 563 f., and Parker, *CAE*). There are good reasons for believing that this cycle was developed in Egypt in the fourth century BC (see Parker, *CAE*). Ptolemy refers to it much later, at *Almagest* VI 2:

Now 25 Egyptian years less 0; 2, 47, 5 days contain approximately [in fact, exactly!] an integer number [309] of [mean synodic] months, and [in 25 years] the mean notions are . . . , [translation and annotation from Ptolemy–Toomer, *PA*, 276]

and Ptolemy's table of conjunctions and oppositions which then follows, together with some calculations, is based on this cycle. Other more accurate but less convenient cyclical calendars, the Metonic cycle and its refinement by Callippus, were described in Section 2.4(b), and an extract from another parapegma is illustrated in Plate 5 and described in Section 7.1(c).

These calendars are precisely the raw material of what my Eudoxus called 'astronomical ratios' (see B_{60} & E_{61}): any sequence that describes the patterns generated by interacting periodic phenomena. My dialogue and these examples both deploy two equivalent ways of describing the lengths of successive lunar months: either as a sequence which describes the succession of new days and new moons, as in the left-hand column of P. Ryl. iv 589, which is a condensed form of my kind of description as

. . . MDDD . . . (29 times) . . . DDMDD . . . (30 times) . . . DDMDD . . . ,

or in terms of the length of the months, as in the right-hand column, for which I introduced the notation { . . . , 29, 30, 30, 29, . . . }. There is not enough text in P. Ryl. iv 589 for us to continue this sequence uninterrupted for very long, but the example of P. Carlsberg 9 is complete. From it we can reconstruct the relationship between this lunar calendar (in which a period of 9125 days is split into 309 mean synodic months) and a civil zodiacal calendar. (It is skilfully argued in Parker, *CAE*, that the lunar calendar was originally closely tied to actual lunar observations, though anomaly and the underlying approximation, of $\frac{9125}{309} = 29 \cdot 5307 \ldots$ for the mean synodic month of 29·53059 days, mean that the fit can never have been exact and that it will drift with the passage of time.) From this, a complete enumeration of the astronomical ratio of day to month that it embodies could be extracted, in the form of a list of the lengths of the 309 successive months.

Our evidence about early Egyptian and Greek astronomy contains many other such examples of enumerations of almost periodic astronomical phenomena; I shall not press this point further, but again refer the reader to Neugebauer, *HAMA*. Let us now look to the next stage.

There are two main directions in which we can turn. The first is to explore, observationally and computationally, the irregularities of the heavens: the anomalies, the erratic behaviour of the planets, the fine details of the observable universe. The tradition that stretches from the astronomers of Babylonia, through Hipparchus and Ptolemy, then the Arabs, up to the sixteenth century and on to today, one of the gigantic intellectual achievements of mankind, is a massive programme of such an exploration. Note how this kind of astronomy has, throughout, been carried out within an arithmetised tradition, and indeed it can scarcely be conceived without the use of such arithmetic for its records and calculations. In this approach the cyclical phenomena which so dominate the early stages of observation atrophy and fall away, just as the original lunar calendars give way to the much more abstract but observationally and computationally convenient solar calendars. But there is a second route which beckons. Since the astronomical data so often presents itself in cyclical form, an understanding of the behaviour of interacting cycles might lead directly to the heart of the phenomena of astronomy themselves. Within the programme I have been developing, this suggests that we look for some *anthyphairetic* understanding of *astronomical* ratios. The first step in such an understanding would be merely to relate these two ways of describing the same thing. Until this first step is achieved, we do not know what further illumination is possible.

What stifles this second enterprise is its subtlety and difficulty. The problem is not hard to formulate (E_{55}–B_{60}), the initial explorations are promising (E_{61-6}), but the next steps, for example the two algorisms described in the dialogue at E_{67} and B_{68}, are far too ingenious for everyday use. (The algorism in E_{67} is found, independently and almost simultaneously, in E. B. Christoffel, OA (1875) and H. J. S. Smith, NCF (1876) and is also described in my AREP; the algorism in B_{68} is due to Christopher Zeeman; see Zeeman, GG, and Series, GMN). I suggest that Plato's and Eudoxus' approach to astronomy may have been to attempt to explore further in this direction. We have little direct evidence about whether they might have achieved any results, such as E_{67} and B_{68}, but these ideas and procedures are very closely related to the techniques of *Elements* V and their applications in Book VI.

We retire from these tentative explorations of this kind of investigation of interacting, almost-periodic phenomena with profound insights but few results beyond the "formal definitions and propositions about ratio and proportion" (E_{57}, my unconventional but defensible description of *Elements* V). The original motivating problems (to understand and manipulate cyclical phenomena via their astronomical and anthyphairetic ratios) recede, unanswered; the formal machinery becomes the theory that survives and takes on an independent life of its own, and a new phase of development begins. For a good illustration of how this formal theory can then be used in geometry, see the proof of *Elements* VI 1.

This is only a sketch of the first steps of a proposal for reconstructing some of the astronomy of Plato's circle, a conceptual framework within which to fit our fragmentary and tentative evidence. (Some of this evidence is negative: for example, we perhaps need to account for the lack of any further astronomy of consequence in Egypt and in the post-Eudoxan tradition.) I have not here begun to introduce any geometrical or kinematic considerations that might enrich and complicate further the point of view, as Glaucon recognises at *Republic* VII, 530b–c (a passage that was quoted at the beginning of this section); for an illustration of what might be involved here, see Riddell, EMES. But I believe that it does offer a plausible interpretation that is compatible with Plato's remarks on astronomy within the Academy, our evidence about Eudoxus' achievements, and other early developments in both Greek and Egyptian astronomy.

4.5 Academic music theory

4.5(a) *Introduction*

Plato's astronomy and music theory (*harmonikē*) are "kindred sciences", both of which should be studied not by experiment and observation but by "ascending to generalised problems". Music theory is the last topic in the curriculum, *Republic* VII, 530d–531c:

[SOCRATES:] We may venture to suppose that as the eyes are framed for astronomy so the ears are framed for the movements of harmony; and these are in some sort kindred sciences, as the Pythagoreans affirm and we admit, do we not. . . . Then, since the task is so great, shall we not enquire of them what their opinion is and whether they have anything to add? And we in all this will be on the watch for what concerns us, [which is] to prevent our fosterlings from attempting to learn anything that does not conduce to the end we have in view, and does not always come out at what we said ought to be the goal of everything, as we were just now saying about astronomy. Or do you not know that they repeat the same procedure in the case of harmonies? They transfer it to hearing and measure audible concords and sounds against one another, expending much useless labour just as the astronomers do.

[GLAUCON:] Yes, by heaven, and most absurdly too. They talk of something they call minims and, laying their ears alongside, as if trying to catch a voice from next door, some affirm that they can hear a note between and that this is the least interval and the unit of measurement, while others insist that the strings now render identical sounds, both preferring their ears to their minds.

[SOCRATES:] You are speaking of the worthies who vex and torture the strings and rack them on the pegs; but—not to draw out the comparison with strokes of the plectrum and the musician's complaints of too responsive and too reluctant strings—I drop the figure, and tell you that I do not mean these people, but those others whom we just now said we would interrogate about harmony. Their method exactly corresponds to that of the astronomer; for the numbers they seek are those found in these heard concords, but they do not ascend to generalised problems and the consideration which numbers are inherently concordant and which not and why in each case.

4.5 Academic music theory 131

[GLAUCON:] A superhuman task.
[SOCRATES:] Say, rather, useful for the investigation of the beautiful and the good, but if otherwise pursued, useless.
[GLAUCON:] That is likely.

Plato distances himself even from the Pythagorean tradition of music theory in a way that leaves us with very little evidence of any kind on which to base our reconstruction; at most we have a single text, the *Sectio Canonis,* which I shall describe, that may give us a later and flawed example of "generalised problems and the consideration of which numbers are inherently concordant and which not and why in each case". Any interpretation must therefore be very speculative; I shall present here a collection of ideas that relate the basic objects and operations of music theory to the development of *logistikē,* and will also include some related topics of wider interest. The exposition will be in another dialogue, between Archytas and the slaveboy, and again I beg the reader to ignore any distortions of chronology or social order that this introduces.

4.5(b) *Archytas meets the slaveboy*

ARCHYTAS$_{70}$: Do you have a few minutes to spare? Ever since Plato started sending me details of your discussions on anthyphairesis, I've been wanting to talk to you. I'd like to see what you think about a suggestion of mine on how music theory may help us to understand more about the behaviour of ratios.
BOY$_{71}$: I'd be delighted. Meno lets me spend as much time as I want on mathematics now.
ARCHYTAS$_{72}$: Have you been able to make any progress with the problem of compounding anthyphairetic ratios? [See B_{32}–S_{35}, with its modification B'_{22-4} in Section 4.2 above.]
BOY$_{73}$: Not much. I can only work out examples by ugly and crude methods that rest on some rather dubious operations, but I can't understand much or prove anything.
ARCHYTAS$_{74}$: Would you tell me more?
BOY$_{75}$: Let's take an example like compounding $\sqrt{2}:1 = [1, \bar{2}]$ and $\sqrt{3}:1 = [1, \overline{1, 2}]$. First work out some convergents of the ratios:

$1:1 < 7:5 < 41:29 < 239:169 < \ldots \sqrt{2}:1 \ldots < 577:408 < 99:70 < 17:12 < 3:2$
$1:1 < 5:3 < 19:1 \ < \ 71:41 \ < \ldots \sqrt{3}:1 \ldots < \ 97:56 \ < 26:15 < \ 7:4 \ <2:1.$

Now multiply together the antecedents and the consequents of any two underestimates and work out the anthyphairesis of the result; and then do the same for two overestimates. We might as well ignore all but the best approximations, so this gives

$$239 \times 71 : 169 \times 41 < ? < 577 \times 97 : 408 \times 56$$
i.e. $$16969 : 6929 < ? < 55969 : 22848$$
i.e. $$[2, 2, 4, 2, 2, \ldots] < ? < [2, 2, 4, 2, 6, \ldots].$$

Hence the new unknown ratio in the middle looks as if it really will be $[2, \overline{2, 4}] = \sqrt{6}:1$. But I can't explain what's going on, let alone prove anything. What's more, I can't really see what it means to compound two commensurable ratios $p:q$ and $r:s$ to give $(p \times r):(q \times s)$. If we add, giving $(p+r):(q+s)$, then we get the basic operation of the *Parmenides* proposition which I can understand much better. For example, it is a mean, while compounding isn't; and it is fundamental to anthyphairesis, which I now understand better and better, but compounding doesn't seem to have much connection with anthyphairesis. If both of the ratios are multiples, $n:1$ and $m:1$, or both are parts, $1:n$ and $1:m$, then compounding involves the product of the *arithmoi* n and m, and it is possible that, in the more general case, it may be an extension of this kind of result. Next, I can use compounding to say something about the ratio of two rectangular numbers [see *Elements* VIII 5], with a similar geometrical result about two rectangles, or about two equiangular parallelograms [see *Elements* VI 23]; but these seem isolated results of no great importance. Finally, while I can prove the particular geometrical relations like $\sqrt{2} \cdot \sqrt{3} = \sqrt{6} \cdot 1$ that are suggested by the manipulation above, there's something curious about the underlying geometry that I don't understand and can't prove. That's a summary of just about everything I can say about compounding.

ARCHYTAS$_{76}$: Have you tried exploring what happens with commensurable ratios?

BOY$_{77}$: No. So I'm glad to have met you because they've told me that you've done a lot with these ratios of *arithmoi*.

ARCHYTAS$_{78}$: Well, my idea is that if you look at this operation in the context of the simplest kind of commensurable ratios, then you might get some help from music theory. Let me explain. When Plato first started to tell me about the need for a mathematical 'definition' of ratio, I couldn't understand what he was talking about. Surely, I then thought, a ratio $b:c$ is just a convenient word for 'the relationship between two lines (or numbers) b and c' or 'the separation between two sounds'; does it really need any more definition than that? And if b can be expressed as m parts of something, and c as n parts of the same thing, then isn't the ratio $b:c$ just $m:n$?

BOY$_{79}$: If you'd said that in Athens, you'd have had people like Socrates asking you awkward but interesting and fruitful questions! For instance, you haven't really said what $m:n$ is.

ARCHYTAS$_{80}$: So they keep telling me, and so I learned from the account of your second meeting with Socrates! Now, although the music theorists don't usually use the word *logos*, but talk instead of *diastēma*, 'intervals', I'll treat musical intervals as if they are ratios. The most important intervals in music are the octave $2:1$, twelfth $3:1$, and double octave $4:1$, and the fifth $3:2$, fourth $4:3$, and tone $9:8$; and all of them except the last are consonant intervals. These all correspond either to multiple ratios $n:1 = [n]$ or to the so-called epimoric ratios $(n+1):n = [1, n]$, so they're particularly simple kinds of anthyphairetic ratios, of a form that is specified by just one number n. I might also add that the next such kind of simple anthyphairetic ratio is $[1, 1, n] = (2n+1):(n+1)$ (where $n \geq 2$, since $[1, 1, 1] = [1, 2]$ is an epimoric ratio), and this doesn't seem to be of any importance in music.

BOY$_{81}$: Much as I am a champion of anthyphairesis, I feel here that there is a more commonplace explanation for this word *epimorion*, 'a part in addition', that it describes the simplest kinds of fractional quantities greater than one, namely $1 + n'$ [see Chapter 7]. However, we can push your anthyphairetic explanation a bit further. Take a general ratio $b:c = [n_0, n_1, n_2, \ldots]$ and try to work out just what it is about b and c that is described by each of the n_ks. You'll see that the connection is very direct in the case of n_0 and n_1, but it starts to get rather remote for n_2 and beyond. For example, it's immediately clear why $[2] = 2:1$, $[1, 3] = 4:3$, and $[2, 3] = 7:3$, etc., but it's rather harder to see why $[2, 3, 4] = 30:13$ without thinking in detail of the subtraction process. Do you see what I mean? So these ratios from music theory are also simple in the sense that the relations between the antecedent, the consequent, and the terms of the anthyphairesis are all absolutely transparent. In this sense, ratios like $[1, 1, n]$ are also getting rather complicated.

ARCHYTAS$_{82}$: Yes, when Plato sent met the account of your second conversation with Socrates, I had to think for a long time to work out how you could compare the ratios just by looking at the terms of the anthyphairesis [see S$_{17-21}$ and S$'_{21}$ in Section 2.3(b)], since I couldn't fathom out what these later n_k meant. Then I saw how each was associated with a remainder that came from either the antecedent or the consequent, according as k was even or odd, just as you said there, and that made all the difference.

Back to music theory: we see that these basic intervals in music theory all correspond to these simplest kinds of anthyphairetic ratios, each of them described by a single number n, in either the fractional or the anthyphairetic sense. Next let's look at what the basic operations of music theory do to these intervals. In music we adjoin or subtract two intervals to make a new interval, rather as lines or other magnitudes are added or subtracted in geometry, and in this way we can generate all the basic intervals of music out of fourths and fifths. For instance:

a fourth and fifth give an octave,	i.e. $4:3$ and $3:2$ give $2:1$,
a fifth and an octave give a twelfth,	i.e. $3:2$ and $2:1$ give $3:1$,
two octaves together give a double octave,	i.e. $2:1$ and $2:1$ give $4:1$,
and a fifth less a fourth gives a tone,	i.e. $3:2$ less $4:3$ gives $9:8$.

Now these operations correspond to compounding the ratios (or the opposite of compounding when the intervals are subtracted) and all of these examples of compounding (but not its opposite) give rise to concordant ratios.

BOY$_{83}$: Are all concordant ratios multiple or epimoric?

ARCHYTAS$_{84}$: An interesting question that I'll leave to one side, since I've now got to my main point. I've seen how Eudoxus' ideas on astronomical ratios have led to mathematical insights into the relation between astronomical and anthyphairetic ratios, and to some elegant mathematical definitions and proofs [see E$_{57}$ and *Elements* V]; is there any possibility that these musical analogies might give some other such insights into properties of ratios? For example, consider the problem of compounding anthyphairetic ratios. Surely it must be simpler to guess or work out what the rule should be by looking at simple examples like

[1, 3] and [1, 2] give [2],
[1, 2] and [2] give [3], and
[1, 3] and [1, 8] give [1, 2],

which are precisely the examples arising from music theory, and perhaps also using some further insights from music theory, than by studying your example

[1, 2, 2, 2, ...] and [1, 1, 2, 1, 1, ...] give [2, 2, 4, 2, 4, ...].

So if we investigate music to see why certain basic intervals are concordant, and how and why new concordant intervals are generated out of this basic set, we may get closer to understanding the operation of compounding ratios.

BOY$_{85}$: I'm not so sure it will help, but it's certainly worth trying. Let me explain further why this particular problem may be of special interest. Plato is looking for some kind of absolute comparison against a standard [see *Statesman* 283c–284d], and I think that ratios might give some mathematical illustration of this, if only we could find how to add and multiply them. For example, our basic results about inequalities between ratios tell us that

$$[1, \bar{2}] < [2, \overline{1, 4}] < [4, \overline{4, 8}] < [5, \overline{1, 1, 1, 10}],$$

but, until we find out how to relate the anthyphairesis of $2b:c$ with $b:c$, we don't have any general way of seeing that the second ratio is *exactly* twice the first, the third is *exactly* three times the first, and so on. Or again, if we could find out how to add and subtract ratios in general then, since we can already compare two ratios for size, we could define ratios of ratios. I don't know quite what this would mean, but I think our hope of understanding would increase if we could first perform the operations and explore their behaviour. So I'm particularly interested in arithmetic with ratios.

ARCHYTAS$_{86}$: What do you mean here by 'arithmetic'? It doesn't seem to have much to do with *arithmētikē*, the study of the *arithmoi*.

BOY$_{87}$: It's part of a more general point of view. Just as *arithmētikē* grows out of the properties of the *arithmoi*—their addition and multiplication, the subtraction of the less from the greater, the unit, oddness, evenness, and primality, and so on—so by analogy I'm coming to think of the study of adding, subtracting, multiplying, etc. of lots of other kinds of things as looking for their own characteristic underlying 'arithmetic'. This arranges the sciences into some kind of hierarchy. For example, all aspects of the arithmetic of anthyphairetic ratios seem problematic while, for geometrical magnitudes, especially for straight lines, addition and subtraction are straightforward, but multiplication seems to be a different kind of operation with its own subtleties. That's connected with what I was saying at the beginning about $\sqrt{2} \cdot \sqrt{3} = \sqrt{6} \cdot 1$, or the general result about $\sqrt{n} \cdot \sqrt{m} = \sqrt{nm} \cdot 1$ for any two *arithmoi* n and m; and there are similar problems about arithmetic with edges of cubes, and perhaps yet other kinds of things.

ARCHYTAS$_{88}$: And in music theory we can also talk about the greater and less, and we can put together and take away musical intervals in an operation that we describe as if it were addition and subtraction, though it seems to be more analogous to multiplication and a more general kind of division than that of addition and subtraction of *arithmoi* or lines; indeed there seems to be nothing in music theory that would correspond to that addition and subtraction. (So, incidentally, we can define the anthyphairesis of musical ratios, just as you were asking above [in B_{85}]. Here's the example of 2:1 and 3:2:

2:1 less 3:2 once leaves 4:3, smaller than 3:2,

3:2 less 4:3 once leaves 9:8, smaller than 4:3,

4:3 less 9:8 twice leaves 256:243, smaller than 9:8,

9:8 less 256:243 twice leaves 531441:524288, smaller than 256:243,

etc.,

so the musical anthyphairetic ratio of the ratio 2:1 and 3:2 is $[1, 1, 2, 2, \ldots]$. I can easily prove that this won't terminate, since the remainder will always be either $3 \times 3 \times \ldots \times 3:2 \times 2 \times \ldots \times 2$, or the other way round, so it can't be 1:1; but I can't say anything more about it. Or we could define a musical version of Eudoxus' astronomical ratios. For example 3:2 could be described by the pattern of

1 2 4 8 16 32 64 128 256 512 1024 2048 ...
1 3 9 27 81 243 729 2187 ...

So now $3:2 = \{\{\dagger 1, 2, 1, 2, 1, 2, 2, 1, \ldots\}\}$. Can you make anything of these manipulations?) On the other hand, music only involves ratios of *arithmoi*, and that sometimes restricts the possibilities; for example, it means that the tone 9:8 cannot be divided into two equal musical intervals. I see how you are saying that these different subjects—geometry, astronomy, music theory, and I'm sure we'll find others—seem to be growing out of a considerable body of overlapping ideas about ratio, and information about one branch seems to feed back into the other topics.

But I'm completely baffled about your remarks about $\sqrt{2} \cdot \sqrt{3} = \sqrt{6} \cdot 1$. What are you talking about there?

BOY[89]: Let me spell that out, since you might be unfamiliar with the definitions and shorthand that I'm using. We choose some assigned line a, and let $b, c, d, \ldots$ be any other lines. Geometrical addition $b + c$ corresponds to adjoining lines, and we can similarly adjoin areas and volumes. Now consider geometrical multiplication of lines. Let $b.c$ denote the rectangle with sides b and c; it's one kind of geometrical multiplication that is inhomogeneous (that is, $b.c$ is a different kind of geometrical magnitude from b and c) and restricted (since we can't define the product of four lines, or two rectangles). There is another kind of geometrical multiplication, where we apply $b.c$ as a rectangle to the assigned line a; the width of this rectangle will be a homogeneous unrestricted product of b and c. This satisfies all the same kinds of properties of multiplication as the multiplication of *arithmoi*, and much more, but it now depends on an arbitrary choice of a.

ARCHYTAS[90]: Surely that's just the fourth proportional to $b, c,$ and a [see *Elements* VI 12]?

BOY₉₁: Yes! And you see here how the assigned line *a* is now analogous to the unit in *arithmetikē*, and how this kind of multiplication of lines is more general than the multiplication of *arithmoi,* since the geometrical fourth proportion always exists, while it doesn't necessarily exist in *arithmetikē*. There are lots of interesting developments relating these two kinds of geometrical multiplication that Theaetetus has started exploring in order to extend the scope of our anthyphairetic knowledge of the dimensions of squares, but it soon gets very complicated. [See *Elements* X, to be described in Chapter 5.] Here, I'll just consider the rectangular multiplication $b.c$. Given b and c, we can define $\sqrt{(b.c)}$ to be the side of the square equal to $b.c$—call it the 'square side' of b and c; it's the mean proportional (or geometric mean) of b and c. Now, in a straightforward example or illustration, I tend to refer to the assigned line and the square on it as the unit line and the unit square, or even omit mention of them altogether, and so just say 'the square side of 2' or '$\sqrt{2}$' for $\sqrt{(2a.a)}$, etc. I've studied closely your amazing geometrical configuration for doubling, tripling, etc. the cube, in connection with the seemingly intractable problem of the dimensions of cubes; you could use this notation there and write $\sqrt[3]{2}$ for the 'cube edge' of $2a.a.a$.

ARCHYTAS₉₂: So this result $\sqrt{2} . \sqrt{3} = \sqrt{6} . 1$ is just a statement about rectangles, that the rectangle with sides $\sqrt{2}$ and $\sqrt{3}$ is equal to the rectangle with sides $\sqrt{6}$ and 1 [see Fig. 4.1]. Or we could say that the homogeneous product of $\sqrt{2}$ and $\sqrt{3}$ is $\sqrt{6}$.

BOY₉₃: Exactly; and I'll explain how this result gives yet another illustration of a profound link between *arithmetikē* and *geomētrikē*. One direct assault on the problem leads to this interesting configuration [see Fig. 4.2], in which we start with the assigned line AB = 1 and construct [using *Elements* II 14] AC' = AC = $\sqrt{2}$ and $\sqrt{AD'} = \sqrt{3}$; then [using *Elements* III 35], we construct AG' = $\sqrt{6}$. Hence we want to show that $AC . AD' = AG' . AB$. If we subtract the rectangle ABPD' from both and apply the standard result on complements about a diagonal [*Elements* I 43; see Fig. 1.2, above], we see that $\sqrt{2} . \sqrt{3} = \sqrt{6} . 1$ means that the point P lies on the diagonal AQ. But I can't prove anything from this figure, though I'm still looking. A more successful proof uses gnomons of a square [see Fig. 4.3]. We start with the unit line AB and unit square ABB″B′, and put unit gnomons round it, in the usual way, thus generating lines AC = $\sqrt{2}$, AD = $\sqrt{3}$, etc. (There's an easy way of doing that by setting AC = AB″, etc., or we can use many other such relations like AD″ = AG.) Then $\sqrt{2} . \sqrt{3}$ is a rectangle with sides AC and AD', while $\sqrt{6} . 1$ is a rectangle with sides AB and AG', for instance. If we subtract the

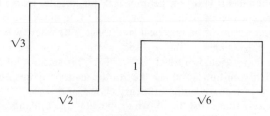

Fig. 4.1

4.5 Academic music theory

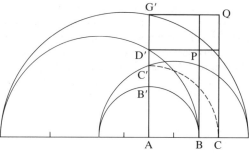

Fig. 4.2

overlapping rectangle ABPD' from both, then again we see that we have to prove that the point P lies on the diagonal AQ. This will hold if the angles PAD' and QAG' are equal. Now these triangles have their sides proportional: we start with a line (AB or AC), construct the square side of three-times the square on these lines (AD' or AG') and construct triangles with these two sides around the right angle (AD'P or AG'Q). Hence these triangles have their sides proportional, and so [see *Elements* VI 5] their angles will be equal. But there's something strange about this kind of argument. To take the easiest possible example, why is it that the angles of any equilateral triangle are always equal to a third of two right angles?

ARCHYTAS[94]: Surely the angles of an equilateral triangle are all equal and therefore, since the sum of . . .

BOY[95]: Yes, but just why is the sum of the angles of a triangle two right angles? The astronomers have started describing geometrical models of the earth and heavens in terms of nested spheres. Suppose we look at equilateral triangles whose edges are great circles on a sphere, precisely the kind of thing these astronomers sometimes consider. We can still talk about angles and triangles, and the angles of a very small equilateral triangle are indeed each still approximately two-thirds of a right angle but, as the triangle grows, so also the angles grow. If one vertex is at a pole and the other two are on the equator, the angles will each be a right angle; then as the triangle expands over the sphere,

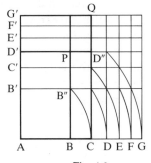

Fig. 4.3

so the angles can grow to almost three and a third right angles each. This example doesn't *prove* anything, but it certainly makes you think. I recall that the first thing Socrates said to me was that a square could be "either larger or smaller", and I was naïve enough to let it pass [see *Meno* 82c]. I wonder what he would have answered if I'd questioned him on this point.

ARCHYTAS$_{96}$: I doubt if even all of your evident mathematical skills could persuade Socrates to answer questions, rather than ask them. Alas, I must be getting on. Farewell.

4.5(c) *Compounding ratios*

Addition and multiplication of common fractions,

$$\frac{m}{n}+\frac{p}{q}=\frac{mq+np}{nq} \quad \text{and} \quad \frac{m}{n}\times\frac{p}{q}=\frac{mp}{nq},$$

are operations so basic to arithmetised mathematics that it may be difficult for us today to conceive of a mathematics of any sophistication in which they are unknown or unimportant. But, as my dialogue explains—see especially B$_{75}$—I believe that this may indeed have been the case for early Greek mathematics. There is a manipulation in the *Elements*, generally described in translation by the word 'compounding', that corresponds to the multiplication of two common fractions, but Euclid does not develop or exploit its properties; and there is nothing that deals with the addition of common fractions. In the next section I shall suggest that the *Sectio Canonis* might have been an attempt, albeit obscure and irretrievably flawed, to explore the behaviour of compounding through the study of music, and music may have been the one place in ancient science and mathematics where the operation arose naturally; the idea behind this proposal is set out in A$_{76-84}$. Then, in Chapter 7, I shall describe our evidence about arithmetical calculations, and argue that a commonly accepted conclusion, that Greek mathematicians conceived and sometimes even expressed fractional manipulations in the form of common fractions, written upside-down, is far from being an established fact. In this section, I shall now discuss some questions of the interpretation of some relevant passages of the *Elements*.

Our assumptions about the role of these manipulations with common fractions within mathematics are pervasive and often not articulated, so I shall start with a brief statement of the point of view against which I am arguing. This is taken from van der Waerden, *SA*, where the interpretations are always set out with exemplary clarity. We read there, with the original emphases:

It is probable that it was *calculation with fractions* which led to the setting up of this theory [sc. *Elements* VII]. Fractions do not occur in the official Greek mathematics before Archimedes, but, in practice, commercial calculations had of course to use them. The reason why fractions were eliminated from the theory is the theoretical *indivisibility of unity*. In the *Republic* (525e), Plato says: "For you are doubtless aware that experts in this study, if anyone attempts to cut up the 'one' in argument, laugh at him and refuse to allow it; but if *you* mince it up, *they* multiply, always on guard lest the one should appear to be not one but a multiplicity of parts." ... When fractions are thus thrown out of the pure theory of numbers, the question arises whether it is not possible to create a mathematical equivalent of the concept fraction and thus to establish a theoretical foundation for computing with fractions. Indeed, this equivalent is found in the ratio of numbers [again, sc. *Elements* VII]. ... The art of computing, including the calculation with fractions, is called 'logistics' by the Greeks [pp. 115–16; also see pp. 49–50].

Contrast this description with an illustration of a Euclidean manipulation with proportions, taken from *Elements* XIII 11. I have replaced the references to the particular lines involved by the letters b, c, d, and e; otherwise the text is taken verbatim from Heath, *TBEE* iii, 462–3:

... therefore, proportionally, as b is to c, so is d to e. And doubles of the antecedents may be taken; therefore, as the double of b is to c, so is the double of d to e. But, as the double of d is to e, so is d to the half of e; therefore also, as the double of b is to c, so is d to the half of e. And the halves of the consequents may be taken; therefore, as the double of b is to the half of c, so is d to the fourth of e. And b_1 is double of b, c_1 is half of c, and e_1 a fourth part of e; therefore as b_1 is to c_1, so is d to e_1. *Componendo* also, as the sum of b_1, c_1 is to c_1, so is [the sum of d and e_1] to e_1; therefore also, as the square on the sum of b_1, c_1 is to the square on c_1, so is the square on [the sum of d and e_1] to the square on e_1.

This is taken from a proposition in which the theory of proportion is applied fluently, as we see from the last two steps and the further passage quoted below. Yet Euclid takes four steps to deduce from $b:c::d:e$ that $2b:\frac{1}{2}c::d:\frac{1}{4}e$, in a context where the fractional quantities $\frac{1}{2}c$ and $\frac{1}{4}e$ pose no problems since they correspond to labelled lines in the figure. Compare this with the manipulation *componendo*, that

$$\text{if } x:y::z:w \quad \text{then} \quad (x+y):y::(z+w):w,$$

(see V Definition 14 and Propositions 17 & 18; this operation is also known as 'composition') and the even more drastic step that

$$\text{if } x:y::z:w \quad \text{then} \quad x^2:y^2::z^2:w^2$$

(see VI 22, with the long note in Heath, *TBEE* ii, 242 ff.) which go through without comment. We see a similar confident manipulation later in the same proof (Heath, *TBEE* iii, 464) where Euclid argues, for different lines b, c, and d:

Let the square on b be equal to that by which the square on c is greater than the square on d; therefore the square on c is greater than the square on d by the square on b. .... And since the square on c is five times the square on d, therefore the square on c has to the square on d the ratio which 5 has to 1. Therefore, *convertendo*, the square on c has to the square on b the ratio which 5 has to 4, ...

Here Euclid justifies the last step by his reference to conversion of ratios (V Definition 16 and Proposition 19):

$$\text{if } x:y::z:w \quad \text{with } z<x \ \& \ w<y \quad \text{then} \quad x:y::(x-z):(y-w).$$

It is not immediately clear how the argument is meant to go; for example, all the magnitudes x, y, z, and w need to be homogeneous for V 19 to be applied whereas here, in the immediate context of XIII 11, they are not. (Incidentally, this passage is the only place I know of where the manuscripts of the *Elements* use numerals rather than words to denote numbers; see Chapter 7, below.) Yet despite the fluency in these apparently more complicated kinds of manipulations, the operations of doubling the antecedents,

$$\text{if } x:y::z:w \quad \text{then} \quad 2x:y::2z:w,$$

and asserting that

$$\text{if } x:y::z:w \quad \text{then} \quad 2x:y::z:\tfrac{1}{2}w$$

seem to require special and individual explanation, though their formal justification follows immediately from V 4. The intuition behind Euclid's treatment does not seem to be based on an understanding of the manipulation of common fractions, where these operations like doubling or halving the antecedent or consequent are straightforward; by contrast, these operations seem intractable for anthyphairetic and astronomical ratios (see B_{85} and Section 9.3(a), below). Conversely, the operations *componendo* and *convertendo* (though not composition) are more straightforward and natural for anthyphairetic and astronomical ratios, but less so for common fractions.

There is nothing in the *Elements* that seems remotely connected with the general addition of two ratios. While composition can be conceived in terms of the addition of $x:y$ and $y:y$ to give $x+y:y$, there is no hint of this point of view, and the closely related manipulations of conversion and of V 12, that

$$\text{if } x:y::z:w \quad \text{then} \quad x:y::(x+z):(y+w),$$

(which is the *Parmenides* proposition in the case of equality) are very emphatically not directly related to the addition and subtraction of fractions, whatever the erroneous feeling among novices that they ought to be; also see the contrast between $(p \times q):(r \times s)$ and $(p+q):(r+s)$ described in A_{75}.

Let me now turn to the compounding of ratios—which is not to be confused with the composition of ratios though, as we shall see, they share the same Greek name. This does correspond to our multiplication of common fractions. The operation appears in *Elements* VI Definition 5 (which is almost certainly interpolated), VI 23, and VIII 5, all of which were quoted at the end of Section 1.2(d); they will also be quoted again below. The procedure, as it appears in VI 23 and VIII 5, but not VI Definition 5, can be described as follows. Given two ratios $b:c$ and $d:e$—and note yet once again, in passing, that Euclid gives no mathematically serviceable definition of ratio in the *Elements* adequate for a proof of VI 1, the *Topics* proposition, though he does, here and elsewhere, appeal to the idea of ratio—if we then write

$$b:c::x:y \quad \text{and} \quad d:e::y:z,$$

then $x:z$ is called the compound of $b:c$ and $d:c$. To bring out the connection with the multiplication of common fractions, we can carry out this operation by writing

$$b:c::b\times d:c\times d \quad \text{and} \quad d:e::c\times d:c\times e$$

(where the products $b\times d$, etc. are interpreted appropriately; this is a version of the *Topics* proposition). Then the compound of $b:c$ and $d:e$ is $b\times d:c\times e$. When b, c, d, and e are lines, as in VI 23, then the lines y and z can be found for any line x, by the basic construction of a fourth proportional at VI 12; but when they are numbers, as in VIII 5, the operation requires the idea of continued proportion, the subject of VIII 4 and, indeed, of most of Book VIII. On the other hand, the garbled definition at VI Definition 5 seems to be striving towards an arithmetised operation, in which the two ratios are 'multiplied together' in some sense.

The role of compounding within the *Elements* is analysed in great detail in Mueller, *PMDSEE* and, to avoid an enormously long excursion through the evidence and its interpretations, I shall restrict myself to quoting some of his conclusions, with which I concur, and refer the reader to this book for the detailed analyses:

It is possible to say that compounding ratios is an analogue of multiplying fractions. The serious question for interpretation is whether compounding should be viewed as a representation of multiplying, i.e., as a device for representing the multiplication of fractions in the language of proportionality. I shall be arguing that compounding should not be viewed in this way [p. 88, on VI Definition 6].

[Euclid's] failure to exploit VIII 5 is a good indication that he does not construe compounding as a representation of multiplying.... Since there is no other candidate for the representation of the multiplication of fractions, it also seems unlikely that Euclid construes duplicating and triplicating as squaring and cubing [fractions] [pp. 92–3, on VIII 11 & 12].

It is not, of course, possible to show that Euclid does not have calculations or something like them in mind when he deals with ratios. But he certainly does not make any such concern explicit, and, as we shall see, the way he proceeds in applying ratios suggests anything but a calculational model. In particular, although he uses the compounding of ratios in later books, he shows no clear sense of its relation to duplicating and presumably, therefore, none of its relation to multiplication [pp. 135–6, on proportion, ratio, and calculation in Book V].

In general the geometric books confirm the impression gained from the arithmetic ones [VII–IX] that Euclid does not construe compounding as multiplication. VI 23 itself is, in a sense, evidence of this fact since the product of the lengths of two sides of a parallelogram does not produce a value of any mathematical significance [p. 154, on VI 23].

Euclid's failure to reduce VI 14 to 23 is one piece of evidence that he does not view compounding of ratios as some form of multiplication. Further evidence is provided by VI 19 and 20 . . . [p. 162, on VI 14, 19, & 20].

[Euclid's] failure [to use the notion of compounding in the solid Books XI and XII] is perhaps the strongest evidence that he does not construe compounding ratios as multiplication [p. 221, on XI & XII; also see pp. 225–6 where the great simplifications that are made possible by using compounding of ratios are described].

Euclid's failure to prove this extension of XII 9 is perhaps some further confirmation of the view that the connections among compounding, multiplying, and volumes were not so immediately clear to him as they are to us [p, 229, on XII 9].

Finally, let us look at the terminology used for some of these operations in the *Elements*. We find that the the verb *suntithēmi*, 'to put together', its associated noun *sunthesis* and adjective *sunthetos*, the verb *sugkeimai* which serves as a passive, and the various participles, are used with an enormous range of meanings. Here are some examples of Heath's translations, with the exact Greek form given in each case. A word from this group can mean the sum, of two numbers:

Let two numbers AB, BC prime to one another be added (*sugkeisthōsan*). I say that the sum (*sunamphoteros*) is also . . . [VII 28],

or of two lines, of which examples abound in Book X; see, for example:

To find two lines incommensurable in square which made the sum (*sugkeimenon*) of the squares on them . . . [X 33].

It can also be used with overtones of multiplication, as in:

A composite (*sunthetos*) number is that which measured by some number [VII Definition 13].

It describes the operation on ratios called composition, or *componendo*:

Composition (*sunthesis*) of a ratio means taking the antecedent together with the consequent as one in relation to the consequent by itself [V Definition 14], as well as that of compounding:

> A ratio is said to be compounded (*sugkeisthai*) of ratios when the sizes of the ratios multiplied together make some (?ratio, or size) [VI Definition 4],

and

> Equiangular parallelograms have the ratio compounded (*sugkeimenon*) of the ratios of their sides [VI 23, also see VIII 5].

It is used in connection with procedures of analysis and synthesis in Book XIII:

> Synthesis (*sunthesis*) is an assumption of that which is admitted (and the arrival) by means of its consequences at something admitted to be true [see Euclid-Stamatis, *EE* iv, 198, and Heath, *TBEE* i, 137–40 & iii, 442–3].

In addition to these technical senses, it is used with the general sense of something being made up of, consisting of, constructed of, etc. We find these same words used elsewhere to describe logical and dialectical procedures; and the word *sunthesis* is employed in grammatical works and literary criticism in very much the same way as 'composition' is used in English.

The precision of terminology that has been developed in parts of modern mathematics, by means of which different operations can be distinguished and compared, is not found in the *Elements*. However, the spirit of such a study of these different operations can be discerned: there is no sense of Euclid confusing addition, multiplication, composition, and compounding, even though they bear the same name. In this sense, my slaveboy's programme, described at B_{87}, is not anachronistic though, as Archytas objects at A_{86}, our use today of 'arithmetic' to describe operations of addition, subtraction, multiplication, and division with very general and different kinds of 'numbers' conveys a very different sense from that of the Greek '*arithmētike*'.

Now let us turn to the objects and manipulations of Greek music theory.

4.5(d)   *The Sectio Canonis*

This is not the appropriate place to attempt a general survey of the role of music in early Greek scientific and moral theory. For this, I refer the reader to the articles in the *OCD*, s.v. music (by Mountford & Winnington-Ingram) and in Sadie, *NGDMM* vii, s.v. Greece, ancient (by Winnington-Ingram), with their references, and the articles and books by Barker, Bowen, and Burkert that will be cited here or listed in the Bibliography.

I wish here to illustrate the very limited comment about music theory set out in A_{78}–A_{84}: that it is conceivable that our understanding of mathematics might be helped by a study of the theory of music, in the way I have suggested that the study of cyclical phenomena in astronomy can help our understanding of *Elements* V, Definition 5. In the event, it will turn out that this contribution from music to mathematics is negligible or non-existent, but nevertheless the idea is worth trying. My illustration will be cast in the form of a description of the *Sectio Canonis*, a treatise on the construction of the musical scale.

The *Sectio Canonis* (Euclid–Heiberg, *Opera* viii (ed. Menge), 158–83, or Jan, *MSG*, 148–66; recent study in Barker, MAESC) is a short treatise on music theory. Its author is unknown and its date is uncertain, but it clearly belongs within a mathematical, as contrasted with empirical, study of music, and this places it broadly within the Pythagorean and Academic traditions, rather than those associated with Aristotle's Lyceum. The propositions that make up the bulk of the work are formulated in Euclidean style, and it is generally included within the Euclidean corpus. A complete translation will appear in Barker, *GMW* ii, and I am grateful to Andrew Barker for providing the following literally rendered and annotated passages. We start with the introduction to the treatise:

If there were stillness and motionlessness, there would be silence; and if there were silence and nothing moved, nothing would be heard. Then if anything is going to be heard, impact (*plēgē*)[a] and movement must first occur. Then since all *phthongoi*[b] occur when some impact occurs, and since it is impossible for an impact to occur unless movement has occurred beforehand—and since of movements some are more closely packed (*pyknoterai*), others more sparse (*araioterai*), and the more closely packed ones make the *phthongoi* higher-pitched, the sparser ones lower-pitched—it follows that some *phthongoi* must be higher-pitched, since they are indeed composed of more closely packed and more numerous movements, while others are lower-pitched, since they are indeed composed of sparser and less numerous movements. Thus those that are higher-pitched than what is right[c] reach what is right [through] being slackened[d] by the taking-away of movement, while those that are too low-pitched reach what is right [through] being tightened[e] by the addition of movement. We must therefore agree that *phthongoi* are composed of parts (*moriōn*), since they reach what is right by addition and taking-away. All things that are composed of parts are spoken of (*legetai*)[f] with respect to one another in a ratio (*logos*) of number, so that it is necessary that the *phthongoi*, too, be spoken of (*legesthai*) in a ratio of number with respect to one another. Of numbers, some are spoken of (*legontai*) in multiple ratio [to each other], some in epimoric, some in epimeric, so that it is necessary that the *phthongoi*, too, be spoken of (*legesthai*) in ratios of these sorts with respect to one another. Of these[g] the multiple and the epimoric are spoken of (*legontai*) with respect to one another under a single name.[h]

Again, of the *phthongoi* we grasp (*gignōskomen*) some as being concordant, others as discordant, and [we grasp] the concordant ones as making a single blend, the one [produced] out of both [*phthongoi*], the discordant ones as not [doing so]. Since these things are so, it is likely[i] that the concordant *phthongoi*, since they make for themselves a single blend of the *phōnē*,[j] the blend [produced] out of both, are among the numbers which are spoken of (*legomenōn*) with respect to one another under a single name, being either multiple or epimoric.[k]

Notes:
Round brackets indicate transliterations of the Greek, and square brackets contain words added to clarify the English version.

[a] *Plēgē*, literally 'a blow'. Also, instead of "impact and movement" one could read "an impact and a movement".

[b] *Phthongoi* are sounds, but in technical writings specifically pitched sounds, sounds each of which is at some one definite pitch: it is commonly used to mean 'note'. The term *psophos* is used for 'sound' in its most general sense, 'noise', whether or not it has definite pitch. The term *phōnē*, literally 'voice', and properly of the human voice, is often used in musical authors to mean 'musical sound', not necessarily of the voice, and not necessarily of a single definite pitch. See the occurrence indicated at note *j* below.

[c] *To deon*, what is right, correct, or proper, or what is required; here, evidently, the pitch that the musician requires.

[d] *Aniemenous*: this relates to the slackening of a string to lower pitch, but in standard jargon the verb is applied directly to the sound or note, in the sense 'to lower in pitch'.

[e] *Epiteinomenous*: the uses of this verb are parallel to those mentioned in the previous note.

[f] *Legetai*, from *legein*, 'to say': this is also the verb cognate with *logos* where *logos* means 'ratio', and this occurrence of it might therefore be translated 'are related', 'are ratioed'. But here the ambiguity, or perhaps conceptual assimilation, is part of the argument, since the fact that numbers standing to one another in multiple or epimoric ratios are 'spoken of with respect to one another under a single *name*' is taken to imply that things in these (but not other) kinds of ratio possess a real unity.

[g] That is, 'these numbers', not 'these ratios' as many commentators have assumed; see the similar passage at the end of the introduction: "the numbers which are said ... ". Also see Barker, MAESC, 2–3.

[h] The word for 'name' here (*onoma*) is not cognate with *logos, legein*, etc., and has no special mathematical usage.

[i] *Eikos*: 'A reasonable hypothesis', 'a fair assumption'.

[j] On the sense of *phōnē* see note *b* above. It seems to be carefully chosen here: the sound in question is a musical one, but it is not a sound of one definite pitch, a 'note', since it is precisely the sound constituted out of the blending of two notes. It is a phenomenon distinct from either of its components, or from the two considered merely as existing simultaneously: it is a 'third thing', but it is not a third note.

[k] The adjectives 'multiple' and 'epimoric' here qualify the *phthongoi*, not, grammatically speaking, the numbers.

The first half of the first paragraph seems to be related to another account of music given by Archytas in his Fragment B1 (Diels & Kranz, *FV* i, 431–5; recent detailed study in Bowen, FEPHS and Huffman, AAF1), except that while Archytas outlines what we might call a 'velocity hypothesis' in which sluggish and weak movements produce low-pitched sounds, quick and powerful movements produce high-pitched sounds, the *Sectio* sets out an 'impulse hypothesis'. The other apparent proponents of the impulse hypothesis seem later in time and associated with the Lyceum (see the pseudo-Aristotelian *Problems* XIX 39, 921a7–31, and *De Audibilibus* 803b27–804a9), though there are difficulties in the interpretation of these passages. For this reason, and in order to avoid overloading my dialogue yet further, I did not have my Archytas suggest developing a theory of the patterns generated by two interacting trains of impulses. These patterns are precisely what I have called astronomical ratios (see E_{55} for an example) and their structure, in which near coincidences are generated and decay, can best be described in terms of the corresponding anthyphairetic ratios. The simplest patterns will be generated by the multiple ratios, and these are in some sense completely static, with a regularly occurring coincidence when the two trains of impulses are in phase. The patterns of the epimoric ratio (familiar from the vernier scale, which exploits the astronomical ratio 10:9) will be almost static. But the pattern exhibited by the less simple ratios $[n_0, n_1, n_2, \ldots]$, with more than two non-zero terms, will have a shifting structure in which hierarchies of almost periodic blocks of terms move in and out of prominence, rather as my Eudoxus described at E_{67}; a simple example of this phenomenon will occur with a ratio like $[1, 2, 1\,0000\,0000, \ldots]$. We could also apply this kind of analysis to the rhythmical structure of music, to explain why the rhythmical signatures described by multiple and epimoric ratios such as 3:2 and 4:3 have a structure that is immediately clear and easily understood, while signatures like 5:3, 8:3, 7:4, etc. are much harder to comprehend by ear. There is some evidence of interest in these rhythmical ratios in Plato, Aristoxenus, and Aristides Quintilianus, but I shall not develop any of these rich ideas here.

In the second half of the introduction, we find what seems to be an argument that the study of *phthongoi*, pitched sounds, should be assimilated into mathematics. The argument is scarcely explicit, precise, or coherent, but it seems to go as follows, with my notes and additions in parentheses: Pitched sounds consist of a succession of impacts, and since the pitch can be varied by adding or taking away movement and therefore impacts (and since each impact is an indivisible unit?), therefore the relation between two pitched sounds can be described as a ratio of two numbers. (Note, in passing, that the argument is specious and the conclusion is false.) Any ratio (taken, as almost always in music theory, the greater to the less) will be either a multiple ratio $n:1$, or an epimoric ratio $(n + 1):n$, or neither of these, an epimeric ratio. (On the use of the word *morion* or *meros*, plural *moria* or *merē*, 'part' and 'parts', see Sections 1.2(c) and 7.1(b). *Epimoric* means, literally, 'a part in addition', *epimeric* means 'parts in addition'.) The multiple and epimoric ratios are described by a single name. (This is enigmatic, and no completely satisfactory explanation presents itself; see the discussion and proposal in Barker, MAESC, 2–3. In A_{80-2}, I suggest a slight addition to

4.5 Academic music theory

Barker's argument, that this single name may be the single number n which is needed to describe the simplest ratios $[n]$ and $[1, n]$, or, arithmetically, the simplest numerical and fractional quantities n and $1 + n'$. These are precisely the multiple and epimoric ratios, and the Greek way of describing each of them will be given below.) Therefore the concordant intervals should then all be found among the multiple and epimoric ratios.

Some such conclusion does seem to have been held by Academic music theorists since later commentators remark, with some disbelief, that while musical practice and the theorists of the Lyceum regarded the octave plus a fourth (2:1 and 4:3, making 8:3) as a concordant interval, the Academic music theorists treated it as a dissonance or ignored it altogether; see Ptolemy–Düring, *HKP* 13. For example, no reference to this interval appears anywhere in the *Sectio Canonis*; also see my Archytas' evasion in B_{83}–A_{84}.

We now turn to the propositions of the *Sectio*. I shall first cite all of their enunciations and discuss briefly the proof of a few of them. The translation is again literal and the following details should be noted: The words *logos* (ratio) and *diastēma* (interval) are distinguished here; frequently they are conflated into 'ratio'. The propositions deal with combinations and decompositions of the multiple intervals (*pollaplasion diastēma*), for example the duple (*diplasion diastēma*), or triple (*triplasion diastēma*), and with the epimoric intervals (*epimorion diastēma*), the hemiolic (*hēmiolion diastēma*, 'half-whole', for 3:2), epitritic (*epitriton distēma*, 'third-in-addition', for 4:3), and the epogdoic (*epogdoon diastēma*, 'eighth-in-addition', for 9:8). The numbers of the propositions are set in square brackets because they are not found in the best manuscripts.

[1] If a multiple interval put together (*suntethen*) twice makes some interval, this interval too will be multiple.
[2] If an interval put together twice makes a whole that is multiple, then the interval will also be multiple.
[3] In the case of an epimoric interval, no mean number, neither one nor more than one, will fall within it proportionally.
[4] If an interval which is not multiple is put together twice, the whole will be neither multiple nor epimoric.
[5] If an interval put together twice does not make a whole that is multiple, that interval itself will not be multiple either.
[6] The duple interval is composed (*sunestēken*) of the two greatest epimoric [intervals], the hemiolic and epitritic.
[7] From the duple interval and the hemiolic, a triple interval is generated.
[8] If from a hemiolic interval an epitritic interval is subtracted (*aphairethēi*), the remainder left is epogdoic.
[9] Six epogdoic intervals are greater than one duple interval.

[10] The octave (*dia pasōn*) interval is multiple.
[11] The interval of the fourth (*dia tessarōn*) and that of the fifth (*dia pente*) are each epimoric.
[12] The octave interval is duple.
[13] It remains to consider the interval of a tone (*toniaion*) to show that it is an epogdoic.
[14] The octave is less than six tones.
[15] The fourth is less than two and a half tones, and the fifth is less than three and a half tones.
[16] The tone will not be divided into two or more equal intervals.
[17] The *paranētai* and *lichanoi* will be found by means of concords, as follows. . . .
[18] The *parhypatai* and *tritai* do not divide the *pyknon* into equal intervals.
[19] To mark out the *kanōn* according to the so-called immutable *systēma*.
[20] It remains to find the moveable notes. . . .

Propositions 1–9 develop some purely arithmetical results about intervals, Propositions 10–18 translate these into statements about musical intervals, and Propositions 19 and 20, which some have argued are interpolations, describe how to mark out a particular kind of scale.

First, consider the terminology. The word *diastēma*, 'distance between' or 'interval', in a very general sense, is often treated by ancient and modern commentators as if it were here synonymous with *logos* 'ratio' (see A_{80}), but the author of the *Sectio* seems to maintain a careful distinction between the two. The mathematical Propositions 1–9 are permeated with the language of ratios, and they explicitly cite some results from some external mathematical theory of ratio or proportion, which are then translated into statements about intervals. See, for example, Proposition 2, which invokes *Elements* VIII 7:

If an interval put together (*sunthen*)[a] twice makes a whole that is multiple, then that interval will also be multiple.[b]

```
  C     B        D
  |-----|--------|
  6    12       24
```

Let there be an interval BC, and let it be that as C is to B so B is to D (*hōs ho* C *pros ton* B, *houtōs ho* B *pros ton* D),[c] and let D be a multiple of C. I assert that B is also a multiple of C. For since D is a multiple of C, C therefore measures D. But we have learned (*emathomen de*)[d] that where there are numbers in [continued] proportion (*analogon*)—however many of them—and where the first measures the last, it will also measure those in between.[e] Therefore C measures B, and B is therefore a multiple of C.

Notes:
[a] *Sunthen*, here and throughout the *Sectio*, has the mathematical sense of compounding, as in *Elements* VIII 5.
[b] The figures show some variations between the manuscripts, and the numbers appended to them may be later additions.

c This is exactly the same terminology for proportions that we find throughout Book VII.

d One of the very occasional explicit back-references in the *Elements*, in X 10, uses this same word, though with the different particle *gar*. See Section 1.2(b) and its Notes, above.

e This result follows immediately from *Elements* VIII 7. The argument will work only for continued proportion (*hexēs analogon* or, from *Elements* VIII 8 onwards, often *sunexēs analogon*): $a:b::b:c$, $b:c::c:d$, etc., the subject of *Elements* VIII. To see why, consider the simplest example of a non-continued proportion $2:3::4:6$. Also see Aristotle, *Nicomachean Ethics*, 1131a30–b16, and the note to *Elements* V Definition 8, in Heath, *TBEE* ii, 131.

We have here an enunciation phrased in terms of intervals between numbers. It is immediately translated into the terminology of proportions, and is established by means of a quoted result. Then, since the conclusion is to be related to the enunciation, there is an implied final translation back into the language of intervals.

Very similar remarks can be made about Proposition 3, in which the results of *Elements* VII 22 and VIII 8 are cited. The proposition is described by Boethius, who attributes it to Archytas, and it is also elsewhere translated by him (see Boethius–Friedlein, *IM* III 11 & IV 2). It has attracted an enormous commentary, and I refer the reader to the description in Heath, *TBEE* ii, 295, and the recent discussions in Burkert, *LS*, 442–7, and Knorr, *EEE*, 212–25 for further details.

Another example of the mathematical style can be found in Proposition 6:

The duple interval is composed (*sunestēken*) of the two greatest epimoric [intervals], that is from the hemiolic and the epitritic.

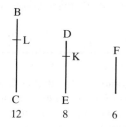

Let BC be hemiolic of DE, and let DE be epitritic of F. I say that BC is duple of F. I took away (*apheilon*) EK, equal to F, and CL, equal to DE. Then since BC is hemiolic of DE, BL is a third part (*triton meros*) of BC, and a half (*hēmisu*) of DE. Again, since DE is epitritic of F, DK is a fourth part of DE, and a third part of F. Then since DK is a fourth part of DE, and BL is a half of DE, DK will therefore be a half of BL. Now BL was a third part of BC: therefore DK is a sixth part of BC. But DK was a third part of F: therefore BC is a duple of F.

Alternatively:

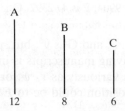

Let A be hemiolic of B, and let B be epitritic of C. I say that A is duple of C. Since A is hemiolic of B, A contains B and a half of B. Then two A's (*duo hoi* A) are equal to three B's. Again, since B is epitritic of C, B contains C and a third of C. Therefore three B's are equal to four C's. Now three B's are equal to two A's and therefore two A's are equal to four C's, and therefore A is equal to two C's: therefore A is the duple of C.

The arguments of both proofs are very basic and long-winded and both seem to avoid any explicit use of the words associated with ratio or proportion. Both styles of argument are found elsewhere in the treatise, so there is no suggestion that either proof is an interpolation. Consider the second proof first; it is an exploration of a particular case of the simple deduction that,

$$\text{if } nA = mB \quad \text{and} \quad mB = pC, \quad \text{then} \quad nA = pC,$$

in which m and p are then reduced to their lowest terms and described. The proofs of Propositions 7 and 8 are similar, and perhaps that of Proposition 9 (cited below) can be viewed as an extension of this method. Now consider the first proof. This is an exploration of operations like: "I take away ... CL, equal to DE [from BC]. Then, since BC is hemiolic of DE, BL is ... a half of DE". This could be an almost explicit explanation of the statement that the anthyphairesis of BC:DE is [1, 2]; or it could also be an almost explicit statement of the arithmetised relation between BC and DE, that $BC = (1 + 2')DE$. Unfortunately for us, these kinds of descriptions are almost identical when applied to the anthyphairetic ratios [1, n] and the fractional quantity $1 + n'$, and nowhere in the treatise is any more complicated kind of such a description explored. (See A_{80}–B_{81}; a similar example arose in Section 2.4(d), where the ratio 4673 2' : 14688 in the *Measurement of a Circle* was described as "greater than threefold and there are remaining 667 2', which is less than the seventh part of 4672 2' ". There, also, a little more information might have made it clear whether the underlying idea was anthyphairetic or fractional.)

The final citation of an external result occurs in Proposition 9:

Six epogdoic intervals are greater than one duple interval.
Let A be one number. Let the epogdoic of A be B, let the epogdoic of B be C, let the epogdoic of C be D, let the epogdoic of D be E, let the epogdoic of E be F, and let the epogdoic of F be G. I say that G is more than the duple of A.

4.5   Academic music theory   151

Since we have learned how to find numbers of epogdoic of one another, the numbers A, B, C, D, E, F, G have been found. A is 26 2144, B is 29 4912, C is 33 1776, D is 37 3248, E is 41 9904, F is 47 2392, G is 53 1441: and G is more than the duple of A.

Here A is 8^6, B is $8^5 . 9, \ldots$, and G is 9^6; but, parenthetically, practically every numeral in the surviving manuscripts is incorrect. In particular, the sampi Ϡ, for 900, appears variously as τ, ψ, or Ϡ; see Chapter 7, below. The reference in the proposition could be to *Elements* VIII 2:

To find numbers in continued proportion, as many as may be prescribed, and the least that are in a given ratio.

Indeed Euclid observes, in a porism, that the extreme terms of this continued proportion will be powers of the terms of the original ratio, just as in this application.

Let us now turn to the music theoretic manipulations in Propositions 10–20; but since these are described and analysed in detail in Barker, MAESC, I can be very brief. We find here that the terminology of ratios is confined to three carefully organised references to the mathematical theory. The first is right at the end of the long Proposition 12, to be quoted below; the second is in Proposition 16:

The tone will not be divided into two or more equal intervals.
It has been shown that it is epimoric. In the case of an epimoric interval neither several mean [numbers] nor one fall within it proportionally. Therefore the tone will not be divided into equal intervals.

and the third is a similar reference in Proposition 18.

The Introduction seems to provide the link between the mathematical theory of Propositions 1–9 and the musical theory of Propositions 10–20, and we can abstract that link into two principles:

The interval between two notes can be expressed in terms of numbers (*arithmoi*), and

All concords are either multiple or epimoric ratios.

As I remarked earlier, the second principle distinguishes Pythagorean and Academic musical studies from those of the Lyceum, and we can elaborate this, in the context of the *Sectio*, as follows. In Proposition 6, quoted above, the treatise investigates the relation

$$(4:3) \text{ and } (3:2) \text{ make } (2:1),$$

and, in Proposition 7, the relation

$$(2:1) \text{ and } (3:2) \text{ make } (3:1)$$

Then the music theoretic consequences are analysed in the comprehensive Proposition 12:

The octave interval is duple.

We have shown that it is multiple; it is thus either duple or greater than duple. But since we showed that the duple interval is composed of the two greatest epimorics, it follows that if the octave is greater than duple it will not be made up of just two epimorics, but of more. But it is made up of two concordant intervals, the fifth and the fourth. Therefore the octave will not be greater than duple: therefore it is duple.

But since the octave is duple, and the duple is made up of the two greatest epimorics, it follows that the octave is made up of the hemiolic and the epitritic, since these are the greatest. But it is made up of the fifth and the fourth, and these are epimoric. The fifth, therefore, since it is greater, must be hemiolic, and the fourth epitritic.

It is clear then that the fifth and octave is a triple. For we showed that the triple interval is generated from a duple and a hemiolic interval, so that the octave and a fifth is also a triple.

The double octave is a quadruple.

But nowhere does the *Sectio* discuss any mathematical or musical implication of the relation

$$(2:1) \text{ and } (4:3) \text{ make } (8:3).$$

Much more serious than this omission is the flaw to be found in the proof of Proposition 11:

The interval of the fourth and that of the fifth are each epimoric.
Let A be *nētē synēmmenōn*, let B be *mesē*, and let C be *hypatē mesōn* [three fixed notes on the scale, each a fourth apart]. The interval AC, being a double fourth, is therefore discordant [a musical fact]: it is therefore not multiple [!; see below]. Thus since the two equal intervals AB and BC when put together make a whole which is not multiple, neither is AB multiple [Proposition 1]. And it is concordant: therefore it is epimoric. The same demonstration applies also the the fifth.

The inference that a discordant interval is not multiple cannot follow from any argument or assumption stated or hinted anywhere in the treatise or elsewhere. Note that the second principle of the Introduction, described above, does *not* state that all multiple and epimoric intervals are concordant; this would ensure the truth of the proposition but it is clearly false, as the examples of $5:1$, $7:1$, and $9:8$ clearly illustrate. There seems to be no explanation of the flaw in this proof, nor any way of recovering the proposition, and since the remaining propositions depend on it, the remainder of the argument collapses.

It may be misplaced effort to look for a sophisticated mathematical explanation for a treatise so obscure and flawed as the *Sectio*, but nonetheless I have tried, in my dialogue. Just as the idea of ratio in an astronomical context might plausibly lead to the profound insight of *Elements* V Definition 5, as I proposed in Section 4, so an exploration of ratio in a musical context might yield some other mathematical insights.

4.5 Academic music theory 153

Moreover, the manipulations of music theory seem to depend fundamentally on the operation of compounding, an operation which seems to pose some serious problems for mathematicians. My purely speculative suggestion, set out in A_{70-84}, is that music theory might plausibly give some help with this problem. It might become apparent only later that the compounding of anthyphairetic ratios is an even more subtle operation, both conceptually and computationally, as I explained in Section 2, above. In the event, the proposal in A_{84} that it may be easier to infer it from simple commensurable examples will turn out to be misguided, as I shall explain in Section 9.3(a).

Let me leave this brief discussion of the *Sectio Canonis* by emphasising a part of my opening description of the treatise: that its author is unknown and its date is uncertain.

4.5(e) Further problems

The final speeches of my dialogue, $B_{89}-A_{96}$, introduce further questions about the connections between geometry and *arithmētike* that I shall only discuss very briefly here. The description in $B_{89}-B_{91}$ of the geometrical operations is included in preparation for the work of the next chapter, and to make the geometrical meaning of a statement like $\sqrt{2} . \sqrt{3} = \sqrt{6} . 1$ quite clear. The symbolism introduced is, of course, intended for the convenience of the modern reader, and there is no suggestion in Greek mathematics of any such developments; all of the symbolic manipulations should be read as shorthand descriptions of geometrical manipulations.

The straightforward construction in B_{93} of the lines involved in $\sqrt{2} . \sqrt{3} = \sqrt{6} . 1$ leads, in Fig. 4.2, to a configuration closely resembling that called the *arbēlos*, the shoemaker's knife, which is found in the Archimedean *Book of Lemmas*, Archimedes-Heiberg, *Opera* ii, 510–25 and Archimedes-Heath, *WA*, pp. xxxii ff. and 301 ff. This treatise is known only in an Arabic translation of what appears to be a later Greek compilation of heterogeneous results. The observation in B_{95} on the angles of an equilateral triangle leads immediately to the concern of Postulate 5 of *Elements* I and what is now called non-Euclidean geometry. Our evidence on this topic from early Greek geometry is very sparse, and I shall not enter the question of its interpretation here beyond observing that, although this postulate is generally described as if its origin lies in problems associated with parallel lines and indeed it is often referred to as the 'Parallel Postulate', the postulate itself is articulated without invoking the language of parallels:

That, if a straight line falling on two straight lines make the interior angles on the same side less than two right angles, the two straight lines, if produced indefinitely, meet on that side on which are the angles less than two right angles.

Indeed it more resembles a converse of *Elements* I 17:

In any triangle two angles taken together in any manner are less than two right angles;

and it can be read as the assertion that it is possible to construct a triangle of any size with angles equal to those of a given triangle. Its formulation seems to be more closely concerned with the properties of similar figures than with parallel lines.

Our earliest treatises on Greek mathematics, *On the Moving Sphere* and *On Risings and Settings* by Autolycus of Pitane, who flourished at the end of the fourth century, are on spherical geometry, and we have later reports of work by Eudoxus on subtle constructions involving spheres. However, we have no explicit evidence that Greek geometers ever considered the spherical model for geometry, as in B_{95}, however attractive the idea may be. Spherical geometry is excluded from the *Elements* by Postulate 2:

To produce a finite straight line continuously in a straight line,

but we do not know whether this was the reason for the inclusion of this postulate. My reason for introducing non-Euclidean geometry is to bring the little set of dialogues themselves in a full circle with my slaveboy's reflection at the end of B_{95}, and so point to a further level of mathematical subtlety underlying Socrates' example of doubling a square in the *Meno*.

4.6 Appendix: The words *logistikē* and *logismos* in Plato, Archytas, Aristotle, and the pre-Socratic philosophers

4.6(a) *Plato*

The references to *logistikos* (the lemma form of *logistikē*) in Brandwood, *A Word Index to Plato,* divide between about twenty-five where the context is explicitly mathematical and about ten with the more general sense of 'intellectual principle' or 'reason'; while *logismos* (excluding ten instances found in the pseudo-Platonic *Definitions*, all of them non-mathematical) divide between about thirty explicitly mathematical references and forty usages with the more general sense of 'rational discussion' or 'reason'. It might also be interesting to check, in the same way, the 1500 or so entries for *logos*.

The relevant entries from the *Index* are reproduced below; they are divided into grammatical categories with lemmata underlined, as in the *Index*. Those items that seem to me to refer to specifically mathematical passages are set in bold-face type. An asterisk indicates some variati d

4.6 Logistikē and logismos

between the manuscripts or editorial emendation, details of which can be found in the *Index*. This list should cover all occurrences of these words in the Platonic and pseudo-Platonic corpus. A similar index to *alogos* and *(ar)rhetos* can be found in the Appendix to Chapter 5.

λογισμός Timaeus 34a.8; *Laws* I 644d.2.
λογισμόν Euthyphro **7b.10**; *Hippias Minor* **367a.9**, **c.5***; *Republic* IV 440b.1, VII **522c.7**, **524b.4**, IX **587e.5**; *Phaedrus* **274c.8**; *Timaeus* 30b.4, 33a.6; *Laws* VII 805a.3; *Epinomis* 974a.2; *Definitions* 412e.2, 416a.1,12.
λογισμοῦ Meno 100b.2; *Phaedo* 66a.1; *Symposium* 207b.7; *Republic* IV 439d.1, 441a.9, IX 586d.2; *Timaeus* 36e.6, 72e.2, 77b.5, 86c.3; *Critias* 121a.7; *Sophist* 248a.11; *Philebus* 21c.5, 52b.3; *Laws* I 645a.1,5; *Epinomis* 981c.4; *Definitions* 413c.3, 415e.6, 416a.18.
λογισμῷ Meno 98a.4; *Phaedo* 79a.3, 84a.7; *Republic* I **340d.4**, IV 431c.6, VI 496d.6, VIII 546b.1, X **603a.4**, 604d.5, 611c.3; *Parmenides* 130a.2; *Phaedrus* 249c.1; *Timaeus* 52b.2, 57e.1, 72a.1; *Laws* III 697e.2, VII 813d.1; *Epistle* VII 340a.2; *Definitions* 412b.4,5, 415d.10.
λογισμοί *Laws* VII **817e.6**, X **896c.9**.
λογισμούς Hippias Minor **367a.1**; *Protagoras* **318e.2**; *Republic* VI **510c.3**, VII **525d.1**; *Theaetetus* **145d.2**; *Statesman* **257a.7**, **b.7**; *Philebus* 11b.8; *Laws* VII **819b.2**.
λογισμῶν Hippias Minor **366c.6**, **367b.7**, **c.2,3,5***; *Hippias Major* **285c.5**; *Republic* VII **526d.8**, **536d.5**; *Timaeus* **47c.2**; *Sophist* 254a.8; *Philebus* **57a.1**; *Laws* VII **809c.4**; *Definitions* **412a.7***.
λογισμοῖς *Philebus* 52a.8; *Laws* X **897c.6**, XII **967b.3**.
λογιστέον (nt.) *Timaeus* 62a.1.
λογιστής *Republic* I **340d.6**; *Statesman* 260a.5.
λογιστικός *Hippias Minor* **367c.4**; *Theaetetus* **145a.7**.
λογιστικόν *Republic* I **340d.3**.
λογιστικῷ *Republic* VII **525b.6**, IX **587d.11**.
λογιστικοί *Euthydemus* **290c.1**; *Republic* VII **526b.5**; *Laws* III **689c.9**.
λογιστική *Charmides* **166a.5**; *Gorgias* **450d.6**, **451c.2,5**; *Republic* VII **525a.9**; *Statesman* **259e.1**; *Philebus* **56e.7**; *Definitions* 411e.10, 414c.3.
λογιστικήν *Gorgias* **451b.5**; *Republic* VII **525c.1**.
λογιστικῆς *Charmides* **165e.5**, **166a.10**; *Hippias Minor* 366c.6; *Definitions* 412b.9.
λογιστικῇ *Statesman* **259e.5**, **260a.10**.
λογιστικόν *Charmides* **174b.5**; *Republic* IV 439d.5, 440e.6,9, 442c.11, VIII 550b.1, 553d.1, IX 571c.4, d.7, 580d.4*, X 605b.5; *Timaeus* 37c.1.
λογιστικοῦ *Republic* IV 440e.8, 441a.5, X 602e.1.
λογιστικῷ *Republic* IV 441a.3, e.4.
λογιστικά *Hippias Minor* 366d.5.

4.6(b) Archytas

Two fragments of Archytas, B3 and 4 in Diels & Kranz, *FV* ii, 436–8, are concerned with *logismos* and *logistikē*. They are cited here in full:

B3. *Logismos,* when discovered, stops strife and increases concord; when it occurs, there is no excess of gain, but there is equality; for by this we settle our disputes. By this, then, the poor take from the powerful, and the rich give to the needy, both sides trusting that through this they will get fair treatment. It is a rule and it prevents men from doing wrong: it stops those who know *logismos* before they do wrong, persuading them that they will not be able to escape notice, when they come to it: and it prevents those who do not know from doing wrong, showing by that very fact [i.e. that they do not know *logismos*] that they are doing wrong. [Translation adapted from Freeman, *APSP,* and Harvey, TKE; see Chapter 6, n. 11, below.]

B4. *Logistikē,* it seems, in regard to wisdom is far superior to all the other sciences, especially geometry, because *logistikē* is able to do more clearly any problem it will . . . and—a thing in which geometry fails—*logistikē* adds proofs and, at the same time, if the problem concerns forms, *logistikē* treats of the forms also. [Translation adapted from Freeman, *APSP.*]

4.6(c) Aristotle

The following entries are taken from Bonitz, *Index Aristotelicus,* which is probably far from complete. Aristotle almost always uses the words in the sense of the 'reasoning faculty'. Passages in bold-face type indicate a context which contains some additional reference, as when this 'reasoning faculty' is associated with the 'rational and irrational principles' of the soul, or, very occasionally, with some technical usage; but none of these passages seem to shed any light on the nature of mathematical *logistikē.* Again, there is a massive entry for *logos,* about fifteen times longer than these entries combined.

λογισμός *Posterior Analytics* **88b12**, 100b7; *Topics* 145b2, b17; *Physics* **200a23**; *On the Soul* 415a8, 433a12, b29; *Metaphysics* 980b28, 1015a33; *Nicomachean Ethics* 1111a34, 1117a21, 1119b10, 1141b14, 1150b24, 1220a1, 1250a11, b13; *Politics* 1312b29, **1322b9**, **1334b24**, 1369b7; *Rhetoric to Alexander* 1429a17; *Fragment* 96 1493b32; *Fragment* 97 1493b37.

λογισταί *Politics* **1322b11**; *Fragment* 406 1546a4, a10, a21; *Fragment* 407 1546a26.

λογιστήριον *Fragment* 406 1564a4.

λογιστικος *Topics* 126a8, a13, 128b39, 129a11 sqq, 134a34, 136b11, 138a34, b2, b13, 145a29, 147b32; *Physics* 210a30; *On the Soul* **432a25**, 433b29, 434a7; *Nicomachean Ethics* **1139a12**, 1182a18, a20, 1249a31, b30, 1250a3, a16; *Rhetoric* **1369a2**.

4.6   *Logistikē* and *logismos*   157

4.6(d)   *Pre-Socratic philosophers*

The index to Diels & Kranz, *Die Fragmente der Vorsokratiker* contains nine columns of entries for *logos,* half a column for *logismos,* and one entry (*Archytas* B4; see above) for *logistikē*. Here are the entries for *logismos,* each with its reference number and volume, page, and line entries. Square brackets denote texts described as spurious; for further details of the classification, see *FV* i, 1. The bold type is as in Section 4.6(a), above.

Anaxagoras	A66	ii 22.4	
Archytas	**B3**	**i 437.7**	(see above)
Democritus	B187	ii 183.10	
	B290	ii 205.19	
	[B302	ii 222.19]	
	[B306	ii 223.32]	
Epicharmus	[B56	i 208.5 f.]	
	[B57	i 208.9]	
Hippias	80A5	ii 256.2	(≡ Plato, *Protagoras* 318e.2)
	86A11	**ii 328.32**	(≡ Plato, *Hippias Major* 285c.5)
Hippocrates of Cos	64C2	ii 67.18	(i.e. the physician)
Pythagoras	9	i 101.2	
Theodorus	4	i 397.25	(≡ Plato, *Theaetetus* 145d.2)

5
ELEMENTS IV, X, AND XIII: THE CIRCUMDIAMETER AND SIDE

5.1 The circumdiameter and side

5.1(a) *The problem*

Let us now turn to the problem of the circumdiameter and side, introduced by my Socrates at S_{41}. It is very similar in spirit to the very fruitful problem of the diagonal and side, discussed in Section 2.2, out of which grew the remarkable discoveries of the dimension of squares in Chapter 3. We take a regular polygon with its circumscribed circle, and try to describe the relation between the circumdiameter and a side. Instead of the side we can take a diagonal; and instead of the diameter, we can take the radius, since doubling or halving a ratio has an effect on the anthyphairesis that seems difficult to elucidate. We can also consider the inscribed circle, or investigate the similar three-dimensional problems associated with an edge or diagonal of a regular polyhedron inscribed in or circumscribed around a sphere.

The geometrical constructions associated with several such problems are described in *Elements* IV and XIII. Most of them lead to lines that are either commensurable or are related as the sides of squares, and they are described as such by Euclid; see *Elements* XIII 12 (equilateral triangle and hexagon), 13 (tetrahedron), 14 (octahedron), and 15 (cube). In these cases the anthyphairesis can be completely described by an additional short and now routine calculation. But three examples, the pentagon, icosahedron, and dodecahedron, give rise to more complicated relations than pure sides of squares. The worst kind of behaviour considered in the *Elements* occurs already with the pentagon, and I shall concentrate exclusively on this example. My geometrical descriptions throughout this chapter will be based on the treatment in Taisbak, *CQ*, to which I shall add an anthyphairetic motivation.

5.1(b) *The pentagon*

Consider the circumradius and side or diagonal of a pentagon. There is no difficulty in evaluating and expressing the relations between these lines using the language and techniques at Euclid's disposal. Take a regular decagon $A_1B_1A_2B_2 \ldots A_5B_5$ inscribed in a circle (see Fig. 5.1), and note that Euclid's basic construction of the pentagon at *Elements* IV 11 does,

5.1 The circumdiameter and side

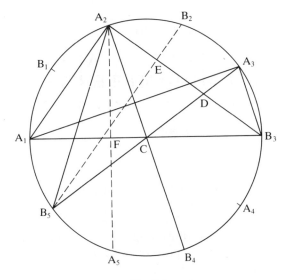

Fig. 5.1

in fact, start from the side of a decagon inscribed in a circle, though Euclid does not bring out this fact. Then the diameters, such as A_1B_3, A_2B_4, and A_3B_5, will intersect in the centre C of the circle. Now most of the angles of the figure can be expressed in terms of the angle subtended at the circumference by a side of the decagon, which is one-twentieth of the circle, or 18°. Hence many 36°–36°–72° isosceles triangles can be found, in particular A_3B_3D (so $B_3D = B_3A_3$ = the side of the decagon) and A_2CD (so $A_2D = A_2C$ = the radius of the circumcircle = the side of the inscribed hexagon). Hence the line A_2B_3 is the sum of the sides of the inscribed hexagon and decagon, and so is divided in extreme and mean ratio; this result is proved at *Elements* XIII 9. Note that since A_2B_5D is also a 36°–36°–72° triangle and B_5EB_2 is perpendicular to A_2EB_3, then $A_2E = ED = q$, where $2q = r$, the radius of the circle. Write $B_3E = p$; then, by XIII 3, one of the block of propositions on the metrical properties of the extreme and mean ratio (see Section 3.5(d), above),

$$B_3E^2 = 5A_2E^2, \quad \text{i.e.} \quad p^2 = 5q^2.$$

Hence
$$B_3A_3 = B_3D = B_3E - ED = p - q,$$
and
$$B_3A_2 \quad = \quad B_3E + EA_2 = p + q.$$

Now, from the right-angled triangle $A_1A_2B_3$,

$$A_1A_2^2 + A_2B_3^2 = A_1B_3^2,$$

so
$$A_1A_2^2 + (p+q)^2 = 16q^2,$$
and hence
$$A_1A_2^2 + p.r = 10q^2 = 10(r/2)^2.$$

Similarly, from the right-angled triangle $A_1A_3B_3$,
$$A_1A_3^2 + A_3B_3^2 = A_1B_3^2,$$
or
$$A_1A_3^2 + (p-q)^2 = 16q^2$$
and hence
$$A_1A_3^2 = p.r + 10(r/2)^2.$$

We, today, express these kinds of results in the arithmetised form:
$$\text{side of pentagon} = r/2\sqrt{(10-2\sqrt{5})}$$
$$\text{diagonal of pentagon} = r/2\sqrt{(10+2\sqrt{5})},$$
but we find nothing of this kind in the *Elements*. It is not that Euclid is unable or unwilling to articulate this kind of result. We can write these formulae as $\sqrt{(10(r/2)^2 \pm r . \sqrt{5(r/2)^2})}$ and describe them, within the idiom of Books II or XIII 1–5, by:

The square on the side [or diagonal] of a pentagon is less [or greater] then ten squares on half the radius by the rectangle contained by the radius and the side of the square equal to five times the square on half the radius.

Nor is it that such a statement is difficult to understand. In Fig. 5.2(a), we start with "ten squares on half the radius" and then, in Fig. 5.2(b), double the base, halve the sides, and subtract "the [upper] rectangle contained by the radius $[r=2q]$ and the side $[p]$ of the square equal to five times the square on half the radius [so $p^2 = 5(r/2)^2 = 5q^2$]". This leaves the lower, shaded rectangle equal to the square on the side of the pentagon; then the side itself can be constructed using *Elements* II 14. A similar procedure, in which $p.r$ is added, not subtracted, will give the square on the diagonal. Nor is it that this kind of result is remote from the kinds of manipulations and evaluations that occur in the *Elements*: We find an exploration of the pentagon at XIII 11 which involves the vertices of the decagon, the projection of a side of the pentagon on an adjacent diameter (for example A_1F in Fig. 5.1), and a line equal to a quarter of the radius, $q/2$ in my notation. Euclid's argument turns on the fact that $A_1A_2B_3$ and A_1FA_2 are similar right-angled triangles, so that

$$A_1A_2^2 = A_1F . A_1B_3, \quad \text{i.e.} \quad \text{side}^2 = (\text{projection}) . (4q).$$

This rectangle appears as the lower, shaded rectangle in Fig. 5.2(c), where it arises from a further doubling of the base and halving of the side of the

5.1 The circumdiameter and side

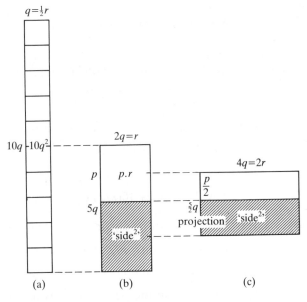

Fig. 5.2

rectangle of Fig. 5.2(b). While the likes of Fig. 5.2 are not generally drawn in the *Elements*, these kinds of figures do seem to follow the Euclidean style of argument much more closely than manipulations of the corresponding algebraic formulae.

For each of the cases of the circumdiameters and sides (or edges) of the pentagon, icosahedron, and dodecahedron, instead of providing something which corresponds to our kind of metrical description, Euclid suppresses all of the numerical constants that give essence to our evaluations and enunciates only qualitative results. For example, at XIII 11, we have:

If in a circle which has its diameter expressible (*rhētos*) an equilateral pentagon be inscribed, the side of the pentagon is the *alogos* ('without ratio') straight line called minor (*elassōn*).

This proposition means that the side of the pentagon can be expressed as the difference of the sides of two squares, where these squares satisfy certain further kinds of very elaborate qualitative, not quantitative, relations of a kind that have been developed, in the *Elements*, in the classification of Book X, and which I shall describe in Section 5.2, below.

5.1(c) *The extreme and mean ratio*

The extreme and mean ratio gives another illustration of Euclid's procedure. If a line of length 1 is cut in extreme and mean ratio, the longer segment has length $\frac{1}{2}(\sqrt{5}-1)$, the shorter $\frac{1}{2}(3-\sqrt{5})$, and their ratio is $\frac{1}{2}(\sqrt{5}-1):1$. While these arithmetised statements might seem to us today to contain some of the essential facts about the extreme and mean ratio, they do not occur in the *Elements*, though there is absolutely no doubt that Euclid could have articulated them, expressed geometrically, among the propositions that he does prove at *Elements* XIII 1–5. Instead this block of results finishes, at XIII 6, with:

If an expressible (*rhētos*) straight line be cut in extreme and mean ratio, each of the segments is an *alogos* straight line, the one called apotome (*apotomē*).

This, again, is a qualitative statement that is, this time, sufficiently simple to be described without further terminology: Each of the segments can be expressed as a difference of two lines which are incommensurable with each other, but such that the squares on these lines are commensurable with each other and with the square on the whole.

5.1(d) *Anthyphairetic considerations*

Let us now try to explore these ratios anthyphairetically. We immediately face two problems: that the evaluation of the anthyphairesis is sometimes rather difficult, beset with theoretical and practical problems, and that the behaviour related by the heuristic search seems to be unpredictable.

The first problem in evaluating the terms of a ratio like $\frac{1}{2}\sqrt{(10 \pm 2\sqrt{5})}:(1 \text{ or } 2)$, using some algorism such as those described in Section 2.3, has already manifested itself in Section 5.2, where the example of $(\sqrt{2}+\sqrt{3}):1$ was cited to illustrate the intractability of algorisms for anthyphairetic arithmetic. I shall illustrate the difficulties in more detail using this slightly simpler ratio. The algorism of Section 2.3(b) depended on some kind of test to identify whether

$$p:q \text{ is } <, =, \text{ or } > \quad (\sqrt{2}+\sqrt{3}):1.$$

Suppose we manipulate boldly, ignoring problems of meaning and interpretation, and all historical and mathematical problems:

$p:q$ is $<$, etc. $(\sqrt{2}+\sqrt{3}):1$ according as p is $<$, etc. $(\sqrt{2}+\sqrt{3})q$,

i.e. according as p^2 is $<$, etc. $(5+2\sqrt{6})q^2$,

provided we can justify these operations, including the arithmetical result that $\sqrt{2} \cdot \sqrt{3} = \sqrt{6}$, on which see B_{93},

i.e. according as $(p^2 - 5q^2)$ is $<$, etc. $2\sqrt{6}q^2$,

i.e. according as $p^4 - 10p^2q^2 + 25q^2$ is $<$, etc. $24q^4$,

although if $p^2 < 5q^2$, some further investigation is needed,

i.e. according as $p^4 + q^4$ is $<$, etc. $10p^2q^2$.

5.1 The circumdiameter and side

TABLE 5.1. Evaluation of $(\sqrt{2}+\sqrt{3}):1 = [3, 6, 1, 3 \text{ or more}, \ldots]$. (Compare Table 2.1.)

Under-estimate	Over-estimate	New estimate $p:q$	p^4	q^4	p^4+q^4	$10p^2q^2$	Under/equal/over
3:1	4:1	7:2	2401	16	2417	1960	over
3:1	7:2	10:3	10000	81	10081	9000	over
3:1	10:3	13:4	28561	256	28817	2704	over
3:1	13:4	16:5	65536	625	66161	64000	over
3:1	16:5	19:6	130321	1296	131617	129960	over
3:1	19:6	22:7	234256	2401	236657	237160	under
22:7	19:6	41:13	2825761	28561	2854322	2840890	over
22:7	41:13	63:20	15752961	160000	15912961	15876000	over
22:7	63:20	85:27	52200625	531441	52732066	52670250	over

The trouble now, as anybody who applies the algorism will find, is that the numbers grow far too quickly for the calculation to be carried very far; the first three terms are calculated in Table 5.1 and, at the next step, the calculation will overflow the capacity of the standard Greek numerals (see Chapter 7). No periodic or regular behaviour presents itself, though we notice incidentally that $(\sqrt{2}+\sqrt{3}):1$ is close to and greater than the ratio of the circumference to diameter of a circle. (The explanation of this approximation can be seen in a hexagon inscribed in, and a square circumscribed around, a circle; see Popper, *OSE* i, Chapter 8, n. 9.)

We attempt these calculations for the same motives that lay behind the calculations of Section 3.3 (see especially Table 3.1), in the hope that they may uncover some plausible hypotheses. After this, we hope that the calculations will fall away, to be replaced by deductive proofs, as happened with the problem of the dimension of squares; compare Sections 3.3–3.6. So let us continue to ignore any historical and mathematical scruples relating to any aspect of this calculation, and pass to the results, to see if any plausible hypotheses would eventually present themselves. For the pentagon, we get:

$$\text{side}:\text{circumradius} = \sqrt{(10-2\sqrt{5})}:2$$
$$= [1, 5, 1, 2, 3, 2, 28, 2, 27, 3, 22, 1, 7, 1, 9, \ldots],$$
$$\text{side}:\text{circumdiameter} = \sqrt{(10-2\sqrt{5})}:4$$
$$= [0, 1, 1, 2, 2, 1, 6, 1, 56, 1, 54, 1, 1, 1, 10, \ldots],$$
$$\text{diagonal}:\text{circumradius} = \sqrt{(10+2\sqrt{5})}:2$$
$$= [1, 1, 9, 4, 1, 1, 1, 2, 1, 1, 2, 3, 7, 1, 2, \ldots], \text{ and}$$
$$\text{diagonal}:\text{circumdiameter} = \sqrt{(10+2\sqrt{5})}:4$$
$$= [0, 1, 19, 2, 3, 6, 5, 1, 1, 1, 3, 2, 1, \ldots].$$

We are again faced with behaviour similar to that encountered in the problem of the dimension of cubes (see Section 4.3), that no regularity seems to be evident. The purely anthyphairetic approach seems blocked.

We can try to investigate this problem from the other side, and attempt to evaluate the ratio of the side or diagonal to circumradius or diameter of the pentagon directly, by geometric arguments such as were given in Section 2.1 for the simpler cases of the diagonals and sides of regular polygons. We can indeed try, but I can see no possibility of progress with the example of the circumcircle of the pentagon. However the case of the extreme and mean ratio is amenable to this approach, for here we can both evaluate the ratio geometrically and explore the construction metrically, and then compare what we find. We then get

$$\tfrac{1}{2}(\sqrt{5}-1) : \tfrac{1}{2}(3-\sqrt{5}) = [\bar{1}].$$

We also notice that this particular expansion is much easier to evaluate directly, since the test for whether

$$p:q \text{ is } <, =, \text{ or } > \tfrac{1}{2}(\sqrt{5}-1) : \tfrac{1}{2}(3-\sqrt{5})$$

reduces to finding whether

$$p^2 \text{ is } <, =, \text{ or } > pq + q^2,$$

and the numbers involved in the arithmetical exploration only grow quadratically. But it is difficult to see, from just this one example, what may be happening in general, what it is that is so special about the relation $(\sqrt{5}-1):(3-\sqrt{5})$ that it yields a purely periodic anthyphairesis. Also the issue is complicated by another phenomenon: all examples of ratios of sides of squares seem to have the form $[n_0, n_1, \ldots, 2n_0]$, and so contain an even term somewhere in their anthyphairesis. Hence the extreme and mean ratio is an example of a periodic anthyphairetic expansion which cannot be expressed as a ratio of sides of squares (see Section 3.5(b)). Can anything more be said about this new class of expressible ratios?

A final series of evaluations will illustrate the problem further. The following calculations, one of which requires a great deal more computational or analytical effort, yield

$$(\sqrt{2}+\sqrt{3}):\sqrt{6} = [1, 3, 1, 1, 15, 1, 1, 1, 19, 2, 4, 1, 1, 3, 47, 1, 3, \ldots],$$
$$(\sqrt{2}+\sqrt{3}):(\sqrt{6}-1) = [2, 5, 1, 6, 4, 2, 2, 2, 106, 12, 2, 1, 1, 2, 2, \ldots],$$
$$(\sqrt{2}+\sqrt{3}):(\sqrt{6}+1) = [0, 1, 10, 2, 1, 1, 1, 15, 9, 5, 2, 6, 5, 4, 17, \ldots],$$

and

$$(\sqrt{2}+\sqrt{3}):(\sqrt{6}-2) = [6, 1, 2742, 1, 1, 1, 3, 1, 160, 1, 1, 2, 5, 3, 1, \ldots],$$

5.1 The circumdiameter and side

but

$$(\sqrt{2}+\sqrt{3}):(\sqrt{6}+2) = [0, 1, \bar{2}].$$

Let us explore to see why this last result may be so much simpler than the rest. I shall consider the reciprocal ratio $(\sqrt{6}+2):(\sqrt{3}+\sqrt{2})$, taken 'the greater to the less', and will work out in detail the geometrical manipulations which simplify this expression. We start from the identities

$$(\sqrt{3}+\sqrt{2}).(\sqrt{3}-\sqrt{2}) + \sqrt{2}.\sqrt{2} = \sqrt{3}.\sqrt{3},$$

so

$$(\sqrt{3}+\sqrt{2}).(\sqrt{3}-\sqrt{2}) = 1.1,$$

and

$$(\sqrt{6}+2).(\sqrt{3}-\sqrt{2}) = \sqrt{2}.1,$$

which correspond to the configurations of Figs. 5.3(a) and 5.3(b). (Note that the latter identity uses manipulations like $\sqrt{6}.\sqrt{3} = \sqrt{18}.1 = 3.\sqrt{2}$, on which again see A_{88}–B_{93}.) Hence we have:

$$(\sqrt{6}+2):(\sqrt{3}+\sqrt{2}) = (\sqrt{6}+2).(\sqrt{3}-\sqrt{2}):(\sqrt{3}+\sqrt{2}).(\sqrt{3}-\sqrt{2})$$

    (by the *Topics* proposition)

$$= \sqrt{2}.1 : 1.1 \quad \text{(by the evaluations above)}$$

$$= \sqrt{2}:1 \quad \text{(by the *Topics* proposition)}$$

$$= [1, \bar{2}].$$

However, this result seems to be very special, and were such a manipulation as this ever explored, it would perhaps be more likely to prompt an investigation of the behaviour of rectangles like $(\sqrt{p} \pm \sqrt{q}).(\sqrt{r} \pm \sqrt{s})$ than of the underlying properties of anthyphairesis.

These manipulations may seem to us to be very artificial and laboured. But when Euclid deals with a similar kind of result at *Elements* X 112–14, his treatment is also notoriously complicated and, if viewed algebraically,

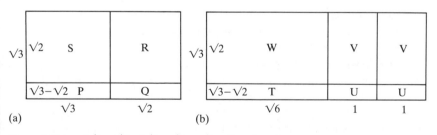

Fig. 5.3(a). $(\sqrt{3}+\sqrt{2}).(\sqrt{3}-\sqrt{2}) + \sqrt{2}.\sqrt{2} = \sqrt{3}.\sqrt{3}$, i.e. $P+Q+R = P+S$, since $S = Q+R$. So $(\sqrt{3}+\sqrt{2}).(\sqrt{3}-\sqrt{2}) = 1.1$. (b) $(\sqrt{6}+2).(\sqrt{3}-\sqrt{2}) = \sqrt{2}.1$, i.e. $T+2U = V$, since $\sqrt{6}.\sqrt{3} = 3.\sqrt{2}$, i.e. $T+W = 3V$, and $\sqrt{6}.\sqrt{2} = 2.\sqrt{3}$, i.e. $W = 2U+2V$, so $T+2U+W = 3V+2U = V+W$.

difficult to follow. He never seems to recognise the relevance and simplification that would follow from using freely the basic identity $(b+c).(b-c) = b^2 - c^2$. Further, there seems to be nothing corresponding to this identity in the configurations of Book II. I shall return briefly to this question at the end of Section 5.2(g) below.

The material I have described so far has provided sufficient clues to permit an informed guess as to what might be happening, but I do not think that much of this information would be accessible in any such coherent form within the early Greeks' arithmetical techniques, while the underlying phenomenon itself, if it were ever guessed, would probably lie outside their techniques of deductive geometrical proof. The result is that the *only* known regular behaviour of anthyphairesis connected with the operations of taking sums, products, and roots (of any order) is that ratios of the form $\pm(p \pm \sqrt{q}):r$ (or, in arithmetised terms, those numbers that arise as solutions of quadratic equations with integers coefficients) will have an anthyphairesis that is eventually periodic,

$$\pm(p \pm \sqrt{q}):r = [n_0, n_1, \ldots, n_k, \overline{m_0, m_1, \ldots m_l}];$$

and, conversely, any such expression will give rise to a ratio that can be expressed in this way. (This converse can be demonstrated geometrically by the final generalisation of the procedure of synthesising ratios in Section 3.5 but, as I observed there, I feel that this also would be highly implausible as a reconstruction of Greek procedures since it involves an extended subsidiary quasi-algebraic computation.) Much more reasonable would be to propose that any such arithmetical heuristic explorations of the problems were abandoned at some early stage since they gave no insight and were computationally intractable.

5.2. Elements X: A classification of some incommensurable lines

5.2(a) *Introduction*

An investigation such as I have described in Section 1 sets up a remarkable contrast. On the one hand, there is the spectacular success of the anthyphairetic exploration of the ratios of sides of squares, $\sqrt{n}:\sqrt{m}$, in which a wide range of arithmetical and geometrical techniques and problems are bound together into some apparently coherent whole. On the other hand, there is an equally spectacular lack of success in the search for any similar regular, predictable anthyphairetic behaviour of any other geometrical ratios, apart from the extreme and mean ratio. Such a contrast might lead to the growth of an attitude in which these sides of squares come to be considered as the basic underlying understandable geometrical objects in terms of which everything else should be

5.2 *Elements* X

described. I propose that Book X of the *Elements* may indeed be a product of such a point of view.

The book opens with four definitions:

[1] Those magnitudes are said to be commensurable (*summetroi*) which are measured by the same measure, and those incommensurable (*asummetroi*) which cannot have any common measure.

[2] Straight lines are commensurable-in-square (*summetroi dunamei*) when the squares on them are measured by the same area, and incommensurable [-in-square] when the squares on them cannot possibly have any area as a common measure.

[3] With these hypotheses, it is proved that there exist straight lines infinite in multitude which are commensurable and incommensurable respectively, some in length only, and others in square also, with an assigned (*protetheisa*) straight line. Let the assigned straight line be called expressible (*rhētos*), and those straight lines which are commensurable with it whether in length and in square, or in square only, expressible, but those incommensurable with it *alogoi*.

[4] And let the square on the assigned straight line be called expressible, and those areas which are commensurable with it expressible, but those which are incommensurable with it *alogoi*, and the straight lines that produce them *alogoi*, that is, in case the areas are squares, the sides themselves, but in case they are any other rectilineal figures, the straight lines on which are described squares equal to them.

I have rendered *rhētos* as 'expressible', but have left *alogos* in transliterated form. This latter word means 'without words' or, if the material of Book X is indeed not just about the lines and areas in themselves, but also their mutual relations, 'without ratio'. These Greek words are almost always rendered into English as 'rational' and 'irrational', respectively. But:

(i) According to Euclid's definitions, if we denote the assigned line by a, then both $3a:2a$ and $\sqrt{3}a:\sqrt{2}a$ are ratios of expressible lines. However, in our arithmetised description of ratios today, the first corresponds to what is universally called a rational number, the second to an irrational number. So the usual translation can lead to mathematical confusions and misunderstandings.

(ii) The use of the words 'rational' and 'irrational' gives the impression that the definitions refer to a dichotomy in which every line is either rational or irrational; but this is not the case in the original Greek. In fact, Euclid makes no attempt in the classification of Book X to consider every possible kind of line; rather, he works outwards from the assigned line to generate new kinds of lines and areas that are described in terms of the lines and areas that have already been considered. He first describes the expressible lines in Definition 3, and then he goes on to describe thirteen different kinds of *alogoi* lines that arise from some very special constructions.

(iii) The word pairs *rhētos/arrhētos* and *logos/alogos,* which do describe dichotomies, are not uncommon from the fifth or fourth century onwards, when they are found in a variety of technical and non-technical contexts. The only surviving possibly mathematical pre-Socratic use of *alogos* is in the title of a book by Democritus, *Peri alogōn grammōn kai nastōn,* and of *arrhētos* is in a later and unreliable report of an aphorism of Lysis, *arithmos arrhētos theos*; both of these will be discussed in Section 8.3(b). An example of Plato's use of these words occurs at *Theaetetus* 201e–202c, where Socrates describes a dream in a remarkable passage many of whose words have mathematical overtones, for example:

... In that way, the elements (*stoicheia*) have no account (*aloga*) and are unknowable (*agnōsta*), but they are perceivable (*aisthēta*); and the complexes (*sullabai*) are knowable (*gnōstai*) and expressible (*rhētai*) and judgeable (*doxastai*) in true judgement (*alētheia doxa*). ...

Given the many problems that are associated with these words it would seem best, in contexts that might involve unfamiliar nuances of specialised technical meanings, to make a special effort to keep track of the original terminology. So I have included an index to the occurrences of *alogos* and *(ar)rhētos* in Plato, Aristotle, and the pre-Socratic philosophers in an appendix to this chapter; but I shall not discuss the important topic of their use and meaning further in this book.

The suggestion of rendering *rhētos* as 'expressible' was made in van der Waerden, *SA,* 168–79; he also translates *alogos* as 'unreasonable'. Another approach is taken in Taisbak, *CQ*: he makes no attempt to translate the words but merely replaces them by neutral, easily remembered alternatives: 'red' for *rhētos,* 'amber' for *mēsos,* and 'obscure' for *alogos,* and this has the additional effect of enhancing his opinion that Book X may be nothing more than a virtuoso display of mathematical reasoning. This question of the motivation of Book X is my reason for introducing this discussion of the book here. It has openly perplexed commentators for the past five hundred years or more, and practically everybody now cites Stevin's description of Book X as the "cross of mathematicians" (see Heath, quoting Loria, *TBEE* iii, 8–9, and for more details of the context of Stevin's remark, Knorr, CM, 41). Recent opinions on the book are:

We are prepared to face the possibility that there was no other point than to entertain [!] us with good logic [Taisbak, *CQ,* 58],

and

The true merit of Book X, and I believe it is no small one, lies in its being a unique [?] specimen of a fully elaborated deductive system of the sort that ancient philosophies of mathematics consistently prized [Knorr, CM, 60].

The classification of Book X is applied in Book XIII, as the examples quoted in the previous section have already illustrated. In this limited sense, the role of the book is explained. But Mueller articulates the incompleteness of this kind of answer:

One would, of course, prefer an explanation that invoked a clear mathematical goal intelligible to us in terms of our own notions of mathematics and which, under analysis, would lead univocally [?] to the reasoning in Book X. Unfortunately Book X has never been explicated successfully in this way, nor does it appear amenable to explication of this sort. Rather, Book X appears to be an expedient for dealing with a particular problem and at the same time a mathematical blind alley [Mueller, *PMDSEE*, 270–1].

Before embarking on a detailed description of Book X, it is worth giving a broad schematic sketch of its contents. With more than 115 propositions, some very long, the book is massive (more than a quarter of the bulk of the *Elements*), monolithic, and forbidding:

Book X does not make easy reading. ... Up to X 28 it goes fairly well, but when the existence proofs start with X 29 ... one does not see very well what purpose all of this is to serve. The author succeeded admirably [?] in hiding his line of thought ... [van der Waerden, *SA*, 172],

Simply put, Book X is a pedagogical disaster [Knorr, *CM*, 59],

and

Danger! Ne s'aventuer dans la lecture de ce livre qu'après une solide préparation [Introduction, by Itard, to the reprinting of Euclid–Peyrard, *OE*, p. xiv].

The subject matter, announced in the First Definitions, quoted at the beginning of this section, is (in)commensurability and (in)commensurability-in-square, and this is the *first time* that this topic of incommensurability has explicitly occurred in the *Elements*. The propositions can be divided into groups:

X 1–18: general properties of expressible lines and rectangles,
X 19–26: medial lines and rectangles,
X 27–35: constructions underlying binomials and apotomes,
X 36–41, 42–7, 48–53, 54–9, 60–5, 66–70, & 71–2: blocks of propositions dealing with each of the six types of additive irrational lines. They are described in X 36–41 and also, in a different geometrical configuration, in the Second Definitions following X 47,
X 73–8, 79–84, 85–90, 91–6, 97–102, 103–7, & 108–10: blocks of propositions, parallel to the previous, dealing with each of the six types of subtractive irrational lines. They are described in X 73–8 and also, in a different geometrical configuration, in the Third Definitions following X 84,
X 111–14: the relations between binomials and apotomes,

X 115: medials of medials, and

X 116 ... : material that is most probably interpolated. It includes two proofs, which will be discussed in Section 8.3, that the diagonal of a square is commensurable with the side.

My account will be based on the interpretation given in Taisbak, *CQ*, Chapter 2, 26–61. This book sets out a purely geometrical account which conforms in detail to the criteria set out above, in Section 1.2, and the author's quirky, idiosyncratic, witty style provides the perfect antidote to the relentless, repetitive monotony of the original. This treatment, like many of Taisbak's mathematical and historical opinions, corresponds to much in Knorr, CM (which is an elaboration of Knorr's earlier discussion in *EEE*, 278–85; also see Knorr, ETB), but while Knorr's descriptions and procedures can generally be translated into geometrical terms, the tendency of an incautious reader will be to interpret his notation arithmetically. A useful, short, clear, but highly arithmetised description is given in van der Waerden, *SA*, 168–72; and a long and detailed account, which develops simultaneously Book X and its application in Book XIII, can be found in Mueller, *PMDSEE*, 251–306, though additional complexity is introduced here both in the elaborate notation that is used and by the way the side and circumdiameter of the icosahedron is made the locus for problems that already arise, in their most acute form, in the much simpler configuration of the side and circumdiameter of the pentagon. I refer the reader to these accounts, especially to Taisbak and Knorr, for further details of the treatment that I shall present here.

No comprehensible description of Book X can avoid a massive reorganisation of Euclid's opaque and unhelpful order of presentation and I shall not dwell on this fact any further. I shall also set aside most of the textual issues: questions of whether some of the lemmas are interpolations or not, whether certain propositions are authentic, how some lacunae should be filled, and discussions of possible authorship. I am mainly concerned to present a geometrical description of the book, as we have it. Such is the nature of the subject matter that it imposes a much more formal approach than I have adopted hitherto.

5.2(b) *Expressible lines and areas*

DEFINITIONS (X Definitions 3–4)   There is an **assigned** line. An **area** is called **expressible** if it is commensurable with the square on the assigned line; a **line** is **expressible** if it is the side of an expressible square.

REMARKS

(i) The definitions and propositions will be cross-referenced to the Euclidean definitions and propositions from which they have been adapted, and they will not be further numbered. Remarks, parenthetic remarks, and other such items will not, unless otherwise indicated, be found in the *Elements*.

(ii) Euclid refers in Definition 3 to the assigned line and its square as 'the expressible'. I shall maintain the distinction between *an* expressible line (or area) and *the* expressible line (or square) by always calling the latter the assigned line (or square), and reserving the letter a (or a^2) for it.

(iii) Euclid's expression *hē to chōrion dunamenē* means 'the side of a square equal to', and it is abbreviated in Heath's translation to 'side'; see his notes on X Definition 4 and Proposition 54, *TBEE* iii, 13 and 119. I shall refer to it as the 'square side' and abbreviate it by '$\sqrt{}$'. I repeat: when it occurs here in the context of Greek mathematics, '$\sqrt{}$' should always be read as 'side' or 'square side', and never as 'square root'. It can be constructed using *Elements* II 14.

(iv) I shall use the geometric shorthand that was introduced in Section 1.2(e), developed thereafter, and described in B_{89} and B_{91}: $a, b, c, \ldots$ denote lines; $b.c$ the rectangle with sides b and c, and b^2 the square with side b; if B is an area, then $\sqrt{B}$ denotes the square side of B; if λ is an *arithmos* then λb (or λB) denotes λ concatenated copies of b (or B)—there is no problem about the ambiguous notation $\lambda b.c$ since clearly, by the configuration of the *Topics* proposition, $(\lambda b).c$ and $\lambda(b.c)$ are equal; and $\sqrt{(\lambda b.b)}$ is abbreviated to $\sqrt{\lambda}b$. Care should be exercised with this last notation, since $\sqrt{\lambda}b$ has no arithmetical sense. For example, $\sqrt[4]{\lambda}b$ has neither arithmetical nor geometric sense and will never be used; instead, the notation $\sqrt{\sqrt{\lambda}b}$ will be used for $\sqrt{(\sqrt{\lambda}b.b)}$, the square side of the rectangle with sides $\sqrt{\lambda}b$ and b. The line a will always denote the assigned line. The notation just described could be extended to include the typical expressible area, μa^2 for a fraction μ, but neither the proofs of the theory, nor the illustrations, need involve anything more than areas of the form λa^2 and their sides $\sqrt{\lambda}a$, where λ is an *arithmos*.

PROPOSITIONS (X 5–16)

(i) Any two expressible areas are commensurable. Conversely anything commensurable with an expressible area is expressible.

(ii) The sum and difference of two expressible areas is expressible.

(iii) Two expressible lines are commensurable if and only if their squares have a ratio equal to that of some square number to some square number. (For example, $\sqrt{2}a$ and $\sqrt{18}a$.)

(iv) If two expressible lines are incommensurable, then their squares will be commensurable, so they can also be described as commensurable-in-square-only. (For example, a and $\sqrt{2}a$.)

(v) A line commensurable with an expressible line is expressible; and a line commensurable-in-square with an expressible line is expressible.

(vi) The sum and difference of two commensurable expressible lines is expressible.

PROOFS Left as an exercise.

PROPOSITION (X 19) The rectangle contained by commensurable expressible lines is expressible.

PROOF See Fig. 5.4, where h and w are the commensurable expressible lines. By the *Topics* proposition, $h.w : h^2 = w : h$. Now w and h are supposed commensurable, so $h.w$ and h^2 are commensurable, so $h.w$ is expressible. QED

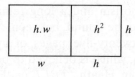

Fig. 5.4

REMARKS

(i) The *Topics* proposition is used almost sixty times in Book X; for a discussion, see Knorr, *EEE*, 259–61.

(ii) The letters h and w stand for height and width. The usual translation of *platos* is breadth, which suggests the letter b, but this will be reserved for later use.

(iii) The height h will later become any expressible line, and it is sometimes of interest to see if there is any difference between the cases when it is commensurable or incommensurable with the assigned line a.

PROPOSITION (X 20) If an expressible area is applied to an expressible height, its width will be expressible and commensurable with its height.

PROOF See Fig. 5.4. If $h.w$ and h^2 are both expressible, then they are commensurable, so h is commensurable with w, so it is expressible.
QED

REMARK There are two basic operations which transform an area into a line: either to take its square side or to apply the area, typically a square or rectangle, to a given height h and to take its width w, as in this proposition. This can be performed using *Elements* I 43–5; see Fig. 1.2, above. These operations are used repetitively throughout Book X.

5.2(c) *Medial lines and areas*

PROPOSITION (X 21) A rectangle contained by incommensurable expressible lines is *alogos*.

PROOF See Fig. 5.4, where h and w are now incommensurable expressible lines. So $h.w$ and w^2 are now incommensurable, so $h.w$ is not an expressible area. QED

DEFINITION (X 21 & 23/24) An **area** is called **medial** (*mesos*) if it is equal to a rectangle contained by incommensurable expressible lines. A **line** is called **medial** if it is the side of a medial square.

REMARKS

(i) Book X contains many unnumbered lemmas, porisms, definitions, and remarks, which will be referred to here by their adjacent propositions. So X 23/24 is a porism and a long remark, some or all of which may be interpolated, to be found between X 23 and 24. The critical edition contains some twenty-five pages of appendix material, Euclid-Stamatis, *EE* iii, 211–36, which is believed to be interpolated and most of which is not translated in Heath, *TBEE* iii. For a brief discussion, see Knorr, *EEE*, 269.

(ii) Euclid does not explicitly define a medial area (see X 21) but he uses the idea from Porism 23/24 onwards.

PROPOSITION (X 22) If a medial area is applied to an expressible height, its width will be expressible and incommensurable with the height.

PROOF Let h be expressible and $h.w$ be medial. Then $h.w = b.c$ where b and c are incommensurable lines. So, by the usual manipulation with proportions, $h:c = b:w$, so, by VI 22, $h^2:c^2 = b^2:w^2$. Since h^2 and c^2 are expressible areas, they are commensurable, so w^2 is commensurable with the expressible area b^2, so it is expressible. But w is incommensurable with h since $h.w$ is not expressible. QED

PROPOSITION (X 23 & 23/24)
  (i) An area commensurable with a medial area is medial.
  (ii) A line commensurable with a medial line is medial.

PROOF See Fig. 5.5.

(i) Let b^2 be (equal to) a medial area, and let c^2 be commensurable with b^2. Apply both to the expressible line h, so $b^2 = v.h$ and $c^2 = w.h$ where v is expressible and incommensurable with h (X 22) and w is commensurable with v. So w is expressible and incommensurable with h, so c^2 is medial.

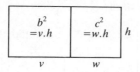

Fig. 5.5

(ii) Let b be medial and let c be commensurable with b; then b^2 is medial and c^2 is commensurable with b^2. So, by what we have just proved, c^2 is medial, so c is medial. QED

REMARKS

(i) Two medial lines may be commensurable, commensurable-in-square-only, or incommensurable-in-square. For example $\sqrt{\sqrt{2}a}$ and $\sqrt{\sqrt{32}a}$, $\sqrt{\sqrt{2}a}$ and $\sqrt{\sqrt{8}a}$, and $\sqrt{\sqrt{2}a}$ and $\sqrt{\sqrt{3}a}$ respectively.

(ii) Two medial areas may be commensurable or incommensurable: in Fig. 5.5, we can have b^2 and c^2 medial, h, v, and w expressible, and v and w either commensurable or incommensurable. For example, $h = \sqrt{2}a$, $v = \sqrt{3}a$, and either $w = \sqrt{12}a$ for the first case, or $w = \sqrt{5}a$ for the second.

(iii) If two medial areas are incommensurable, it makes no geometrical sense to say that they are 'commensurable-in-square'. If we give an arithmetised interpretation, in terms of the square and fourth roots of fractions, we see that medial lines are of the form $\sqrt[4]{\lambda a}$, and medial areas are $\sqrt{\lambda a.a}$, where λ is not a perfect square, and so any two medial areas are indeed arithmetically commensurable-in-square. It is possible to articulate such a result geometrically, using X 22: If two incommensurable medial areas are applied to an expressible height, then their breadths will be medial and commensurable-in-square-only. The fact that Book X contains no suggestion of this kind of result is yet another indication that its procedures are not being conceived in any arithmetised form.

PROPOSITION (X 24) The rectangle contained by commensurable medial lines is medial.

PROOF We adapt the proof of X 19 above. If h and w are commensurable medial lines, then $h.w$ and h^2 are commensurable and h^2 is medial. So $h.w$ is medial. QED

EXAMPLE The lines $\sqrt{\sqrt{2}a}$ and $\sqrt{\sqrt{32}a}$ contain the medial area $\sqrt{8a.a}$.

PROPOSITION (X 25) The rectangle contained by medial lines that are commensurable-in-square-only is either expressible or medial.

REMARK Euclid's proof can be broken into conceptually simpler steps

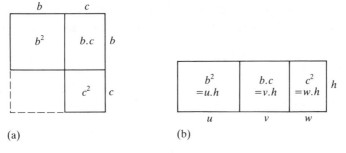

Fig. 5.6

by inserting the following:

LEMMA (X 53/54) A rectangle is the mean proportional between the squares on its sides.

PROOF OF THE LEMMA See Fig. 5.6(a), in which $b^2:b.c = b:c = b.c:c^2$, by two applications of the *Topics* proposition. QED

PROOF OF THE PROPOSITION See Fig. 5.6; let b and c be medial lines, commensurable-in-square-only, and apply b^2, $b.c$, and c^2 to an expressible line h, as in Fig. 5.6(b). Then, by X 22, u and w are expressible and each is incommensurable with h; but since

$$u:w = h.u:h.w = b^2:c^2,$$

they are commensurable with each other. Now, by the Lemma, $b.c$ is the mean proportional between b^2 and c^2, so v will be the mean proportional between u and w, so $v^2 = u.w$, so v^2 is expressible, so v is expressible. Finally, since v can be either commensurable or incommensurable with h, $v.h = b.c$ can be either expressible or medial. QED

REMARKS

(i) This proposition is illustrated at X 27 and 28 by what are, in effect, abstractions of the following two examples: the medial lines $\sqrt{\sqrt{2}a}$ and $\sqrt{\sqrt{8}a}$ are commensurable-in-square-only, and contain a rectangle equal to $2a.a$, which is expressible, while the medial lines $\sqrt{\sqrt{6}a}$ and $\sqrt{\sqrt{24}a}$ are also commensurable-in-square-only but they contain a rectangle equal to $\sqrt{12}a.a$, which is medial.

(ii) The medial is Euclid's first example of an *alogos*; see X 21. However, in his description of Book X, Taisbak does not class the medials among the *alogoi*; see his Definition 216 at Taisbak, *CQ*, 35.

EXERCISES

(i) If an expressible area is applied to a medial height, its width is medial and commensurable-in-square-only with the height.

(ii) If a medial area is applied to a medial height, its width is medial and either commensurable or commensurable-in-square-only with the height.

(iii) The sequence of lines $\sqrt{2}a$, $\sqrt{\sqrt{2}a}$, $\sqrt{\sqrt{\sqrt{2}a}}$, ... is an infinite sequence of *alogoi* lines, no two of which are either commensurable, or commensurable-in-square. (This result appears as X 115, very probably a later interpolation.)

5.2(d) *Sums and differences*

PROPOSITION (X 36 & 73) Neither the sum nor the difference of two incommensurable expressible lines is expressible.

PROOF See Fig. 5.7(a), on the left for sums, the right for differences. Let x and y be incommensurable expressible lines, with x greater than y, and write

$$q_+ = x + y, \qquad q_- = x - y.$$

Then

$$q_+^2 = x^2 + 2xy + y^2, \qquad q_-^2 + 2xy = x^2 + y^2.$$

Now $x^2 + y^2$ is expressible and $x.y$ is medial; hence $2x.y$ is medial and so is incommensurable with $x^2 + y^2$. Hence neither q_+^2 nor q_-^2 can be commensurable with the expressible area $x^2 + y^2$, so neither is expressible. QED

REMARK The notational advice of writing q_+ and q_- is very useful for exhibiting the parallel cases of addition and subtraction, and it will be frequently adopted, sometimes implicitly, in many subsequent propositions and comments. But although the arithmetised manipulations may be presented in a uniform way in both cases, the geometrical meaning is embodied in distinct though parallel figures. Figures 5.7(a) and (b) set out such a basic set of parallel figures, on the left for addition, on the right for subtraction, and the lowest figure 5.7(c) is common to both columns. The details of these figures will be explored and exploited in the subsequent propositions.

PROPOSITION (X 26) Neither the sum, nor the difference, of two medial areas is expressible.

PROOF by contradiction. Suppose x^2 and y^2 are (equal to) two medial areas such that $x^2 \pm y^2$ is expressible. Apply these areas to an expressible height h (see Figs. 5.7(a) and (b)), so that

$$x^2 = h.u, \quad y^2 = h.v, \quad \text{and} \quad (x^2 \pm y^2) = h.(u \pm v).$$

Then u and v are each expressible and incommensurable with h (X 22), while $u \pm v$ is expressible and commensurable with h (X 20). So $(u \pm v)$ and u are incommensurable expressible lines whose difference or sum v is also expressible, which contradicts the previous proposition. QED

REMARK Euclid proves only the case of the difference of two medial lines in X 26, and part of his proof is duplicated later, in the proof of X 36.

EXERCISES
(i) Prove that neither the sum, nor the difference, of two expressible lines is medial. (This result is not found, in this simple form, in the *Elements*, but it does follow from later results, as is pointed out in the remarks at X 72/73 and 111/112.)

(ii) Prove that neither the sum nor the difference of two incommensurable medial areas is medial.

BASIC PROPOSITION ON ADDITION (Assembled out of VI 28, part of X 17, and X 59/60.) Let $q = x + y$. If q^2 is applied to a line h, $q^2 = h.w$, then w can be written

$$w = b + c \quad \text{with} \quad b > c, \quad h.b = x^2 + y^2, \quad \text{and} \quad h.c = 2x.y.$$

Conversely, given $w = b + c$ with $b > c$, then q, the square side of $w.h$, can be written

$$q = x + y \quad \text{with} \quad x^2 + y^2 = h.b \quad \text{and} \quad 2x.y = h.c.$$

PROOF See Fig. 5.7$_+$. If $q = x + y$, then $q^2 = x^2 + y^2 + 2x.y$ (Fig. 5.7(a$_+$)). Now construct adjacent rectangles $h.u = x^2$, $h.v = y^2$, and $h.c = 2x.y$ (Fig. 5.7(b$_+$)). Since $2x.y + (x - y)^2 = x^2 + y^2$ (Fig. 5.7(a$_-$)), we have $2x.y < x^2 + y^2$, so $b < c$. Conversely, given

$$h.w = q^2 \quad \text{with} \quad w = b + c \quad \text{and} \quad b > c,$$

we first find how to split b into $u + v$ with

$$h.u = x^2, \quad h.v = y, \quad \text{and} \quad x.y = \tfrac{1}{2}h.c.$$

Since $x.y$ is the mean proportional between x^2 and y^2 (see X 53/54, above) so $\tfrac{1}{2}h.c$ must be the mean proportional between $h.u$ and $h.v$, so $c/2$ must be the mean proportional between u and v, so $u.v = (c/2)^2$. Hence we must solve the geometric problem $u + v = b$, $u.v = (c/2)^2$. This is precisely the problem of elliptical application of areas, introduced in

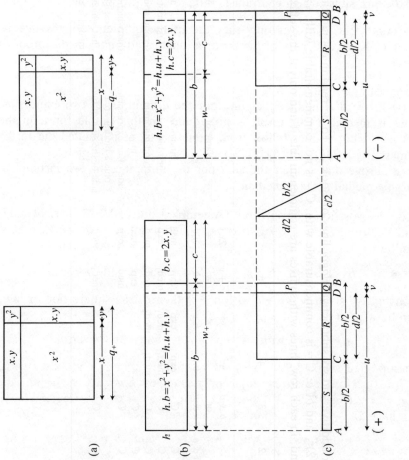

Fig. 5.7

TABLE 5.2. The six additive and six subtractive *alogoi* lines of Book X. To be read in conjunction with Fig. 5.7: $u+v=b$, $u.v=(c/2)^2$, and $b\mathcal{C}c$; C = is commensurable with, $\mathcal{C}$ = is not commensurable with; exp = expressible, med = medial, and, in each row, lines with suffixes are commensurable if and only if the suffixes are equal.

column row	1	2	3	4	5	6	7	8	9
Class of $w=b\pm c$ $b\mathcal{C}c$	$h\&b$	$h\&c$	$b\&d$ or $u\&v$	x^2 = $h.u$	y^2 = $h.v$	x^2+y^2 = $h.b$	$2x.y$ = $h.c$	Name of $g_+=x+y$	Name of $q_-=x-y$
1	C	$\mathcal{C}$	C	exp	exp	exp	med	binomial	apotome
2	$\mathcal{C}$	C	C	med$_1$	med$_1$	med$_1$	exp	first binomial	first apotome of a medial
3	$\mathcal{C}$	$\mathcal{C}$	C	med$_1$	med$_1$	med$_1$	med$_2$	second binomial	second apotome of a medial
4	C	$\mathcal{C}$	$\mathcal{C}$			exp	med	major	minor
5	$\mathcal{C}$	C	$\mathcal{C}$			med	exp	side of an expressible plus a medial area	that which produces with an expressible area a medial
6	$\mathcal{C}$	$\mathcal{C}$	$\mathcal{C}$			med$_1$	med$_2$	side of the sum of two medial areas	that which produces with a medial area a medial whole

Book X at 16/17 and 17, which starts:

If there are two unequal straight lines [b and c] and to the greater [b] there be applied a [rectangle] equal to the fourth part of the square on the less [i.e. $u.v = \frac{1}{4}c^2$] and deficient (*elleipon*) by a square [i.e. $u + v = b$] and . . . .

The solution of the problem has already been given by Euclid at VI 28. It requires a condition (*diorismos*), stated there, which here means that $(c/2)^2$ must not be greater than $(b/2)^2$, and so is precisely the hypothesis of our proposition. We therefore follow the procedure of VI 28, as set out in Fig. 5.7(c): Bisect AB = b at C, construct the square on BC = $(b/2)^2$ and the square with side CD = $d/2$, where $(d/2)^2 = (b/2)^2 - (c/2)^2$, so that $b/2$ is the hypotenuse of the right-angled triangle with sides $b/2$, $c/2$, and $d/2$. But then

$$P + Q + R = BC^2 - DC^2 = (b/2)^2 - (d/2)^2 = (c/2)^2$$

while

$$P + Q + R = S + R \quad \text{since} \quad P + Q = S,$$

and hence S + R is the required elliptical application of $(c/2)^2$ to AB. So AD = u, DB = v is the required splitting of AB = b. Finally, we construct $x^2 = h.u$ and $y^2 = h.v$ on the same diagonal, as in Fig. 5.7(a_+), and verify immediately that $q = x + y$ gives the required decomposition of the square. QED

BASIC PROPOSITION ON SUBTRACTION Let $q = x - y$. If q^2 is applied to a line h, $q^2 = h.w$, then w can be written

$$w = b - c \quad \text{with} \quad b > c, \qquad h.b = x^2 + y^2 \quad \text{and} \quad h.c = 2x.y.$$

Conversely, given $w = b - c$ with $b > c$, then q, the square side of $w.h$, can be written

$$q = x - y \quad \text{with} \quad x^2 + y^2 = h.b \quad \text{and} \quad 2x.y = h.c.$$

PROOF See Fig. 5.7_. The details are left as an exercise.

REMARKS

(i) The operations of passing back and forth between a rectangle of the form $h.(b \pm c)$ and an equal square of the form $(x \pm y)^2$ are used repetitively X 54–9 and 60–5 (for sums) and X 91–6 and 97–102 (for differences). They are fundamental to an understanding of Book X; see B_{89-91} in Section 4.5(b).

(ii) The line $d = \sqrt{(b^2 - c^2)}$ (see Fig. 5.7(c)) plays an important role in the theory. Euclid expresses the sentence 'the square on b is greater than the square on c by the square on d' by '*he b tēs c meizon dunatai tēi d*' (see the note in Heath, *TBEE* iii, 43). Following Taisbak, I shall refer to d as the 'square difference' of b and c.

5.2 Elements X

(iii) This is an appropriate place to describe an arithmetised formula that translates these basic propositions and which has even been called "the quintessence of Book X" (see Pappus–Thomson & Junge, *CPBXEE*, 23 & Knorr, *CM*, 64 n. 32). To maintain the parallel, write $b = \beta h$, $c = \gamma h$, so $d = \delta h$ where $\delta^2 = \beta^2 - \gamma^2$, and now interpret β, γ, and δ, as real numbers or algebraic symbols. Then

$$\sqrt{(\beta \pm y)} = \sqrt{(\tfrac{1}{2}\beta + \tfrac{1}{2}\sqrt{(\beta^2 - \gamma^2)})} \pm (\tfrac{1}{2}\beta - \tfrac{1}{2}\sqrt{(\beta^2 - \gamma^2)})$$
$$= \sqrt{(\tfrac{1}{2}\beta + \tfrac{1}{2}\delta)} \pm \sqrt{(\tfrac{1}{2}\beta - \tfrac{1}{2}\delta)}.$$

For one proof of this identity write

$$\sqrt{(\beta + \gamma)} = \xi + \eta \quad \text{and} \quad \sqrt{(\beta - \gamma)} = \xi - \eta,$$

where $x = \xi h$ and $y = \eta h$. Then

$$2\xi = \sqrt{(\beta + y)} + \sqrt{(\beta - y)} \quad \text{and} \quad 2\eta = \sqrt{(\beta + \gamma)} - \sqrt{(\beta - \gamma)}.$$

Therefore

$$4\xi^2 = 2\beta + 2\sqrt{(\beta^2 - \gamma^2)} \quad \text{and} \quad 4\eta^2 = 2\beta - 2\sqrt{(\beta^2 - \gamma^2)},$$

and hence

$$\xi = \sqrt{(\tfrac{1}{2}\beta + \tfrac{1}{2}\delta)} \quad \text{and} \quad \eta = \sqrt{(\tfrac{1}{2}\beta - \tfrac{1}{2}\delta)},$$

as required.

It is possible to express these manipulations geometrically in the style of Fig. 5.7; but this derivation, even expressed geometrically, is very different from Euclid's. (Note how this proceeds by manipulating $\sqrt{(\beta + \gamma)} = \xi + \eta$ and $\sqrt{(\beta - \gamma)} = \xi - \eta$ simultaneously, while the Euclidean treatment keeps the two cases strictly separate, parallel, and independent.) It is possible to derive the two arithmetised results separately in the particular case where β and γ each have the form of the square root of a fraction, and ξ and η belong to some restricted class of real numbers, but this kind of derivation seems to have little to do with Euclidean techniques. So the correspondence between Euclid's manipulations and these formulae seems to be much less close than is sometimes asserted or assumed.

PROPOSITION (17 & 18) With the same notations as the previous propositions, u is commensurable with v if and only if b is commensurable with the square difference d.

PROOF See Fig. 5.7(c); $u = b/2 + d/2$, $v = b/2 - d/2$, and hence $b = u + v$, $c = u - v$. QED

REMARK We see from this that $x^2 = \tfrac{1}{2}h.(b + d)$, $y^2 = \tfrac{1}{2}h.(b - d)$, and $x^2 - y^2 = h.d = h.\sqrt{(b^2 - c^2)}$, but these geometrical identities do not seem to appear in Book X.

5.2(e) *Binomials and apotomes*

DEFINITIONS (X 36 & 72) The sum of two incommensurable expressible lines is neither expressible nor medial; it will be called a **binomial** (*ek duo onomatōn*). Similarly the difference of two incommensurable expressible lines is neither expressible nor medial; it will be called an **apotome** (*apotomē*).

PROPOSITION (X 60 & 97) If the square on a binomial (or apotome) is applied to an expressible height, its width is a binomial (or apotome).

PROOF See Fig. 5.7. If x and y are expressible lines, then x^2 and y^2 are expressible areas, so their sum is expressible, so $h.b$ is expressible, so b is expressible and commensurable with h. If now x and y are incommensurable expressible lines, then $x.y$ is medial, so $2x.y = h.c$ is medial, so c is expressible and incommensurable with h. So b and c are incommensurable expressible lines. Hence $b + c$ is a binomial, and $b - c$ is an apotome. QED

REMARK This result is summarised in the top line of Table 5.2, reading from right to left: columns 9 and 10 imply 4, 5, 6, and 7, in that order. Columns 1, 2, and 3 will now be explained.

CENTRAL QUESTION When does the converse of this result hold?

This is answered by:

PROPOSITION (Converse of X 54 & 91) If the square side of the rectangle with width $w = b + c$, a binomial (or $b - c$, an apotome), and height h is a binomial $x + y$ (or an apotome $x - y$), then h, b, and d will be mutually commensurable. Such a binomial is called a **first binomial** (or **first apotome**) in the classification below.

PROOF If $(b \pm c).h = (x \pm y)^2$, as supposed, then $x^2 = h.u$ and $y^2 = h.v$, as in Fig. 5.7. So $h.u$ and $h.v$ are expressible, so h, u, and v are mutually commensurable, so, by X 17 and 18, h, b, and d are mutually commensurable. QED

REMARKS
(i) This result is not explicitly given, but it does seem to underly the organisation of Book X.

(ii) These last two propositions apply directly to the extreme and mean ratio. We take the construction in *Elements* II 11:

To cut a given [expressible] line [h] so that the rectangle contained by the whole and one of the segments [$h.s$] is equal to the square on the remaining segment [g^2].

Here s is the lesser segment, g is the greater. Then, by these last two propositions, if g is an apotome, s will be an apotome, and hence s must be a first apotome. Now we check that g is, indeed, an apotome. This follows immediately from XIII 1, which proved that $(g+h/2)^2 = 5h^2$, so $(g+h/2)$ and $h/2$ are incommensurable expressible lines whose difference is g.

This proof is found in *Elements* XIII 6. Euclid enunciates that s and g are apotomes, but he also deduces within the proof, as above, that s is a first apotome with respect to h. In fact, the subclassification of the binomials and apotomes, to be given below, is internal to the mechanisms of Book X and it depends on the particular choice of h in a way that I shall describe.

ABBREVIATIONS I shall sometimes write 'is commensurable with' as C, and 'is not commensurable with' as $¢$.

DEFINITIONS (X 47/48 & 84/85) Let, as now usual, h be an expressible line, b and c be incommensurable expressible lines with $b > c$, and $d = \sqrt{(b^2 - c^2)}$ be their square difference. Then there are the following six mutually exclusive classes of binomials $b+c$ and apotomes $b-c$:

1st:	hCb	&	$h¢c$	&	bCd
2nd:	$h¢b$	&	hCc	&	bCd
3rd:	$h¢b$	&	$h¢c$	&	bCd
4th:	hCb	&	$h¢c$	&	$b¢d$
5th:	$h¢b$	&	hCc	&	$b¢d$
6th:	$h¢b$	&	$h¢c$	&	$b¢d$

In the kth class, $b+c$ will be called a kth *apotome* and $b-c$ a kth *binomial* with respect to h.

REMARKS

(i) The six classes of binomials depend on the particular choice of the assigned line h. Clearly, if h is replaced by a line commensurable with it, the classes are not changed relative to this new height; but if it is incommensurable, the classes will be permuted. (Any permutation of the first three classes can be induced, and then classes 4, 5, and 6 will be permuted in the same way.)

(ii) It is usually proposed that Euclid's constructions, namely:

PROPOSITION (X 48–53 & 85–90) To find the kth binomial and apotome

are demonstrations that each of the six possibilities can occur; but that would be satisfied by a list of examples such as I shall give below. In fact Euclid describes the form of every example in every class, though one of the constructions, in the second lemma at X 28/29, is inadequate. (This observation is due to Knorr, CM, 57–8, which should be consulted for further details.) I shall not summarise Euclid's proofs, which depend on the construction of X 27–35, but will simply enumerate six examples. The first and fifth examples are based on the lesser and greater segments of the line $2h$ cut in extreme and mean ratio, analysed above, and the fourth on the diagonal and side of a pentagon inscribed in a circle of radius $2h$, as will be explained below.

1st:	$b = 3h$	$c = \sqrt{5}h$	$d = 2h$.
2nd:	$b = \sqrt{12}h$	$c = 3h$	$d = \sqrt{3}h$.
3rd:	$b = \sqrt{8}h$	$c = \sqrt{6}h$	$d = \sqrt{2}h$.
4th:	$b = 10h$	$c = \sqrt{20}h$	$d = \sqrt{80}h$.
5th:	$b = \sqrt{5}h$	$c = h$	$d = 2h$.
6th:	$b = \sqrt{10}h$	$c = \sqrt{8}h$	$d = \sqrt{2}h$.

(iii) It would be possible to extend this kind of classification further by splitting each of classes 4, 5, and 6 into two: if $b\mathcal{C}d$, then we can have either $c C d$ or $c \mathcal{C} d$. All of the examples given above illustrate the case $c C d$; the following three have $c \mathcal{C} d$.

4th:	$b = 3h$	$c = \sqrt{7}h$	$d = \sqrt{2}h$.
5th:	$b = \sqrt{6}h$	$c = h$	$d = \sqrt{2}h$.
6th:	$b = \sqrt{5}h$	$c = \sqrt{3}h$	$d = \sqrt{2}h$.

The fact that Euclid does not explore these possibilities illustrates further how the subclassification of binomials and apotomes is more than a gratuitous formal exercise, and how it is not important in its own right but is rather an intermediate step towards a further construction which we shall now explore.

(iv) This description of kth binomials and apotomes is incorporated into rows 1–6, columns 1–3 of Table 5.1, which we now consider in close detail.

5.2(f)   *The six additive and subtractive alogoi lines*

REMARKS ON TABLE 5.2

(i) Table 5.2, which must be read in conjunction with Fig. 5.7, summarises much of the information about the six additive and six subtractive *alogoi* lines of Book X. Note the following further abbreviations and notations: 'exp' for expressible, 'med' for medial, and medial lines within a row are commensurable if and only if they have the same suffix.

(ii) The following immediate implications between the columns of the table have already been established:

$$1 \Leftrightarrow 6 \quad \text{and} \quad 2 \Leftrightarrow 7 \quad (X\,19\text{–}22), \quad \text{and}$$
$$4\ \&\ 5 \Rightarrow 6 \text{ in rows 1–3.}$$

(iii) Euclid's opaque treatment of the subject corresponds to reading the rows of the table from right to left, starting (in X 29–35, 36–41, & 73–8) with the constructions and definitions of the strange group of lines described in columns 8 and 9. We shall follow the different and much clearer process of exploring the table from left to right, row by row, and finishing by proving that lines q_+ and q_- give twelve new kinds of *alogoi* lines. Note, then, that b and c are always incommensurable expressible lines, divided into the six classes relative to a given expressible line h. Columns 4–7 then refer to the constructions of Fig. 5.7.

PROPOSITION (row 1: X 36, 54, 74 & 91) If $b+c$ is a first binomial (or $b-c$ is a first apotome) with respect to h, then the square side $x+y$ of $h.(b+c)$ is a binomial (or the square side $x-y$ of $h.(b-c)$ is an apotome).

PROOF Columns 1–3 specify the condition for the first binomial or apotome. The results in row 1, columns 4–7 follow immediately. So x and y are expressible (columns 4 and 5) and incommensurable (column 7), so q_+ is a binomial and q_- is an apotome. QED

PROPOSITION (row 2: X 37, 55, 75 & 92) If $b \pm c$ is a second binomial or apotome then x and y are medial, commensurable-in-square-only, and contain an expressible rectangle. Such a line $x+y$ is called a **first bimedial**, and $x-y$ is a **first apotome of a medial**.

PROOF Since h and b are incommensurable (column 1) $h.b = x^2 + y^2$ is medial (column 6). Since u and v are commensurable (column 3) $h.u = x^2$ and $h.v = y^2$ will be commensurable (columns 4 and 5); neither square can be expressible since their sum is medial, so x^2, y^2, and $x^2 + y^2$ are medial and mutually commensurable. So x and y are medial and commensurable-in-square-only, since $x.y$ is expressible (X 24). QED

PROPOSITION (row 3: X 38, 56, 75, & 93) If $b \pm c$ is a third binomial or apotome, then x and y are medial, commensurable-in-square-only, and contain a medial rectangle. Such a line $x+y$ is called a **second bimedial**, and $x-y$ is a **second apotome of a medial**.

PROOF We deduce that x^2, y^2, and $x^2 + y^2$ are medial and mutually commensurable, as in the previous proposition, but $x^2 + y^2 = h.b$ and $2x.y = h.c$ are incommensurable because b and c are, as always, incommensurable. Finally, since $x.y$ is medial, x and y are commensurable-in-square-only. QED

PROPOSITION (row 4: 39, 57, 76, & 94) If $b \pm c$ is a fourth binomial or apotome, then x and y are incommensurable-in-square, and such that $x^2 + y^2$ is expressible and $x.y$ is medial. Such a line $x+y$ is called a **major**, and $x-y$ is a **minor**.

PROOF   Clear from row 4.

REMARKS

(i) In rows 4–6, the lines x and y cannot be further described within the classification of Book X. The next remark will illustrate one kind of case; others can easily be devised.

(ii) Write $b = \beta h$, $c = \gamma h$. I described earlier, in Section 5.2(d), the arithmetised identity

$$\sqrt{(\beta \pm \gamma)h} = \sqrt{(\tfrac{1}{2}\beta + \tfrac{1}{2}\delta)h} \pm \sqrt{(\tfrac{1}{2}\beta - \tfrac{1}{2}\delta)h}, \quad \text{where} \quad \delta = \sqrt{(\beta^2 - \gamma^2)}.$$

Let us apply this to the pentagon inscribed in a circle of expressible radius $r = 4h$. We then have

$$\text{diagonal} = \sqrt{(10 + 2\sqrt{5})h} = \sqrt{(5 + 2\sqrt{5})h} + \sqrt{(5 - 2\sqrt{5})h} = x + y$$
$$\text{side} = \sqrt{(10 - 2\sqrt{5})h} = \sqrt{(5 + 2\sqrt{5})h} - \sqrt{(5 - 2\sqrt{5})h} = x - y,$$

so $x = \sqrt{(5 + 2\sqrt{5})h}$, $y = \sqrt{(5 - 2\sqrt{5})h}$, and hence $x^2 + y^2 = 10h^2$ is a rational area and $x.y = \sqrt{5}h^2$ is a medial area. Therefore the diagonal is an example of a major line, and the side of a minor line.

Euclid's proof of this result, given in XIII 11, contains nothing of this kind of arithmetised reasoning. He argues (also see Fig. 5.1) that since $A_1A_2B_3$ is equiangular with A_1FA_2, $A_1A_2^2 = A_1F . A_1B_3$ (see X 32/33). Now a metrical evaluation, which he does in detail (some parts of this were quoted and discussed in Section 4.5(c)), shows that A_1F is a fourth binomial with respect to the expressible diameter A_1B_3; hence the side A_1A_2 is a minor. He does not consider the diagonal.

PROPOSITION (Row 5: X 40, 58, 77, & 95)   If $b \pm c$ is a fifth binomial or apotome, then x and y are incommensurable-in-square and such that $x^2 + y^2$ is medial and $x.y$ is expressible. Such a line $x + y$ is called the **side of an expressible plus a medial area**, and $x - y$ is **that which produces with an expressible area a medial whole**.

PROOF   Clear, even to the description of the names, from row 5.   QED

PROPOSITION (Row 6: X 41, 59, 78, & 96)   If $b \pm c$ is a sixth binomial or apotome, then x and y are incommensurable-in-square and such that $x^2 + y^2$ and $x.y$ are both incommensurable and medial. Such an $x + y$ is called the **side of a sum of two [incommensurable] medial areas**, and $x - y$ is **that which produces with a medial area an [incommensurable] medial whole**.

PROOF   We need only verify that $x^2 + y^2 = h.b$ and $2x.y = h.c$ are incommensurable; but b is, as always, incommensurable with c.   QED

REMARK   The next cycle of results works up to the result that the twelve

PROPOSITION (X 36–42, 72/73, & 111/112) Each of the lines $x \pm y$ of columns 8 and 9 is *alogos*, and none of them is medial.

PROOF Consider first rows 1, 2, 4, and 6; in each case $q^2 = (x^2 + y^2) \pm 2x.y$ is the sum or difference of an expressible and a medial area (columns 6 and 7), so it is neither expressible nor medial.

Now consider rows 3 and 6. Here q^2 is the sum of two incommensurable medial areas, so it is neither expressible nor medial (X 26 and the exercise on incommensurable medial lines, above). QED

PROPOSITION (X 42–7 & 79–84) Each of the lines $q = x \pm y$ can be divided in only one way into a sum or difference of appropriate kinds of lines x and y.

We use:

LEMMA (X 41/42) If $x + y = x_1 + y_1$ with $x > y$, $x_1 > y_1$, and $x > x_1$, then $x^2 + y^2 > x_1^2 + y_1^2$.

PROOF Set $x + y = 2s$, $x - y = 2d$, and $x_1 - y_1 = 2d_1$. Then $x = s + d$ and $x_1 = s + d_1$ and so, since $x > x_1$, $d > d_1$. Now, by II 5,

$$x.y + d^2 = s^2 \qquad x_1.y_1 + d_1^2 = s^2,$$

so $x.y < x_1.y_1$. Hence, since

$$x^2 + y^2 + 2x.y = x_1^2 + y_1^2 + 2x_1.y_1,$$

we see that $x^2 + y^2 > x_1^2 + y_1^2$. QED

PROOF OF THE PROPOSITION Consider first the case of the binomial. Suppose $q = x + y = x_1 + y_1$, where x and y are pairs of incommensurable expressible lines with $x > y$ and $x_1 > y_1$. If x and x_1 are not equal, we may suppose that $x > x_1$. Then

$$x^2 + y^2 + 2x.y = x_1^2 + y_1^2 + 2x_1.y_1$$

and, by the lemma, $(x^2 + y^2) > (x_1^2 + y_1^2)$. Hence

$$(x^2 + y^2) - (x_1^2 + y_1^2) = 2x_1.y_1 - 2x.y.$$

But the left-hand side is expressible, while the right-hand side is not (X 26).

A similar argument applies to an apotome. It also applies to any q in rows 2, 4, and 5, since one of the entries in columns 6 and 7 is an expressible area and the other is a medial. However, the following argument, given by Euclid for rows 3 and 6, will in fact deal with all cases in rows 2–6:

Let $q = x \pm y$ be any of these other *alogoi* lines from columns 8 and 9, rows 2–6, and suppose that

$$q = x \pm y = x_1 \pm y_1 \quad \text{with} \quad x > y, \quad \text{and} \quad x > x_1 > y_1.$$

Express

$$q^2 = h.w \quad \text{where} \quad w = b \pm c, \quad h.b = x^2 + y^2, \quad h.c = x.y, \quad b > c,$$

and

$$q^2 = h.w_1 \quad \text{where} \quad w_1 = b_1 \pm c_1, \quad \text{etc.}$$

Then

$$h.w = h.b \pm h.c = h.b_1 \pm h.c_1 \quad \text{with} \quad b > b_1,$$

so

$$w = b \pm c = b_1 \pm c_1 \quad \text{with} \quad b > b_1.$$

But this gives two different decompositions of the binomial or apotome w, which is impossible. QED

PROPOSITION (X 66 & 103) Let $w = b + c$ be a kth binomial (or $w = b - c$ a kth apotome) with respect to h, and let w_1 be commensurable with w. Then w_1 is a kth binomial (or a kth apotome) divided in the same ratio as w.

PROOF Let $w : w_1 = b : b_1$; then it will also be equal to $(w - b) : (w_1 - b_1)$, i.e. $c : c_1$. Hence, if w and w_1 are commensurable, so are b and b_1, c and c_1, and d and d_1. Hence w_1 is a binomial of the same class as w. QED

PROPOSITION (X 66–70 & 103–7) Any line commensurable or commensurable-in-square with any of the *alogoi* lines of columns 8 and 9 is the same kind of line, and is divided in the same ratio.

PROOF If $q_1 = x_1 \pm y_1$ is commensurable or commensurable-in-square with $q = x \pm y$, then $q_1^2 = h.(b_1 \pm c_1)$ is commensurable with $q^2 = h.(b \pm c)$. So $b_1 \pm c_1$ is commensurable with $b \pm c$, so is the same class of binomial or apotome as $b \pm c$ so, by the very procedure we have been describing, q_1 will be the same kind of irrational as q, and will be divided in the same ratio. QED

REMARK Euclid does not adopt this kind of proof. Instead he uses repetitively the argument of the previous proposition. Perhaps for this reason, he does not consider the commensurable-in-square case of this proposition, though he does appear to use it in XIII 18. (This observation is due to Mueller, *PMDSEE*, 283, 299; also see Knorr, CM, n. 38.)

PROPOSITION (X 111) No binomial is commensurable with any apotome.

PROOF by contradiction. Suppose $q = x + y = x_1 - y_1$ where x and y and x_1 and y_1 are incommensurable expressible lines. Write $q^2 = w.h$, where h is expressible; then $w = b + c = b_1 - c_1$, where $b + c$ is a first binomial and $b_1 - c_1$ is a first apotome. Hence b and b_1 are both commensurable with h, and c and c_1 are both incommensurable with h. But, since $(b_1 - b_2) = c_1 + c$, then $c = (b_1 - b_2) - c_1$, and the expressible line c has been written as an apotome $(b_1 + b_2) - c_1$, which is impossible.   QED

CONCLUSION (X 111/112) The thirteen classes of *alogoi* lines: medial, binomial, apotome, first bimedial, ... are all disjoint; and any line that is commensurable or commensurable-in-square with a line in a given class is also in that class.

There is some mopping-up of associated results to be done. First the descriptions in columns 8 and 9 are, in fact, perfectly general:

PROPOSITION (X 71–2 & 108–10)
(i) The square side of a sum (or difference) of a rational and medial area will only give one of the additive (or subtractive) *alogoi* lines of rows 1, 2, 4, and 5.
(ii) The square side of a sum (or difference) of two incommensurable medial areas will only give one of the additive (or subtractive) *alogoi* lines of rows 3 and 6.

PROOFS Left as exercises.

Finally, the binomials and apotomes are related by the results:

PROPOSITION (X 112–13) If an expressible area h^2 is applied to a binomial (or apotome) height h, its width will be an apotome (or binomial) of the same class with respect to h.

PROPOSITION (X 114) If $b_1 + c_1$ is a binomial, $b_2 - c_2$ is an apotome, and $b_1 : b_2 = c_1 : c^2$, then the rectangle $(b_1 + c_1).(b_2 - c_2)$ is expressible.

REMARK I shall not prove these results in detail here. To us, they are simple variations on the basic identity

$$(b + c).(b - c) = b^2 - c^2$$

around which simple proofs, and even simple geometric proofs, can be constructed and generalised to the other *alogoi* lines; I encourage the reader to explore these. Yet Euclid's proofs are most unwieldy, though perhaps not to the extent that some commentators have asserted. (Contrast the discussion in Heath, *TBEE* iii, 246, which also touches on the issues of the authenticity of these results, with the attractive treatment in Taisbak, *CQ*, 58–61.) They offer yet more evidence that

Euclid does not seem to conceive the material of Book X in any arithmetised or algebraic way.

5.2(g) *The scope and motivation of Book X*

I shall set out my proposals in some theses and antitheses which culminate in my principal thesis.

(i) Euclid's treatment throughout the *Elements* is purely geometrical, with no undertones of arithmetised mathematics.

I have already illustrated this thesis in many places throughout this book, in particular at several places in the discussion of *Elements* X, and it should be unnecessary to dwell on it further here.

(ii) Euclid's treatment in *Elements* II and X is two-dimensional, based on squares and rectangles, and he nowhere shows any interest in giving any similar results for cubes or other three-dimensional figures.

This thesis is a truism, so I will bring out its implications more prominently. The basic results of a three-dimensional exploration would be straightforward: the *Topics* proposition in three dimensions; the cubic version of X 9, which would follow from VIII 27 in the same way that the square version is apparently based on VIII 26; the three-dimensional analogue of the results of Book II, such as the geometrical identity

$$(x+y)^3 = x^3 + 3x^2.y + 3x.y^2 + y^3$$

illustrated in Fig. 5.8; etc. We could then define a new class of 'cubic expressibles' by taking the edges of cubes commensurable with the assigned cube, then define new kinds of 'cubic' medials, binomials, and apotomes, and begin to explore their properties. Such manipulations suggest themselves immediately, and the early results are attractive exercises in the kinds of reasoning found throughout the plane *and solid* geometry of the *Elements* and in our other fragmentary evidence of early Greek mathematics. Yet, with the sole exception of the three-

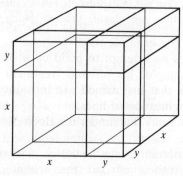

Fig. 5.8. $(x+y)^3 = x^3 + 3x^2.y + 3x.y^2 + y^3$.

dimensional *Topics* proposition at XI 25, none of these results is found in the *Elements*. For example, while the general idea of II 1 is repeated in the particular cases of II 2 and 3, and has indeed already been silently assumed in the proof of I 47, nowhere do we find its three-dimensional analogue articulated; nowhere do we find the result of Fig. 5.8 set out. The contrast is striking: the explorations of Books II and X are strictly restricted to two-dimensional phenomena.

(iii) The subject of Book X is *not* the manipulation or classification of quadratic surds, or of the roots of quadratic or biquadratic equations with integer or fractional coefficients.

This statement is a reformulation of thesis (1), above. The notes in Heath, *TBEE* iii, provide very interesting and useful discussions of textual and terminological matters concerning Book X, but his mathematical analysis is, I believe, irrelevant. The same comment goes for the analyses in Pappus–Thomson & Junge, *CPBXEE*, 17–32 and Euclid–Stamatis, *EE* iii, pp. xi–xv. The perplexities that follow from such kinds of arithmetised and algebrised descriptions are raised several times in Mueller, *PMDSEE*, Chapter 7.

(iv) The implied underlying theme of the book is some qualitative description of the ratios of certain kinds of lines, and words *rhētos* and *alogos* are indeed referring specifically to the *ratios* of *lines*.

The best way of establishing this thesis is to appeal to Euclid's text, and look at the role played by Book X in Book XIII, where it is used to classify other lines arising in the extreme and mean ratio (XIII 6), the side and circumdiameter of the regular pentagon (XIII 11), and the edge and circumdiameter of the icosahedron (XIII 16) and dodecahedron (XIII 17). This cycle of results on the five regular solids is then summarised in XIII 18 and, in the course of this proof, we read:

> The said sides, therefore, of the three figures, I mean the pyramid, the octahedron, and the cube, are to one another in expressible ratios (*logois rhētois*). But the remaining two, I mean the side of the icosahedron and the side of the dodecahedron, are not in expressible ratios either to one another or to the aforesaid sides; for they are *alogoi*, the one being minor and the other an apotome.

Also, Euclid makes no attempt to build up a systematic classification of *areas*; see, for example, his offhand treatment of the medial area in X 21–3. The areas that are named are introduced only with a view to developing the classification of lines.

(v) The idea of ratio that underlies Book X is not arithmetised but anthyphairetic.

Consider first arithmetised mathematics, mathematics in which the idea of number is extended beyond the *arithmoi* to include fractional quantities, and then further to include more and more general kinds of numbers, and where this general kind of number is then used to define and manipulate ratios. The striking dichotomy here is between fractions, the 'rational numbers' of modern mathematics, which describe ratios of *arithmoi,* and the rest, the 'irrational numbers', which cannot be described in finite terms by ratios of *arithmoi.* This dichotomy pervades arithmetised mathematics and so now dominates our thinking about ratios; for example, within basic arithmetised mathematics, we can perceive little qualitative difference between irrational square roots and irrational cube roots. This dichotomy is not suggested anywhere by any of the terminology of Book X and does not correspond to any of the classifications found there.

Now consider anthyphairetic mathematics. Those ratios that can be now completely understood and described in finite terms by the *arithmoi* include the ratios of the sides of commensurable squares, that is the ratios of expressible lines $\sqrt{m} : \sqrt{n}$, and some sporadic examples like the extreme and mean ratio. If we then wish to extend outwards from these accessible, expressible lines, it would be natural to start with those lines that arise from the simplest operations of adding, subtracting, and squaring pairs of expressibles. This seems to correspond closely to what we find in Book X, which starts from the expressible lines and goes on to construct medials, apotomes, and binomials, which are truly baffling from an anthyphairetic viewpoint. These new lines are profoundly *alogoi,* without reason, and Book X is then one of the very few successful attempts to find order in a corner of this mathematical jungle. Sections 9.2(d) and 3(b) will describe briefly the modern and largely unsuccessful further attempts at this kind of exploration.

5.3 Appendix: The words *alogos* and *(ar)rhētos* in Plato, Aristotle, and the pre-Socratic philosophers

For an explanation of this list, see the Appendix to Chapter 4.

5.3(a)   *Plato*

ἄλογον (m.) *Timaeus* 42d.1.
ἀλόγους *Republic* VII **534d.5**.
ἄλογος (f.) *Definitions* 414c.7, 416a.23.
ἄλογον *Theaetetus* **201d.1**; *Timaeus* 47d.4.
ἀλόγου *Timaeus* 28a.3; *Laws* III 696e.1.
ἀλόγῳ *Republic* IX 591c.6; *Timaeus* 69d.4.
ἄλογον (nt.) *Gorgias* 465a.6, 496b.1, 519e.3; *Phaedo* 62b.2, c.6, 68d.12*;
    Symposium 202a.6; *Republic* X 609d.9, 11; *Parmenides* **131d.2, 144b.3**;

5.3   *Alogos* and *(ar)rhetos*

Theaetetus **199a.3**, **203d.6**, **205c.9**, **e.3**; *Timaeus* 51e.4; *Sophist* 219e.4, 238c.10, e.6, 239a.5, 259a.1; *Philebus* 55b.1; *Axiochus* 365e.5.
ἀλόγου *Philebus* 28d.6.
ἄλογα *Protagoras* 321c.1; *Theaetetus* **202b.6**, **203a.4**, **b.6**; *Sophist* 241a.5*, 249b.1.
ἀλογώτερον (nt.) *Charmides* 175c.7; *Gorgias* 519d.1.
ἀλογώτατον (nt.) *Apology* 18c.8.
ἀλογώτατα *Philebus* 55c.3.
ἀλόγως *Gorgias* 501a.6; *Republic* IV 439d.4; *Phaedrus* 238a.1; *Timaeus* 43b.2, e.3, 53a.8; *Laws* II 669d.4, IX 875b.8.

ἄρρητον (f.) *Laws* VII 788a.3.
ἀρρήτων *Republic* VIII **546c.5**.
ἄρρητον (nt.) *Sophist* 238c.10, e.6, 239a.5; *Laws* VI 754a.4; *Alcibiades* I 122d.2.
ἄρρητα *Hippias Major* **303c.1**; *Symposium* 189b.4; *Sophist* 241a.5*; *Laws* VII 793b.3, 822e.2.
ἀρρήτων *Hippias Major* **303b.7**.

ῥηταί *Theaetetus* 205d.9.
ῥητάς *Theaetetus* 202b.7.
ῥητῶν *Republic* VIII **546c.5**.
ῥητόν *Theaetetus* 205e.7; *Epistle* VII 341c.5.
ῥητά *Hippias Major* **303b.8**; *Republic* VIII **546c.1**, *Laws* VII 817d.3; *Epistle* VII 341d.5.
ῥητοῖς *Symposium* 213a.2; *Laws* VIII 850a.7.

5.3(b)   Aristotle

ἄλογος *Posterior Analytics* **76b9**; *Physics* 188a5, 212b30, b34, 252a24; *On the Heavens* 289a6, b34, 291b13; *Meteorology* 355a21, a35, 362a14, 366a9; *On the Soul* **432a26**; *Prophesying by Dreams* 463a23; *Generation of Animals* 722b30, 734a7; *Problems* 954b35; *On Indivisible Lines* **968b18**; *Metaphysics* 999b23, 1002a29, 1046b2, **1083a8**; *Nicomachean Ethics* 1095b15, 1102a18–1103a10, 1111b1, b13, 1117b24, 1120b18, 1172b10; *Magna Moralia* 1198a17; *Eudeman Ethics* 1218a29; *Politics* 1334b18, b21; *Oeconomica* 1343b13; *Rhetoric* 1369a4, 1370a18; *Poetics* 1460a13, 1461b14.

ἀρρήτως *Fragment* 40 1481b6.

ῥητός *On Indivisible Lines* **968b15**, **b18**; *Metaphysics* 1017b1, b3; *Nicomachean Ethics* 1162b26, b31, 1163a5.

5.3(c) *Pre-Socratic philosophers*

ἄλογος

Democritus	⎰ **A33**	ii	**91.20** ⎱	(See Sections 5.2(a)
	⎱ **B11p**	ii	**141.24** ⎰	and 8.3(b))
	A105	ii	109.35	
	A116	ii	111.25	
	B164	ii	177.1	
	B292	ii	206.6	
Empedocles	B136	ii	367.4	
Heraclitus	A16	i	147.35, 148.19 and 35	
Leucippus	A22	ii	77.1	
Parmenides	A45	i	226.2	
Philolaus	B11	i	412.11	
Protagoras	C1	ii	296.23	(≡ Plato, *Protagoras*, 321c)

ἄρρητος

Lysis	**4**	i	**421.5**	(See Sections 5.2(a) and 8.3(b))

PART TWO

EVIDENCE

> The papyrologist works with an ancient script which he transforms into a modern script, with the language denoted by the script, and with the meaning conveyed by the language. This linguistic trinity—script, language, and meaning—is fixed irremovably at the centre of his activity, and it is an indivisible trinity.
>
> H. C. Youtie, *The Textual Criticism of Documentary Papyri*, p. 16.

Chapter 6 speaks for itself, and needs no further introduction. Chapter 7 describes a wide range of different kinds of texts—scientific and commercial, advanced and elementary, abstruse and pedagogic—whose common element, examined in great detail here, is their treatment of numbers and, especially, fractions. I wish to illustrate how the implications of a kind of arithmetical manipulation might change as this text is recopied with different conventions of writing or printing, or translated, or interpreted in a different mathematical context. All the examples chosen relate to the question: Did the Greek mathematicians of the fifth and fourth centuries BC have at their disposal the manipulations of common fractions such as $p/q + r/s = (ps + qr)/qs$, $p/q \times r/s = pr/qs$? I believe that we have no good evidence on which to argue that they did. Of course, insofar as fractional quantities were manipulated correctly, they did use procedures that were equivalent to our operations with common fractions, but their different conceptions and operations may have given rise to different intuitions and abstractions about the underlying mathematics.

Another purpose of Chapter 7 is to set in higher relief the non-arithmetised nature of early Greek mathematics; in this, the chapter is an extension of arguments introduced in Section 4.5, above. However, I nowhere consider the important and vast subject of Ptolemaic astronomy, which is a blend of Babylonian arithmetical and Greek geometrical methods, but only note, in passing, that we have no Greek evidence of any use of or interest in Babylonian sexagesimal arithmetic before the second century BC.

6
THE NATURE OF OUR EVIDENCE

6.1 *ΑΓΕΩΜΕΤΡΗΤΟΣ ΜΗΔΕΙΣ ΕΙΣΙΤΩ*

Plato, it is said, placed an inscription over the door to the Academy: "Let no one unskilled in geometry enter". Since my aim is to propose a new and central role for Plato in the development of early Greek mathematics, this well-known and often-cited story provides an epitome of the main theme of this book.

Let us, however, pause a moment and enquire about the evidence for the story. Plato (*c.* 429–347 BC) founded his school sometime around 385 BC, and the 'early Academy',[1] the fourth-century period of existence under its first three directors or scholarchs, Plato, his nephew Speusippus, and Xenocrates, forms the shadowy background to the mathematical developments discussed here. The Academy is then commonly believed to have continued in existence for almost a millennium, directed by a 'golden chain' of scholarchs who kept alive and developed the Platonic tradition, until the closure of the Athenian schools by an order promulgated by the emperor Justinian in AD 529, the same year that St Benedict founded the monastery of Monte Cassino (thus, coincidentally, providing historians with a clear and symbolic break between pagan antiquity and medieval Christianity). Such is the almost universal impression to be gained from both scholarly and popular accounts alike, yet the evidence for every remark of substance in the previous sentence is found, under scrutiny, to be either lacking or distorted, and this whole line of interpretation is unfounded. This is not the place to attempt a summary of this scrutiny and details of alternative interpretations,[2]

---

[1] See Cherniss, *REA*. This short and erudite book is essential reading for anybody interested in the scanty evidence concerning the early Academy and what has been made of it by modern scholars.

[2] See Cameron, *LDAA*, Glucker, *ALA,* and Lynch, *AS*. These works are each fascinating and authoritative syntheses of a wide range of different kinds of evidence. Cameron poses and discusses the following questions: "What was Justinian's motive? Did he give the last push to a tottering edifice, or destroy a thriving intellectual centre? Indeed, did he actually succeed in destroying anything at all? What did the philosophers do on their return?" (p. 7). Lynch's book assembles the literary, archaeological, epigraphical, and legal evidence concerning the Athenian schools in general, and so has much of importance to say about the Academy. Glucker's massive re-evaluation of the context of later Platonism must, if it is accepted, modify or put into question practically everything previously written on the subject. His work is set out in an equally massive book of some 450 densely argued pages, with a second instalment promised, directed at a readership so different from what I

beyond the following general conclusions. It now appears that the Academy fell into disuse after the directorship of Philo, who went to Rome,[3] a refugee from the Mithridatic wars, sometime around 85 BC, and soon after this the tradition of philosophical schools in Athens languished. Then there was a general revival of interest in Plato's philosophy and mathematics in the second century AD, in Syria and Asia Minor, at which time a few philosophers may have migrated back to Athens. However, Athens does not seem to have become once more celebrated for its philosophical activity before the fifth century,[4] when the study of Plato's work again flourished in a prosperous and active school directed successively by Plutarch of Athens, Syrianus, Proclus, Marinus, Hegias, and Damascius; this school appears to have survived Justinian's edict; and its next scholarch, Simplicius, appears to have continued to write,

---

envisage here that I dare to suggest the following page references to his main conclusions for the interested but non-specialist reader: pp. 88–97 (Philo and Antiochus); 138–9, 145, and 153–8 (on the revival of Platonic studies in Asia Minor and Syria in the second century AD, but its absence from Athens until the revival of the neo-Platonic school of Athens in the fifth century AD); 226–55 (the school's property, including a summary of recent archaeological investigations); 296, 306–15, and 321–2 (the 'Golden Chain' of scholarchs and Proclus' mystical attitude to Plato's 'divine philosophy'); 322–9 (a summary and extension of Cameron, LDAA, on the 'end' of the Athenian schools); and 364–79 (the succession of the Peripatetic, Stoic, and Epicurean schools and the attitude of the Athenian citizens to the philosophical activity within their city). My only unease about these investigations concerns their neglect of the role of mathematics within philosophy throughout the period with which they are concerned. Compare, for instance, the conclusions of Cherniss, quoted at the beginning of Chapter 4, with the only three allusions to mathematics I could find in Glucker's book: two passing references to Theon of Smyrna on pp. 136 and 212, and the following passage on p. 260: "From our passage [Plutarch, *De E apud Delphos*, 387 f.] we learn that he [Plutarch] had already studied some mathematics—a subject with which, at the dramatic date of the dialogue [AD 66/7] he was somewhat too passionately involved". This disdain, or distaste, for mathematics is general. See, for example, the review of Cherniss, *REA* by Solmsen (*Classical Weekly* 40 (1946), 164–8, on 168): "This [mathematics] was the kind of work to which Plato kept the members of the Academy with inexorable strictness, refusing to discuss with them, especially with the younger men, the *more serious philosophical problems* connected with his theory of ideas" (my emphasis); and it would be easy and profitless to point to scores of similar instances.

According to Cameron, Monte Cassino is now believed to have been founded in AD 530!
[3] Glucker's stronger conclusion, that Philo died in Rome, is queried in Sedley, EA; see especially n. 2.
[4] Here, for example, is one anecdotal piece of evidence for the dearth of intellectual activity in Athens, in a letter from the neo-Platonic philosopher Synesius of Cyrene to his brother, written around AD 396: "... And as for Athens: A curse on the accursed boatman who brought me here! Why, present-day Athens has nothing worth venerating except for the famous names of its places. It's just like the skin that is left from a sacrificed victim: a remainder of a life that once was. That's the condition of things now that philosophy has gone from here. All that remains for us is to travel around and wonder at the Academy, the Lyceum, and—by Zeus—the painted Stoa which gave its name to the philosophy of Chrysippos, though it's now no longer painted since the proconsul took away the panels on which Polygnotes of Thasos displayed his art. In our time Egypt has received and continues to cherish the creative works of Hypatia, while Athens—which was formerly the dwelling place of the wise—now has only bee-keepers to make her famous" (*Epistle* 136, quoted in Lynch, *AS*, 195 f.).

have students, draw his emoluments, and even practise his pagan religion—albeit in reduced circumstances and more circumspectly—into the second half of the sixth century. But though this Athenian neo-Platonic school consciously revived and emulated the tradition of Plato's Academy, it never seems to have been called 'The Academy' by its contemporaries. In fact this name, with its modern connotations, is an obstacle to the solution of what is characterised, in Cherniss, *The Riddle of the Early Academy,* as the more genuine problem of understanding Plato's Academy:[5]

What, then, did Plato really do in his Academy? . . . 'Academy' and 'Academic' are terms which men of formal training, who speak as their own the modern European tongues, have been pleased to apply to themselves and their organisations. Is it not wonderful, therefore, that by a more or less unconscious retrojection modern scholars have attached the particular significance which 'Academy' has in their own milieu to the garden of Plato's which was situated in the suburb northwest of Athens called 'Academia' after a mythical hero. . . . The external evidence for the nature of the Academy in Plato's time is extremely slight . . . [*REA,* 61 f.].

The same point is made in Glucker, *Antiochus and the Late Academy*:

For the ordinary educated Roman of [Cicero's] age, the place is just what it appears to be to a visitor to Athens—a gymnasium. To us, influenced as we are by the Platonic tradition, the word 'Academy' has come to mean an institution of learning, a learned society, or at least a place of theoretical ('academic') education. In ancient Athens, the Academy was first and foremost a public park dominated by its gymnasium, and the connection between it and Plato's school was only one of the numerous historical reminiscences in an area rich in history. (In order to obtain the right perspective, one can add that the excavations in the area of the Academy have not proved to be a tourist attraction in our times, although they are mentioned in most guidebooks to Greece.) It is in such a light that Pausanias, in what is essentially a tourist guide [*Description of Greece,* I xxix–xxx], describes the area.[6] He proceeds to describe in detail monuments like the enclosure to Artemis nearby and the small shrine of Dionysus Eleutherus . . . . He lists the graves of the various war heroes and of distinguished statesmen and philosophers . . . . He [describes] the altar to Eros built by Charmus . . . ; the altar of Prometheus in the Academy itself; the altar to the Muses—most probably the one dedicated by Plato and decorated by Speusippus (Diogenes Laertius, *Lives of the Eminent Philosophers* IV, 1); an altar to Athena and an olive tree. It is only then that he mentions that not far from the Academy is a memorial to Plato, in a place where a god indicated to Socrates in a dream that Plato was to

---

[5] The 'riddle' of Cherniss' title is to understand why some modern scholars need to propose an 'unwritten doctrine' to account for the discrepancy between Aristotle's account of Plato's theory of idea-numbers and what is found in Plato's own writings.

[6] Pausanias lived during the second century AD. The *OCD* observes: "He loves all religious and historical remains . . . on which he writes plainly and honestly. His accuracy herein is confirmed by existing remains".

be the greatest of philosophers. Pausanias narrates the dream in great detail, but says nothing to the effect that Plato was buried on what was his own estate near the Academy (if that was the case), or of the very existence of such an estate. He passes on immediately to a description of the tower of Timon the misanthrope, whose property, as we know from other sources, was near Plato's school. . . . The traveller in Attica was shown Plato's grave, but he was not told of the connection between the Academy and Plato himself or his school [*ALA*, 244 f.].

To return, then, to the inscription. It is most unlikely that the archaeological record could now be of any help in verifying or refuting the proposal that Plato himself had any part in putting it up[7] so we have, instead, to rely on references to it by ancient writers;[8] but these cannot be traced back any earlier than the fourth century AD. The earliest such reference occurs in an oration written in AD 362 by the emperor Julian the Apostate, who mentions an inscription over the entrance to Aristotle's classroom and alludes to another over Plato's, without giving any details of their contents. No other reference to Aristotle's inscription is known, and the first details of the wording of Plato's inscription come from an anonymous scholiast[9] who has been identified as probably the fourth-

---

[7] The site of the Academy has been buried under the sprawl of twentieth-century Athens without ever being properly excavated, and the scattered reports of the partial excavations have never been brought together into any kind of comprehensive work; see Whycherly, PAPS, especially Part II, 2–10. A full discussion of the archaeological evidence, with references to some fifty archaeological reports, can be found in Glucker, *ALA*, 237–46; he concludes: "The one subject on which the excavations have not so far shed any new light is precisely the issue which Mr. Aristophron hoped to clear up with the aid of these excavations: the connection between the Academy and the school of philosophy established in that area by Plato. . . . As to the history of Plato's school, we are at present still largely at the mercy of our literary sources. And the literary sources hold their peace. Or rather, they behave as though any connection between the Academy and the philosophical school founded by Plato was restricted to the 'Classical' period of the school, between Plato and the pupils of Carneades and Clitomachus [i.e. up to the end of the second century BC]". Lynch, *AS*, 17–31, also deals with the archaeological evidence, but mainly with reference to the Lyceum, which was to the east of Athens.

[8] My summary of these sources is based on Saffrey, AME (= *ΑΓΕΩΜΕΤΡΗΤΟΣ ΜΗΔΕΙΣ ΕΙΣΙΤΩ, Une inscription légendaire*). Also see Riginos, *Platonica, The Anecdotes Concerning the Life and Writings of Plato,* in which the sources of 148 anecdotes (including this) are examined; very few can be traced back to the first century AD or before.

[9] A scholium is a marginal note in a manuscript (for some examples, see Plate 4) and its author is the scholiast. The scholium reads:

εἰ δὲ ἡ γεωμετρία ἐπεγέγραπτο δὲ ἔμπροσθεν τῆς διατριβῆς τοῦ Πλάτωνος ὅτι ΑΓΕΩΜΕΤΡΗΤΟΣ ΜΗΔΕΙΣ ΕΙΣΙΤΩ. ἀντὶ ἀνίσος καὶ ἄδικος. ἡ γὰρ γεωμετρία τὴν ἰσότητα καὶ τὴν δικαιοσύνην ζητεῖ.

(There had been inscribed at the front of the school of Plato "Let no one who is not a geometer enter'. [That is] in place of 'unfair' or 'unjust': for geometry pursues fairness and justice).

Andrew Barker, who has provided this and the following translation, notes: "The piece is riddled with ambiguities. 'No one who is not a geometer' might be paraphrased as 'no one who is not geometrically minded'. 'No one unskilled in geometry' doesn't quite work here,

century orator Sopatros, in an annotation of a manuscript of Aelius Aristides. Sacred places sometimes had inscriptions such as "Let no unfair or unjust person enter"; and this scholium implies that the author of Plato's inscription has substituted *ageōmetrētos*, 'ungeometrical', for *anisos kai adikos,* 'unfair or unjust', in the normal formula; also the scholium seems to indicate that the inscription was not put up by Plato. The story is repeated and used by the sixth-century Alexandrian neo-Platonic philosophers Philoponus, Olympiodorus, Elias, and David to impute a variety of different motives to Plato in having the inscription put up in the first place. Finally the most commonly used standard source for the story is the twelfth-century Byzantine Johannes Tzetzes (described in the *OCD* as a "copious, careless, quarrelsome, Byzantine polymath") in his *Book of Histories* (or *Chiliades*), a review of Greek literature and learning in 12,674 verses.[10] Thus the evidence for the story is late and of doubtful provenance.

In support of the evidence, it can be added that the word *ageōmetrētos* is found in technical use during the fourth century BC (Aristotle, *Posterior Analytics* 77b12–13: "One should therefore not discuss geometry among those who are *ageōmetrētoi*, for in such a company an unsound argument will pass unnoticed."), so the inscription is philologically acceptable; Julian's account of the inscription seems to allude to a well-known story, so the anecdote could have been in circulation before his time (but how much earlier?); and there can be no doubt that Plato regarded mathematics as being extremely important. However, that the inscription corresponds so well with Plato's own opinions can be used as an

---

since the sequel suggests that *ageometrētos* indicates a disposition as much as a skill. *Anisos* 'unfair', literally 'unequal', is used with something of the sense of 'antiegalitarian'. In this usage, someone who is *isos* pursues equality as an end or ideal, while someone who is *anisos* rejects it—see Aristotle, *Nicomachean Ethics* 1129a32–b1. 'Unfair' or 'unjust': the scholiast may possibly mean 'unfair or unjust', or again 'unfair and unjust'. In such a context *kai* may mean either 'and' or 'or', and in the latter case there is no way of telling which arrangement of inverted commas best expresses the scholiast's intention."
[10] Tzetzes, *Chiliades VIII*, 974–7:

Πρὸ τῶν προθύρων τῶν αὐτοῦ γράψας ὑπῆρχε Πλάτων·
Μηδεὶς ἀγεωμέτρητος εἰcίτω μου τὴν cτέγην·
Τουτέcτιν, ἄδικος μηδεὶς παρειcερχέcθω τῇδε·
Ἰcότης γὰρ καὶ δίκαιόν ἐcτι γεωμετρία.

(On the front of his doorway Plato had written 'Let no one who is not a geometer enter my house.' That is, 'Let no one who is unjust come in here', for geometry is equality and justice.) Andrew Barker notes: "Similar comments to those on Sopoatros' scholium apply here. In both passages I would take *adikos* to mean 'unjust' rather than the more general 'unrighteous', since this facilitates the parallel with geometry, which presumably has to do with 'due proportions' and the like; but the more general senses can't be ruled out if 'righteousness' is being understood, for instance, within a broad conception of virtue as a mean. In any case, writers were not always conscious of the ambiguity, or sometimes appear to play on it deliberately—as (arguably) does Plato himself."

argument against its authenticity: it could be a later elaboration of Plato's words, for example at *Gorgias* 508a: "Geometric equality is of great importance among gods and men".[11] The main argument against the authenticity of the story is the fact that no direct evidence can be found earlier than the fourth century AD, and 747 years separate the tentative date of the Academy's foundation and the date of Julian's oration. There can also be added a further kind of negative evidence *ex silentio*: Aristotle (384–322 BC), who spent from his eighteenth to thirty-seventh year in the Academy, leaving on Plato's death in 348/7 BC, was very critical of the mathematical bias of Academic philosophy. For instance, in a discussion of Plato's theory of ideas (*Metaphysics* 992a32–b1) he writes: "Mathematics has come to be identical with philosophy for modern thinkers, though they say it should be studied for the sake of other things", and a little later, he adds that, in this approach, "The whole study of nature has been annihilated".[12] It is therefore curious that Aristotle nowhere points to the inscription as a flagrant and defiant advertisement by Plato of the very tendency of which he, Aristotle, is so unequivocally critical.

I have dwelt on this story of Plato's inscription at such length because it provides a convenient way of introducing what we find, time and time again, when we study the spectacular developments in mathematics that took place in the period which culminated in the compilation of Euclid's *Elements* and *Data* at the end of the fourth century BC: the evidence for many stories cannot be traced back any earlier than a period five, six, seven, or even eight hundred years after the event, and we often have little indication of whether the stories are authentic, plausible, or misleading. It is now worth pursuing this question even further and exploring the form in which the surviving evidence, early or late, reaches us.

6.2 Early written evidence

Suppose we take a substantial text relating to the early developments of Greek mathematics—Euclid's *Elements*, for example. What, precisely, are we considering?

---

[11] A bizarre, though possible, explanation of the background of 'geometric equality' at *Gorgias* 508a, has been proposed in Harvey, TKE: that this was part of an intellectual argument developed by the aristocratic Archytas and Plato to discredit democracy. In place of the simple political formulation that all citizens are equal, which loosely corresponds to the application of an *arithmetic* proportion in which each term stands at an equal distance from its neighbour, one should rather reward each according to his value, and this is better described by a *geometric* proportion, in which each term bears the same relationship to its neighbour. Harvey even proposes that *logismos*, in Archytas' fragment B3, might mean geometric proportion, but this surely cannot be the case for Plato's *logismos*; see Section 4.2. Also see de Ste. Croix, *CSAGW*, 413–4.

[12] Also see Cherniss, *REA, passim*, and especially p. 68.

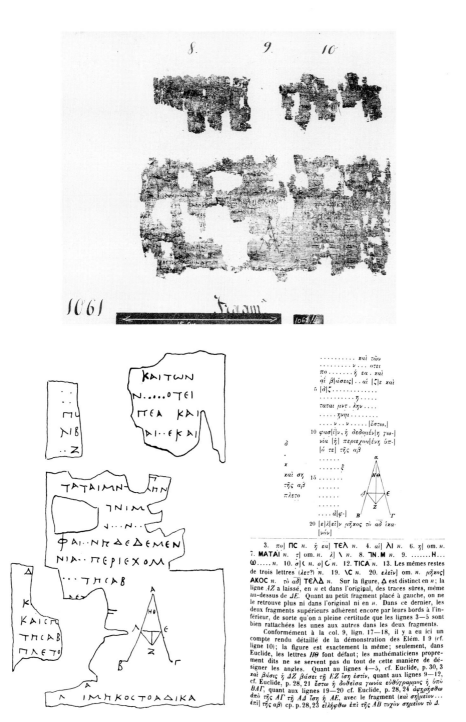

1. P. Herc. 1061: photograph of columns 8, 9, and 10; and drawings and transcription of column 9 in J. L. Heiberg, Quelques papyrus traitant de mathématiques, *Oversigt over det Kgl. Danske Videnskabernes Selskabs Forhandlinger* 2 (1900), 164–5. See Section 6.2.

XXIX. Euclid II. 5.

8·5 × 15·2 cm.

Fragment from the bottom of a column, containing the enunciation, with diagrams, of Euclid II. 5, and the last words of the preceding proposition.

From the character of the handwriting, which is a sloping rather irregular informal uncial, this papyrus may be assigned to the latter part of the third or the beginning of the fourth century. Diaereses are commonly placed over syllabic ι and υ. Iota adscript is not written. The corollary of Proposition 4 seems to have been omitted, while the two lines illustrating the division εἰς ἴσα καὶ ἄνισα in Proposition 5 are not found in ordinary texts. Otherwise the papyrus shows no variants from the text of the Oxford edition of 1703 or that of Peyrard, beyond the mistake of τετραγωνου for τετραγώνῳ in l. 9, and the spelling μετοξυ for μεταξύ in l. 6.

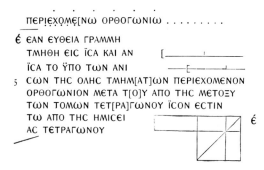

```
     ΠΕΡΙΕΧΟΜΕ[ΝΩ ΟΡΘΟΓΩΝΙΩ . . . . . . . . .
   έ ΕΑΝ ΕΥΘΕΙΑ ΓΡΑΜΜΗ
     ΤΜΗΘΗ ΕΙC ΪCΑ ΚΑΙ ΑΝ
     ΪCΑ ΤΟ ΫΠΟ ΤΩΝ ΑΝΙ
 5   CΩΝ ΤΗC ΟΛΗC ΤΜΗΜ[ΑΤ]ΩΝ ΠΕΡΙΕΧΟΜΕΝΟΝ
     ΟΡΘΟΓΩΝΙΟΝ ΜΕΤΑ Τ[Ο]Υ ΑΠΟ ΤΗC ΜΕΤΟΞΥ
     ΤΩΝ ΤΟΜΩΝ ΤΕΤ[ΡΑ]ΓΩΝΟΥ ΪCΟΝ ΕCΤΙΝ
     ΤΩ ΑΠΟ ΤΗC ΗΜΙCΕΙ
     ΑC ΤΕΤΡΑΓΩΝΟΥ
```

5. ΤΗC Ο corrected from ΠΕΡΙ.   6. l. μεταξύ.   9. l. τετραγώνῳ.

1. If the reading is correct—and though the traces of letters after ΠΕΡ are scanty, there seems to be no alternative—the corollary of Prop. 4 was omitted. After ΟΡΘΟ-ΓΩΝΙΩ, too, there would not be room for more than about nine letters, so ὅπερ ἔδει δεῖξαι must have either been omitted or, more probably, abbreviated.

2-3. The shortness of these lines indicates that there were two horizontal strokes in the margin, the first showing the division into equal, the second that into unequal parts. The first is entirely broken away, and only the left-hand part of the second is preserved.

2. P. Oxy. i 29: photograph of text; and complete publication in B. P. Grenfell & A. S. Hunt, *The Oxyrhynchus Papyri* i (1898), 58. See Section 6.2.

Fr. (a)

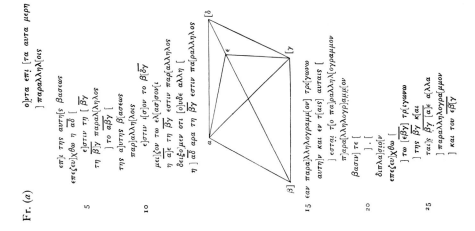

    ο]ντα επι [τα αυτα μερη
    ] παραλληλ[οις

    επι] της αυτη[ς βασεως
5 επεξευ]χθω η $\overline{αδ}$ [
    ε]στιν τη [$\overline{βγ}$
    τη $\overline{βγ}$ παραδ[λληλος
    ] το $\overline{αβγ}$ [
    της α]υτης βασεως
    παρ[αλλη]λοις
10 ε]στιν ι[σ]ον το $\overline{β[δγ}$
    μει]ζον τω ελ[ασσον[ι
    η $\overline{αε}$ τη $\overline{βγ}$ εστιν παρ[αλληλος
    δει]ξομεν οτι [ο]υδε αλλη [
    η] $\overline{αδ}$ αρα τη $\overline{βγ}$ εστιν πα[ραλληλος

15 εαν παρα]λληλογραμμ[ον] τρι[γωνω
    αυτη]ν και εν τ[αις] αυταις [
    ] εσται τ[ο παραλλη[λογραμμον
    π]αρα[λληλογρα]μμ[ον
    βασιν] τε [
20   ].[
    διπλα]σ[ιο]ν
    επεξευ]χθω
    ]τω [εβγ] τρι[γωνω
    ]της $\overline{βγ}$ κ[αι
    ταις $\overline{βγ}$ [α]ε α[λλα
25   ]παραλληλογρα[μμον
    ] και του εβ[γ

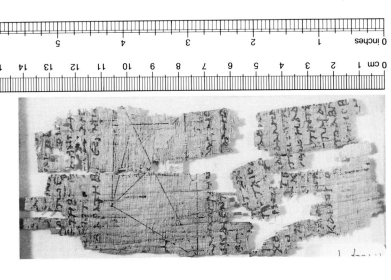

3. P.Fay. 9: photograph of text; and transcriptions in B. P. Grenfell, A. S. Hunt, & D. G. Hogarth, *Fayûm Towns and their Papyri* (1900), 96–9. See Section 6.2.

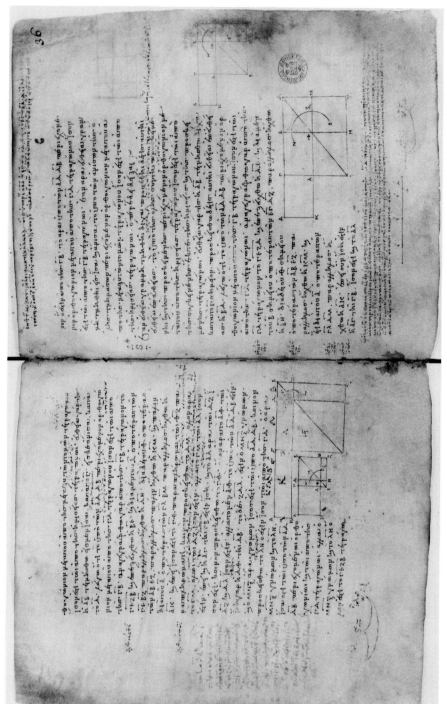

4. Bodleian manuscript D'Orville 301, folios 35ᵛ and 36ʳ: *Elements* II 5 and 6. See Section 6.3.

Fr. (b), 2nd hand. Col. iv.

55 [ἡ] νὺξ ὡρῶν ιγ ίβ΄ μ΄ ε΄, ἡ δ᾽ ἡμέρα ιβ΄ ε΄ λ΄ϟ΄.
[ι]ς Ἀρκτοῦρος ἀκρώνυχος ἐπιτέλλει,
[ἡ] νὺξ ὡρῶν ιββ΄ ί ε΄ μ΄ ε΄, ἡ δ᾽ ἡμέρα ιαθ΄ ί λ΄.
[κ]ς Στέφανος ἀκρώνυχος ἐπιτέλλει
[κ]αὶ βορέαι πνείουσιν ὀρνιθίαι, ἡ νὺξ
60 [ὡρ]ῶν ιβ∠λ΄, ἡ δ᾽ ἡμέρα ιαγ΄ ί λ΄. Ὄσιρις
[π]εριπλεῖ καὶ χρυσοῦν πλοῖον ἐξά-
[γε]ται. Τῦβι ⟨ε⟩ ἐν τῶι Κριῶι. κ ἰσημερία
[ἐα]ρινή, [ἡ] νὺξ ὡρῶν ιβ καὶ ἡμέρα ιβ,
[κ]αὶ ἑορ[τ]ὴ Φιτωρώιος. κζ Πλειάδες
65 [ἀκ]ρώνυχ[οι] δύνου[σ]ιν, ἡ νὺξ ὡρῶν ιαβ΄ ϟ΄ ϟ΄,
[ἡ] δ᾽ ἡμέρα [ι]βί λ΄ μ΄ ε΄. Μεχεὶρ ς ἐν τῶι
[Τ]αύρωι. Ὑάδες ἀκρώνυχοι δύνουσιν,
[ἡ] νὺξ ὡρῶν ιαΖί λ΄ ε΄,

55. l. ίε΄ for ίβ΄.   57. ίε΄ corr. from ε΄.   65. ϟ΄ corr.   68. l. λ΄ϟ΄ for λ΄ε΄.

5. P. Hib. i 27: photograph of column 4 and its transcription, both from B. P. Grenfell & A. S. Hunt, *The Hibeh Papyri* i (1906), 138–57 and Plate VIII. See Section 7.1 (c).

1847 [1036].   30 B.C. - 14 A.D.

ε ∠ d η ∠ d **²** *Γεω(μετρία) γῆ(ς) Καλλιε(δόντος) διὰ Πετεχ(ώνσιος)* σ⌐ αρ^χ *Λιβ(ὸς)* **³** d' η' ι'ϛ' λ'β' $\frac{\angle\ d'\ \eta'}{a}$ η ι'ϛ' [λβ] / ἄ(ρο)ν(ραι) d' ι'ϛ'. *Βο(ρρᾶ) ἐχ(ομένη)* ει^ας λ'β'. **⁴** *Ἄλ(λη) ἐχο(μένη) ἐν τῶι ἐδά(φει)* ∠ d'η' $\frac{a\ \angle\ \eta'}{\iota'\varsigma'}$ a λ'β' / ∠ d λ'β' [.'.']. **⁵** *Ἄλ(λη) ἐχο(μένη)* ∠ η' $\frac{\angle\ d'\ \eta'}{\angle\ d'}$ ∠/ d' η' ιϛ'. **⁶** *Ἄλ(λη)* δ $\frac{\angle\ d'\ \eta'\ \iota\varsigma'\ \lambda'\beta'}{a\ \eta'\ \lambda'\beta'}$ ο / δ d' **⁷** *Ἄλ(λη) ἐχ(ομένη)* δ ∠ d $\frac{\angle\ d\ \iota'\varsigma'\ ..}{\angle\ d'\eta'\ \iota'\varsigma'\ \lambda'\beta'}$ / γ ∠ d η' λ'β' **⁸** / ἄ(ρο)ν(ραι) θ ∠ d' η' [ι'ϛ']. **⁹** *Βο(ρρᾶ) ἐχ(ομένη) διὰ Λυσιμάχ(ου)* **¹⁰** ε $\frac{\angle\ d'\ \eta'\ \lambda'\beta'}{a\ d'\ \eta'\ \lambda'\beta'}$ ο / ε ∠ d λ'β'. **¹¹** *Ἄλ(λη) ἐχο(μένη)* δ $\frac{\angle\ d'\ \eta'}{\angle\ d'\ \eta'\ \lambda'\beta'}$ ο / γ d' η' λ'β'. **¹²** / θ ∠ η' ι' ϛ'.

3 Au quatrième côté du quadrilatère, le η n'est pas surmonté du trait qui indiquerait qu'il s'agit de 1/8. Il faut cependant comprendre 1/8 et non 8, en raison du produit : 1/4 et 1/16 d'aroure.   7 Il manque un quatrième nombre aux dimensions des côtés données pour le calcul de la superficie. Le signe o (= ὁμοίως ?) indique que le quatrième côté considéré est égal au premier.   9 *Λυσίμαχος* : le même sans doute que dans W.O. 760 et 761.

2 Je ne comprends pas σ⌐ = σα( ). αρχ = probablement ἀρχ(ή).

6. O. Bodl. ii 1847: photograph of text; and transcription in J. G. Tait, & C. Préaux, *Greek Ostraca in the Bodleian Library at Oxford* ii (1955), 308. See Section 7.1 (d) and 8.1.

7. P. Lond. ii 265 (p. 257), columns 2 and 3. See Section 7.3 (c).

8. M.P.E.R., N.S. i 1, column 6.
See Section 7.3 (d).

Euclid himself is an elusive figure.[13] We know nothing of him from contemporary references, and the only thing that can be hazarded with any degree of confidence is that he may have lived some time around 300 BC, since the general theory of proportion to which Aristotle appears to allude, at *Posterior Analytics* I.5, 74a17–25, appears in polished and complete form in Book V of the *Elements*. He most probably lived before Archimedes, since Archimedes refers to earlier, possibly Euclidean, works,[14] though the only explicit citation, in *Sphere and Cylinder* I, 3: "... by the second [proposition] of the first of the [books] of Euclid", is uncharacteristic of Archimedes' style, and so is very probably a later addition to the manuscript. A stronger argument for dating some kind of Euclidean text to around 300 BC will be described below, but the evidence for this will also not be without its complications.[15]

Nine or ten works are attributed to Euclid, of which the *Elements*, *Data*, *Division of Figures*, *Phaenomena*, *Optics*, and *Sectio Canonis* survive in one form or another. Of these the most varied and important is the *Elements*, which comprises thirteen books containing, in all, more than 133 definitions and 465 propositions, occupying more than 20,000 lines of text.[16] It deals with a variety of topics in a recognisably varied range of mathematical and literary styles. Euclid is the compiler, not the author, of the work: he is believed to have taken source works by other mathematicians and edited them, adapting and rearranging the material, perhaps even inserting new material of his own, to make the complete treatises. The original source works cannot have been written very long after the change in Athenian intellectual and cultural activity from an oral to a written tradition; this appears to have taken place in the sixth, fifth, and fourth centuries, BC, and a range of explanations has been proposed to account for this development.[17] Let us now consider what

---

[13] See the *DSB* for articles on Euclid (by I. Bulmer-Thomas) and the transmission of the *Elements* (by J. Murdoch; this deals mainly with the Arabic and Latin traditions that I shall not consider here).

[14] In *Sphere and Cylinder* I 2 and 6; *Quadrature of the Parabola* 3, *Conoids and Spheres* 3; and *Method* I and II; these passages are cited in the testimonia in Euclid–Stamatis, *EE* i, pp. xii–xiii and some are excerpted in Thomas, *SIHGM* ii, 50 f. (for *Sphere and Cylinder* I 2) and 224 f. (for the *Method*). However, there are indications that Archimedes may have learned some mathematics from pre-Euclidean sources; see Knorr, APEPT.

[15] The suggestion that Euclid was the pen-name of a group of mathematicians or the name of a teacher, used collectively by his students, was put forward in Itard, *LAE*, 11. The fact remains that we have no facts on this subject. For a recent discussion of the difficulties posed by Proclus' attempt to locate Euclid (*CFBEE* 68, lines 10 ff.) see Fraser, *PA* ii, 563, n. 82; other ancient references are discussed in notes 84–91 and the text to these notes.

[16] A 'line of text' means a line of the critical edition, Euclid–Heiberg, *Opera*. The text and scholia of the *Elements*, vols. i–v, without Latin translation and with added prefatory material, has been reissued in Euclid–Stamatis, *EE*; but see notes 34, 40, 58, and Chapter 7 n. 1, below.

[17] See, for example, for one of the most popular of the explanations, Hall, *CCT*, 26–7: "Up to the end of the sixth century BC Greek literature is in this state of ceaseless flux, and is exposed to all the dangers of a tradition that is practically oral. And then the change comes

actual writings survive, and in what form. Here I shall only lead up to consideration of the Greek Euclidean tradition; there is also, up to the tenth century AD, a debased Latin tradition; and also a rich but largely unexplored Arabic tradition which plays an important role in the transmission of Greek mathematical ideas.[18]

Inscribed, as opposed to written, texts will scarcely enter into our discussion. The most common materials on which writing is found are pieces of pottery or stone (called 'ostraca') or wooden writing boards or wax tablets, for ephemeral work,[19] and, for more enduring records,

---

swiftly and suddenly with the birth of a new form of literature, not local nor occasional nor professional as the older forms had been, but Pan-Hellenic in its appeal, although it sprang from a single city-state. This new form was Attic Tragedy, which never lost the hold which it rapidly obtained over the Greek race in all quarters of the ancient world. The enthusiasm for Tragedy created a reading public, since but few Greeks could hope to see the masterpieces of the great dramatists performed in Athens. Thus an impulse was given to the production of books which ends in the growth towards the end of the fifth century of an organised book trade with its centre in Athens". This from a book published in 1913; nobody today would speak with such confidence. Here, for example, is Turner, *ABFFCBC*, 16: "Let us . . . attempt to trace the steps by which books came into common use in Athens. The target before such an inquiry is a correct appreciation of the part played by the written word in the revolution which took place in the techniques of thought during the fifth century BC. It is a target against which Wilamowitz scored a brilliant and authoritative miss [in *Einleitung in die griechischen Tragödie* (1907), Chapter 3]. He laid it down that the texts of the Greek tragedies were the first books. This positive statement I hope to demonstrate as false; while his rigorous definition of a book as 'what is published by its author through the medium of an organised book trade for the benefit of an expectant public' is to import from the nineteenth century a criterion which obscures understanding of the fifth." Also see the mythical story told by Plato of the invention of "*arithmos te kai logismos*, geometry and astronomy, not to speak of draughts and dice, and above all writing" at *Phaedrus* 274d–275b, quoted in Section 1.2(e), and the notes to that section. For a further discussion, which includes the oriental background against which Greek culture is set, see Pfeiffer, *HCS*, Chapter 2. The earliest evidence for the development of the Greek alphabet out of the Phoenician is found in the eighth and seventh centuries BC and comes from graffiti on pottery; for a summary and references, see Coldstream, *GG*, 295–302 and 382.

[18] For references concerning the Arabic and Latin traditions, see note 13 and Toomer, *LGMWAT*. For further details of the role of writing in ancient Greek culture, see Turner, *GP*; I shall rely heavily on this and other works by Turner in what follows. Wherever possible, I shall refer to papyrus texts by the standard method adopted by papyrologists: an abbreviation for the published collection or publication, followed by the publication number; the abbreviations are expanded in *GP*, Chapter IX or the Introduction to Liddell, Scott, & Jones, *GEL*. If a text has no such publication number, but appears in the second edition of Pack, *GLLTGRE*, its reference number there will be given and, for anything particularly important to our topic here, further details of its publication. In this way, I hope that every relevant cited text can be easily located, even by somebody who has never previously ventured into the papyrological literature. The Euclidean texts will be included in a forthcoming *Corpus dei Papiri Filosofici*, to be published by the Società Italiana per la ricerca dei Papiri greci e latini in Egitto.

[19] See Diogenes Laertius-Hicks, *LEP*, III 37: "Some say that Phillip of Opus copied out the *Laws* [of Plato], which were left on waxen tablets." For illustrations of tablets, complete with the schoolboys' writing exercises, see Turner, *GMAW*, Plate 4, and *PW*, Plate 6; other examples will be cited in the next chapter. Tax-collectors' receipts and calculations were often made on ostraca; see Plate 6 for an example.

6.2 Early written evidence

papyrus rolls: the pith of the papyrus plant was fashioned into oblong sheets which were then immediately joined by the maker to form rolls about thirty feet long. The height of the rolls varied greatly, and might be altered by cutting, but a typical range of heights for literary texts would be eight to twelve inches.[20] An exceptionally narrow roll, P. Mich. iii 146, only four inches high, will be cited in the next chapter; it is a pocket-sized ready reckoner. The writing was set out, column by column, side by side on the roll, in capital, unaccented letters without gaps between words, but with the words occurring at the end of a line divided between lines according to strictly observed rules.[21]

SURVIVINGEARLYGREEKPAPYRIARESETOUTJUSTLIKETHISAN
DINSCRIPTIONSARESOMETIMESCARVEDBOUSTROPHEDONICAL
LYBACKWARDSANDFORWARDSONALTERNATELINESINDIVERS
DIALECTSINALIVINGCHANGINGVARIETYOFSTYLESASSOCIATED
WITHLEGALADMINISTRATIVELITERARYSOCIALANDTECHNICAL
SUBJECTSANDINANEVOLVINGANDVARYINGSTYLEOFHANDS ~ ~

Accents, marks to resolve ambiguous word divisions, and even indications of change of speaker in a play, were only slowly and irregularly introduced during the Alexandrine period, and it is not before the ninth-century Byzantine manuscript tradition that written Greek approached its now standard form, with upper and lower case letters, spaces separating words, and an elaborate array of accents and breathings; this development will be described in the next section. The most common type now used for printed Greek is based on the beautiful Greek handwriting of the Cambridge scholar Richard Porson (1759–1808) which itself was based on this Byzantine script; it was designed and first used by Cambridge University Press for his posthumously published editions of Euripides.[22] For an example of a fount designed to correspond more closely to early Greek script see Plate 2, a reproduction from volume i of *The Oxyrhynchus Papyri*; this experiment was gradually abandoned after a few years, and the editors of the series now use

---

[20] I am here only discussing Greek texts (i.e. texts written in Greek). Papyrus as a writing material is much older: around 3000 BC, at Saqqara, a blank roll was placed in the tomb of the Vizier Hemaka for the use of the dead man. See Černý, *PBAE*; Cockle, RCP; Lucas, *AEMI*; and Lewis, *PCA*.

[21] See Turner, *GMAW*, 19–20. One rule, about division between doubled consonants, may have forced a scribe in an example which follows to cram in an extra μ above the line when splitting the word γραμ|μη; see Turner, Fowler, Koenen, & Youtie, EEI, on lines 27–8 of P. Mich. iii 143.

[22] See Sandys, *HCS* ii, 424–30. There is a facsimile of Porson's handwriting in Watson, *The Life of Richard Porson, M.A.*, opposite p. 260. (Porson was celebrated for his witticisms and epigrams; this is a Greek epigram boldly rendered into English: "The Germans in Greek | Are sadly to seek; | Not five in fivescore, | But ninety five more: | All save only Herman, | And Herman's a German.")

Porson's Greek types, with a recently introduced and increasingly popular variation, which I have adopted, of a sigma in the form of a *c*, without any distinction between medial and final forms.[23] (In fragmentary unspaced texts, such a distinction may be problematic or impossible to determine.)

Papyrus rolls are fragile and will tear, if handled too roughly; they are inconvenient, as any microfilm user will appreciate: Athenian vase-paintings show readers getting into difficulties with a twisted roll, and the aged Verginius Rufus broke a hip while trying to collect up one he had dropped;[24] but they continued to be the main vehicle for permanent records until between the second and sixth centuries AD, when a gradual process of 'codification', of transcription of manuscripts into codices, was carried out. (A codex is a forerunner of the modern book: flat sheets of material are gathered, folded, and stitched. The word derives from the Latin name of a hinged set of wooden tables. Early codices were made from either papyrus or parchment.[25])

Do we then have a papyrus roll containing an early manuscript of the *Elements*? That would be far too much to expect. Papyrus will rot unless kept perfectly dry, though it can be preserved in a damp climate if it is baked at a high temperature until it is carbonised, and then kept undisturbed—like burned paper, it is then useless for everyday purposes, and the writing is frequently illegible, but it is in a stable and relatively indestructible state. This is certainly the main reason why only two pieces of papyrus have been found in Greece, both of them in tombs. The oldest dates from the second half of the fifth century, and so is perhaps the earliest surviving Greek papyrus; it comes from a recently uncovered tomb between Athens and Sounion which contained a male skeleton, a papyrus roll, a bronze pen with a split nib, the remains of a tortoise shell lyre, and two oil flasks; "The papyrus, a shapeless, flattened mass, will need a miracle to be preserved and unrolled, as it did to survive 2500 years".[26] The second piece, a line-by-line commentary on an Orphic cosmological poem, was found in 1963, in Derveni, north of Thessaloniki; it has been preserved through being baked during the burial rites.[27] These two texts, together with a cache found in Herculaneum, to

---

[23] In fact, on the insubstantial evidence cited in the previous note, Porson mainly used the *c* sigma; but, curiously, there is one σ, in Τεύτονοσ, where we now would use ς.
[24] See Turner, *GP*, 7.
[25] The origins and development of the Greek codex are still far from understood, and literary texts on papyrus rolls are still occasionally found in the ninth century. See Roberts & Skeat, *BC*, who also discuss the transition from papyrus to parchment.
[26] Report of *The Times*, 25 May 1981, supplemented by Cockle, RCP, 147.
[27] There is a very interesting preliminary account of the discovery of the Derveni papyrus, followed by a most illuminating transcript of a discussion, in Kapsomenos, OPRT; for a photograph, brief discussion, and bibliography, see Turner, *GMAW*, no. 51. The text still has not had its definitive publication; see Turner, Tsantsanoglou, & Parássoglou, ODP.

be described below, may be the only written (again, not inscribed) Greek texts that have not passed through Egypt in the course of their transmission down to us.

No surviving Greek papyrus found in Egypt can be confidently dated by handwriting, context, or other internal evidence to before Alexander's conquest of Egypt, though a roll of *Persae* by Timotheus, Pack no. 1537, was found, in 1902, at Abusir, in a sarcophagus that has been dated by some, on archaeological evidence, to the middle of the fourth century, though this is disputed.[28] For a long time, the oldest document that could be securely dated on internal evidence was P. Eleph. 1, a marriage contract of 311 BC from Elephantine Island, Assuan.[29] This may now have been supplanted by a boldly written notice found in the Sacred Animal Necropolis at Saqqara which reads: "Of Peukestas. No one is to pass. The chamber is that of a priest. (ΠΕΥΚΕΣΤΟΥ | ΜΗΠΑΡΑΠΟΡΕΥΕΣ-ΘΑΙΜΗ | ΔΕΝΑΙΕΡΕΙΩΣΤΟΟΙΚΗΜΑ)", or, more colloquially, "By order of the commander-in-chief Peukestas. Out of bounds to troops. Ritual areas." This notice can perhaps be dated, on the evidence of the name of Peukestas, to around 330 BC.[30]

Let us now turn to technical and Euclidean texts. Perhaps our earliest scientific text is P. Hib. i 27, a parapegma related to Eudoxus' calendar; this gives a very clear and important illustration of early numerical practice, and it will be described and reproduced in the next chapter (see

---

[28] For a bibliography, see Pack no. 1537.

[29] The text and translation is in Hunt & Edgar, *SP* i, 2–5, and it is worth quoting in full: "In the seventh year of the reign of Alexander, son of Alexander, the fourteenth year of the satrapship of Ptolemy, in the month Dius. Marriage contract of Heraclides and Demetria. Heraclides takes as his lawful wife Demetria, Coan, both being freeborn, from her father Leptines, Coan, and her mother Philotis, bringing clothing and ornaments to the value of 1000 drachmae, and Heraclides shall supply to Demetria all that is proper for a freeborn wife, and we shall live together wherever it seems best to Leptines and Heraclides consulting in common. If Demetria is discovered doing any evil to the shame of her husband Heraclides, she shall be deprived of all that she brought, but Heraclides shall prove whatever he alleges against Demetria before three men whom they both accept. It shall not be lawful for Heraclides to bring home another wife in insult of Demetria nor to have children by another woman nor to do any evil against Demetria on any pretext. If Heraclides is discovered doing any of these things and Demetria proves it before three men whom they both accept, Heraclides shall give back to Demetria the dowry of 1000 drachmae which she brought and shall moreover forfeit 1000 drachmae of the silver coinage of Alexander. Demetria and those aiding Demetria to exact payment shall have the right of execution, as if derived from a legally decided action, upon the person of Heraclides and upon all the property of Heraclides both on land and on water. The contract shall be valid in every respect, wherever Heraclides may produce it against Demetria, or Demetria and those aiding Demetria to exact payment may produce it against Heraclides, as if the agreement had been made in that place. Heraclides and Demetria shall have the right to keep the contracts severally in their own custody and to produce them against each other. Witnesses: Cleon, Gelan; Anticrates, Temnian; Lysis, Temnian; Dionysius, Temnian; Aristomachus, Cyrenaean; Aristodicus, Coan."

[30] See Turner, CCOS.

Section 7.1(c) and Plate 5). Our earliest glimpse of Euclidean material will be the most remarkable for a thousand years: six fragmentary ostraca containing text and a figure, Pack no. 2323, found on Elephantine Island in 1906/7 and 1907/8 and published, with photograph and commentary, in Mau & Müller, MOBS. These texts are early, though still more than 100 years after the death of Plato (they are dated on palaeographic grounds to the third quarter of the third century BC); advanced (they deal with the results found in *Elements* XIII, 10 and 16, on the pentagon, hexagon, decagon, and icosahedron); and they do not follow the text of the *Elements* (though they are clearly connected with the material of these propositions). So they give evidence of someone in the third century BC, located more than 500 miles south of Alexandria, working through this difficult material; and the fact that the texts are on potsherds and are not the received text of the *Elements* suggests that this may be an attempt to understand the mathematics, and not a slavish copying or learning of the material. For example, the text jumps straight from the middle of Proposition 16 (Heath, *TBEE* iii, on p. 483) to the porism found at the end, and it contains nothing on the classification of the side as a minor line. These fragments are the evidence, mentioned earlier, which corroborates assigning Euclid to c. 300 BC,[31] but one complication in this proposal is that we cannot be sure of their relationship to the received text of the *Elements*: it is not inconceivable that the fragments represent a preliminary version of the final text, not the other way round, though they provide, in themselves, slender evidence for such a proposal.

One of the first and most spectacular discoveries of papyrus was at Herculaneum: a library of some 1800 rolls that had been buried in boiling volcanic ash and rain during the eruption of Vesuvius in AD 79. The town was rediscovered when a well was being dug in the gardens of a house on the lower slopes of Mount Vesuvius in 1752, and it has subsequently been excavated. The publication of these texts became a prestige project commissioned by the Bourbon Kings of the Two Sicilies and, between 1802 and 1806, the Prince of Wales (later King George IV) sent his chaplain John Hayter to supervise the unrolling of the papyri; then, in 1819 and 1820, he sent Sir Humphry Davy (later President of the Royal Society) and the Reverend Peter Elmsley (later Camden Professor of Ancient History at Oxford) to make chemical experiments on the papyri and to decipher them.[32] Initially great expectations were aroused that their contents might unlock the secrets of antiquity, but unfortunately

---

[31] The first reference to these ostraca in connection with the problem of locating Euclid appears to be Fraser, *PA* ii, 558, n. 43.
[32] See Davy, SOEPFRH. A survey of recent restoration work on the papyri, especially by A. Fackelmann, is given in Cockle, RCP, 158. There is a recent bibliography of the Herculaneum papyri, Gigante, *CPE*; and Cavallo, LSSE, contains sixty-four remarkable plates of photographs, by T. Dorandi, of fragments of the collection.

very little of the material proved to be legible, and the books in the library are mostly restricted to Epicurean philosophy. One of the papyrus rolls (P. Herc. 1061) does provide us, however, with a second glimpse of the *Elements*, in an essay by Demetrius Lacon. The title page of the roll, apart from the author's name and a few letters at the end, has broken away and much of the text that follows is either missing or illegible. Generally the state of the texts is such that photographs show little of use (see Plate 1); but as the various papyrus rolls from Herculaneum have been unrolled or broken open, layer by layer, sometimes inevitably being destroyed in the process, they have been drawn, and some drawings and a transcription of column 9, from Heiberg, QPTM, are also given in Plate 1.[33] Column 8 lines 9–17 cites *Elements* I Definition 15, the definition of a circle; column 9 cites *Elements* I 9, the proposition on bisecting an angle, and its figure and a summary of the proof are given in column 10; column 9, lines 12–15 cites *Elements* I 3, a proposition that is used in the proof of I 9; and finally column 11 gives the enunciation and figure of I 10. The text itself appears to be a defence of an Epicurean doctrine that allowed the possibility of an infinite divisibility of a line, against criticism by Polyaenus, and is believed to date from around 100 BC.

Our next Euclidean fragment[34] comes from a dramatic and extensive find of papyri at Oxyrhynchus, in January 1897, by B. P. Grenfell and A. S. Hunt, in one of a series of expeditions for papyri sponsored by the Egypt Exploration Fund (later to become the Egypt Exploration Society). Oxyrhynchus was a large Graeco-Roman town in the desert to the west of and above the Nile valley, adjacent to the modern Arab village Behnesa, about 120 miles south of Cairo. A series of rubbish mounds yielded an enormous quantity of papyrus: "The flow of papyri soon became a torrent it was difficult to keep pace with",[35] and indeed about half of the papyri found there still remain to be transcribed and published. Publication is still taking place in the series *The Oxyrhynchus Papyri*, 53 volumes to 1986; and volume i, edited by Grenfell and Hunt and published in 1898, included a fragment containing the enunciation of *Elements* II 5, together with a roughly drawn figure (see Plate 2). This fragment is dated by Grenfell and Hunt to the third or fourth century AD, but Sir Eric Turner has revised this dating in some notes that he very kindly provided me on the published Euclidean fragments from Egypt.

---

[33] Heiberg's conclusions have not all been accepted by later scholars; see, for example, De Falco, *EDL*. Further work on this fragment, by T. Dorandi, is in progress, in connection with the *Corpus dei Papiri Filosofici* (see n. 18, above).
[34] The next three texts are unreliably reprinted in Euclid–Stamatis, *EE* i, Appendix II.
[35] See Turner, *GP*, Chapter 3 and James, ed., *EEEES*, for full and exciting accounts of this and other discoveries. This quotation is from Grenfell & Hunt's own account, cited in *GP*, 29.

Of this, he wrote[36] (with my additions in square brackets):

P. Oxy. i 29. Height 8.5 cm, width 15.2 cm.
A *kollesis* (a pasted join between two sheets of papyrus, the join being made by the manufacturer at the time the roll was put together) runs through the middle of the column of writing. Thick and poor quality papyrus. Writing in a single broad column. Lower margin about 3 cm, broken above. The *kollesis* and the fact that this is a II 5 show that the fragment is almost certainly from a longer (? complete) manuscript of Euclid [to which can also be added the evidence of the broken letters in the first line: these probably come from the end of either the enunciation, or the conclusion of II 4]. An empty space at the left (with no trace of the preceding column) of at least 3 cm shows that this was a generously laid out manuscript. But the figure is drawn without rule, free-hand, and carelessly [and is unlabelled].

The handwriting is a medium to large rounded capital, roughly bilinear, fairly quickly written, clear to read but with no claim to be regarded as calligraphic. I should date it confidently to the end of century i AD/early century ii, say AD 75–125. ... [Here follows a discussion of palaeographical features, supporting this date; see note 37.]

Note that this text was one of the first finds of Grenfell & Hunt at Behnesa (Oxyrhynchus). They published in a great hurry, and no doubt included this piece to illustrate the wide range of subjects on which papyri could throw light. Their palaeographical (especially dating) framework was still to be worked out. By 1902 I do not think that the dating offered would have been 'third or fourth century'.

Their transcription is very good. Of line 1 only περ is certainly read. The traces are compatible with the restoration [of the enunciation or conclusion of II 4], but a guess of a different word could be verified, i.e. stated to be acceptable or not. Between lines 1 and 2, a paragraphus indicated that a new section begins: also see the cursive ε' = number 5 in left-hand margin and the enlarged initial letter of εαν.

I doubt whether the restoration of the figure at the end of lines 4 and 5 is correct [also see the discussion below], because on this restoration there is no reason why in line 2 the writing after γραμμα should not continue to the right, instead of resuming at the beginning of line 3. Perhaps the figure was something quite different. All there is to go on is the short portion of a horizontal line visible more or less accidentally, since the papyrus is broken off above line 4, περιεχομενον.

---

[36] Personal communication. Turner's query about the dating was noted in Pack, *GLLTGRE*, no. 368.

[37] Turner's notes continue: "The verticals of ι τ η γ ν π often terminate at the foot with left-pointing serifs, υ is widely spread. The hand is rounded and fairly upright, but the writing takes an upward slope to the right. Note the τ in μετοξυ (and elsewhere), the frequent cursive ε, and the enlarged initial ε in εαν. Some parallels for the handwriting, in general, in dated documents are: P. Oxy. 378 (BM 809) col. ii (Domitian); P. Ryl. 597; P. Amh. 66 (AD 124); P. Lond. 298 (AD 124); and among undated documents: P. Oxy. 1793 (Callimachus); P. Ryl. 484 (assigned to this period); P. Lit. Lond. 132 (Hyperides). There are tremata on initial ῖ, ϋ. One final ν is extended to fill the line (this is not significant for the date)."

6.2 Early written evidence

The two 'errors' in Greek are superficial: τετραγωνου is probably attracted into the genitive by the preceding ημιςειας. In μετοξυ the *o* is not an error but the use of an alternative form of the word common in the later koine.

I have the impression that there *may* be ink (one or two letters?) at the top left of the figure level with line 9 and at the lower right. But my photograph has gone brown, and these may not really be ink.

The text of II 5 reads, in Heath's translation:

If a straight line be cut into equal and unequal segments, the rectangle contained by the unequal segments of the whole together with the square on the straight line between the points of section is equal to the square on the half.

There is a space to the right of lines 1 to 4, and traces of a horizontal line on line 4, invisible on the photograph in Plate 2: these led Grenfell and Hunt to propose that the space was taken up with a figure showing two lines: 'a straight line cut into equal segments' and 'a straight line cut into unequal segments'. As Turner points out above, and because such a figure is quite unprecedented in our early evidence of the text, this is implausible. The question of the figure is discussed by Heiberg who also disagrees with this restoration, but I cannot understand fully his alternative proposal.[38] An alternative reconstruction, which I have not seen discussed in the literature, is as follows. As was noted above, the broken words in line 1 could come either from the enunciation (*protasis*) or the conclusion (*sumperasma*) of II 4, since the conclusion is a word-for-word repetition of the enunciation.[39] This fragment does not appear to come from a scribe's copy of the work; with its quickly written hand on poor quality papyrus, it is more likely to be working notes by somebody who

---

[38] Heiberg, QPTM, 148: "La figure, elle aussi, est identique à celle de nos manuscrits; il est étrange qu'elle ne paraisse pas porter trace de lettres et qu'elle soit placée immédiatement après la protase: dans nos manuscrits mathématiques, la figure se trouve régulièrement à la fin du théorème. Peut-être cette place insolite fournit-elle la raison pour que l'έ ajouté désigne la figure comme appartenant à ce théorème. Les éditeurs expliquent le peu de longueur des lignes 2–4 en admettant qu'il y a eu à leur droite une figure, soit deux lignes droites divisées pour exemplifier les mots εἰς ἴςα καὶ ἄνιςα; ils ajoutent qu'on ne voit les traces que de la ligne inférieure. Mais une figure de ce genre est tout à fait inouïe dans nos manuscrits mathématiques et on ne peut plus inutile, en sorte que cette explication est sujette à caution. On devrait plutôt penser que la fin nécessaire du théorème précédent (II 4) ὅπερ ἔδει δεῖξαι, qui, selon les éditeurs, a rempli, sous quelque forme abrégée, le reste de la ligne 1, a eu la distribution que voici: ὅπερ ἔδει a eu sa place à la ligne 1, tandis que δεῖξαι s'est trouvé au-dessous, devant les ligne 2–3; alors la barre qui suit la ligne 4 a pu être mise comme signe de séparation." But what is meant by "δεῖξαι s'est trouvé au-dessous, *devant* les lignes 2–3"? There is an ample margin to the left which contains no trace of writing.

[39] For this reason the conclusion is often omitted from editions and translations of the *Elements*; see how Heath generally finishes his translations with "Therefore etc." from I 4 onwards, as explained in his note to that proposition. Ancient authors, editors (perhaps even Euclid), or scribes did the same: the conclusion is often omitted in Books VII and IX, or cut short, from Book X onwards, with the words καὶ τὰ ἑξῆς (and the rest); see Heath, *TBEE* i, 58.

understood the material. This opinion is corroborated by the roughly drawn figure and, in particular, by the way in which the figure is unlabelled and the diagonal line is undoubled: the diagonal enters only in the construction of the figure and does not play any part in the proof, and an unlabelled figure would provide poor help for understanding the text of Euclid's proof. The text could have been written by somebody who wrote down the enunciation, drew the figure, and then worked through the proof doubling the lines on the figure as he did so, but without writing down any of the details of the proof or labelling the figure; observe the generous space immediately below the last line and the second paragraphus at the end of the fragment. (The proof is a very long and elementary verification of the associated geometrical construction; see the discussion of Book II in Chapter 3.) This would explain why the figure came immediately after the enunciation, rather than at the end of the proof, why the figure is unlabelled and the diagonal is undoubled, and it would also explain line 1 and the gap to the right of line 1 to 4: these could have contained, in similar style, the last line of the *enunciation* of II 4 and the associated figure for this proposition; and the fragment could have been part of a manuscript of notes by someone working through Book II of the *Elements*.

The next text, P. Fay. 9, is indeed fragmentary; its photograph and transcription are reproduced in Plate 3. It was edited and published in 1900 in Grenfell, Hunt, & Hogarth, *FTP,* along with other papyri found in the Fayûm, a region some fifty miles north of Oxyrhynchus.[40] It is assigned to the latter half of the second century AD and it contains part of *Elements* 39 and 41. The difficulties of this fragment arise from the fact that no single complete line of text survives, and when the surviving letters are compared with the received text, the lines of the fragment seem to contain a very irregular number of letters: between nearly corresponding points on the first ten lines of the fragment, there are 32, 39, 81, 26, 41, 38, 36, 40, and 23 letters in the received text, to be described in the next section.[41] It is worth quoting part of the editors' discussion:

The general tendency of the fragment is towards compression; and some agreements with the manuscript called p [= Paris gr. 2466] are noticeable. The irregularities of the text followed by the papyrus extended to the order of propositions. Proposition 39 is immediately succeeded by Proposition 41. Proposition 40 was either omitted or else placed in some other position. It is noticeable that the diagram of Proposition 39 is drawn at the end of the

---

[40] The text is reprinted in Euclid–Stamatis, *EE* i, 188 f., but there is a displacement of its vertical alignment and some of the spacing is incorrectly reproduced, and the final two lines of fragment (b) should read:] ταυταις α] | πα]ραλληλ[ο]ς [.

[41] These figures, which are based on my own counts, differ slightly from those given by the editors.

6.2 Early written evidence

demonstration, instead of, as is usual, at the beginning. [Incorrect: the diagrams in the manuscripts usually come at the end of the demonstrations. See note 37 and Plate 4.]

An easy explanation of the eccentricities of this fragment would be obtained if it could be supposed that it did not form part of a regular book, but was merely an imperfectly remembered exercise. But this is not a satisfactory view. The words are correctly spelled, and the handwriting, though not of the regular literary type, is by no means ill formed. Its date is apparently the latter half of the second century AD. The papyrus was found with a number of documents belonging to the reigns of Antoninus, Marcus, and Commodus.

Sir Eric wrote of this fragment:

I am inclined not to dismiss this text as yet. At least it deserves further scrutiny, but it is going to be difficult to do it. The photographs from Mount Holyoke show that it has come to pieces (of course, perhaps it was always in pieces[42]), and it has been put into the frames just any how. The placing of the pieces doesn't agree with Grenfell & Hunt's publication, and they are obviously wrong in their relation to each other.

Grenfell & Hunt's date is probably correct enough. On page 97 they point out that the papyrus does not seem to correspond with the manuscripts, and leave it at that; I wonder if this is sufficient.

The reason why I think it would be interesting to know more is in the drawing of the figure, and its two surviving letters. It seems to be that these letters are written by the original scribe, but with a different, sharper pen (perhaps with the one with which he ruled the lines of the figure.) In that case he was taking some trouble (using a rule), and perhaps these indications are enough to suggest that we have part of a real manuscript for the *Elements* here, and its difficulties and divergencies deserve further investigation.

The challenge of matching the fragment to the text had already been taken up by Heiberg, who also gave an extended discussion of the question of the missing Proposition 40. Unfortunately his proposals are too long and detailed to quote or summarise here, but his overall conclusion is that this fragment, as he reconstructs it, diverges greatly from our text as received from other sources; but that this in no way justifies rejecting the text, in view of the character of the writing.[43]

---

[42] Miss Anne Edmonds, Librarian of Mount Holyoke College, where the papyrus is now kept, has kindly sent me copies of correspondence with organisers of the Egypt Exploration Fund that indicate that the fragment was already badly broken up when it was received by the college in 1903/4: "The last [papyrus] may be a puzzle or lesson to some mathematicians who can read Greek. I fear that it is very much more mutilated now than it was when the account of it was written" (Letter from W. W. Goodwin, June 12th, 1904).

[43] Heiberg, PE. After a brief discussion of P. Herc. 1061 and P. Oxy. i 29 on pp. 47–48, his discussion of this text and his proposed reconstruction follows on pp. 48–53. Here is his general conclusion: "Bei dieser Reconstruction bin ich so wenig wie möglich von unseren Hss. abgegangen, und wenn auch das positive hier und da unsicher bleibt, steht das negative Ergebniss, wozu schon die Herausgeber gelangt sind, unumstösslich fest, dass der

Therefore the fragment remains as a tantalising unresolved puzzle: that the surviving papyrus fragment that most closely resembles a professional scribe's copy differs so from the received text.

The final published Euclidean papyrus is a fragment containing the first ten definitions of Book I of the *Elements*, P. Mich. iii 143, published by F. E. Robbins in Winter, Robbins, & others, *PUMC* iii, 26–8. Sir Eric's notes began: "I have compared the photograph and text as published. At a number of points I would quarrel with the transcript (all would need to be checked with the original) . . . ." The subsequent reexamination has led to a republication of the text in which the original editor's assessment that it "might well be the attempt of an older youth to reproduce from memory definitions imperfectly learned" has now been reversed to "One might guess that the piece was written by a schoolmaster, a grammaticus, making sure he has the text which he wanted to dictate to his students or on which he intended to base his next lesson"! For further details and a photograph, see the republication, Turner, Fowler, Koenen, & Youtie, EEI.

Such is the extent of the published Euclidean papyrus fragments: material corresponding to about sixty complete lines of Heiberg's text, plus fragmentary information about a further sixty lines.[44] Moreover, neither the earliest evidence in the Elephantine ostraca, nor that in the fragment that most resembles a professional scribe's copy, P. Fay. 9, follows the received text. In addition to these fragments on papyrus, there is a parchment palimpsest from the seventh or eighth century, British Library Add. gr. 17211, which consists of five partially legible parchment leaves containing parts of Book X and Book XIII Propositions 14 (but there numbered 19), which was collated by Heiberg in his edition of the *Elements*.[45]

There is a persistent and widespread belief that the lack of surviving texts from this early period is due to the burning of the library of Alexandria by the Arabs after their conquest in AD 642. The evidence

---

Papyrus sehr stark von unserem Text abweicht, der allgemein als vortrefflich gilt. Man könnte daher versucht sein, wie es auch geschehen ist, das ganze wegzuwerfen als eine verwilderte Ueberlieferung. Die Herausgeber haben an ein 'imperfectly remembered exercise' gedacht, verwerfen aber mit vollem Recht diesen Gedanken angesichts der correcten Buchstabirung und des ganzen Schriftcharakters. Dass wir ein wirkliches Buch vor uns haben, bestätigt die Normalzeile, und es kommen noch andere Umstände hinzu, wodurch die einfache Verwerfung dieser unserer ältesten handschriftlichen Quelle ganz unmöglich wird" (p. 50).

[44] At least one more Euclidean text can be expected from the Oxyrhynchus collection.

[45] A palimpsest is a parchment (as here) or other medium that has been washed or rubbed clean to make way for another text; sometimes parts of the lower text can still be read. Here the *Elements* was the earlier text, replaced by a text of Homer; the legible parts of the original are published in Heiberg, PEE. There is a brief general account of palimpsest discoveries in Reynolds & Wilson, *SS*, 174 ff.

connected with this story is very well and authoritatively presented in Butler, *The Arab Conquest of Egypt* (1902), 401–26, and the supplemented second edition by P. M. Fraser (1978) adds a summary of subsequent work, which does not significantly alter Butler's conclusions.[46] I can do no better than to quote two passages, and refer the reader to this book for a full discussion.

The story as it stands in Abû 'l Faraj is well known and runs as follows. There as at this time a man, who won high renown among the Muslims, named John the Grammarian. He was an Alexandrian, and apparently had been a Coptic priest, but was deprived of his office owing to some heresy by a council of bishops held at Babylon. He lived to see the capture of Alexandria by the Arabs, and made the acquaintance of 'Amr, whose clear and active mind was no less astonished than delighted with John's intellectual acuteness and great learning. Emboldened by 'Amr's favour, John one day remarked, 'You have examined the whole city, and have set your seal on every kind of valuable: I make no claim for aught that is useful to you, but things useless to you may be of service to us.' 'What are you thinking of?' said 'Amr. 'The books of wisdom,' said John, 'which are in the imperial treasuries.' 'That', replied 'Amr, 'is a matter on which I can give no order without the authority of the Caliph.' A letter accordingly was written, putting the question of Omar, who answered: 'Touching the books you mention, if what is written in them agrees with the Book of God, they are not required: if it disagrees, they are not desired. Destroy them therefore.' On receipt of this judgement, 'Amr accordingly ordered the books to be distributed among the baths of Alexandria and used as fuel for heating: it took six months to consume them. 'Listen and wonder', adds the writer [*ACE*, 401–2].

After twenty pages of discussion, Butler concludes:

It may not be amiss to briefly recapitulate the argument. The problem being to discover the truth or falsehood of the story which charges the Arabs with burning the Alexandrian Library, I have shown—
(i) that the story makes its first appearance more than five hundred years after the event to which it relates;
(ii) that on analysis the details of the story resolve into absurdities;
(iii) that the principal actor in the story, viz. John Philoponus, was dead long before the Saracens invaded Egypt;
(iv) that of the two great public Libraries to which the story could refer, (a) the Museum Library perished in the conflagration caused by Julius Caesar, or, if not, then at a date not less than four hundred years anterior to the Arab conquest; while (b) the Serapeum Library either was removed prior to the year 391, or was then dispersed or destroyed, so that in any case it disappeared two and a half centuries before the conquest;
(v) that fifth, sixth, and early seventh century literature contains no mention of the existence of any such Library;
(vi) that if, nevertheless, it had existed when Cyrus set his hand to the treaty

---

[46] Additional references, for which I am indebted to P. M. Fraser, are Furlani, GFIBA and SIBA, and Meyerhoff, JGPAAM.

surrendering Alexandria, yet the books would almost certainly have been removed—under the clause permitting the removal of valuables—during the eleven months' armistice which intervened between the signature of the convention and the actual entry of the Arabs into the city; and

(vii) that if the Library had been removed, or if it had been destroyed, the almost contemporary historian and man of letters, John of Nikiou, could not have passed over its disappearance in total silence.

The conclusion of the whole matter can be no longer doubtful. ... One must pronounce that Abû 'l Faraj's story is a mere fable, totally destitute of historical foundation.

My only concern in this matter has been to establish the truth, not to defend the Arabs. No defence is necessary: were it needful, it would not be difficult to find something in the nature of an apology. For the Arabs in later times certainly set great store by all the classical and other books which fell into their hands and had them carefully preserved and in many cases translated. Indeed they set an example which modern conquerors might well have followed. ... [*ACE* 424–6].

For the answer to the question posed at the beginning of this section, let us now turn to our most important source of information about Greek literature and science: Byzantine and medieval manuscripts from the ninth century AD onwards.

6.3 The introduction of minuscule script

Two developments in writing appear to have had an important effect on the survival and transmission of Greek manuscripts. The first, the change from the papyrus roll to the codex form, was described briefly in the last section;[47] the second, to be described here, was the development of a new script.[48] Although the further developments in the substrate of writing—the change from papyrus to parchment and the introduction of paper by the Arabs, who learnt the process from Chinese prisoners of war taken at Samarkand, perhaps in AD 751—were ultimately to prove very important in the spread of written and, later, printed books, in the short run they were less influential.

Up to the beginning of the ninth century AD, almost all Greek literary works continued to be written in variants of the unspaced capital script illustrated in the preceding section. This seemed to be slow to write, to be difficult to read, and to require a generous area of papyrus or parchment. Early administrative and financial documents used a wide range of abbreviated, cursive scripts that can be very difficult to decipher; for an example, see Plate 6. Various uses of cursive scripts for literary texts had not previously led to the emergence of a distinctive, stable, and

---

[47] See, especially, n. 25.
[48] My account will be very heavily based on Lemerle, *PHB*, Reynolds & Wilson, *SS*, and Wilson, *SB*.

6.3 Minuscule script

widely employed script; but then, sometime around AD 800, in circumstances that are still far from clear, a new kind of compact, cursive script took over and thereafter came to dominate the production of literary texts. This script soon came to employ majuscule and minuscule letters, ligatured groups of letters, spaced words, and a full range of accents and breathings; and very much more text could be fitted onto the page. It is now generally referred to as 'minuscule' script, and almost all of the Greek texts that survive have been through a process of minusculisation; indeed this was, in most cases, necessary for their preservation, as very little survives in the old script.

The first precisely dated manuscript in minuscule, Leningrad gr. 219, copied in AD 835, is a Gospel named after the Archimandrite Porphoryrij Uspenskij; another text which contains a few leaves of minuscule, and may date from 813–20, is Leiden B.P.G. 78, a copy of Ptolemy's *Handy Tables* with a commentary by Theon. Since the minuscule script in both of these examples does not look primitive or experimental, it may have been developed as much as fifty years earlier. The other scientific manuscripts that survive from this early period illustrate further how the old capital script still continued in use (Ptolemy's *Almagest* survives complete in the older style in Paris gr. 2389 and in minuscule in Vatican gr. 1594, and the old capital form is often retained for the headings of books and chapters that are otherwise written in minuscule), and how the earliest minuscule versions might not necessarily be the best (Euclid's *Elements* is found first in a beautiful minuscule manuscript dated to AD 888, Bodleian D'Orville 301 (see Plate 4), then a better text is found in the slightly later minuscule Vatican gr. 190; these texts will be referred to by the letters B and P respectively[49]). Another ninth-century manuscript, Vatican gr. 204, contains, in the following order, an important collection of scientific texts:[50] Theodosius, *Sphaerica*; Autolycus, *On the Moving Sphere*; Theon's recension of Euclid's *Optics* with *Prolegomena*; Euclid, *Phaenomena*;[51] Theodosius, *On Habitations, On Nights and Days, and On Days and Nights* II; Aristarchus, *On the Sizes and Distances of the Sun and Moon*; Autolycus, *On Risings and Settings* I and II;

---

[49] It is editorial practice to associate letters, called sigla, with the manuscripts used in a critical edition of a text. The *conspectus siglorum* for Euclid–Heiberg, *Opera*, is P = Vatican gr. 190 (first used by Peyrard for his French translation of 1804 and 1809); B = Bodleian D'Orville gr. 301; F = Florence Laurenziana gr. 28.3; V = Vienna phil. gr. 103; b = Bologna gr. 18 and 19; p = Paris Bibliothèque Nationale gr. 2466; q = Paris gr. 2344; and L = London Add. gr. 17211. For more details of these manuscripts, see vol. v of *Opera*, reprinted in Euclid–Stamatis, *EE*, v_1 (in Latin) or Heath, *TBEE* i, Chapter 5 (in English). Further Euclidean manuscripts are described in Heiberg, PE.

[50] This is taken from the description, due to Menge, given in Aristarchus–Heath, *AS*, 325 f.

[51] There is a photograph of an opening of this text in Neugebauer, *HAMA* iii, Plate VIII, and a very interesting discussion of text figures in general, and those in this manuscript in particular, in ii 751–5.

Hypsicles, *On Rising-times*;[52] Euclid, *Catoptrica*; Eutocius, *Commentary on Books I-III of Apollonius' Conics*; Euclid, *Data*; Marinus, *Commentary on Euclid's Data*; and *Scholia to Euclid's Elements*, incomplete at the end.

Two people in the ninth and early tenth century were important in the preservation and transmission of Greek scientific texts: Leo the Philosopher (c. 790–after 869), Archbishop of Thessaloniki 840–3 and later head of the school established under the emperor Bardas; and Arethas (c. 850–?), Archbishop of Caesarea in Cappadocia.[53] Leo's interest in mathematics and astronomy led him to acquire a copy of some works of Archimedes, Apollonius' *On Conic Sections*, some works on mechanics by the unknown authors Marcellus and Quirinus, a geometrical treatise by Proclus, and a work by Theon; we know this[54] from some obscure epigrams he wrote, preserved in the *Greek Anthology* IX 200, 202, and 578, which do not permit closer identification. He certainly must have possessed a copy of Ptolemy's *Almagest* though not, as is commonly supposed,[55] the manuscript Vatican gr. 1594, and he wrote a note on *Elements* VI 5 that is found, in the later scribe's hand, in the Euclid manuscript B;[56] but none of the books owned by Leo are known to survive. Eight volumes of the library of Arethas have survived and been identified through his marginal notes. The earliest is our earliest surviving complete manuscript of the *Elements* (see Plate 4); it was written by the scribe Stephanus, possibly one of the most accomplished of Byzantine calligraphers, in AD 888, and it cost fourteen gold pieces, at a time when salaries of civil servants started at seventy-two gold pieces a year, though they could rise in exceptional circumstances to 3500. Arethas also commissioned an important copy of Plato which contains all the main dialogues except the *Republic, Laws,* and *Timaeus*: Bodelian E. D. Clarke 39, written by John the Calligrapher in 895 and costing twenty-one gold pieces. These are the only two manuscripts of his that bear directly on our interests here, though it has been suggested that a copy of Pappus' *Collection* (Vatican gr. 218) might have belonged to his library, on the flimsy grounds that its minuscule script hints at an origin in

---

[52] This is the earliest surviving *Greek* text to divide the circle into 360 degrees, or to show any other influence of Babylonian sexagesimal arithmetic; see Chapter 7, below.

[53] For full biographies, see Wilson, *SB*, Chapters 4 (Leo) and 6 (Arethas), and the *DSB* (Leo).

[54] Except for information about the Archimedes manuscript. Leo may have owned the lost archetype of most of our surviving manuscripts of Archimedes' works; see Chapter 7 n. 45 and text.

[55] See Wilson, TBS, where it is pointed out that the identification is based on a note on folio 263^v, but this is written in a much later hand, of the thirteenth century or afterwards, and is of no particular significance.

[56] See Euclid–Heiberg, *Opera* v, 714 (= Euclid–Stamatis, *EE* v$_2$, 341): *Hupomnēma scholion eis tas tōn logōn sunthesis te kai aphairesin Leontos.*

Constantinople and the subject matter coincides with one of Arethas' interests.[57]

It is evident that most of our texts of the *Elements* derive from an edition made by Theon of Alexandria in the middle of the fourth century AD. The scribe of the manuscript P appears also to have had a pre-Theonine manuscript to which he also referred, and this enables us to get a glimpse of an earlier recension of the text; this manuscript was used by Peyrard for a French translation of parts of the *Elements* of 1804, and it forms the most important text for Heiberg's critical edition, in his *Euclidis Opera Omnia* (1883–1916). The other texts he used and the textual procedures he followed are described, in Latin, in his *Prolegomena Critica*, (Euclid-Stamatis, *EE* v_1, pp. xvi–lxxxix), and the evidence from some of the papyrus fragments and further Byzantine manuscripts is discussed in his later article 'Paralipomena zu Euklid'.[58] All of this material is described in Heath's admirable translation, *The Thirteen Books of Euclid's Elements* (= *TBEE*), Chapters 5 and 6.

The photograph of an opening, folios 35 verso and 36 recto, of the manuscript B, Bodleian D'Orville 301, is given in Plate 4; it contains *Elements* II Proposition 5, starting in the middle of the word ὀρθογώνιον of the enunciation (compare this with the same text on Plate 2), and Proposition 6. Observe the wealth of scholia, some copied by the scribe from his exemplar, some added later, the interlinear annotations, and the many figures. Another facsimile from B, of folio 45 verso, *Elements* III 4 and the beginning of 5, can be found as a frontispiece to Heath, *TBEE* i, with brief description on p. xi.[59]

To summarise and answer the question which opened Section 2: our sources before AD 888 tell us about approximately 1 per cent of the eventual text of the *Elements,* and possibly a similar amount of other scientific and literary texts; almost all of the surviving Greek material comes in the form of Byzantine or medieval manuscripts. On the other hand, the wealth of documentary papyri, dealing with all manner of financial, administrative, legal, and personal matters have vastly increased our understanding of everyday aspects of life in Graeco-Roman Egypt.

---

[57] See Wilson, *SB*, 129–30: "We must not imagine Arethas as being rich enough to commission or acquire all the important or calligraphic manuscripts produced in his lifetime, nor should we think so ill of the culture of the capital as to suppose him the only man of his generation willing to explore the abstruse areas of mathematics".

[58] Beware; the bibliography of Euclid–Stamatis, *EE* i, pp. xxxvii–xxxix does not contain Heiberg's very important article PE, and still less does the reprinted text incorporate any of the material therein; the existence of this article is only noted on *EE* v_1, p. xv. (Concerning this page, also see Chapter 7, n. 1, below.)

[59] For a photographic reproduction of a complete mathematical manuscript, see Heron–Bruins, *CC* i; this manuscript is our only surviving copy of Heron's *Metrica*.

In the next chapter I shall examine in detail part of one small family of Byzantine manuscripts of a short mathematical work, Archimedes' *Measurement of a Circle,* and describe some other kinds of texts, both scientific and commercial, found on papyrus.

NOTE ADDED IN PROOF: There is a detailed description of Vatican gr. 218, together with a new critical edition and translation of part of its contents and essays on matters of reconstruction and interpretation, in the recently-published Pappus of Alexandria, *Book 7 of the Collection,* ed. and tr. A. Jones, Springer-Verlag, 1986.

7

NUMBERS AND FRACTIONS

7.1 Introduction

7.1(a) *Numerals*

Almost the only notation for numbers (i.e. numerals) found in the earliest and best manuscripts of Euclid's works (apart from the *Sectio Canonis*) is used to label the propositions;[1] with the exception of four numerals in *Elements XIII* 11,[2] those few numbers that are used in the text are written out as words. This is an extreme case; most mathematical texts, and all manner of financial, administrative, and legal documents, employ a notation that will underly the main topic of this chapter. But first I shall start by severely delimiting and emphasising the scope of my discussion here:

(i) The Greeks employed two different kinds of numerals, the so-called acrophonic and alphabetic systems. There are wide regional variations in the acrophonic system; the earliest and best-represented version in our evidence, that found in Attica, is based on a variety of different forms of the six elementary symbols: $I = 1$, Π (for πέντε; the right-hand leg of the pi is shorter than the left) $= 5$, Δ (for δέκα) $= 10$, H (for ἑκατόν) $= 100$, X (for χίλιοι) $= 1000$, and M (for μύριοι) $= 1\,0000$, together with four compound symbols, also found in a variety of

---

[1] The definitions are not numbered in the best manuscripts of the *Elements*, and they are grouped together in blocks connected by δέ. For example, in Book I, Definitions 1–3, 4–6, 7–10, 11–12, 15–18, and 19–22 are so connected; for a discussion, see Turner, Fowler, Koenen, & Youtie, EEI, 19–20. The numbers that occur in the manuscripts of the (possibly pseudo-Euclidean) *Sectio Canonis,* Proposition [9] (here the propositions are unnumbered) require some emendation for them to make sense; see Section 4.5(d). This, we shall see, is typical. There are abundant numerical illustrations in the later scholia to the *Elements,* one of which has entered the text of some manuscripts; see Euclid–Stamatis, *EE* ii, 237–8. In particular, there is a frequently cited later annotation on folio 32^v of B which consists of the Arabic numerals for 1 to 9 (see Vogel, BIZB, which reproduces facsimiles of the text, and Wilson, MP, 401–4 for an informed discussion); unfortunately the description of this in Euclid–Heiberg, *Opera* v, p. xix has been omitted from Euclid–Stamatis, *EE*. This missing text reads, insofar as it can be reproduced here: "*De notis numeralibus arabicis, quae in scholiis Vindobonensibus maxime in libro X occurrunt, hoc tantum commemorabo, scholia illa manu V^b,* h.e. *sine dubio saec XII, exarata esse. pro numero 5 usurpatur 0, nostrum vero 0 punctum est vel ^{O}; prorsus similes sunt series numerorum in B fol. 32^V* (ad initium libri 11) *m. rec.* [see below] *et in b ad II, 1 m. rec.* ψῆφος ἰνδική [see below]." The omitted passages reproduce the form of these Arabic numerals; examples of these can also be seen in many of the scholia in Book X, for example nos. 152, 155, 170, 183, 184, and many from no. 356 onward.

[2] This passage was quoted in Section 5.5(c), above.

forms, which denote 5, 500, 5000, and 5 0000. So much is well known and frequently described,[3] but what is less frequently pointed out is that, with the exception of a few early papyrus accounts recently found at Saqqara, of which only one has so far been published,[4] and occasional archaic usages in literary stichometry, the acrophonic system is only found on inscriptions, and has little relevance to mathematics.[5] I shall consider exclusively the alphabetic system.

(ii) We must differentiate clearly between texts which are copies of earlier texts, such as Byzantine manuscripts, and texts which may be originals or which, if copies, more closely reflect early practice, such as some material found on papyrus. One objective of this chapter will be to emphasise the problem of assessing the effect of the later modification of the conceptions and notations, especially in the descriptions of fractional quantities, during the ninth to sixteenth centuries, the period during which most of the surviving texts were copied. But note that I shall follow here the almost universal practice of presenting any Greek words or texts in modern printed form, as they have been transcribed by their editors,[6] and will only occasionally refer to the wide diversity of different styles which are found in the original manuscripts.

(iii) In scientific computations from the time of Hipparchus and Hypsicles (c. 150 BC), and especially in Ptolemy's *Almagest* and its commentaries, the Greeks employed an alphabetic version of the sexagesimal system found earlier in Babylonian cuneiform clay tablets that date back almost to 2000 BC. But there is no trace whatsoever of sexagesimal numbers in any Greek text from before the second century BC, and I shall not consider them further here.[7]

---

[3] See, for example, Heath, *HGM* i, 29 ff., Thomas, *SIHGM* i, 41 ff., etc.
[4] Turner, FODMS. Also see Section 6.2, n. 30 and its text.
[5] For descriptions of the acrophonic system, based on a thorough familiarity with the epigraphical evidence, see Guarducci, *EG* i, 417–28; Tod, *AGNS*; and the more specialised Johnston, *TGV*, Chapter 6, and Lang, *NNGV*. Concerning stichometry—the labelling or counting of lines or verses in a text—see Tod, *AGNS*, 32 f. (= GNN 130 f.), Turner, *GP* 94 f., McNamee, *AGLPO*, 122, and, for examples, P. Ryl. iii 540, P. Oxy. x 1231 and xlii 3000, and P. Herc. 1151. These uses of acrophonic numerals are not unlike the way we have retained Roman numerals in stylised contexts such as labelling the books of the Bible (as in 'I Kings' for 'the first book of Kings') or on clock dials (which consistently employ the form IIII, now elsewhere very unusual).
[6] So one text, P. Lond. ii 265, in Section 7.3(c), below, will be given without diacritical marks, while others have them added or corrected by their editors. For a not entirely successful attempt to use characters that more closely resemble early script, see Fowler & Turner, *HP*. The editors of *The Oxyrhynchus Papyri* gradually abandoned their original practice of employing the specially created fount that can be seen in Plate 1 for their diplomatic transcriptions (i.e. transcriptions as close as possible to the written text).
[7] The earliest surviving text which contains anything resembling the sexagesimal system is Hypsicles, *Anaphorikos* (*On Rising-times*); see Hypsicles–De Falco, Krause, & Neugebauer, *AG*. (This manuscript is in Vatican gr. 204; see Chapter 6, n. 52 and text.) Also there are later secondary reports that Eratosthenes (c. 250 BC) divided the circumference of a circle into sixty parts; the Keskinto inscription in Rhodes, *IG* xii(1) 913, has a division of

7.1 Introduction 223

(iv) Any convenient and easily remembered sequence of labels or marks can be used to stand for the sequence

first, second, third, fourth, fifth, ... ;

for example we can, and often still do, use the letters of the Greek alphabet, $\alpha, \beta, \gamma, \delta, \varepsilon, \ldots$. Sometimes, as for example in labelling the books of the *Iliad* and *Odyssey*, the next letter in this sequence is ζ, as we might expect; in this case there will be no mathematical overtones, and we have a pure letter labelling system. Our earliest evidence about numerical notations is of such kinds of usage, and this also will not concern us here;[8] ordinal arithmetic will not become a part of mathematics before the nineteenth century.[9]

The main topic of this chapter will be the use of the alphabetic numerals and the associated system for treating fractional quantities, in school, commercial, and scientific texts. I shall regard it as very important to pin down as many as possible of the various points of the discussion to illustrative examples which, in many cases, will then be reproduced and transcribed in the plates, and translated and described in the text.

The standard Greek alphabet consists of twenty-four letters. In the alphabetic numeral system,[10] this is supplemented by three letters:[11]

vau or *digamma*, originally written ⊂ or F, which later, in minuscule, is called *stigma* and is transcribed as ς, a letter not unlike, but not

---

the circle into 360°; and the Antikythera Mechanism of 80 BC has brass circular scales divided into 360° (see Price, *GG*). Numerical practice is rarely consistent, and different systems are often mixed together, even within a single context. For discussion and references, see Neugebauer, *HAMA* ii, 590 ff. and 698 ff.

[8] See Tod, LLGI. For examples of confusion between ζ as the sixth letter of the Greek alphabet and the numeral for seven, see Keaney, ETTHP, 296 n. 2, and Sharples, NETDP, 144.

[9] Landau's joke at the beginning of his *Grundlagen der Analysis* (1930) could scarcely have been understood before the twentieth century: "'Theorem 1', 'Theorem 2', ..., 'Theorem 301' are simply labels for distinguishing the various theorems, ... and are more convenient for the purposes of reference than if I were to speak, say, of 'Theorem Light Blue', 'Theorem Dark Blue', and so on. Up to '301', as a matter of fact, there would be no difficulty whatever in introducing the so-called positive integers. The first difficulty—overcome in Chapter 1—lies in the totality of the positive integers 1, ..., with the mysterious series of dots after the comma, in defining the arithmetical operations on these numbers, and in the proofs of the pertinent theorems" (Landau, *FA*, p. v).

[10] The best general account of this alphabetic system I know is Smyly, EAGL; this article is informed with Smyly's experience in transcribing difficult numerical papyri, but note that all of his comments on common fractions, p/q, are based entirely on examples drawn from minuscule manuscripts. Also Coulton, TUGTD, 74–89, Neugebauer, *ESA*, Chapter 1, and de Ste. Croix, GRA, are highly recommended. The Phoenician antecedents of the Greek alphabet are described in Coldstream, *GG*, 295–302 and 382.

[11] For further details of these extra letters, see Liddell, Scott, & Jones, *GEL*, s.vv. F (between ε and ζ), M, ϟ (both between π and ρ), Σ, δίγαμμα, and κόππα.

identical with, the final form of sigma in Porson's Greek types;
 koppa ϟ, or ϙ, and
 san or ssade ϡ, later, in minuscule, called *sampi* and written ϡ.
These twenty-seven letters were used in three groups of nine to represent the unit digits:

	α	β	γ	δ	ε	ϛ	ζ	η	θ
for	1	2	3	4	5	6	7	8	9,

the tens:

	ι	κ	λ	μ	ν	ξ	ο	π	ϟ or ϙ
for	10	20	30	40	50	60	70	80	90,

and the hundreds:

	ρ	ς	τ	υ	φ	χ	ψ	ω	ϡ or ϡ
for	100	200	300	400	500	600	700	800	900.

An extra place of thousands was often originally indicated by a hooked or looped line extending above the letter, but this notation gave way, in the fourth century AD, to an oblique stroke below and to the left, right, or under the letter.[12] The notation of a lower left stroke was later adopted in minuscule manuscripts, and thousands are now usually transcribed as either a lower case letter with a lower left accent, or as an upper case letter: ͵α or Α to ͵θ or Θ for 1000 to 9000. The digits may be written in any order in the course of a calculation, but the answer to the calculation or a number in isolation is almost always written in decreasing order of digits, left to right. So, for example 5324 may appear in an edited texts as ͵ετκδ or $\overline{Ετκδ}$, where the bar over the numeral is a sign, also generally adopted around the fourth century AD (though found not infrequently before then, especially to indicate dates; see Section 7.1(c), below, for an example), and then used in minuscule manuscripts, that is used to distinguish a number from a word or from the letters used to label a figure.[13] The evolution of the letter san into sampi caused some

---

[12] In the Achmîm Mathematical Papyrus (no. 12 in the catalogue in the Appendix) the scribe employs all three usages to denote thousands; see, for example, the Table Paléographique, Planche 1; also see Milne & Skeat, *SCCS*, 62–4. There are valuable discussions of the papyrological evidence concerning thousand, myriad, and fraction indicators in Brashear, *CTAL*, 215 n. 2, and MS.

[13] In one manuscript of Euclid, the scribe has confused the use of the alphabet to label the figure with its use as numerals, and so has translated the letter labels on the figure as Arabic numerals; see Heiberg–Euclid, *Opera*, supplement (= vol. x), ed. Curtze, pp. xvi–xxvii. Numbers can also be confused with words; for example the scribe of Heron's *Metrica* at one place copied the text μγ γ́ λῆμμα (...43 3'. Lemma...) as 43 3' 28' 40' 41', in the notation to be described in the next section; see Heron-Bruins, *CC* i, folio 77^r line 16 or

confusion; we shall see a typical example where we find $\overline{,\varepsilon\tau\kappa\delta}$ where we would expect $,\overline{\varepsilon\lambda\lambda\delta}$, 5924, and confusion between λ and λ or ψ is also common.

Larger numbers, tens of thousands up to tens of millions, were originally indicated by writing these letters over a mu (for $\mu\upsilon\rho\iota\acute{\alpha}\delta\varepsilon\varsigma$, myriads or ten thousands); then, from the second century AD onwards, other abbreviations and arrangements appear. This system ran out at a myriad myriad, 'ten thousand times ten thousand', the proverbial multitude that no man could number (see Daniel 7:10 and Revelation 5:11).[14] It is now easy for us to see how, in principle, to continue this procedure automatically up to a myriad blocks each contining a myriad (a myriad raised to the power of a myriad) and this appears to be the basis of the system described by Apollonius in a book that is now lost, but which is referred to in Pappus, *Collection* II (excerpted in Thomas, *SIHGM* ii, 352–7). Archimedes' *Sandreckoner* describes an exponential system for arbitrarily large numbers (excerpted in Thomas, *SIHGM* ii, 198–201) in which the process just described—which gives numbers of what he calls the first period—is then repeated to give numbers of the second period.[15] Since Archimedes takes as the limit of his basic system of numbers not the myriad, but the myriad myriad, this yields a myriad myriad raised to the power of a myriad myriad, then raised again to the power of a myriad myriad. At this point he stops his description and gives a theoretical analysis of his procedure.

This interest in very large numbers is clearly theoretical and restricted to mathematics. The only place outside the works of Archimedes, the

---

Heron–Heiberg, Schöne, & others, *Opera* iii, 50. There is no guarantee that an edited and printed text follows the practice of the scribe of the manuscript in its handling of these matters; see how the bars over numerals representing dates are omitted in Plate 5, lines 56, 58, 62, 64 and 66. Also, the numerals that occur in *Elements* XIII 11 (see the opening paragraph, above) are written in the manuscripts B and P, at least, in exactly the same way as the labels on the diagrams: in majuscules in B, minuscules in P, both distinguished by a superior bar; but one cannot tell this from Heiberg's text.

[14] So this would be a suitable number to use in my modified S'_{21}, in Section 2.3(b), just as I did in the last paragraph of E_{67}, in Section 4.4(b). A similar device can be used in manipulating anthyphairetic ratios in a calculating machine, which has the same finite capacity for integers. I have imitated the feature of counting by myriads by splitting transcriptions of numbers that arise in Greek contexts into blocks of four, rather than the now conventional blocks of three or five.

[15] Since Archimedes does not describe how the precise value of a number is got by adding together the various orders of the different periods, it might be more accurate to say that Archimedes' system describes the size of, or an approximation to, a number and this is how he uses it, to obtain an estimate of the number of grains of sand that would fill the universe. Contrast the way we say 'the solution of Archimedes' *Cattle Problem* is $7.76\ldots \times 10^{206544}$', in which the relation between the approximate and exact values is more straightforward. (The number itself completely fills $46\frac{1}{4}$ sheets of computer print-out, as reproduced in Nelson, SACP, 165–76!)

now lost book of Apollonius, and Diophantus[16] that I know where there is any attempt or need to express a number bigger than a myriad myriad is a very corrupt passage in Theon of Smyrna–Hiller, *ERMLPU*, 126–7, where Theon estimates the volume of the earth. Our only known manuscript, Venice Marciana gr. 303, of this half of the treatise (pp. 120–205) was made in the fourteenth or fifteenth century by a scribe who could not read Greek numerals, so almost all of them are corrupt and have had to be restored by the editor.[17] On the other hand, there is little sign of any difficulty in commercial documents with manipulations which involve numbers in the tens or hundreds of millions, and such large numbers are not uncommon.

7.1(b) *Simple and compound parts*

Now let us turn to fractions. First, the word: I shall always use it with its general meaning of "a numerical quantity that is not an integer" (*OED*, s.v. fraction). All other meanings will be qualified: a *common*, or *vulgar* fraction for p/q, which is *proper* or *improper* according as p is less than q or not; when $p = 1$, the common fraction is a *unit* fraction; a *decimal* fraction for the decimal expression of a fraction, which may then be *rounded* or not, *periodic* or not, and *terminating* or not; a *sexagesimal* fraction; a *complex* fraction, for a common fraction that has a fraction for its numerator or denominator; a *continued* fraction; and so on. Note that fractions, being 'numerical quantities', belong to what I have called the arithmetised style of mathematics, and the non-arithmetised approach may often be signalled by the use of the alternative terminology of ratios.[18]

I shall argue in this chapter that we have no evidence for any conception of common fractions p/q and their manipulations such as, for example, $p/q \times r/s = pr/qs$ and $p/q + r/s = (ps + qr)/qs$, in Greek mathematical, scientific, financial, or pedagogical texts before the time of Heron and Diophantus; and even the fractional notations and manipulations found in the Byzantine manuscripts of these late authors may have been revised and introduced during the medieval modernisation of their minuscule script. Among the thousands, possibly the tens of thousands, of examples of fractions to be found in *contemporary* Egyptian (hieroglyphic, hieratic, and demotic), Greek, and Coptic texts, all but a few

---

[16] See Diophantus–Heath, *DA*, 47 (but Heath's example can only be found in the apparatus to Diophantus' text, Diophantus–Tannery, *Opera* ii, 323.8, and there are further corruptions too involved to be pursued here). For further comments on Diophantus' notation, see n. 28, below. There is an isolated exception to my statement in a table of squares which goes up to $1\,0000 \times 1\,0000$; see no. 68 in the Appendix to this chapter.
[17] See Smyly, NTS, for a discussion.
[18] See the example of Aristarchus, *On the Sizes and Distances of the Sun and Moon*, in Section 7.3(b), below.

7.1   Introduction

isolated examples in five texts (P. Lond. ii 265 (p. 257); M.P.E.R., N.S. i 1; and three demotic papyri published in Parker, *DMP*), all to be described in detail in Section 7.3, below, use throughout the following 'Egyptian' system for expressing fractions:

We take the basic sequence of the *arithmoi*:

$$\text{two, three, four, five}, \ldots,$$

represented in Greek by the letters $\beta, \gamma, \delta, \varepsilon, \ldots$, and convert it into the sequence

$$\text{half, third, quarter, fifth}, \ldots,$$

where, after the exceptional cases of the first few terms, for which special symbols (to be described below) are assigned, the derived symbol is got by appending to the numeral a long line, straggling upwards, now generally called a 'fraction indicator' and usually transcribed as an accent, $\acute{\gamma}, \acute{\delta}, \acute{\varepsilon}, \ldots$, or a prime, $\gamma', \delta', \varepsilon', \ldots$; clear examples can be seen in Plates 5 and 6. Later a variety of strokes, bars, and dots are found as fraction indicators.

This sequence: half, third, quarter, fifth, ... is often confused with the nearly homonymous sequence of ordinal numbers: first, second, third, fourth, fifth, ..., and its terms are often called unit fractions. The universal Greek way of referring to them is as μέρος or μόριον, 'part' (plural μέρη or μόρια, 'parts'). The two Greek words are used interchangeably, sometimes both in the same passage; and they are often translated as 'fractions' or even 'denominators',[19] but I shall always refer to the $\angle, \acute{\gamma}, \acute{\delta}, \acute{\varepsilon}, \ldots$ more accurately as 'parts', and distinguish, where necessary, between simple parts[20] (i.e. isolated members of the sequence) and compound parts (to be described below). One reason why I wish to avoid the name of unit fractions is because of the overtones of the notation it suggests, $1/q$, which immediately introduces associations with the common fraction p/q; these associations are completely absent from the Greek description and notation. I shall instead write the translated sequence of parts as $\bar{2}, \bar{3}, \bar{4}, \ldots, \bar{10}, \bar{11}, \ldots$, in imitation of early Greek practice, or as the typographically more convenient and understandable $2', 3', 4', \ldots, 10', 11', \ldots$, or as 3rd, 4th, ..., 10th, 11th, ...,[21] and I

---

[19] See, for example, *Timaeus* 36a–b, or Heron–Heiberg, Schöne, & others, *Opera* v, 96.4 and 12. (Liddell, Scott, & Jones, *GEL*, s.vv. μέρος and μόριον, cite the apparently incorrect *Stereometrica* 2.14 and 16 and translate, following Heiberg's *index verborum*, as 'the denominators of fractions'.) Euclid refers to an *arithmos* and its corresponding *meros* as being ὁμώνυμος, 'called by the same name', in *Elements* VII 37–39.

[20] A simple part is also called an aliquot part; see the *OED*, s.v. aliquot, which cites Billingsley on *Elements* V Definition 1: "This ... is called ... a measuring part ... and of the barbarous it is called ... an aliquot part".

[21] The notation $\bar{2}, \bar{3}, \bar{4}, \ldots$, based on Egyptian practice, is also used, but this conflicts with the later scribal notation for distinguishing a number or figure label from the text, or for differentiating one kind of number from another. See n. 13, above, and the text to that note.

recommend strongly that they be read 'the half, the third, the quarter or the fourth, ...'[22] and not 'one-half, one-third, one-quarter or one-fourth, ...'.

The sequence of parts starts with $\acute{\beta}$, 'the two parts', τὰ δύο μέρη, an expression for 'two-thirds'.[23] This, to begin with, was the standard letter beta with the usual fraction indicator (see Plates 5 and 6, to be described below, and the detailed analysis of this point in Fowler & Turner, HP), and it is found later in a range of variants and forms; each text may show its own version. However extraordinary it may seem to us, it is an incontrovertible fact that the sequence of parts starts with two-thirds, and this term has exactly the same status as the other parts in the symbolic expression of fractions. (This is an additional reason for avoiding the name 'unit fractions' for the sequence $\acute{\beta}, \angle, \acute{\gamma}, \acute{\delta}, \ldots$, since $\acute{\beta}$ is not a unit fraction.) I shall transcribe all variants of this symbol as $\acute{\beta}$, and translate it as $\tilde{3}$ or 3″. The symbol $\acute{\delta}$ is often found in a special form in which one side of the Δ is extended to form the fraction indicator, and this is sometimes transcribed as d or d′; see for example, the transcription in Plate 6. The various signs for the half may or may not incorporate a fraction indicator which then may or may not be shown on the transcribed text; here they will all be transcribed as $\angle$.

More complicated fractions than simple parts are expressed as sums of an integer and *different* simple parts;[24] I shall refer to them as *compound parts*. They are laid out as follows: first come the digits of the integer part of the number, written out in decreasing order as was explained above; this is followed by the successive parts, in order (i.e. order of increasing 'denominator'), the digits of each part (i.e. 'denominator') being written, as usual, in decreasing order.[25] Some pairs of symbols may be ligatured into a composite symbol, for example $\angle\acute{\delta}$ for $\frac{3}{4}$ (not unlike an upside-down β, and sometimes printed as such) and $\acute{\beta}\varsigma$ for $\frac{5}{6}$.

Let us now look at two texts that will illustrate many of these features.

---

[22] Curiously, just as we can never say 'second' for 'half' so the Greeks could never write $\acute{\beta}$ (see below); and just as we have the alternatives 'quarter' and 'fourth', so they had an optional special form for $\acute{\delta}$. However, Greek has a special terminology for epimoria (described in Section 4.5(b) (see A_{60} & B_{61})) and their reciprocals (see the next note) which has no idiomatic translation.

[23] This principle is sometimes said to be general, with τὰ τρία μέρη, the three parts (sc. of four) for three-quarters, and so on, but I cannot cite any Greek examples, and only two-thirds has an associated symbol. Also see n. 83, below.

[24] The only example I know of an expression that involves repeated parts is in a schoolboy's exercise, P. Louvre [inv.] MND552K side A, described at no. 17 in my catalogue in the Appendix.

[25] The only Greek example I know where the parts are not arranged in decreasing order is P. Oxy. xii 1446, a survey of land and rents. The fractional quantity $\gamma\varsigma\acute{\rho}\nu\acute{\iota}\beta$ (3 6′ 150′ 12′) occurs twelve times and $\delta\acute{\delta}\nu\tau\iota\beta$ (interpreted as 4 4′ 50′ 300′ 12′) occurs once. The editors suggest that the anomaly may have been caused by the later addition of an extra 12′ artaba of wheat to the rent of the land. Also see n. 88, below.

7.1 Introduction

7.1(c) *P. Hib. i 27, a parapegma*

Many of the features so far described are illustrated in the early papyrus text, P. Hib. i 27; see Plate 5. In addition to its very full original publication in Grenfell & Hunt, *HP* i, 138–57, there are the easily accessible descriptions of its astronomical content in Neugebauer, *HAMA* ii, 599 f. & 686 ff., and of its numerical features in Fowler & Turner, HP, so my account here can be very brief. This text is perhaps our earliest physically surviving Greek scientific or semi-scientific text; it comes from the cartonnage of a mummy and is dated, from internal evidence of the text, to about 300 BC, and this date seems to be corroborated by palaeographical analysis. It opens with a fragmentary and elementary introduction, believed to be based on Eudoxus' astronomical system, and this is followed by a parapegma (see Section 4.4(c)) of which column 4, lines 55–68, is reproduced and transcribed here in Plate 5. The translation of this part of the text is:

55 The night is 13 12′ 45′ hours, the day 10 3″ 5′ 30′ 90′.
56 16th, Arcturus rises in the evening,
57 the night is 12 3″ 15′ 45′ hours, the day 11 9′ 10′ 30′.
58 26th, Corona rises in the evening,
59 and the north winds blow which brings the birds, the night
60 is 12 2′ 30′ hours, the day 11 3′ 10′ 30′. Osiris
61 circumnavigates, and the golden boat is brought
62 out. Tybi 5th, the sun enters Aries. 20th, spring equinox,
63 the night is 12 hours, and the day 12 hours
64 and the feast of Phitorois. 27th, Pleiades
65 set in the evening, the night is 11 3″ 6′ 90′ hours,
66 the day 12 10′ 30′ 45′. Mechir 6th, the sun enters
67 Taurus. Hyades set in the evening,
68 the night is 11 2′ 10′ 35′ hours, ...

As is typical with numerical texts, there are several errors which here can be corrected with some confidence. They are noted in the editors' apparatus:

55 for 13 12′ 45′ read 13 15′ 45′
68 for 11 2′ 10′ 35′ read 11 2′ 10′ 30′ 90′

and the editors have restored some parts, including two smudged numbers on lines 57 and 65 where the scribe has illegibly altered the text.

Note the following features:

(i) The distinctive long straggling fraction indicators.

(ii) The letter β in $\acute{\beta}$ for $\frac{2}{3}$ (e.g. end of line 55); it is the same as the β in $\iota\beta$, 12′ (e.g. beginning of line 57) or the letter β (e.g. line 59).

(iii) The distinction between stigma (fragmentary examples in lines 56 and 58; complete example in line 66) and sigma (no examples occur as numbers; the letter occurs in lines 56 and 58, with no distinction between upper and lower cases, and initial, medial, or final forms). The written letters are completely different, and the transcribed printed letters, in this well-designed fount, are slightly different.

(iv) The letter koppa (e.g. at the end of line 55).

(v) The way the dates, the ordinal numbers, are distinguished throughout by a superior bar (not indicated in the transcription), but how the *arithmoi* are not otherwise distinguished from the text.

It is now standard editorial practice to modernise the notation of fractions. For example $\iota\beta\acute{\epsilon}\lambda$ at the end of line 55 would now almost always be silently translated as $10\frac{41}{45}$, as was done in Grenfell and Hunt's original edition. There is no harm whatsoever in this as long as it is done consciously; this kind of translation makes texts accessible and comprehensible. But we must remain aware that translations like this can alter the significance of the text, and that similar translations may have occurred in some texts during their long period of transmission and revision, down to our own time.

7.1(d)   O. Bodl. ii 1847, *a land survey ostracon*

There is evidence of Egyptian land surveys from at least 3000 BC, and many detailed Greek examples from the Ptolemaic period, dating from the second century BC onwards, have been discovered and published. They vary from short calculations on potsherd or stone ostraca to long papyrus rolls setting out the survey of a whole village. There are, broadly speaking, two kinds of record: one giving the owner, location, dimensions, and area of each field, while the other, the annual return, gives the area (but not the dimensions), the crop under cultivation, the annual tax paid, and sometimes details of roads, drains, and canals which pass through the land.[26] A typical short example of the first kind of record, O. Bodl. ii 1847[27] from Thebes in Egypt, dated to between 30 BC and AD 14, is given with its transcription in Plate 6. These rapidly written, cursive, highly abbreviated texts present enormous problems to the uninitiated reader, rather as happens today with doctors' prescriptions. However, most abbreviations in this text have been successfully resolved, the added letters being put in parentheses (though, on the bottom line of the apparatus, the second editor notes that one abbreviation in line 2 has defeated her while another is uncertain).

The text reads, in translation:

1   5 2' 4' 8 2' 4'

---

[26] For a discussion of land surveys in general, and the papyrological evidence in particular, see Crawford, *KEVPP*.
[27] Also known as O. Tait; it is published in Tait, Préaux, & others, *GOBLO* ii, 308.

7.1 Introduction 231

2 The land measurement of the (land) of Kalliedon (farmed) by Petechonsis, beginning(?) on the west.

3 4' 8' 16' 32' $\dfrac{2'\ 4'\ 8'}{1}$ 8' 16' / 4' 16' arouras. Adjacent to the north 32'.

4 Another (field) adjoining the bottom, 2' 4' 8' $\dfrac{1\ 2'\ 8'}{16'}$ 1 32' / 2' 4' 32'.

5 Another (field) adjoining 2' 8' $\dfrac{2'\ 4'\ 8'}{2'\ 4'}$ 2' / 4' 8' 16'.

6 Another (field) 4 $\dfrac{2'\ 4'\ 8'\ 16'\ 32'}{1\ 8'\ 32'}$ ○ / 4 4'.

7 Another (field) adjoining 4 2' 4' $\dfrac{2'\ 4'\ 16'\ \ldots}{2'\ 4'\ 8'\ 16'\ 32'}$ / 3 2' 4' 8' 32'.

8 / 9 2' 4' 8' [16'] arouras.

9 Adjacent on the north, (farmed) by Lysimachos

10 5 $\dfrac{2'\ 4'\ 8'\ 32'}{1\ 4'\ 8'\ 32'}$ ○ / 5 2' 4' 32'.

11 Another (field) adjoining 4 $\dfrac{2'\ 4'\ 8'}{2'\ 4'\ 8'\ 32'}$ ○ / 3 4' 8' 32'.

12 / 9 2' 8' 16'.

The basic calculation, which occurs in lines 3, 4, 5, 6, 7, 10, and 11 is as follows. The lengths a, b, c, and d of the four sides of a field (with a opposite c, b opposite d), measures in schoinia and half, fourth, eighth, sixteenth, and thirty-second parts of a schoinion (the word means 'measuring cord', and 1 schoinion = 100 cubits, 1 cubit = elbow to knuckle, approximately) are written out $a\dfrac{b}{d}c$. If two sides of a field are equal, then the second will be replaced by an ○, perhaps for *homoion*, or a dot, or even omitted (see lines 6, 7, 10, and 11, and the editor's note to line 7). In the case of a triangular field, *ou* (for *outhēn*, nothing) would replace one side; there are no examples here. The oblique line / is one of the standard symbols in documentary papyri for equality; it is usually described as an abbreviation for the gamma of *ginetai* or *ginontai*, 'becomes', but is more probably derived from an Egyptian sign, which is later often Hellenised into a gamma; then, in the Byzantine period, this symbol disappears completely, and is replaced by much more explicit biliteral or triliteral abbreviations which continue into the minuscule manuscripts.[28] The area of the field follows, given in arouras (1 aroura = 1 square schoinion, equal to approximately 0·676 acres, or 0·25

---

[28] The text M.P.E.R., N.S. i 1, to be described below in Sections 7.3(d) and 4(a) uses variously this sign, the biliteral abbreviation ¡†, or the word *ginetai*. On these abbreviations, see Blanchard, *SAPDG*, 29 ff, especially notes 9–11 and the text to these notes, and McNamee, *AGLPO*. Differences between the abbreviations in the papyrological traditions and those found later in Byzantine minuscule texts are very common. For example, consider the abbreviations for subtraction: throughout the Byzantine minuscule manuscripts of Diophantus and once in Heron (at *Metrica* III, 7; see Heron–Bruins, *CC* i, folio 103ʳ, line 5 for a photograph), we find a sign similar to a rounded vertical arrow, ↑. I am grateful

hectares), and it always appears to be calculated by multiplying together the averages of the two pairs of opposite sides,

$$\text{nominal area} = \tfrac{1}{2}(a+c) \times \tfrac{1}{2}(b+d) = \tfrac{1}{4}(ab + ad + bc + cd),$$

and approximating the result to within the thirty-second or less of an aroura. Look, for example, at line 3: We have $a = 4'\ 8'\ 16'\ 32'$, $c = 8'\ 16'$, so

$$\tfrac{1}{2}(a+c) = 8'\ 8'\ 16'\ 64'\ = 4'\ 16'\ 64',$$

while $b = 2'\ 4'\ 8'$, $d = 1$, and so

$$\tfrac{1}{2}(b+d) = 2'\ 4'\ 8'\ 16'.$$

Now do a long multiplication sum in what is, in effect, binary representation to get

```
  2'   4'   8'   16'
× 4'        16'         64'
─────────────────────────────────
  8'  16'  32'  64'
       32'  64'  128'  256'
                 128'  256'  512'  1024'
─────────────────────────────────
  8'  16'  16'  32'   64'   128'  512'  1024'
```

and simplify the carries to give $4'\ 32'\ 64'\ 128'\ 512'\ 1024'$. We do not

---

to Alain Blanchard for the following sketch of the general development of the earlier Greek symbol for 'from which': "My impression is that there must again be a break between the papyrological and medieval periods. The papyrological forms, derived from the demotic ↙, *wp.t* (for this word see Mattha, *DO*, 156, no. 190, line 2; Erichsen, *DG*, 85; and Pestman, *GDZA*, 76–8) are relatively faithful to their prototype, with different kinds of deformations. There are various kinds of examples: ↙, U.P.Z. ii 157.9, 242 BC and M.P.E.R., N.S. i 1 [see Section 7.3(d), below]; , P. Petrie ii 39c.9, iii BC, to P. Lond. i 99 (p. 158).28, iv AD; and ↙, P. Berl. Leihg. 13.15, beginning of ii AD. The demotic word appears to be used for itemising, with subtracting as a secondary meaning; the Greek expression seems to be ἀφ' ὧν, 'from which'. Despite some reinforcement of this sign at the end of its evolution (for example, ↙, O. Bodl. i 1957.3, ii AD? and ↙, P. Lond. ii 755 verso (p. 221).38, iv AD), the sign did not survive beyond the Roman period. In the Byzantine period it is replaced by a more explicit abbreviation $\alpha\varphi$ (e.g. P. Sorb 61.16,, V AD? & P. Lond iv 1419.1396, after AD 716); or $\alpha\varphi^\omega$ (e.g. P. Lond iv 1413.221, AD 716–21); or $\alpha\varphi^\circ$ (e.g. P. Lond iv 1419.43, after AD 716)."

The possibility that the sign ⸋ might be an introduction during the Byzantine minusculisation was dismissed by Tannery, the editor of Diophantus' works, in SSCG, 208–9: "A la vérité, comme on ne le rencontre que dans ce passage [Heron, *Metrika* III.7] ... il pourrait n'y avoir là qu'une abréviation byzantine; cependant cette dernière hypothèse n'est guère vraisemblable." Tannery's argument is that the manuscript of Heron was copied more than two centuries earlier than those of Diophantus, at a time when Diophantus was not being studied; and Diophantus' procedures had been too little vulgarised for them to have been the source of this sign. But, one can object, these are far from the only possibilities. Also see Diophantus–Heath, *DA*, 41-4.

know just how the taxman has performed this calculation, which here has been rounded up to 4′ 16′, an approximation in his favour of 256′ 1024′ of an aroura.

It is easy to show that the area of the field is actually $\frac{1}{4}(ab \sin ab + ad \sin ad + bc \sin bc + cd \sin cd)$, so the taxman's formula will *always* overestimate the area unless the field is rectangular. Sometimes it seems that a field may be split up into pieces, to reduce the error; and sometimes a correction, called a 'difference of measurement' (*diaphoron schoinismou*) may be noted, though this correction is itself likely to be incorrect, and the tax is still calculated on the uncorrected area.[29] There are also frequent arithmetical mistakes; see, for example, line 7, where the answer to the multiplication

$$4\ 2'\ 4' \times 2'\ 4'\ 8'\ 64' = 4\ 8'\ 16'\ 32'\ 128'\ 256'$$

is given by the scribe as 3 2′ 4′ 8′ 32′, and no readjustment of the doubtful numbers seems to lead to any approximation to the scribe's result, although the editor points out that the absence here of either a number, or an °, or an *ou* in the right-hand place may indicate a scribal omission of a dimension. The entry on line 11 also seems to contain an error. But since some annual tax returns seem purely notional, with the same minutely detailed account being returned in consecutive years, perhaps it is irrelevant to pick on such small details as these.[30]

This text illustrates a feature of commercial documents, that there are conventions restricting the range of permitted parts. In recording and manipulating distances and areas, as here, fractions are always expressed in halves, fourths, eighths, sixteenths, . . . of schoinia and arouras; while taxes and rents are expressed in equivalents of artabas of wheat (see Section 7.3(c), below), where the permitted parts are 2′, 4′, 8′, 5′, 10′, 3″, 3′, 6′, 12′, 24′, ...[31] This means that approximations are necessary— for example, it is not possible to express the third of an aroura of land, or the seventh of an artaba of wheat exactly, within these restrictions—and these approximations are sometimes crude and usually in the taxman's or landlord's favour.

The earliest surviving land registers such as I have described here date

---

[29] Sir Eric Turner commented on a preliminary version of this, in a letter written in April 1983, a week before his death: "I wonder whether *diaphoron schoinismou* means 'tax-payment for measurement' rather than 'difference of measurement'. There are many taxes under the name of *diaphoron*. Even more in the pocket of a king or inspector!"
[30] See de Ste. Croix, GRA, 59: "Mistakes in operating with fractions are particularly common in papyri. But it is sometimes difficult to tell how far deficiencies observable in ancient arithmetical calculations are due to the nature of the scripts used and how far to the comparative indifference of the Greeks and Romans to extreme precision in such matters as the calculation of interest on loans." Indeed, extreme precision is unattainable within the conventions of commercial practice; see the next paragraph.
[31] For more details and references, see Fowler, NFA.

from the second century BC, but some evidence for a land survey in Egypt can be traced back to at least 3000 BC. The approximate formula for the area occurs in Problems 51–3 of the Rhind Mathematical Papyrus, where it is applied to isosceles triangles and trapezia, though many commentators conceal the fact that it is an approximation by adapting the translation, and rendering 'side' as 'height'. After the conquest of Egypt by Alexander in 331 BC, the Greeks took over the existing Egyptian administration, including the land survey, merely imposing Greek as the official language for virtually all administrative and financial documents that were of concern to them. On the other hand, the Greeks never seemed to have operated a land survey in Greece itself, though there are suggestions in Proclus' *Commentary on the First Book of Euclid's Elements* that they did use the same procedure for calculating area. I shall return to this topic, in a discussion of Egyptian land surveys and the stories about the development of Greek mathematics, in Section 8.1.

7.2 Tables and ready reckoners

7.2(a) *Division tables*

The practice of expressing fractions as sums of parts raises two distinct questions: how were these expressions manipulated, and how were they calculated? I shall deal mainly with the first question,[32] and will here describe some of the extensive published primary evidence most, though perhaps not all, of which derives from school and commercial texts.

A preliminary grammatical observation is that the definite article plays a more important role in Greek than in many modern languages, and the sequences of the *arithmoi* and *merē* should really be thought of as

> the duet, the trio, the quartet, the quintet, ...

and

> the half, the third, the quarter, the fifth, ...

Since Greek is an inflected language, it is often through the case that the relationship of one word to another is expressed; and since the higher numbers, after four, are indeclinable, it is the article which will carry this information. So a phrase like 'the quintet' (nominative plural) must be '$οἱ\ ε$', '$αἱ\ ε$', or '$τὰ\ ε$' according as the implied noun is masculine, feminine,

---

[32] My attitude will be analogous to that taken by most users, even most sophisticated mathematical users, of mathematical tables and calculating machines. Few people pause to think how the numbers they use were actually calculated. And this information is usually far from simple; I suspect that any trick or intuition that shows hope of being useful will be exploited, then as now.

7.2 Tables and ready reckoners

or neuter; and this implied presence enhances the concreteness of the number system. Also, while the *arithmoi* take plural articles, as in this illustration, the *merē* take neuter singular articles, for example τὸ ε [μέρος], the 5th part.

Now consider, for example, our common fraction $\frac{12}{17}$. This would be expressed in a division table (such as will be described below[33]) as

$$\tau\tilde{\omega}\nu \quad \iota\beta \quad [\tau\grave{o} \quad \iota\zeta] \quad \angle\,\overline{\iota\beta}\,\overline{\iota\zeta}\,\overline{\lambda\delta}\,\overline{\nu\alpha}\,\overline{\xi\eta}$$
$$\text{of the} \quad 12 \quad [\text{the 17th is}] \quad \tfrac{1}{2}\,\tfrac{1}{12}\,\tfrac{1}{17}\,\tfrac{1}{34}\,\tfrac{1}{51}\,\tfrac{1}{68}$$

for what we would write as

$$\frac{12}{17} = \frac{1}{2} + \frac{1}{12} + \frac{1}{17} + \frac{1}{34} + \frac{1}{51} + \frac{1}{68}.$$

These expressions are far from unique (we could here have, among infinitely many others, 2′ 6′ 34′ 102′, or 3′ 4′ 12′ 51′ 68′ 204′) and they may be written out using different arrangements of diacritical marks or spaces in the original text, or they may be presented differently by a modern editor, with the more convenient and understandable format ∠ ιβ′ ιζ′ λδ′ να′ ξη′ now being the most common. Our ancient evidence clearly indicates that most such expressions would be found in tables that would either have been memorised or consulted.

I shall call these tables 'division tables', rather than their more usual but misleading names 'tables of fractions' or 'multiplication tables'. Many of them have been found and published. An excerpt from a typical table reads:

ἔνατα	ninths
τῆς α τὸ θ̅ θ̅	of the 1 the 9̅ [is] 9̅
τὸ θ̅ χξς β̅	[of the 6000] the 9̅ [is] 666 3̅
τῶν β ϛιη	of the 2 [the 9̅ is] 6̅ 1̅8̅
τῶν γ γ̅	of the 3 [the 9̅ is] 3̅
τῶν δ γ̅θ̅	of the 4 [the 9̅ is] 3̅ 9̅
τῶν ε∠ιη	of the 5 [the 9̅ is] 2̅ 1̅8̅
τῶν ς β̅	of the 6 [the 9̅ is] 3̅
τῶν ζ β̅θ̅	of the 7 [the 9̅ is] 3̅ 9̅
τῶν η∠γιη	of the 8 [the 9̅ is] 2̅ 3̅ 1̅8̅
τῶν θ α	of the 9 [the 9̅ is] 1
τῶν ι αθ̅	of the 10 [the 9̅ is] 1 9̅
τῶν κ βϛιη	of the 20 [the 9̅ is] 2 6̅ 1̅8̅
and so on up to	
τῶν ϥ ι	of the 90 [the 9̅ is] 10
τῶν ρ ιαθ̅	of the 100 [the 9̅ is] 11 9̅
τῶν c κβϛιη	of the 200 [the 9̅ is] 22 6̅ 1̅8̅

---

[33] This example comes from P. Mich. iii 146, no. 20 in my catalogue of division tables, in the Appendix. Also see the following n. 34.

ἔνατα	ninths
and so on up to	
τῶν Τ ρ	of the 900 [the 9̓ is] 100
τῶν Α ριαθ	of the 1000 [the 9̓ is] 111 9̓
τῶν Β ϲκβϛίη	of the 2000 [the 9̓ is] 222 6̓ 1̓8̓
and so on up to	
τῶν Θ Α	of the 9000 [the 9̓ is] 1000
τῶν μ̇ Α ριαθ	of the 10000 [the 9̓ is] 1111 9̓

A catalogue of published tables is given in the Appendix to this chapter, and hereafter I shall refer to the tables by their serial numbers in this catalogue; this example comes from no. 20, except that I have standardised and corrected the use of the fraction indicators.[34]

This table is in what I shall call 'full' format, which lists expressions up to 1 0000 *n*th parts; some tables stop at the entry 'of the *n* [the *n̓* is] 1', in what I shall call 'abbreviated' format. This example starts with the 'heading' *enata,* ninths, and it contains an 'initial entry' "[of the 6000] the 9̓ is 666 3̓" of a kind that is believed to have been used in financial calculations and to date from the time (up to the third century AD at the latest, and possibly earlier) when 6000 drachmae, equalling 1 talent, was the monetary standard. Although inflation and change from a 'gold', then a 'silver', standard rendered this entry obsolete, it continued to be copied into some of the tables for another five hundred years at least. There is also some variation in the wording of the tables. For example, the first line may be written in words:

$$\tau\tilde{\eta}\varsigma\ \mu\iota\tilde{\alpha}\varsigma\ \tau\grave{o}\ \theta\ \theta,$$

and the initial entry may be found as either the first or second line, and it may appear as:

$$\tau\grave{o}\ \theta\ \dot{\alpha}\rho\iota\theta\mu\tilde{\omega}\nu\ \chi\xi\varsigma\beta,$$

as:

$$\tau\grave{o}\ \theta\ \psi\acute{\eta}\varphi\omega\nu\ \chi\xi\varsigma\beta$$

or as:

$$\tau\grave{o}\ \theta\ \dot{\varepsilon}\xi\alpha\kappa\iota\varsigma\chi\iota\lambda\acute{\iota}\omega\nu\ \chi\xi\varsigma\beta.[35]$$

The actual expressions themselves vary from table to table; for example the Egyptian Rhind Mathematical Papyrus, no. 2, copied around 1650 BC from an older archetype, contains a table of tenths, in an extremely

---

[34] There is a photograph of this part of this table in Knorr, TFAEG, 145. The scribe has put fraction indicators on all letters occurring in the answers, irrespective of whether they are *arithmoi* or *merē*.

[35] There are discussions of these features by Brashear, Crawford, and Sijpesteijn in the references to nos. 11, 24, 25, and 39, of the catalogue.

7.2 Tables and ready reckoners 237

concise format, laid out exactly as in the following translation:[36]

10'	3" 30'
5'	3" 10' 30'
5' 10'	3" 5' 30'
3' 15'	
2'	
2' 10'	

while the other two published tables of tenths are as follows[37] (I have again listed the example from the Rhind Mathematical Papyrus for comparison):

	Rhind papyrus (no. 2)	P. Mich. 146 (no. 20)	Achmîm papyrus (no. 12)
1	10'	10'	10'
2	5'	5'	5'
3	5' 10'	5' 10'	4' 20'
4	3' 15'	3' 15'	3' 15'
5	2'	2'	2'
6	2' 10'	2' 10'	2' 10'
7	3" 30'	2' 5'	2' 5'
8	3" 10' 30'	3" 10' 30'	3" 10' 30'
9	3" 5' 30'	2' 3' 15'	2' 3' 15'

Since we have here, at different places in these three tables, agreement of any two against the third, we cannot easily argue for a continuous textual tradition: the tables must have been frequently recomputed, when occasion demanded.

These tables are very common. Some examples, like nos. 7, 12, and 20, are systematic and extensive ready reckoners; others, like no. 38, are less apparently coherent collections; many are isolated single examples. Some seem crude, in the sense that their entries give more cumbersome expressions, by any criterion, than are necessary (see, for example, no. 9. Several entries of this table need correction and can be simplified, for instance 'of 6 the 31' is 12' 20' 31' ⟨62'⟩ 155' 186'' can also be expressed 'of 6 the 31' is 6' 62' 93''). Many of these texts are school exercises. In all, the Appendix contains details of 45 publications containing 146 individual division tables, distributed as follows:

3"	2'	3'	4'	5'	6'	7'	8'	9'	10'	11'	12'	13'	14'	15'	16'
14	8	9	10	5	4	11	7	8	6	7	6	5	7	6	5

17'	18'	19'	20'	23'	24'	25'	29'	30'	31'	48'	49'	90'	150'
5	3	3	2	1	4	2	1	1	1	2	1	1	1

---
[36] See the facsimile and transcript in Plate 33 of Chace, *RMP*. Unfortunately the photograph of this part of the text was not reproduced in the abridged reprint of this book.
[37] The tables of tenths in nos. 6, 7, and 14 of the catalogue are incomplete or not transcribed.

Any discussion of fractional calculations in Greek antiquity should take place against a background of awareness of this published material.

7.2(b) *Multiplication and addition tables*

One feature of the Greek alphabetic system is that the tables for multiplication by $2, 3, 4, \ldots, 9$, by $20, 30, 40, \ldots, 90$, and by $200, 300, 400, \ldots, 900$ are completely different, so this makes a complete multiplication table to be very long.[38] Broadly speaking, three different formats are found for these tables; the first, which resembles the format of a division table, is to set out a table of 1-times, 2-times, ..., 9-times, then sometimes continuing with 10-times, 20-times, ..., 90-times, 100-times, 200-times, ..., 900-times, 1000-times, 2000-times, ..., 9000-times, 1 0000-times. A typical table, for example the 80-times, is then given as:

πα	π	80	1	80
πβ	ρξ	80	2	160
πγ	cμ	80	3	240
etc.				

In the second format, each entry is commuted and repeated. The following illustration of the 8-times table suggests how such a table might have been read:

ηα	η	*oktakis mia estin oktō*	eight-times one is eight
αη	η	*hapax oktō estin oktō*	once eight is eight
ηβ	ις	*oktakis duo estin hekkaideka*	eight-times two is sixteen
βη	ις	*dis oktō estin hekkaideka*	twice eight is sixteen
etc.			

(We have some instances of such multiplications described in Socrates' encounter with the slaveboy, *Meno,* 81e–85d; for example, at 87d, we find both *dis duoin* and *duoin dis* in consecutive speeches.) The need to

---

[38] Many discussions of the consequences of this lack of place-value in Greek arithmetic have no basis in our knowledge, now, of Greek numerical practice. I recommend Smyly, EAGL (see n. 10, above) for an informed opinion: "I am convinced that the [Greek] notation, though it may be inferior in some respects to that now in use, by no means deserves the unlimited contempt which has been heaped upon it" (p. 515). Also de Ste. Croix, GRA, 55–6: "The habitual arrangement of figures in columns in our notation is not an intrinsic virtue of that notation but, on the contrary, an incidental defect, due, somewhat paradoxically, to the combination of its two greatest virtues, namely place-value and the small number of symbols it employs; but ... this very defect (the necessity for the arrangement of figures in columns) has, paradoxically again, provided a very useful stimulus towards the evolution of the advanced concepts of debit and credit (positive and negative entries) ... which was an essential preliminary to a coordinated system of book-keeping by double or even single entry"—and so eventually, one might add, perhaps to negative numbers.

7.2   Tables and ready reckoners   239

commute each entry surely arises not from ignorance of the commutativity of multiplication, but from the need to recognise automatically the different sounds of the cardinal and repetition numbers.

The third format (which occurs in conjunction with the second format in one of our tables, no. 51) has the following form:

$$
\begin{array}{llll}
\eta\alpha & \eta & 8\ 1 & 8 \\
\eta\iota & \pi & 8\ 10 & 80 \\
\eta\rho & \omega & 8\ 100 & 800 \\
\eta\alpha & ,\eta & 8\ 1000 & 8000 \\
\hline
\eta\beta & \iota\varsigma & 8\ 2 & 16 \\
\eta\kappa & \rho\xi & 8\ 20 & 160
\end{array}
$$

etc., up to multiplication by 9, 90, 900, and 9000.

Aristotle makes a reference to multiplication tables at *Topics* 163b24–29:

For just as in geometry it is useful to have been trained in the elements, and in arithmetic to have a ready knowledge of the multiplication table up to ten times helps much to the recognition of other numbers which are the result of multiplication, so too in arguments it is important to be prompt about first principles and to know your premisses by heart.

From this we can infer that we cannot immediately dismiss the evidence I have been describing as being relevant only to the training of scribes and accountants; Aristotle's audience clearly is aware of the basics of this kind of material, though he implies that they may not be fluent in its practice.

Tables of additions, in the same kinds of formats as multiplication tables, are also found and are listed in the Appendix to this chapter. Most of these examples of multiplication and addition tables seem to be more in the nature of systematic school exercises rather than ready reckoners intended for consultation.

7.2(c)   *Tables of squares*

In contrast to our evidence about Babylonian mathematics, relatively few examples of other kinds of mathematical tables have been found. This may be a consequence of the random way in which the papyrus texts have been preserved; on this we can only speculate, but such speculation should be moderated by the fact that a substantial amount of other material, for example astronomical[39] and astrological data, has been preserved on papyrus. Setting aside very fragmentary texts and marginal

---

[39] For astronomical material, see Neugebauer, APO, the inspiration for my catalogue in the Appendix.

annotations whose interpretation is often very doubtful, the only examples known to me of systematic tables other than addition, multiplication, division, and astronomical tables are the five tables of squares listed in the Appendix to this chapter, of which no less than three are early, dating from the third to first centuries BC. This is insufficient evidence to support anything more than the most tentative conclusions, so I shall do no more than remark on the importance of squares of the *arithmoi* within the mathematical reconstructions in Part One of this book.

7.3 A selection of texts

7.3(a) *Archimedes' Measurement of a Circle*

Consider, now, what is apparently the most blatant piece of evidence against my thesis that common fractions are not found in early Greek mathematics: Archimedes' *Measurement of a Circle*.[40] A typical description of part of this short treatise (Heath, *HGM* ii, 50 ff.) is as follows:

[Proposition 3 proves] that the ratio of the circumference of any circle to its diameter (i.e. π) is less than $3\frac{1}{7}$ but greater than $3\frac{10}{71}$. ... The calculation starts from a greater and a lesser limit to the value of $\sqrt{3}$ which Archimedes assumes without remark to be known, namely

$$\frac{266}{153} < \sqrt{3} < \frac{1351}{780}.$$

Let us contrast this description with our received text. Take first the enunciation of the proposition; here, with a word-for-word literal translation, we have:

παντὸc κύκλου ἡ περίμετροc τῆc διαμέτρου τριπλαcίων
Of every circle the perimeter the diameter three-times

ἐcτὶ καὶ ἔτι ὑπερέχει ἐλάccονι μὲν[41] ἢ ἑβδόμῳ μέρει
is and further it exceeds by less than a seventh part

τῆc διαμέτρου, μείζονι δὲ ἢ δέκα ἑβδομηκοcτομόνοιc
of the diameter, but by more than ten seventy-first [parts]

and, in the English word-order

The perimeter of every circle is three-times the diameter and further it exceeds by

---

[40] The critical edition of the text is in Archimedes–Heiberg, *Opera* i, 232–43, with the commentary from the early sixth century AD by Eutocius in *Opera* iii, 228–61. The Greek text, without critical apparatus, is reproduced in Thomas, *SIHGM* i, 316–33. The medieval Latin translations are in Clagett, *AMA*; see especially vol. ii. Also see n. 44, below. On the chronological problems associated with Eutocius, see the *DSB*, s.v. Eutocius.
[41] The Greek expression μέν ... δέ ... merely requires the translation 'but' later in the sentence.

7.3   Selection of texts   241

[a line] less than a seventh part of the diameter but greater than ten seventy-first [parts of the diameter].

Archimedes clearly does *not* here consider the ratio of circumference to diameter as a numerical quantity. Also the phrase 'ten seventy-first parts' seems to refer to the line got by concatenating ten copies of the seventy-first part of the diameter, rather than the common fraction $\frac{10}{71}$. Moreover, the ratio of 10 to 71 is capable of a far wider range of interpretations than as a common fraction, as has already been illustrated many times in this book. In particular, see Sections 2.2(b) and 4(d), where some mathematical features of Archimedes' calculation are analysed; here I shall consider purely textual issues.

Next, Archimedes' treatment of the lower bound to $\sqrt{3}:1$ reads, in translation:

Let there be a circle with diameter $A\Gamma$ and centre E, and let $\Gamma\Lambda Z$ be a tangent and the angle $ZE\Gamma$ one third of a right angle. Then EZ has to $Z\Gamma$ the ratio that 306 has to 153, and $E\Gamma$ has to ΓZ the ratio [greater than] that which 265 has to 153.

Here we have ratios, not fractions.

Now turn to the proof of the proposition. One feature can be seen immediately by consulting the admirable translation in Thomas, *SIHGM* i, 316–33.[42] Look at the parallel Greek and English texts on pp. 324 and 325: six lines of Greek text give rise to more than twenty-four lines of translation. Most of this translation is set in square brackets; it does not come from Archimedes' text (Archimedes–Heiberg, *Opera* i, 232–43) but is freely adapted from Eutocius' long commentary (ibid. iii, 228–61), and it is these additions that make the translation intelligible to us. Now look in detail at a typical passage in this translation, *SIHGM* i, 329.19–28 = *Opera* i, 242.2–5. With the interleaved material from Eutocius and elsewhere deleted, we read:

$$A\Theta : \Theta\Gamma < 5924\tfrac{3}{4} : 780$$

$$< \tfrac{4}{13} . 5924\tfrac{3}{4} : \tfrac{4}{13} . 780$$

$$< 1823 : 240.$$

Therefore $A\Gamma : \Gamma\Theta < 1838\tfrac{9}{11} : 240.$

---

[42] The following corrections and minor amendments to the translation should be noted:
p. 329   line 11. For '= 1351' read '< 1351'.
        line 13. For 'Hence' read '[Hence'.
        line 23. On '$\tfrac{4}{13} . 5924\tfrac{3}{4} : \tfrac{4}{13} . 780$': this is not really a translation, but a very helpful editorial resolution of a complicated piece of text, to be quoted below.
p. 331   line 3. For '24' read '240'.
        line 6. This line '$\tfrac{11}{40} . 3661\tfrac{9}{11} : \tfrac{11}{40} . 240$' is another resolution of even more complicated text.

The corresponding Greek text, with a literal translation, is as follows:

1   ἡ $A\Theta$ ἄρα διὰ τὰ αὐτὰ   πρὸc
     The [line] $A\Theta$ then for the same [reason] in relation to

2   τὴν $\Theta\Gamma$ ἐλάccονα λόγον ἔχει ἢ   ὃν $\overline{,\varepsilon\kappa\delta}$ $\angle$ δ'
     the [line] $\Theta\Gamma$ a less   ratio has than that [ratio] which   5924   2′   4′

3   πρὸc $\overline{\psi\pi}$ ἢ   ὃν $\overline{,\alpha\omega\kappa\gamma}$
     [has] in relation to 780 or [less than] that [ratio] which   1823

4   πρὸc $\overline{c\mu}$·   ἑκατέρα   γὰρ ἑκατέραc
     [has] in relation to 240, each one of the two for[43] of the other [line]

5    $\bar{\delta}$ $\iota\gamma'$ ὥcτε   ἡ $A\Gamma$ πρὸc   τὴν $\Gamma\Theta$
     [is] $\bar{4}$ 13′, so that the [line] $A\Gamma$ in relation to the $\Gamma\Theta$ [is less]

6   ἢ   ὃν $\overline{,\alpha\omega\lambda\eta}$ $\bar{\theta}$ $\iota\alpha'$ πρὸc $\overline{c\mu}$
     than that [ratio] which   1838 $\bar{9}$ 11′ [has] in relation to 240.

This text can be pursued back to the manuscripts using the apparatus of the critical editions by Heiberg and Clagett.[44] All of the surviving Greek texts and Latin translations of Archimedes, except for those found in the palimpsest rediscovered by Heiberg in 1906, to be described briefly below, derive from an archetype compiled in the ninth or tenth century which may have belonged to Leo the Philosopher[45] and which seems to have contained, in addition to Archimedes' works, the commentary on Archimedes by Eutocius and a book by Heron. This codex disappeared in the sixteenth century, but its contents have been reconstructed by Heiberg from four surviving Greek copies made from it between 1450 and 1564; this reconstructed manuscript is Heiberg's and Clagett's manuscript A.

The autograph copy survives of a careful literal Latin translation of parts of A made in 1269 by the Flemish Dominican monk, William of Moerbeke; it contains the translation of all of the Archimedean books in A except *The Sandreckoner,* and of all of Eutocius' commentaries except for that on *Measurement of a Circle*; this is manuscript B.[46] The

---

[43] This particle always comes second in a Greek phrase and first in English: 'for each one of the two [lines] is $\bar{4}$ 13′ of the [corresponding] other line'.

[44] The critical apparatus of Archimedes–Heiberg, *Opera,* must be collated with that in Clagett, *AMA* ii₃, 397–9. The textual details in Archimedes–Heath, *WA,* should be ignored since they are based on Heiberg's first edition of 1880–1; there is a summary based on the second edition in Heath, *HGM* ii, 25–7 and a further account in Clagett, *AMA* ii₁, 54–78.

[45] So Archimedes–Heiberg, *Opera* iii, pp. x ff. and xxii f., Heath, *HGM* ii, 25 f., and Wilson, *SB,* 83; but Clagett, *AMA* ii₁, 55 is noncommittal. Clagett gives details of its probable contents: Archimedes, *Sphere and Cylinder* I and II, *Measurement of a Circle, Conoids and Spheroids, Spiral Lines, Plane Equilibria* I and II, *Sandreckoner, Quadrature of a Parabola*; Eutocius, *Commentaries on Sphere and Cylinder, Measurement of Circle, Plane Equilibria*; and Heron, *On Measures.*

[46] Clagett calls this manuscript O.

translation was corrected by Moerbeke himself in passages, distinguished by a blacker ink, that I shall refer to as B^1. Then, in the fifteenth or sixteenth century, an unidentified hand added further annotations, though none of these are found in *Measurement of a Circle*. Finally the manuscript was acquired by one Andreas Coner in 1508, who added many more, often perceptive, annotations[47] which will here be denoted by B^2. It appears that Moerbeke found difficulty in translating the numbers in *Measurement of a Circle*: sometimes he wrote the corresponding Greek text in the margin, leaving a gap in the translation and then, as B^1, he subsequently inserted the missing text.[48] The marginal Greek annotation was then erased, but these and other erased portions can now be read under ultraviolet light and they give more information about the lost archetype, manuscript A. However, since some of the numbers in A were already wrong or meaningless, this procedure did not ensure an arithmetically correct result, and Coner has clearly worked through the manuscript, correcting and amplifying it further. Clagett (*AMA* ii$_1$, 55 f.) proposes that Moerbeke may not have translated Eutocius' commentary on *Measurement of a Circle* and *The Sandreckoner* because of his difficulties with Greek numerical practice.

The third Archimedes manuscript, a palimpsest in rather poor condition rediscovered by Heiberg in Constantinople in 1906, is a tenth-century copy that has subsequently been erased to make way for a theological text of the twelfth or thirteenth century.[49] Fortunately the original text is still legible, and Heiberg was able to decipher most of it with the aid of a magnifying glass. He recovered, in decreasing order of completeness, our second Greek version of the extant *Sphere and Cylinder* I and II, *Spiral Lines, Measurement of a Circle,* and *Plane Equilibria*; the first Greek texts of *Floating Bodies,* hitherto known only in a Latin translation made by Moerbeke from another Greek manuscript,[50] now lost; the preface and two propositions of the

---

[47] Heiberg dated these annotations to the fifteenth century (*Opera* i, pp. v–vi), although he described Coner's acquisition of the manuscript in 1508 and his subsequent annotations (*Opera* iii, pp. xliii f.). Also Heiberg apparently frequently confused the corrections in the second and third hands; see Clagett, *AMA* ii$_1$, 63 n. 7, 64 n. 8, 68, 73; and ii$_3$, *passim*.

[48] Since the main difference between B and B^1 lies in the colour of the ink, I presume that it must sometimes be difficult to distinguish between them. In any case, the distinction between them is not very great since they are both assigned to Moerbeke, and it is significant to us here only as an indication of his difficulties with Archimedes' text.

[49] This manuscript is described and its text of the *Method* is edited in Heiberg, ENA; also see Archimedes–Heiberg, *Opera* i, pp. v–vi and iii, pp. lxxxv–xc, and the supplement to Archimedes–Heath, *WA*, after p. 326. It was formerly in the Library of the Metochion of the Holy Sepulchre, Istanbul but now is in a private collection, except for one leaf, identified by N. G. Wilson, which has strayed to Cambridge University Library, Add. 1879.23; see Wilson, *SB*, 139.

[50] Heiberg calls this lost manuscript ß, Clagett calls it B. It is described in Archimedes–Heiberg, *Opera* iii, pp. liv–lvii and Clagett, *AMA* ii$_1$, 58–60. It was last mentioned in a catalogue of 1311.

Stomachion, which deals with a kind of puzzle; and, most valuable of all, the new and unexpected *Method.* This palimpsest, which is not directly related to A, is called manuscript C; and those parts of it where Heiberg could detect the presence of text that he was unable to read are identified as (C).

The following variations in the passage above are found in the different manuscripts.

line 2: $,\overline{\varepsilon\lambda\kappa\delta}\ \angle\ \delta'$
A had $,\varepsilon\tau\kappa\delta\ \varepsilon'\ \delta'$
B^1 has   5324 $\bar{3}\ \bar{4}$, in an unsuccessful attempt to correct the irregular[51] ε'
B^2 has the correct[52] 5924 $\frac{1}{2}\ \frac{1}{4}$
C has $,\varepsilon\rho\kappa\delta$
Eutocius' commentary has the correct $,\overline{\varepsilon\lambda\kappa\delta}\ \angle\ \delta'$

line 4: $\overline{c\mu}$
A had $\overline{cv}$
B has   250
B^2 has   240
C has $\overline{c\mu}$
Eutocius has $\overline{c\mu}$

line 5: $\bar{\delta}\ \iota\gamma'$
A had $\bar{\delta}\ i\gamma'\ \alpha'$ according to Heiberg's reconstruction but Moerbeke
    transcribed the Greek $\bar{\Delta}\ \iota\gamma'\ \alpha'$, now erased, in the margin of B
B is erased, and is presumably illegible
B^2 has $\frac{4}{13}$ in the erasure
(C) is illegible, but has space for $\bar{\delta}$ [or $\bar{\Delta}$?] $\iota\gamma'\ \alpha'$[53]
Eutocius has $\bar{\delta}\ \iota\gamma'$

line 6: $\bar{\theta}\ \iota\alpha'$
A omitted the $\iota\alpha$
B has θ', erased, in the margin
B^1 has $\bar{9}$, followed by a gap
B^2 has the correct $\frac{9}{11}$
(C) has no space for the $\iota\alpha$, and
Eutocius has $\bar{\theta}\ \iota\alpha'$, but he interprets it[54] as 9' 11'

These variations illustrate how those parts of the passage that seemed to suggest some kind of manipulation of fractions have all but disappeared as we trace the text back through the surviving manuscripts. For

---

[51] Recall that the conventional way of writing this would be $\delta\varepsilon$ (4' 5'), in decreasing order of size of the parts.
[52] The word 'correct', here and elsewhere, refers of course to the generally accepted interpretation of the passage.
[53] The details of C are understandably sometimes left ambiguous by Heiberg.
[54] Eutocius' commentary gives a long, long multiplication for (1138 9' 11')^2, evaluated using a mixture of fractional notations, whose answer, and with it the inequality he is purportedly verifying, is fudged. See Archimedes–Heiberg, *Opera* iii, 252–3.

example, the fraction $\frac{4}{13}$ on line 5 exists in the meaningless form $\bar{\delta}$ or $\bar{\Delta}$ (?) $\iota\gamma'\ \alpha'$ in A; it is illegible in C; and it is only found explicitly and correctly in a sixteenth-century addition to the Latin translation B; also the operation of multiplying by $\frac{4}{13}$ is described in Greek by a phrase 'hekatera gar hekateras...' which is capable of a range of alternative interpretations. However, some evidence does survive from this examination: if Eutocius interpreted $,\alpha\omega\lambda\eta\ \bar{\theta}\ \iota\alpha'$ as 1838 9' 11', then this implies a misinterpretation of some such notation that he found in his manuscript; and there is also the corrupted evidence of a similar kind of notation on line 5.[55]

An even greater complication, which I shall not describe here, pervades the next passage in the text, whose translation appears in Thomas, *SIHGM* i, 331.2–7, as:

$$AK:K\Gamma < \tfrac{11}{40} \cdot 3661\tfrac{9}{11} : \tfrac{11}{40} \cdot 240 < 1007:66,$$

and I have suggested, in note 42, that the expression in the middle of this inequality should really be placed in square brackets as an editorial interpretation.

These variant readings and later additions and corrections indicate a range of editorial and scribal interferences that have occurred since the now lost manuscript A was compiled. We have no way of assessing the extent of corruption already introduced in the transmission from Archimedes' own version up to manuscript A, but other features indicate that this could be considerable. For example, the order of the propositions is now clearly incorrect: the first two propositions were unnumbered in A, but the second of these uses the result proved in the third proposition, which is numbered $\gamma' = 3$; and the second half of the proof of this third proposition is erroneously numbered $\delta' = 4$.[56] Then Heron refers, at *Metrica* I 37, to a further proposition on the relationship between the area of a sector of a circle and the arc it subtends, and he says that it is in his version of the text; such a result is now only to be found in a translation from Arabic into Latin made by Gerard of Cremona.[57] Also Archimedes wrote in a Doric dialect, and some of his works still preserve this feature but all traces of dialect have been purged from the *Measurement of a Circle*. Further, the treatise has no preface or introductory postulates, unlike almost all of Archimedes' other works.

Heiberg's careful assessment of the text is: "In general, in the whole of this short work, the type of language suffers from so careless a brevity that one recognises the hand of an excerptor rather than that of

---

[55] The reader is also encouraged to explore the textual details of the three references to $\tfrac{10}{71}$ at the conclusion of the proposition.
[56] See Archimedes–Heiberg, *Opera* i, variant readings to 232.1, 234.18, 236.7, and 140.12.
[57] See Claggett, *AMA* ii$_1$, 5, 32, 47, and 57.

Archimedes" (Archimedes–Heiberg, *Opera* i, 233 n. 3), an opinion that does not presuppose that the original text was meant as an elementary exercise, intended for popular consumption. The bolder common opinion, that Archimedes' *original* text was meant for popular consumption, I find hard to accept. Note the subtleties of Proposition 1, which launches immediately into a very summary description of an exhaustion argument of the type found in *Elements* XII 2, based on the bisection arguments of X 1; and note also the difficulties that Moerbeke, a translator of enormous experience with Greek philosophical and scientific works, had with his translation of Proposition 3. Our easy understanding, today, depends on the clarifications, commentaries, explanations, and corrections made by Eutocius, Coner, and other editors and inserted into the received text.[58] Now observe how all of these corrections have been made from within an arithmetised tradition of mathematics. This is very strikingly illustrated by Eutocius' long commentary, a detailed verification of many of the steps, executed throughout in compound parts, though occasionally expressed in the manuscripts in an abbreviated notation that resembles common fractions. I suggest that Archimedes' original intention may have been different; even that the work may originally have been the first steps in the anthyphairetic exploration of the ratio of circumference to diameter of a circle. And I tried to show, in Sections 2.2(b) and 2.4(d), how some of the mathematical features of the calculation which cause problems when interpreted arithmetically fit very naturally within an anthyphairetic context.

7.3(b) *Aristarchus' On the Sizes and Distances of the Sun and Moon*

Aristarchus' calculation in his *On the Sizes and Distances of the Sun and Moon* has survived in a less disturbed state than Archimedes' calculation. As usual, all of our sources are minuscule manuscripts and, of these, the oldest and best, and perhaps the ultimate source of all the other surviving manuscripts, is the beautiful ninth-century Vatican gr. 204, described briefly in Section 6.3. A Latin translation by Valla was published in 1488, another by Commandinus in 1572, and the Greek text was first published by John Wallis in 1688; it is not quite clear which manuscript they used. There is a modern critical edition in Aristarchus–Heath, *AS,* with a very useful introduction and commentary, to which I refer the reader. The numbers in the different manuscripts seem to be free of corruption, though there are minor variations in the way they are expressed.

---

[58] I recommend any reader who remains unconvinced to delete all editorial additions, to replace the numbers in the text by those in A, and then to try to understand the text without using any modern notations and techniques. Also Clagett's comment (*AMA* ii$_1$, 51) that "Coner made significant corrections of these numbers, no doubt by consulting the *Commentary* of Eutocius" should be expanded slightly since, as we have seen, Coner also made corrections that are not found, or are incorrect, in Eutocius' commentary.

7.3 Selection of texts

We now find it natural and obvious to express Aristarchus' calculation in terms of manipulations of common fractions, but Aristarchus actually employs two distinct procedures:

(i) To begin with, in Proposition 4 and part of 7, simple parts are expressed as such. Sometimes they are written out in words, as in Hypothesis 6:

That the moon subtends one fifteenth part (πεντεκαιδέκατον μέρος) of the zodiac,

sometimes they are expressed as numerals, as in Proposition 4:

But a 15' [part] (ιε' [μέρος]) of a sign is a 180' (ρπ') [part] of the whole circle of the zodiac.

(ii) Compound parts are never used, with two very minor exceptions to be described below. Instead anything that is not a simple part is, right from the start, expressed as a ratio. For example:

The diameter of the sun has to the diameter of the earth a ratio greater than that which 19 has to 3, but less than that which 43 has to 6 [Introduction and Proposition 15].

When this language of ratios is in use, it is not mixed with the language of multiples or parts, even in the most obvious cases. See, for example, Proposition 7:

But it was proved that GE has to EH a ratio greater than that which 15 has to 2; therefore, *ex aequali,* FE has to EH a ratio greater than that which 36 has to 2, that is, than that which 18 has to 1.

Also see the two passages quoted in Section 2.4(c).

We appear to have two distinct approaches: the notation of simple parts, which is treated as an extension of the *arithmoi,* and a notion of ratio, which is never confused with any kind of numerical quantity. While there is no difficulty in translating those passages that are expressed in simple parts, style i, into ratios, style ii, the translation in the opposite direction is less evident. So the simple parts seem here to be playing the role of an informal shorthand, not a developed language that can be used freely or manipulated arithmetically.

The first exception to this description is found, isolated, in Proposition 7:

... the angle GBE is a fourth part (τέταρτον μέρος) of a right angle. But the angle DBE is a 30th part (λ' μέρος) of a right angle; therefore the ratio of the angle GBE to the angle DBE is that which 15 has to two (τὰ ιε πρὸς τὰ δύο): for, if a right angle be regarded as divided into 60 (ξ) equal [parts], the angle GBE contains 15 (ιε) of such [parts], and the angle DBE contains two (δύο) [parts].

The language is curiously tentative ("be regarded as divided"), and seems to indicate that though the operation of taking the ratio and simplifying it mildly is not a standard manipulation, yet it is not difficult to fit it within a formally correct procedure.

The second exception is found throughout Proposition 11:

The diameter of the moon is less than the two 45' (δύο με' in Vat. gr. 204; δύο τεσσαρακοστόπεμπτα in Wallis), but greater than the 30' (λ' in Vat.; τριακοστόν in Wallis), of the distance of the centre of the moon from our eye,

and this quantity, 'two 45'', runs through the first half of the proposition. Again, there is no difficulty in adapting the argument to the alternative language of ratio, at some cost in length.

Despite these exceptions, I do not think that we have the evidence here to justify seeing anything which corresponds to the manipulation of general fractions. And again, as I illustrated in Section 2.4(c), one of the most problematic features of the calculation, the derivation of the inequalities in Propositions 13 and 15, fits very naturally within the context of an anthyphairetic ratio theory, but it seems to need an anachronistic kind of explanation in arithmetised mathematics, especially if it is interpreted in terms of inequalities between fractional quantities.

7.3(c) *P. Lond. ii* 265 (*p.* 257)

This papyrus, published in Kenyon, Bell, & others, *GPBM* ii, 257 ff. is not typical of the thousands of financial documents that have been found, but it provides a good example for illustrating the style and procedures of a straightforward and repetitive calculation; Plate 7 is a photograph of columns 2 and 3, lines 22–67.[59] The main reason for including it here is that it is sometimes cited as evidence for the Greek use of a notation for common fractions.[60] The papyrus is tentatively assigned, on palaeographical grounds, to the first century AD, and it concerns the conversion between different standards of artabas, a measure of wheat which was also used as a unit for expressing all manner of commercial transactions. This text deals with six different sizes of artabas,[61] called *dromōi*,

---

[59] The text starts with a column of demotic, which does not appear to have been published, and which is not mentioned in the edition of the Greek text. Then Greek column 1 contains lines 1–21; 2 22–45; 3 46–67; 4 68–91; 5 92–109; 6 110–30; 7 131–50; and 8 151–62. The verso contains a census return in Greek, also unpublished.

[60] For an influential example, see Heath, *HGM* i, 44, quoted in n. 92, below.

[61] Much as today there are Imperial pints (= 20 fluid ounces: 'A pint of water weighs a pound and a quarter'), the US pint (= 16 fluid ounces; 'A pint's a pound the world around'), the Scotch pint (= 3 Imperial pints), and other local variants. A suggestive modern translation of artaba might be 'bushel'. Similar conversions can be found in P. Mich iii 145.

chalkōi, anēlōtikōi, Philippou, Gallou, and *Hermou*,[62] and it contains twenty-nine examples of conversions of 625 artabas of one standard into another. There are potentially thirty possible different such conversions; three of the conversions in the text are performed in two ways, to illustrate slightly different methods, and part of the text is missing. Other parts of the text are fragmentary but can generally be restored with confidence because of the repetitive nature of the calculations and the stylised verbal formulae describing the operation. In the examples that follow, I shall omit most of the editorial details of restorations and difficult readings, except where they involve the arithmetical calculations or the 'fractional' notation, our main concern here. The scribe has made several errors and approximations, some of which I shall describe. The text contains two standard abbreviations: ō, for artabas; and /, one of the common signs for equality.[63] The names of some of the different artabas, and occasional other words, are also abbreviated.[64]

A typical sum,[65] lines 25–9, with a word-for-word translation, reads:

25 δρομωι   ō   χκε   αι   Γαλλ∠   ποcαι
   Dromōi   artabas 625 the   *Gallou*, how many

       προc   κγ $\tilde{β}$ ⟨ταιc   ρ   ō⟩[66]
   at the rate of 23 $\tilde{3}$ ⟨to the 100 artabas⟩?

26 ωc   δει   ποιηcαι   ποιει   τac   ō   χκε   επι   τac
   As   you should   do.   Do   the   artabas 625   to   the

27 κγ $\tilde{β}$ του   διαφορου   / $\mu\hat{δ}ψϟ\overset{α}{α}\tilde{β}$
   23   $\tilde{3}$ of the difference /   14791 $\tilde{3}$.

---

[62] The words *dromōi, chalkōi,* and *anēlōtikōi,* all in the dative, are adjectives agreeing with a noun to be understood, probably *metron. Phillippou, Gallou,* and *Hermou* are genitives referring either to persons or, less probably, to places. But these grammatical points are unimportant for my argument here.
[63] See n. 28, above.
[64] I am very grateful to T. S. Pattie for examining the original of this papyrus, and for confirming the following corrections to the published text:
Line 40: There is a fraction indicator above the eta of *ρκη* but no traces over the rho and kappa. But there is a hole over these letters and very little space between this line and the line above. There seems no doubt that the scribe conceived this as *ρκη*, with fraction indicators.
Line 48: Parts of an alpha with fraction indicator and a theta can clearly be read, so ']$\overset{α}{θ}$ must be restored.
Line 139: There are no fraction indicators over the first iota alpha, but the second iota has them.
Line 147: The papyrus is broken off after the zeta, so ∠ could easily be restored.
Line 160: Restore [νθ ια
Line 162: Restore [φ α $\overset{ια}{ι}$
All of these corrections have been made to the quotations, below.
[65] This particular example is chosen because it contains more fractional manipulations than others and its text is reproduced on Plate 7.
[66] This phrase is frequently omitted. The conversion rate means that $123\frac{2}{3}$ *Gallou* are equal to 100 *dromōi* artabas.

28 τουτων   το   ρ̂   /   ρμζ   ∠ γ́ ί𝛽   ταυτας
   Of these the   100th /   147   2́ 3́ 1́2́.   These
29 προσθες   ταις   ō̄   χκε   /   ψο𝛽   ∠ γ́ μ́𝛽   ⟨δεδεικται⟩[67]
   add   to the artabas 625 /   722   2́ 3́ 4́2́.   ⟨Q.E.D.⟩

or, in a more liberal translation,

Given 625 *dromōi* artabas, how many in *Gallou* will there be, at a rate of 23 3″ per 100 artabas? (I.e. 123 3″ *Gallou* equals 100 *dromōi*.) How you should go about it. Multiply 625 by the 23 3″ which represents the difference / 14791 3″. Divide by 100 / 147 2′ 3′ 12′. Add the 625 / 771 2′ 3′ 12′. Q.E.D.

At six places only in the surviving text (on lines 40, 48, 49, 65, 66, and 139) there is something resembling our notation for common fractions.[68] Let us examine each of these examples, starting with lines 38–40:

38 χαλκωι   ō̄   χκε   αι   δρομωι   ποσαι
   Chalkōi artabas 625 the   *dromōi*, how many?
39 ως   δει   ποιησαι   επει   εστιν   το   διαφορον
   As   you should   do.   Since   it is   the   difference
40 προς   κη   ταις   ρ   ō̄   εστιν $\overset{\rho\bar{\kappa}\bar{\eta}}{\rho}$ μορια   ο
                                                 $\overset{1́2́8́}{100}$
   at the rate of 28 to the 100 arbatas, it is 100 parts which
41 γινεται /   ∠ d λ𝛽   λα𝛽ε   ουν   το   ∠ d λ́𝛽   των
   become   /   2́ 4́ 3́2́. Take   therefore   the   2́ 4́ 3́2́   of the
41/2 ō̄   χκε   /   υπη   d ⟨λ𝛽⟩   τοσαυται   χαλκωι[69]   δεδεικται
     artabas 625 /   488   4́ 3́2́;   as many   in *dromōi*.   Q.E.D.

Here 128 *chalkōi* artabas are equal to 100 *dromōi*; so the conversion proceeds by multiplying by $\frac{100}{128} = \frac{25}{32}$. But note particularly how, before this operation can be carried out, the fraction has to be converted into a sum of parts, $\frac{100}{128} = 2'\ 4'\ 32'$. This complex operation of taking the 2′ 4′ 32′ part of some quantity then recurs in three subsequent calculations, on lines 71, 102, and 130, where it is treated just like a normal division. The

---

[67] The scribe wrote μ́𝛽 in an evident error for ί𝛽, which is given correctly on the previous line. On the previous line he has omitted an accent from the iota and many such accents are omitted elsewhere; I shall restore them all in all subsequent passages. All but two of the sums finish with *dedeiktai*, 'it has been shown'.
[68] The 'fractional notation' is also implausibly reconstructed by the editors on line 160, and it should have been reconstructed on line 162. See n. 64, above.
[69] On this line, the scribe has omitted the fraction λ́𝛽 and written *chalkōi* for *dromōi*.

example on line 102 contains an error;

101/2 5 | ō χπζ ∠ | τουτων | το | ∠ d ĺβ | / | δρομωι | ō | φλς |
|---|---|---|---|---|---|---|---|
| artabas 687 2́ | of these | the | 2́ 4 3́2 | / | dromōi | artabas | 536 |

∠ d ίς ĺβ
2́ 4 ί6 3́2

where the answer should be 537 16' 32' 64'. These calculations with 2' 4' 32' are straightforward and this kind of manipulation, using only the parts 2', 4', 8', 16', ..., would have been familiar from land measurements; see Section 7.1(d), above.

Now look at the fourteenth sum, lines 63–7. Here the scribe abbreviates the procedure and leaves the answer in a non-standard form:

63 ανηλωτικωι   ō   χκε   αι   χαλκωι   ποσαι
    Anēlōtikōi   artabas 625   the   chalkōi,   how many?

64 ως δει ποιησαι λαβε το κά̄ των ō χκε
    As you should do. Take the 21st of the artabas 625

65 / κθ κά̄ ίς ταυτα αφελε απο των ō χκε
    / 29 2́1 ί6. These take away from the artabas 625.

66 λοιπαι   φ϶ε   κά̄ [ε]   τοσαυται   χαλκωι
                 2́1
    There are left 595 5,   as many   in chalkōi.

67 δεδεικται
    Q.E.D.

Here there are 100 chalkōi artabas to 105 anēlōtikōi, so 20 chalkōi to 21 anēlōtikōi; hence the conversion proceeds by subtracting a 21st part of the anēlōtikōi. When the conversion factor is n to $(n + 1)$ or $(n + 1)$ to n, the calculation will be simplified, since it can proceed by subtracting a $(n + 1)$th part, as here, or by adding an nth part, as in the next example. The ratios $(n + 1)$ to n were well known from financial calculations of interest, and they have a special name, *epimorios logos,* with the particular examples of *epitriton* (4 to 3), *epipempton* (6 to 5), *ephekton* (7 to 6), *epogodoon* (9 to 8), and *epidekaton* (11 to 10) being of special importance.[70]

---

[70] See Sections 4.5(b) (A_{60} & B_{61}) and 5(d), and Burkert, *LS,* 439; but Burkert's argument that these epimoric ratios appear in the musical fragments of Archytas and Philolaus as borrowings from the financial calculations of everyday speech seems very doubtful.

A similar procedure is followed in lines 46–9:

46 χαλκωι   ō   χκε   αι   Φιλιππ⌐   ποcαι
   Chalkōi artabas 625 the Philippou, how many?

47 ωc   δει   ποιηcαι λαβε το   ιά   των   ō   χκε
   As you should do. Take the 11th of the artabas 625

                ι ά
48 / [vς] θ   ταυτας   αφελε   απο   των   ō   χκε
              ί ί
   / 56   9. These   take away from the artabas 625.

              ι ά
49 /   φξη   β   τοcαυται   Φιλιππ⌐   δεδεικται
              ίί
   /   568   2   as many   in Philippou.   Q.E.D.

In these last two calculations, the scribe has performed two arithmetical operations with the 'fractional' notation: $1 - \frac{16}{21} = \frac{5}{21}$ and $1 - \frac{9}{11} = \frac{2}{11}$. He performs a similar but more elaborate manipulation in the next example, and then gives an approximation to the result. So here, finally, are lines 136–41; I have left in the editorial details to indicate just how much of this part of the text is editorial restoration:

136 [Γαλλ⌐   ō   χκε   αι   Φι]λιππ⌐   [ποcαι]
    Gallōu artabas 625 the Philippou how many?

137 [ωc   δει   ποιηcαι]   ταυτα[c   χαλκιcον]
    As you should   do.   These convert to chalkōi

138 [ωc προκειται] / [χ]αλκωι   ō   [χμβ ∠ γ́ κ̄δ̄]
    as   set out   /   chalkōi artabas 642 2̂ 3̂ 24̂.

                                 ι ά
139 τουτων   [το]   ια / νη   ε   [ταυτας αφελε]
                                 ίί
    Of these the   11th / 58   5.   These take away

140 απο   των   [χ]αλκωι   ō   χμβ   [∠ γ́ κ̄δ̄ /]
    from the   chalkōi artabas 642   2̂ 3̂ 24̂ /

141 Φιλιππ⌐   [ō]   φπδ γ́ ίβ   [δεδεικται]
    Philippou artabas 584 3̂ 12̂   Q.E.D.

More than half of the calculations proceed by passing first to chalkōi artabas, and then into the required standard; the process is described by the verb χάλκιcον ('chalkise'). In this case, the conversion from Gallou to chalkōi was performed in the 23rd sum, lines 122–6, and it contains a mistake and a skilful approximation: 64687 ÷ 100 is given[71] and subsequently used as 642 2′ 3′ 24′, where it should be 646 2′ 3′ $\frac{11}{300}$, closer to 646 2′ 3′ 27′. The mistake, in a division by 100, is less obvious in Greek than in Arabic notation:

των $\overset{\varsigma}{\mu}$,δχπζ τουτων το ρ̇ / χμβ ∠ γ̇ κδ, instead of χμς ∠ γ̇ κδ.

In line 139, $\frac{5}{11}$ is a reasonable and straightforward approximation for 4 2′ 3′ 24′ ÷ 11, since 4 2′ 3′ 24′ is already close to 5, and this difference will then be divided by 11. In line 141, 3′ 12′ is a reasonable and straightforward approximation for 2′ 3′ 24′ − $\frac{5}{11}$ since $\frac{5}{11}$ is less than 2′ by 22′, close to 24′. The approximations are not always so adept; for example, in lines 120–1, the result of 68570 ÷ 104 is given as 661 12′, where the correct fractional result is $\frac{3}{52}$, which is between 18′ and 17′.

The arithmetical operation in the papyrus that generates the fractions is division, and these divisions are always expressed exactly as we find them in division tables, for example as on lines 28, 102, 64, 47, and 139 in the examples quoted above. The following divisions are performed in the papyrus; the examples cited above are marked with an asterisk:

Division by	On lines
4	35
6	22
10	96
11	47*, 76, 139*, 160
20	12, 44, 134, 154
21	64*
100	4, 6, 15, 28*, 33, 126
104	61, 93, 120, 149
128	40*
207	55, 84, 111
2′ 4′ 32′	41*, 71, 102*, 130

The 'fractional' notation is associated *only* with the divisions by 11, 21, and 128; I have quoted in detail all of these divisions except for that on line 76 (where 595 4′ ÷ 11 is approximated as 54 11′ with an error of 44′ which arises from ignoring the fractional part 4′ in the dividend) and in line 160 (where, in text that is lost and has been completely restored, 650 ÷ 11 = 59 11′ is then subtracted from 650 to leave 591 $\overset{ii}{10}$). Of the other divisions, those by the more commonplace numbers, 4, 6, 10, 20, and 100, are *all* evaluated in a straightforward way,[72] while the seven

---

[71] In fact this part of the text has to be reconstructed, but the scribe subsequently clearly uses the wrong answer.

[72] With one error on division by 100 (see the comment on lines 137–8, above), and one approximation in the division by 20 on line 134, in a passage that has been completely restored by the editors (642 2′ 3′ 24′ ÷ 20 / 32 6′).

divisions by the more unusual numbers 104 and 207 are *all* expressed in conventional simple and compound parts.

Section 7.2 described the extensive evidence we have concerning division tables, the ready reckoners and exercises that deal with the expressions of divisions as sums of parts. These division tables are sometimes quite extensive, though few would contain tables for division by 11 or 21 (see how the sets of tables in nos. 7 and 14 omit these) and none would contain tables for division by 104 or 207. I propose that the 'fractional' notation may plausibly have been used here by the scribe in the following circumstances: By convention, the final answer to any sum should always be presented in a standard form, as simple or compound parts.[73] Whenever possible, this expression should be precise, but approximations were permitted; and the exact value for all commonly occurring divisors could be got from memory or tables. The scribe did not have the tables for division by 11 and 21 to hand, or he could not remember the answers though he was aware that these tables existed; so where the sums required divisions by 11 or 21, the scribe approximated the answer or left a shorthand abbreviation in the text, to be expanded later. Where the scribe knew that the tables did not exist, for example for division by 104 and 207, he always approximated the result. And I think that a close examination of the papyrus leads to one firm overall conclusion: that we have no good evidence here for the notation or manipulation of anything similar in conception to common fractions, beyond simple subtractions like '1 minus the 21st of 16 leaves the 21st of 5', which are conceived throughout as simple or compound parts. I shall return to this kind of explanation at several places below.

7.3(d)   *M.P.E.R., N.S. i* 1

This text[74] is a collection of thirty-eight easy problems in three-dimensional metrical geometry very much in the style of the Heronian *Stereometrica*; it has been edited with full mathematical commentary in Gerstinger & Vogel, *MNWPER* i, 11–76. It was dated by its editors, on palaeographic grounds, to the second half of the first century BC, but this date has now been revised by W. E. H. Cockle and H. Harrauer to the first century AD, at the earliest.[75] Plate 8 is a photograph of column

---

[73] Analogous conventions apply today to the presentation of answers to school mathematical problems.

[74] Also known as P. Vindob. [inv.] 19996 and P. Rainer (N.S.) i 1.

[75] I am very grateful to Walter Cockle and Hermann Harrauer for answering my many queries about this text and for examining, respectively, a photograph and the original. Several of the details described below, particularly of the use of fraction indicators in the 'fractional notation', are not included in the *editio princeps* and are not discernible on the photograph, reproduced here as Plate 8; but they can be securely read on the original. Also there is a revised reading in column 6, line 10: for $\tau . \zeta$ now read $\pi\zeta$. These corrections have

6,[76] which contains seven of the nine examples of the general 'fractional notation' that occur in this papyrus, and the remaining two examples (in Problem 10) arise in exactly the same context as examples in Problem 12, as will be described below. Hence a discussion of column 6 will cover all occurrences of the notation in this papyrus. The column starts in the middle of Problem 12 (whose beginning is lost), and Problem 13 begins in the middle of line 3; this is indicated by a forked paragraphus or *diple obelismene*[77] under the beginning of line 3. There is another such paragraphus, not indicated by the editors, under the beginning of line 11, marking the end of Problem 13.

The seven instances of the 'fractional notation' in this column are as follows:

Problem 11: $\tau\nu\varsigma$ $\bar{\overset{\varepsilon}{\alpha}}$ is used as a label to its figure, not legible on Plate 7. This is an unusual way of writing the part 5', which in any case here needs emendation to yield anything meaningful.

Problem 12: $\overset{\acute{\varepsilon}}{\beta}$ and $\overset{\acute{\varepsilon}}{\delta}$ (twice) occur in the text, in lines 2 and 3, and $\kappa\alpha\delta^{\gamma\varepsilon}$ (in a doubtful reading) occurs as a label to its figure, but this also needs emendation to make mathematical sense. The epsilons in line 2 are written cursively while that in line 3 is written in two strokes, and all of them have fraction indicators.

Problem: 13: $\overset{\acute{\varepsilon}}{\beta}$ and $\overset{\acute{\kappa}}{\gamma}$ occur in the text, in lines 9 and 10. The epilson is written in two strokes, and both the epsilon and the kappa have fraction indicators.

Here now is a transcription and literal translation of column 6, together with its text figures and their labels, in which all abbreviations have been expanded. Note, in particular, that ⁄ʌ is $\lambda o\iota\pi\acute{o}\nu$, '(there is) left'; ⋋ is $\hat{\omega}\nu$, 'of which';[78] and $\gamma\acute{\iota}\nu\varepsilon\tau\alpha\iota$ is written in full or abbreviated as / or ⌡. Further remarks on the abbreviations will be found in Section 7.4(a), below.

---

been incorporated in the transcription, below. Further discussion can be found in the edition of the verso, M.P.E.R., N.S. xv 151, which contains an extensive addition table in a later, schoolboy's hand; see no. 56 in the Appendix.

[76] There is a photograph of column 10, Problems 24 and 25 and their figures, in the *editio princeps* and another photograph in Weitzmann, *ABI* of columns 12 (part), 13, and 14 (part), Problems 28–32 and the figures 12b and c, and 13.

[77] See Turner, *GMAW*, 14–5.

[78] For details of the second abbreviation, see n. 27, above. On the first, A. Blanchard notes: "Here again there is borrowing from the demotic ⌐, *sp,* rationalised into a lambda by the Greeks and then, since the Greek abbreviations contain at least two letters, into ⁄ʌ, ⁄ʌ, and even later, into $\lambda o\bar{\iota}$.

256   Numbers and fractions   7.3

[End of problem 12:]

1   (λοιπὸν)   α   ὧν   [πλε(υρὰ) ⊪] α   [τηλικα(ύτη)] ἡ κάθετος
... there is left 1 of which   a side   / 1. So much [is] the height.
εἶτ[εν ἀ]ναμέτρει τρίγω(νον) ἰσόπλευρ(ον)
Then   measure   a triangle,   equilateral,

2   οὗ   αἱ πλε[υ]ρα[ὶ] ἀνὰ   ιβ γίνεται ξδ β´
of which the sides   each [are] 12. It makes 64 2´,
(ὧν)   τὸ γ´ ⊪ κα δ´   ταῦτ' ἐπὶ τὴν κάθε-
of which the 3rd / 21 4´. This to the height

3   τον α / κα δ´ τηλικα(ύτη) [⟨ἡ⟩ cτ]ήλη
    1 / 21 4´. So much is the block.

ἐὰν δ(οθ)ῆι ἄλλη{ι} τρ[ίγ]ωνο(c) ἡμιτελὴc ἡ κορυ
[Problem 13:] If there is given another triangle, half-complete, the top

4   φῇ ἀνὰ   γ ἡ δὲ βάc[ιc] [ἀν]ὰ   ιη τὰ δὲ κλίματα ἀνὰ   ια
    each [is] 3, the base each [is] 18, the slopes each [are] 11.
ἄφελε τὴν κορυφ(ὴν)
Take away the top,

5   τὰ γ ἀπὸ τῶν ιη [(λοιπὸν) ιε τὰ ιε] ἐφ'ἑα(υτὰ) ⊪ cκε
    the 3 from the 18, there is left 15. The 15   to itself / 225,
ὧν τὸ γ´ ⊪ οε καὶ τὰ ι ἐφ' ἑα(υτὰ) ⊪ ρ[
of which the 3rd / 75, and the 10 to itself / 100.

6   ἀπὸ τούτων ἄφελε τὰ [οε] ... (λοιπὸν) κε
    From these take away the 75   there is left 25,
ὧν πλε(υρὰ) ⊪ ε τηλικα(ύτη) ἡ κάθετος εἶτεν
of which a side / 5. So much is the height. Then

7   cύνθες τὰ γ καὶ τὰ [ιη] ⊪ κα [ ] (ὧν) τὸ ∠´ ⊪ ι ∠´
    put together the 3 and the 18 / 21 of which the 2´ / 10 2´.
ἀναμέτρηcον τρίγω(νον)   οὗ   αἱ
Measure   a triangle, of which the

8   πλευραὶ ἀνὰ   [ι ∠´] ⊪ [μ]ζ ∠´ [d] εἶτεν ἄφελε τὰ γ ἀπὸ
    sides each [are] 10 2´ / 4 7 2´ 4´. Then take the 3 from

7.3 Selection of texts 257

τῶν ιη [(λοιπὸν)] ιε   ὧν   τὸ ∠ ⊹ ζ ∠̣
the 18, there is left 15 of which the 2́ / 7 2́.

9 ἐγμέτρει   ἄλλο [τ]ρίγ(ωνον) ἰςόπ[λ]ευρ(ον)   οὗ   αἱ πλευ(ραὶ)
   Measure out another triangle, equilateral,   of which the sides

   ἀνὰ   ζ ∠ / κε έ̣   ὧν   τὸ γ̇ / η β̇ ταῦτα
   each [are] 7 2́ / 25 5́, of which the 3rd / 8 2́. These

10 πρόσθες τῆι πρώτηι ἄ[λ]λο ἐχθήςει τοῖς πζ ∠ δ/ νϛ γ̇ ταῦτ' ἐπὶ τὴν
   add to the first, elsewhere set-out, to the 87 2́ 4́ / 56 3́. This to the

11 κάθετον ε / cπ ∠ d τόϲων [τ]ῶν ποδῶν ἐϲτὶν ἡ πυραμίϲ
   height   5 / 280 2́ 4́. Of so many feet is the pyramid.

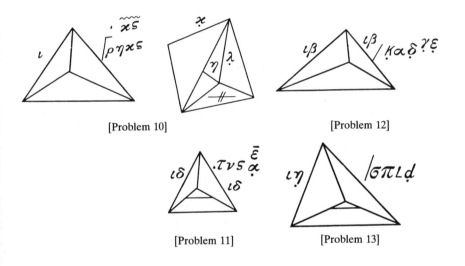

[Problem 10]   [Problem 12]

[Problem 11]   [Problem 13]

Line 2: For ξδ, 64, read ξβ, 62.

Lines 2 & 3: For κα έ̣ 5́, 21 4, read κα ζ̣, 21 7, or κ δ̣ 5́, 20 4.
Line 4: For ια, 11, read ι, 10.
Line 10: For πζ, 87, read μζ, 47.
For notes on the labels on the figures see the *editio princeps*.

The scribe is applying the same procedures for calculating pyramids that we find set out in the Heronian *Stereometrica*; see Heron–Heiberg, Schöne, & others, *Opera* v, 137–61 or Heron–Bruins, *CC*, folios 55[r]–61[r]. The following notes will explain what is happening, and the reader is referred to these texts and commentaries for further details:

(i) The area of an equilateral triangle is (approximately) the 3rd plus the 10th of the square on the side; the scribe performs this operation, without explanation and with errors and approximations, on lines 2, 7–8, and 9, and each such operation will be analysed below. Heron's explanation of this procedure, for the case of an equilateral triangle of side 30 (Heron–Heiberg, Schön, & others, *Opera* V, 148.25–6 or Heron–Bruins, *CC*, folio 58^r.19–20), is:

εὑρεθήϲεται τὸ ἐμβαδὸν οὕτωϲ· τὰ λ̄ ἐφ' ἑαυτά· / ↘. τούτων τὸ γ̄ καὶ τὸ ῑ· / τϟ̄.

The base will be found thus: the 30 to itself / 900. Of these the 3rd and the 10th / 390.

(ii) The square on the circumradius of an equilateral triangle is the 3rd of the square on the side (*Elements* XIII 12); see line 5.

(iii) The height of these regular pyramids is calculated using the right-angled triangle of inclined edge, height, and circumradius of the base; see lines 5–6.

(iv) The volume of a pyramid is the third of the base, multiplied by the height; see lines 2–3.

(v) The procedure for evaluating the volume of a truncated pyramid is clearly described in lines 7–11.

Every occurrence of the 'fractional notation' in the text is associated with an evaluation of the area of an equilateral triangle, as was described above, and none of these operations is explained by the scribe:

Line 2: When the side is 12, its square is 144, of which the 3rd and 10th is 48 and 14 and the 10th of 4, viz. 62 and the 5th of 2. The identical operation was performed correctly in Problem 10, column 5.3–4, but here, in Problem 12, the scribe writes 64 2̇ $\overset{5}{}$ instead of 62 2̇ $\overset{5}{}$.

Line 8: When the side is 10 2', its square is 110 4', of which the 3rd and 10th is 36 2' 4' and 11 and the 10th of 4', which the scribe approximates as 47 2' 4' by neglecting the 40'.

Line 9: When the side is 7 2', its square is 56 4', of which the 3rd and 10th is 18 2' 4' and 5 and the 10th of the 4th of 25, which is 23 2' 4' and the 8th of 5, which is 24 4' 8'. The scribe makes an error and gets 25 5', which he then divides by 3 to give 8 2̇ $\overset{5}{}$. He then adds this to 47 2' 4' to give 56 3̇ $\overset{20}{}$; the fractional part of this can be evaluated by adding the 20th of 8 to the 20th of 10 and of 5.

All other calculations in the text are carried out in the usual simple and compound parts. The text contains divisions by 3, 5 and its multiples, and various powers of 2. The examples of the 'fractional notation' *only* occur in the divisions into 5th parts or its multiples, and *every* such division by 5 or its multiples introduces the notation. Also *every* such 'denominator' has a fraction indicator. Thus an explanation such as was offered for the examples in P. Lond. iii 265 again seems to apply: that the scribe used the notation only when he did not know, could not find, or did not want to break off to compute, a table of division by 5ths or 10ths.

Fractions, expressed in parts, also arise in this papyrus when square roots are quoted: without any explanation, in Problem 14 the square root of 66 3″ is taken to be 8 6′, and in Problem 29 that of 200 as 14 7′. Also the rule for evaluating the area of an equilateral triangle is based on an approximation of $\frac{1}{4}\sqrt{3}$ as 3′ 10′. All of these are good approximations.[79]

7.3(e) *Demotic mathematical papyri*

The small corpus of known mathematical works written in demotic comprises eight texts: five of them, containing in all seventy-two short problems, published in Parker, *DMP*; two in Parker, DMPF and ME; and the division tables, no. 6 in the catalogue in the Appendix below.[80] All of my examples here will be taken from *DMP*, and will be referred to by the problem numbers therein. Readers are directed to this book for photographs, transcriptions, and full commentaries.

Here I shall consider only the demotic treatment of fractions. As in Greek, there is no specific word for 'fraction', and the demotic word *r*3 means 'part' or 'fraction', or even 'number' or 'amount', according to context.[81] The treatment of fractions follows the familiar pattern: there

---

[79] The anthyphaireses are as follows: $\sqrt{200}:\sqrt{3} = [8, \overline{6, 16}]$, $\sqrt{200}:1 = [14, \overline{7, 28}]$, and $\sqrt{3}:4 = [0, 2, \overline{3, 4}]$ where $[0, 2, 3, 4] = 13:30$, or $\sqrt{3}:1 = [1, \overline{1, 2}]$ where $[1, 1, 2, 1, 2, 1] = 26:15$. See Chapters 2 & 3; here again we have another example of unexplained fluency with square root operations.

[80] The inventory numbers, locations, and dates of these texts are as follows. In Parker, *DMP*: P.(dem.) Cairo J.E. [inv.] 89127–30, 89137–43 verso; iii BC, perhaps from Hermopolis: Problems 1–40. P.(dem.) B.M. [inv.] 10399, Ptolemaic, later than P. Cairo, provenance unknown: Problems 41–52. P.(dem.) B.M. [inv.] 10520, early(?) Roman, provenance unsure, possibly Memphis: Problems 53–65. P.(dem.) B.M. [inv.] 10794, uncertain date, unknown provenance: Problems 66–7, the division tables no. 4 in my catalogue. P.(dem.) Carlsberg [inv.] 30, probably ii AD, from Tebtunis: Problems 68–72. In Parker, DMPF: P.(dem.) Griffiths Institute [inv.] I E.7. In Parker, ME: P.(dem.) Heidelberg [inv.] 663, Ptolemaic, perhaps ii or i BC, provenance unknown: three fragments of problems on a trapezoidal field.

Also see Neugebauer, *ESA*, 91: "One large demotic text was found in Tûna el-Gebel, according to *The Illustrated London News* 104 (1939) and *Chronique d'Égypte* 14 No. 28 (1939) 278. No information about this text could be obtained". (Hermopolis West is modern Tûna el-Gebel; this is surely P. Cairo, above.) I would like to thank R. A. Parker and John Tait for further explanations of some of this material.

[81] I was tempted to change the word 'fraction' into 'part' in several of the passages cited below, but have resisted this blind interference with the translations. See, for Egyptian numerals and fractions, Erichsen, *DG*, 694–706, and Gardiner, *EG*, 191–203.

are special signs for 3″, 2′, 3′, and 4′, and thereafter the parts are indicated by a fraction indicator, a small stroke, which may often be omitted, attached to the normal numeral.[82] A ligatured 3″6′ is used to represent $\frac{5}{6}$, but no surviving text seems to contain a similar ligatured 2′4′ for $\frac{3}{4}$, common in Greek documentary papyri.[83]

While the operation of division, 'the nth of m', is described in Greek by very minor variations, for example in the word order, of the basic phrase 'τῶν m τὸ $\hat{n}$', there is a much wider range of expressions in demotic. Here are the translations of some examples:[84]

Problems 66 & 67, the division tables, are expressed thus:

> The way of taking 150′ to 10. Viz.[85]
> 1 result 150′
> 2 result 90′ 450′ (sic)
> 3 result 60′ 300′ (sic)

Problem 13: Carry 15 3″ into 100,
Problem 3: You shall take the number 47 to 100,
Problems 9, 10, & 46–52: You shall say: '4 is what of 5?'. Result, its 3″ 10′ 30′.

Six problems in *DMP*, numbers 2, 3, 10, 13, 51, and 72, coming from three different manuscripts, show the special feature that the results of divisions are not expressed in parts, but employ some more general 'fractional notation'. I shall start with a complete citation of the translation of Problem 3, and with brief observations on these other examples. I have adjusted the quotations to conform to the style of editorial indications used here (described in the index to the plates, p. xxi), and the notation for parts (see n. 79, above); otherwise the quotations are word-for-word from *DMP*:

Problem 3:
1 If it is said to you: '[Carry 15 3″ into 100]',
2 you shall subtract 6 fr[om 100: remainder 94]
3 You shall say: '[6] customarily go[es into 94 (with the) result 1]5 3‴'.
4 Bring it (to the) number 3′: result 47.
5 You shall take the number 47 to 100.
6 47 to 1

---

[82] In the passages cited below, I have replaced Parker's translations $\frac{2}{3}, \frac{1}{2}, \frac{1}{3}, \frac{1}{4}, \ldots$ by 3″, 2′, 3′, 4′, ...

[83] This ligatured 3″ 6′ is very common, and the expression 2′ 4′ seems to be avoided; for example, in Problem 23, the sum of 1, 2′, 4′, and 8′ is then expressed as $1\frac{5}{6}$ 30′ 120′, and 2′ 4′ only seems to occur in the fragmentary and uncertain readings in Problems 33 and 70, while Problem 13 uses the alternative expression 3″ 12′. Also see n. 88, below.

[84] See the discussion in Parker, *DMP*, 8 for further details and examples. Even within the text of the single scribe of Problems 1–40, we find a wide variety of different equivalent expressions.

[85] This word 'viz.' is the translation of ' / ', the transliteration of ⸗. See n. 28 and its text, above.

7.3 Selection of texts

7   94 to 2
8   Remainder 6: result $\frac{6}{47}$
9   Result $2\frac{6}{47}$.
10 You shall reckon it 3 times: result $6\frac{18}{47}$.
11 You shall say: '$6\frac{18}{47}$ is its number.'
12 You shall reckon $6\frac{18}{47}$, 15 3″ times.
13 $6\frac{18}{47}$ to 1
14 $63\frac{39}{47}$ to 10
15 31 2′ $\dfrac{19\ 2'}{47}$ to 5.
16 Total 95 2′ $\dfrac{11\ 2'}{47}$ to 15.
17 $4\frac{12}{47}$ to 3″
18 Total 99 2′ $\dfrac{23\ 2'}{47}$ to 15⟨3″⟩.
19 Its remainder is 2′, which is equivalent to $\dfrac{23\ 2'}{47}$
20 Total 100.

These fractions, all with 'denominators' 47, are indicated here by underlining the 'numerator'. For example, in line 15,

$$31\ 2'\ \dfrac{19\ 2'}{47} \text{ is written } 47\ \underline{2'\ 19}\ 2'\ 31$$

(read from right to left!). This same system applies to Problems 2, 3, 10, and 15, all from the same papyrus. In Problems 51 and 72, below, the 'denominators' are underlined; for example, in Problem 51, $\frac{5}{11}$ is written $\underline{11}$ 5.

Problem 2: A fragmentary part of the text, very similar to Problem 3, based on dividing 100 by 17 3″ and involving divisions by 53.

Problem 10: A problem on division of cloth that involves division by 11, in which the complex expression 1 2′ $\dfrac{1\ 2'}{11}$ appears. Problems 8 and 9 are very similar; they involve division by 6 (giving an answer $\frac{5}{6}$, written as the ligatured 3″ 6′) and by 5 (see the discussion below).

Problem 13: A fragmentary and obscure problem on the division and valuation of pieces of cloth, in which, without any explanation, a further valuation at a rate described by the complex division 3′ 15′ $\dfrac{7\ 2'\ 10'}{131}$ is introduced. This yields a set of complex expressions involving divisions by 131.

Problem 51: The last of a sequence of 6 problems. The previous five problems all ask: given the fraction n' which is added to 1, determine what fraction of $1\,n'$ must be subtracted from it to give 1 again. The answer $(n+1)'$ is verified in each case, for $n = 5, 6, 7, 8,$ and 9. Then Problem 51 asks: given the fraction $\frac{5}{6}$ which is added to 1, determine what fraction of 1 must be subtracted from it to give 1 again. Two of these problems will be quoted in full below.

Problem 72: A fragmentary and obscure part of a problem involving proportions of gold, silver, copper, and lead. Without explanation in the surviving fragment, division by 76 is introduced and then not expressed in parts.

Problems 13 and 72 seem to be so fragmentary and obscure as to defy elucidation, and I shall not consider them further here. Take now the following parallel extracts, from Problems 9 and 10:[86]

Problem 9	Problem 10
You shall say: '4 is what (fraction) of 5?'	You shall say: '3 is what (fraction) of 11?'
Result, its[87] 3″ 10′ 30′. Add it to 4: result 4 3″ 10′ 30′.	Result $\frac{3}{11}$. Add it to 3: result $3\frac{3}{11}$.
You shall reckon 4 3″ 10′ 30′,	You shall reckon $3\frac{3}{11}$
5 times: result 24. . . .	2′ times: result 1 2′ $\frac{1\ 2'}{11}$ . . . .

In the left-hand column, the calculation involving division and multiplication by 5 is carried out in compound parts; in the right-hand column, division by 11 is left in its raw state, but it is clearly conceived in simple and compound parts. This is further corroborated by the corresponding extract from the similar Problem 8, which starts:

You shall say: '5 is what fraction of 6?' Result, its $\frac{5}{6}$. . . . ,

in which the $\frac{5}{6}$ is written as the ligatured 3″ 6′, again a compound part.

Now consider Problem 51. This follows exactly the same stereotyped verbal formula as the previous five problems 46–50, as can be seen in the following parallel quotations.

Problem 48	Problem 51
1 You are told: '7′ is (the) addition: what is (the) subtraction?'	You are told: '$\frac{5}{6}$ is (the) addition: what is (the) subtraction?'
2 The way of doing it. Viz.	The way of doing it. Viz.
3 You shall add 7′ to 1: result 1 7′.	You shall add $\frac{5}{6}$ to 1; result $1\frac{5}{6}$.
4 You shall say: '7′ is what (fraction) of 1 7′?' Result, its 8′.	You shall say: '$\frac{5}{6}$ is what (fraction) of $1\frac{5}{6}$?' Result, its $\frac{5}{11}$.

---

[86] The editorial apparatus, apart from indications of editorial additions, has been omitted from these and most of the following examples.

[87] I do not know whether the inclusion of the possessive article in this phrase in Problems 8 and 9 (demotic "$r\ p3y\,.\,f\ldots$", translated "Result, its ..."), but its exclusion from 10 (demotic "$r\ldots$", translated "Result ...") is significant.

5 You shall say: '7' is (the) addition; 8' is (the) subtraction.'	You shall say: '$\frac{5}{6}$ is (the) addition; $\frac{5}{11}$ is (the) subtraction.'
6 To cause that you know it. Viz.	To cause that you know it. Viz.
7 You shall subtract 8' from 1: remainder 3" 12' 8'.	You shall subtract $\frac{5}{11}$ from 1: remainder $\frac{6}{11}$.
8 The fraction 3" 12' 8'. Viz.	The fraction $\frac{6}{11}$. Viz.
9 Its 7' (is) 8'. You shall add it (8') to it (3" 12' 8'): result 1 again.	Its $\frac{5}{6}$ (is) $\frac{5}{11}$. You shall add it ($\frac{5}{11}$) to it ($\frac{6}{11}$): result 1 again.

In each of Problems 46–50, the subtraction in line 7 is expressed as a compound part, with the feature that, in all of the cases other than the one quoted, these parts involve the ligatured 3" 6', translated as $\frac{2}{3}$. These subtractions are as follows:

Problem 46: $1 - 6' = 3"\ 6'$
Problem 47: $1 - 7' = 3"\ 6'\ 42'$
Problem 48: $1 - 8' = 3"\ 12\ 8'$ (see above[88])
Problem 49: $1 - 9' = 3"\ 6'\ 30'\ 45'$
Problem 50: $1 - 10' = 3"\ 6'\ 15'$
Problem 51: $1 - \frac{5}{11} = \frac{6}{11}$ (see above)

It seems clear from this block of examples that again, while the scribe prefers, whenever he can, to express the divisions as compound parts, here also, as in Problem 10, division by 11 defeats him. Similarly, in Problems 2 and 3, division by 53 and 47 leads to expressions which he also cannot express in parts. In all of these cases, he resorts to abbreviations for expressing the incomplete divisions; and it is these abbreviations that we interpret as a notation for common fractions.

This explanation is again consistent with the proposed explanation of the previous texts, that, while an expression in parts is the preferred form of answer, in cases for which tables are not available, the divisions may very occasionally be left in an incomplete state. I shall now draw these conclusions together and indicate some of their wider implications.

7.4 Conclusions and some consequences

7.4(a) *Synthesis*

The evidence for the proposal that early Greek mathematicians conceived and used manipulations of common fractions is said to come from two sources: from the medieval manuscripts of scientific treatises, and

---

[88] Parker observes (*DMP*, 61, note to line 16): "For no discernible reason the scribe uses the clumsy fraction 3" 12' 8', with 8' in incorrect order after 12', instead of the more convenient $\frac{5}{6}$ 24'." I suggest that 3" 12' may be the standard expression for the 4th of 3. So to subtract 8' from 1, 1 is mentally decomposed into 3" 12' 4', then 8' is subtracted from 4' to give 8'; hence the form of the scribe's answer. Also see n. 83, above.

from contemporary texts on papyrus which are mainly of a commercial and pedagogic nature; and the proposal is often reinforced by a more or less unconscious retrojection of our own ways of thinking, which usually manifests itself in wonder that anyone of any mathematical insight could ever be so obtuse as to reckon in unit fractions.[89]

However, the evidence of calculations in the papyri displays almost unanimous evidence against this proposal. The very few instances there that can be cited as illustrating notions for common fractions appear, on closer scrutiny, more probably to be abbreviations of unresolved descriptions of divisions that are still conceived as sums of unit fractions, and all can be more naturally explained as relaxations of stylistic conventions about how these divisions should be evaluated and expressed. Possibly the abbreviations did then evolve into our conceptions of common fractions, and certainly the practice and popularity of common fractions developed, particularly among Italian mathematicians, from the ninth or tenth centuries onwards. These new fractional notations and conceptions may then have been adopted by the scribes and readers of the medieval manuscripts, and so infiltrated and corrupted the evidence to be found there; on this more work needs to be done.[90]

Of course, the manipulations of fractions expressed as unit fractions are (arithmetically) equivalent to the same manipulations when expressed as common fractions; but they will be conceived differently in the two systems. Some operations will come very close to each other; for example, the manipulation of 'the nth of m' into the equivalent 'the knth of km' is parallel to the identity $m/n = km/kn$, but very far from it in conception; and a frequent step in our evidence is the manipulation 'the nth of m plus or minus the nth of p is the nth of $(m \pm p)$', equivalent to $m/n \pm p/n = (m \pm p)/n$. But reciprocation gives an example of an operation which manifests itself very differently in the two systems; for an illustration, the relation between $\frac{7}{11}$ and $1\frac{4}{7}$ is much more evident than that between 3' 7' 51' and 1 2' 14'. Just one example of some operation such as the addition, subtraction, multiplication, or division of two

---

[89] In this section, only, I shall sometimes use the common but, I believe, misleading description 'unit fractions', to ensure that my proposals are clear to everybody. But I encourage any sympathetic reader to conceive these unit fractions, in Greek style, as the half, the third, the quarter, ... and their combinations, that is as *meros ē merē*, part or parts.

[90] For example, it would be interesting to know if there is any systematic difference between contrasting early manuscripts, like Paris gr. 2389 and Vatican gr. 1594, which contain the same numerical data (here the *Almagest*) but expressed differently, in the old capital and the new minuscule scripts. Also see the notes in Diophantus–Sesiano, *BFSDA*, on the treatment of numbers and fractions in the newly discovered Arabic translation of part of the *Arithmetica*, especially on pp. 37–42 and 437 s.v. *juz'*. This translation was probably made from an earlier exemplar than the surviving minuscule Greek manuscripts (see p. 67).

7.4 Conclusions

fractional quantities, expressed directly as something like, 'the nth of m multiplied by the qth of p gives the nqth of mp' and *clearly unrelated, by context, to any conception in terms of simple and compound parts*, could be fatal to my thesis that we have no good evidence for the Greek use or conception of common fractions. I know of no such example.[91]

We can take my explanation further. First consider the form of the abbreviation, in which the stereotyped phrase '$τῶν\ m\ τὸ\ n$' is written $\overset{\acute{n}}{m}$ with, as is often expressed, 'the numerator and denominator reversed'.[92] The practice of writing one letter above another is very common in the abbreviations found in Greek papyri. See, in Plate 9, how practically every work in O. Bodl. ii 1847 is abbreviated in this way. For example, at the beginning of lines 4 and 5, $Ἀλ(λη)\ ἐχο(μένη)$ is written $\overset{λ}{α}\overset{o}{εχ}$, while, in line 7, the same expression is written $\overset{λ}{α}\overset{χ}{ε}$; and, in Plate 8, in the first line of column 6 of M.P.E.R. i 1, $τρίγω(νον)$ is written $τριγ\overset{ω}{\ }$ and $ἰσόπλευρ(ον)$ is written $ἰσοπλευρ\overset{\diagup}{\ }$.[93] It may be that it was this common practice of the scribes of documentary papyri which led to the

---

[91] For a recent, useful, detailed study of Greek and Egyptian fractional techniques, see Knorr, TFAEG. But what are presented there as manipulations of common fractions $\frac{m}{n}$ are clearly manipulations of the descriptions 'the nth of m' which are conceived throughout the texts under discussion in terms of unit fractions. See pp. 148–51 on the Achmîm Mathematical Papyrus, for example on p. 150: "Although the problems are invariably expressed in terms of unit-fractions and the final solutions are given in this same mode, the actual execution of the arithmetic operations first introduces their conversion to terms of the form $\frac{a}{b}$" (but here $\frac{a}{b}$ invariably abbreviates the expression $τῶν\ a\ τὸ\ b$); and "Thus, the scribes in the arithmetic tradition of late antiquity present themselves in the art of manipulating unit-fractions as *virtuosi* in the art of manipulating unit-fractions; yet their very methods reveal this to be a superfluous art. How was it possible that they failed to perceive this, embellishing these techniques and so distracting from the teaching and development of the more general techniques of [common] fractions employed within their computations?" (The answer here is surely that the scribes were actually conceiving, teaching, and developing unit fraction techniques, and had little or no conception of common fractions.)

[92] See, for an influential example, Heath, *HGM* i, 43 f.: "The most convenient notation of all [for common fractions] is that which is regularly employed by Diophantus, and occasionally in the *Metrica* of Heron. In this system the numerator of any fraction is written in the line, with the denominator *above* it, without accents or other marks (except where the numerator or denominator itself contains an accented fraction); the method is therefore simply the reverse of ours, but equally convenient. In Tannery's edition of Diophantus a line is put between the numerator below and the denominator above ... but it is better to omit the horizontal line (cf. $\overset{ρκη}{ρ} = \frac{100}{128}$ in Kenyon's Papyri ii, No. cclxv.40, and the fractions in Schöne's edition of Heron's *Metrica*)." In fact, these letters $ρκη$ and all of the other examples in P. Lond. ii 265 and M.P.E.R., N.S. i 1 incorporate fraction indicators; see nn. 64 and 75, above. *Every* known instance of this abbreviation in contemporary papyri has been discussed in Sections 7.3(c), (d) and (e), above.

[93] There are about thirty different such abbreviations in M.P.E.R., N.S. i 1; they are listed in Gerstinger & Vogel, *MNWPER*, 47. For abundant other early examples of these abbreviations, see McNamee, *AGLPO* and the index of symbols and abbreviations in P. Lugd. Bat. xxi (= Pestman, *GZA*). The whole question of abbreviations in papyri is very fully discussed in Blanchard, *SAPDG*; see n. 28, above.

writing of numerical abbreviations by superposition. For example:

(i) In the alphabetic system of numerals, $\overset{\alpha}{\mu}$, $\overset{\beta}{\mu}$, etc. denote 1 myriad (1 0000), 2 myriads, etc.

(ii) In describing amounts of money measured in the χαλκοῦc, one-eighth of an obol, $\overset{\alpha}{\chi}$, $\overset{\beta}{\chi}$, etc. is very often used for 1 *chalkous*, 2 *chalkōi*, etc.

(iii) In describing amounts of wheat, or its equivalent, measured in the χοῖνιξ, one-thirtieth of an artaba, the same notation $\overset{\alpha}{\chi}$, $\overset{\beta}{\chi}$, etc. is very often used for 1 *choinix*, 2 *choinikes*, etc.

Moreover, in the 'fractional notation' the 'denominator' would tend to be placed above the 'numerator' in such an abbreviation since this 'denominator' is written with its long fraction indicator which proclaims its different grammatical and conceptual status from that of the 'numerator'. As this fraction indicator would get in the way of anything written too close and above it, it would be much more conveniently situated on top. Further, in these examples of abbreviation by superposition, the later part of the abbreviation is usually written on top, and the same goes for the 'fractional' abbreviation of 'τῶν m τὸ ń' as $\overset{ń}{m}$.

A further argument in support of my thesis can be drawn from the full-format division tables, described in Section 7.2(a), above. Consider how such a table might be used to evaluate a division like the 9th of 842. From the table, we can read off that the 9th of 800 is 88 2' 3' 18', of 40 is 4 3' 9', and of 2 is 6' 18'. when these are added, we get 92 2' 3' 3' 6' 9' 18' 18'; but the repeated parts now have to be removed, by several more steps of working. However, it does not take much insight to see that we can greatly simplify the evaluation as follows: the 9th of 800 is 88 2' 3' 28', and, also from the table, we recognise the fractional expression here as being the 9th of 8; similarly the 9th of 40 is 4 and the 9th of 4. Hence the 9th of 842 is 88 + 4 and the 9th of 8 + 4 + 2, so 93 and the 9th of 5, so 93 2' 18'. I cannot believe that some such simplification was not used in practice; but that it seems never to have been expressed in the tables is yet another indication of the absence of any notation for it, and so the absence of any conception of division in terms other than as sums of parts.

Now consider the contrast between the rather rigid forms of the Greek expressions for division, all very minor variations of 'τῶν m τὸ ń', that are found in the wealth of Greek papyrus texts,[94] and the wealth of different expressions that have been found in the small handful of demotic texts; and the associated contrast of the minute proportion of

---

[94] My catalogue in the Appendix gives only a small part of this evidence, that dealing with systematic tables.

7.4 Conclusions

Greek uses of a 'fractional notation' in surviving calculations with the very much greater proportion of demotic occurrences. The abundance of our school exercises in division indicate how the very rigid Greek style for expressing division seems to have been drummed into schoolchildren and thereafter to have determined their conception of fractions while, in demotic, our evidence seems to indicate a much looser sense of style. My explanation of the occurrences of the 'fractional notation' concerns this imposed sense of style: that the results of properly executed calculations should always be presented in standard form, as expressions in unit fractions, either to be recalled from memory, or looked up in tables, or evaluated, or approximated. If this sense of style was relaxed, so would the necessity for this final, often inconsequential, step be removed, and so unresolved phrases for 'of m the nth' could begin to appear, expressed or abbreviated in a variety of ways. The adoption of the demotic script rather than hieratic for Egyptian mathematical texts may well have left scope for the development of new summary notations for stereotyped common operations like division. In this way, it is perhaps not too surprising that our small clutch of surviving demotic texts should contain so many different ways of describing division; but I do not think this, in itself, gives evidence for an early familiarity with common fractions, since the expressions themselves always seem still to be conceived there in terms of unit fractions.

Another way of expressing this is to view division as an implied question which anticipates some conventional reply, thus 'the nth of m is . . .', or 'the quotient (literally *quotiens*, how many times?) of m by n is . . .'. Answers can be given in many different ways—division with remainder, unit fractions, sexagesimal numbers,[95] decimal numbers, anthyphairetic ratios, astronomical ratios, etc.—and only the context can determine what will be acceptable. Common fractions correspond to the sophisticated step of ignoring this interrogative aspect, and realising that the question itself can be treated as its own answer. In order to see that this step is meaningful and useful, we need to be convinced that the questions themselves can be freely manipulated; for example that 'the nth of m is . . .' and 'the qth of p is . . .' can be added to give 'the nqth of $(mq + np)$ is . . .'. Modern mathematics has shown, time and time again, how difficult such a step of treating a question as its own answer can be, however natural it might subsequently become, and how productive it may be of new concepts and developments. I do not believe

---

[95] We have the example of the completely different reaction of another ancient culture in the Babylonian need for and use of reciprocal and multiplication tables to resolve divisions in their arithmetised sexagesimal mathematics; see Neugebauer, *ESA*, 31–4 and 50. Observe how there is simply no need for a Greek reciprocal table to correspond to these Babylonian examples!

that this step was taken by Greek mathematicians in their calculations with fractions.

The contempt that is often expressed for the system of unit fraction calculations is, I believe, not entirely justified. Unit fraction expansions can be evaluated and manipulated by a wealth of algorisms[96] and they can convey some information, for example of magnitude and approximation, as efficiently as any other representation.[97] The problems that do arise with unit fraction arithmetic are often exaggerated and distorted: on the one hand, unit fraction arithmetic is quite feasible, as our evidence of ancient commercial practice demonstrates abundantly, while on the other hand the problems inherent in any formally correct and complete description of arithmetic with real numbers are far greater than many mathematicians seem to believe.[98] I do not deny that the introduction first of common fractions, then of decimal fractions, introduced a dramatic new fluency and confidence into mathematics, with profound consequences. I only wish to insist that the old unit-fraction representation does not suffer from quite as many disadvantages as is often supposed.

Much more could be said on the *merē* in Greek mathematics. For example, I have just alluded to the problem of the computation of the entries in the division tables, and there is also the issue of Euclid's use of the word *merē*, especially at the beginning of *Elements* VII. These issues must be deferred to another place. Here I shall finish with my final dialogue, a little fantasy to top off this heavy chapter.

7.4(b) *The slaveboy meets an accountant*

ACCOUNTANT$_{97}$: I was out at the garden of Academe on business, and I heard Plato and his friends discussing some of your ideas about mathematics. Then one of my scribes had to copy out your dialogue with Socrates, so I got to read

---

[96] To begin to discuss this issue properly would double the length of this chapter. See n. 32, above.

[97] For example, if sums and differences of unit fractions are permitted, the expression $\pi = 3 + 7' - 791' - 3748549' + \ldots$ conveys the information about the size and best approximations to π as succinctly and directly as any other method; more directly, for example, than the anthyphairesis $\pi = [3, 7, 15, 1, 272, 1, 1, 1, \ldots]$. But the anthyphairesis can contain a lot more additional information.

[98] Of course a formal description of arithmetic with rational numbers is straightforward and so, in this sense, the unit fraction description of rational numbers does introduce unnecessary problems. But the theoretical description of arithmetic with decimal fractions is less straightforward than is generally thought; see the remarks in Sections 1.2(e), 4.2, and 9.3(a), below, and the discussion in Fowler, FHYDF. On the practical difficulties with sexagesimal arithmetic, see Neugebauer, *HAMA* ii, 590 ff. ("The further one moves away from the productive period of Greek astronomy towards the didactic phase the more one finds long and clumsy explanations of sexagesimal operations. Such is the case already in the Commentaries to the Almagest by Pappus and Theon. Byzantine treatises are full of such trivia.") and pp. 968 ff.

7.4 Conclusions

that, and I have a copy here. Could we go through it together, with you taking the part of Socrates?

BOY$_{98}$: All right, though it seems a curious thing to do. Here goes. (Reading:) Tell me, sir, what is the relationship of size between this heap of sixty stones, and that heap of twenty-six stones?

ACCOUNTANT$_{99}$: Do you mean the number of times the smaller goes into the larger?

BOY$_{100}$: Try it.

ACCOUNTANT$_{101}$: It goes more than twice, but less than three-times.

BOY$_{102}$: Can you be more precise?

ACCOUNTANT$_{103}$: It goes twice with eight stones left over.

BOY$_{104}$: Those 'eight stones' aren't related to anything now.

ACCOUNTANT$_{105}$: I just omitted to say that they're still in relation to the other heap of twenty-six stones.

BOY$_{106}$: Go on.

ACCOUNTANT$_{107}$: Give me time; I can go on to describe that relationship. So let's say: twice at the first step, and now I'll tell you about the relationship between eight stones and twenty-six stones.

BOY$_{108}$: Go on.

ACCOUNTANT$_{109}$: (Putting the script to one side:) Those eight stones are less than the third and more than the fourth of the twenty-six stones; so the relationship is: first step, twice; second step, the fourth; and then a bit more.

BOY$_{110}$: (Aside:) I see what he's up to; this could be interesting. But what happens to *arithmētikē* if part of a stone is still called a stone? Does it still make sense? (To the accountant:) Can you be more precise?

ACCOUNTANT$_{111}$: I can work out how much more it is as follows: I need to take the 4th of 1 away from the 26th of 8, which is the 13th of 4; so I need to take the 4 × 13th of 13 away from the 13 × 4th of 4 × 4, which leaves the 52nd of 3, which is less than the 17th and more than the 18th. So the ratio of sixty to twenty-six is expressed by: first step, twice; second step, the fourth; third step, the eighteenth, and a bit more which I can easily work out in the same way to be the four hundred and sixty-eighth exactly. Or, as we learned at school to say and write it:

of 60 the 26th is 2 4' 18' 468'.

There are lots more equivalent ways of describing this same ratio, some much more convenient than this: for example, a preliminary division with remainder and then a look at my ready reckoner gives:

of 60 the 26th is 2 4' 26' 52'.

What's wrong with all that? It seems just as logical and convenient as your anthyphairetic ratios, so why did Socrates dismiss it so contemptuously at the end of S$_{35}$?

BOY$_{112}$: There's the awkward operation of cutting up the unit that the philosophers don't like, though if the mathematicians find something that interests them, they might nevertheless still go ahead. You also have a problem in deciding when two ratios are equal; so you can define ratio, but its not entirely clear how you can go on to describe proportionality! But let's see what might

be involved in a theory of these 'accountant's ratios'. For a fully developed theory we'll have to investigate:
  (i) Evaluating these kinds of ratios (cf. $S_1 - B_{14}$),
  (ii) Describing equality and order (cf. $S_{17} - B_{22}$),
  (ii) Translating backwards (cf. $B_{26} - S_{29}$), and
  (iv) Arithmetic (cf. $B_{32} - S_{35}$),

and, on the way, we hope to pick up insights about yet undiscovered phenomena or appealing new ways of handling mathematical ideas and proofs. Then, we must relate this new definition to the other definitions of ratio.

ACCOUNTANT$_{113}$: But before you start your abstruse investigations, and explain why some of these things are so difficult and so unilluminating, I'd just like to say that I've got an office full of scribes who are calculating with these kinds of expressions all the time, and they don't seem to have many problems in getting by.

BOY$_{114}$: Wait a minute. Instead of doing a reciprocal subtraction process, we could always subtract from the same larger term. Take that ratio of eight to twenty-six that we have to describe [see A_{107-11}]: we could say eight goes into *twenty-six* three-times with two left over; two goes into *twenty-six* thirteen times. Work that out and I think you'll find

of 60 the 26th is 2 3' less (3.13)'.

Think about that, and you'll see that if you don't want subtractions like that "less 39'" to occur you will have to overshoot at each step: four-times eight overshoots twenty-six by six; five-times six overshoots twenty-six by four; seven-times four overshoots twenty-six by two; and thirteen-times two is twenty-six. So now

of 60 the 26th is 2 4' (4.5)' (4.5.7)' (4.5.7.13)'.

Do you see what's happening?

ACCOUNTANT$_{115}$: I give up! Will you people never stick to the point?

BOY$_{116}$: I wonder if we can relate anthyphairetic ratios to these kinds of accountant's ratios. . . .

7.5 Appendix: A catalogue of published tables

See Section 7.2, above, for a general description of the tables in this appendix.

These lists contain all the published examples I have been able to locate in Egyptian (hieratic and demotic), Coptic, and Greek. They are identified as being on papyrus (P.), ostraca (O.), writing tablets (WT.), waxen wooden tablets (WWT.), and as graffiti (G.). The lists are arranged alphabetically by language and script; museum inventory numbers are distinguished from catalogue numbers; where the text appears in a standard edition (like *The Oxyrhynchus Papyri, Papyri in the University of Michigan Collection,* etc.) no further publication details are given, but otherwise they are given in detail (and if a reference that occurs here is not cited elsewhere, it will be given here in full and not listed in the bibliography); the usual name of each text is set in bold type; if the text appears in Pack, *GLLTGRE* (second edition), its reference number there is given; where a date or provenance is assigned or known, it is given; a note about the publication may be added; and each entry finished with a brief description.

I have restricted this list to systematic examples of tables; there are many other published examples of arithmetical exercises involving calculations with, or verifications of, these kinds of expressions which have not been included.

The following abbreviations are used:

ff = full format, up to the nth part of 1 0000.
af = abbreviated format, up to the nth part of n.
ie = initial entry, the nth part of 6000.
BASC = *Bulletin de la Société d'Archéologie Copte.*
ZPE = *Zeitschrift für Papyrologie und Epigraphik.*
RA = *Revue Archéologique.*

7.5(a) *Division tables*[99]

Egyptian, hieratic:
1. **P. Kahun 8,** in Griffith, *HPKG*. A fragmentary collection of mathematical problems and a '$2/n$-table', with verifications, for $n = 3, 5, 7, \ldots, 21$, identical with the expressions in the Rhind Mathematical Papyrus; photograph.
2. **Rhind Mathematical Papyrus** (= P. B.M. [inv.] 10057 and 10058, and a fragment in the Brooklyn Museum), in Chace, *RMP*; copied c.1575 BC by Ahmose from an archetype some 300 years older. An extensive collection of 86 mathematical problems, a table of 10', in af no ie, quoted above, and verifications of nth parts of 2 for $n = 3, 5, 7, \ldots, 99$, and 101; photographs and facsimiles. This

---

[99] A preliminary version of this table was published as Fowler, TP; I have here revised my proposal there to refer to these as 'tables of parts', and now suggest the more conventional and, I hope, acceptable 'division tables'. I would like to thank the many scholars, especially R. S. Bagnall, B. Boyaval, W. Brashear, A. Bülow-Jacobsen, P. Cauderlier, W. E. H. Cockle, B. R. Goldstein, L. Koenen, P. Mertens, P. J. Parsons, T. S. Pattie, R. Pintaudi, P. J. Sijpesteijn, and Sir Eric Turner, who have provided information and references. More material will be published in R. Pintaudi & P. J. Sijpesteijn, *Tavolette Lignee e Cerate della Biblioteca Apostolica Vaticana,* with Appendices, *Quatre cahiers scolaires* by P. Cauderlier, containing tablets from a private collection, now in the Louvre, and *A tablet from the Pierpont Library* by R. S. Bagnall. I am grateful to these editors for sending me advance notice of these very interesting texts. Yet more examples are quite possible from the unpublished residues of other papyrological collections around the world, or from excavations.

text, especially the '2/n-table', has provoked an enormous literature (bibliography to 1929 in Chace, *RMP,* abridged and extended randomly in the 1979 reprint), but with only occasional references to other division tables.

3. **O. Sen-Mūt 153,** in W. C. Hayes, *Ostraka and Name Stones from the Tomb of Sen-Mūt (No. 71) at Thebes* New York: Metropolitan Museum, 1942. Fragments of a table of 7' with auxiliary red numbers; photograph.

Egyptian, demotic:

4. P. B.M. [inv.] 10794, Problems 66 and 67 in **Parker,** *DMP,* 72–3. Tables of 90' and 150', with entries from 1 to 10; photograph.

5. O. Private collection, in **G. Belli & B. Costa, Una tabella arithmetica per uso elementare scritta in demotico,** *Egitto e Vicino Oriente* 4 (1981), 195–200. Table of 2'; photograph.

6. P. Unidentified example, in **E. Revillout,** *Mélanges sur la métrologie, l'économie politique, et l'historie de l'ancienne Égypte,* Paris: Maisonneuve, 1895, pp. lxix–lxxiii. (Revillout describes it as "un papyrus mathématique qui m'a été communiqué par l'Exploration Fund".) Tables of 7', 8', ..., 14', and 15' in af.

Coptic:

7. P. B.M. MS 528, in **W. E. Crum,** *Catalogue of Coptic Manuscripts in the British Museum,* London, 1905, and J. Drescher, A Coptic Calculation Manual, *BASC* 13 (1948–9), 137–60; c. AD 900. A palimpsest parchment codex; the later Coptic text contains multiplication tables (see no. 47, below), mathematical exercises, and division tables for 2', 3″, 3', 7' [*sic*], 4', 5', 6', 8', 9', 10', 12', 15', 16', 20', 24', and 48', in af with ie; only the first few entries of the tables for 2', 15', and 48' are transcribed. Described as "identical with those in the Greek mathematical papyrus of Achmîm", no. 12, below.

8. P. Cambridge University Library [inv.] T.-S. Ar 39:380, in **B. R. Goldstein & D. Pingree, More Horoscopes from the Cairo Geniza,** *Proceedings of the American Philosophical Society* 12 (1981), on 186–9. Two folia containing multiplication tables (see no. 48, below) and tables for 12' 24', and 48', in ff; photograph.

9. **O. Crum 480,** in W. E. Crum, *Coptic Ostraca from the Collections of the Egypt Exploration Fund, the Cairo Museum, and Others,* London, 1902. A fragmentary potsherd table of 31' in af with ie. Crum was unable to identify its purpose in 1902 (but see no. 7, above). There is a

	description in K. H. Sethe, *Von Zahlen und Zahlworten bei den Alten Ägyptern,* Strassbourg: Trübner, 1916, pp. 71–2, but some of his restorations, including the ie, are clearly wrong; see W. Brashear, Quisquiliae, *BASC* 26 (1984), 19–22.
10.	P. Strasbourg Bibliothèque Universitaire [inv.] 4110, same publication as no. 8, above, on 176–7. Table of 24′ in ff, a marginal annotation to an astrological text; photograph.
11.	O. **Wadi Sarga 24–28** (= *Coptica* iii 24–28). Five fragmentary potsherds with brief traces of division tables, four with ie (on this, see W. Brashear, Quisquiliae, *BSAC* 26 (1984), 19–22) for 7′ (3 examples), 11′, 25′, 49′, and one unidentified table.

Greek:

12.	**Achmîm Mathematical Papyrus** (= P. Cair. [inv.] 10758), Pack 2306, in J. Baillet, *Mémoires de la Mission Archéologique Française* ix 1, Paris: Leroux, 1892; vii AD. Leather-covered papyrus codex containing division tables for 3″, 3′, 4′, ..., 8′, 9′, and 10′ in ff and for 11′, 12′, ..., 19′, and 20′ in af, all with ie, and mathematical exercises; complete facsimile.
13.	P. Berol. [inv.] 21296, in **W. Brashear, Greek Papyri: Fractions and Tachygraphy,** *Anagennesis* 3 (1983), 167–77; ii BC. Fragments of tables of 3′ and 4′ in ff; photograph.
14.	WT. Bodleian Gr. Inscription [inv.] 3019 in **P. J. Parsons, A School-Book from the Sayce Collection,** *ZPE* 6 (1970), on 142–3; late iii AD. Seven wooden school tablets containing a wide variety of exercises in Greek and Coptic; three sides contain division tables for 2′, 3′, 4′, 5′, 6′, 8′, 9′, 10′, and 12′, in ff. Only the table for 2′ is transcribed.
15.	**P. Freib i 1**; also see W. Brashear, Greek Papyri: Fractions and Tachygraphy, *Anagennesis* 3 (1983), 167–77; ii or i BC. Fragments of a table of 3″.
16.	WT. Louvre [inv.] AF 1196^1, AF1196^2, AF 1196^3, AF 1197^1, in **B. Boyaval, Tablettes mathématiques du musée du Louvre,** *RA* (1973), 243–60 and Le Cahier de Papnouthion et les autres cahiers scolaires grecs, *RA* (1977), 215–30; P. Cauderlier, Cinq tablettes en bois au musée du Louvre, *RA* (1983), 259–80; and W. Brashear, Corrections à des tablettes arithmétiques du Louvre, *Revue des Études Grecques* 97 (1984), 214–17 (all publications must be consulted; photographs in *RA*); vi AD. Parts of a wooden schoolbook containing multiplication (see no. 54, below) and division tables:

AF 1196¹ side B: table of 9' in ff with ie;

AF 1196² side B: table of 3", with ie and verifications (or possibly derivations; see nos. 26 & 44, below) of each successive entry, up to 900. Also see AF 1196³ side B, below;

AF 1196³ side A: table of 11' in af with ie, and multiplication tables;

AF 1196³ side B: on the basis of three legible letters, Cauderlier proposes a restoration of the continuation of AF 1196² side B, a table of 3" with verifications from 1000 up to 1 0000;

AF 1197¹ side A: table of 14' in af with ie;

AF 1197¹ side B: table of 17' in af with ie.

17. WWT. Louvre [inv.] MND 552, MND 552h, i, k, and l; same publications by B. Boyaval as the previous entry; v or vi AD. Wax tablets containing a collection of school exercises, some arithmetical:

MND 552k: Headed and dated, "Papnouthion son of Iboïs, Mechir 21st" (incorporating a corrected reading by P. Cauderlier; personal communication), a table of 24', in af with ie, with six alternative entries for 2 (of 2 the 24' is 12', or 15'60', or 20'30', or 30'30'60' [sic; this is the only example I know of a repeated part], or 45'48'60'80'90', and one lost entry).

MND 552i: Again headed and dated Mechir 21st, a fragmentary table of 25' with ie, again with duplicated but largely illegible expressions for the 25' of 2.

18. WT. Louvre [inv.] MND 551dl, in **B. Boyaval, *CRIPEL* 2** (1974), 270–1 and W. Brashear, Trifles, *ZPE* 56 (1984), on 64–5; iv–vi AD. Wooden school tablet with fragmentary table of 3" in ff.

19. **P. Mich. iii 134**, Pack 2309; ii AD. Fragments of tables of 23' and 29', and mathematical problems.

20. **P. Mich. iii 146**, Pack 2310; iv AD. Long narrow papyrus roll (about 106.75 × 9.2 cm) broken at the beginning, incomplete at the end, containing tables for 7' (fragmentary), 8', 9' (quoted above in Section 7.2(a)), and 10' in ff, then 11', 12', 13', ..., 17', 18', and 19' (heading εννεακαιδεκατα to a blank column) in af, all with ie and headings.

21. **P. Mich. iii 147**, Pack 2310; early ii AD. Fragments of a table of 4'.

22. **P. Mich. xv 686**; ii–iii AD. Multiplication table (see no. 55, below) and fragmentary division tables including a table of

7.5 Catalogue of tables 275

30′ increasing by halves.

23.   O. Mich. [inv.] 9733, in **H. C. Youtie, A Table of Fractions**, *ZPE* 18 (1975), 17–19; first half of iii AD. Potsherd table of 3″ in ff with ie.

24.   WT. **(P.) Michael. 62**, Pack 2308; vi AD? Wooden tablet containing tables of 2′, 3″, 3′, and probably 4′ (now illegible) in ff with ie and, on the other side, mathematical problems. More complete publication in D. S. Crawford, A mathematical tablet, *Aegyptus* 33 (1953), 222–40.

25.   WT. **M.P.E.R., N.S. xv 154**; vii AD. Wooden tablet containing multiplication tables (no. 59, below) and a table of 2′, in ff with ie τὸ ∠ ἑξακιϲχιλίων ,Γ; photograph.

26. **M.P.E.R., N.S. xv 156**, first published in P. J. Sijpesteijn, Wiener Mélange, *ZPE* 40 (1980), on 97–8, and re-edited in W. Brashear, *Enchoria* 12 (1984), 1–6 and here; v–vi AD. Fragment containing only multiplications, but very possibly coming from a table of 9′ with verifications (or calculations) similar to those in Louvre AF 1196² side B and P. Würzburg K 1024 (nos. 16 & 44, here); photograph.

27. **M.P.E.R., N.S. xv 158**; iii AD? Minute fragment of a table of 3″; photograph.

28. **M.P.E.R., N.S. xv 159**, first published in P. J. Sijpesteijn, Wiener Mélange, *ZPE* 40 (1980), on 98–9; iii or iv AD. Table of 3″ in ff followed by a calculation with money; photograph.

29. **M.P.E.R., N.S. xv 160**; ii AD. Tables of 2′ and 3′ in ff; photograph.

30. **M.P.E.R., N.S. xv 161**; vii–viii AD. Tables of 2′ and 4′ in ff with ie, but not written out in tabular form—the only such example known to me; photograph.

31. **M.P.E.R., N.S. xv 162**; first half of ix AD. Fragment of a table of 3″ in ff with ie; photograph.

32. **M.P.E.R., N.S. xv 163**; vi AD. Fragment of tables of 3′ and 4′; photograph.

33. **M.P.E.R., N.S. xv 164**; vi–vii AD. Fragment of tables of 3′ and 4′ in ff; photograph.

34. **M.P.E.R., N.S. xv 165**; vi–vii AD. End of table of 5′ and table of 3″, in ff with ie,; photograph.

35. **M.P.E.R., N.S. xv 166**; ii BC. End of a table of 4′ in ff; photograph.

36. **M.P.E.R., N.S. xv 167**; ii AD. Fragments of tables of 7′, 8′, and 9′ in ff with headings; photograph.

37. **M.P.E.R., N.S. xv 168** verso; vii AD. Fragment of a table of 14′; photograph.

38. **M.P.E.R., N.S. xv 169** recto and verso; viii AD. Fragment of tables of 13', 14', 11', 17', and 19'; photograph.
39. WT. Moen [inv.] 602, in **P. J. Sijpesteijn, A Wooden Tablet in the Moen Collection**, *Chronique d'Égypte* 56 (1981), 97–101; vi or vii AD. Half of a wooden tablet containing part of a table of 7' in ff with ie.
40. **P. Oxy. xxxiii 2656**, first published in E. G. Turner, *New Fragments of the Misoumenos of Menander*, Bulletin of the Institute of Classical Studies Supplement 17 (1965), on 18–19; iv or v AD. Tables of 13', 14', 15', 16', and perhaps 17' and 18', in af with ie, headed τρεicκαιδεκατα, etc. These tables are found in a papyrus codex that otherwise carried a lost play of Menander; there would have been space for tables of 10' to 19'. Only the table of 13' is transcribed. The republication as P. Oxy. xxxiii 2656 omits the tables.
41. **P. Oxy. xlix 3456**; iii or iv AD. Tables of 7' and 8' in ff, followed by metrological definitions. Only part of each table is transcribed.
42. WT. University College [inv.] 36114, Pack 2312, in **H. Thompson, A Byzantine Table of Fractions**, *Ancient Egypt* 1 (1914–15), 52–4. Wooden school tablet containing tables of 15' and 16' in af with ie; facsimile. The editor misunderstood the initial entry, but it can clearly be guessed from his drawing and I have checked the original.
43. WT. Würzburg [inv.] K 1014, in **W. Brashear, Holz- und Wachstafeln der Sammlung Kiseleff**, *Enchoria* 13 (1985), 13–23 and Tafel 1–13; vi–vii AD. Four writing tablets containing school exercises in multiplication and division. Tables of 6', 3″, 5', 3' (in that order) in ff with ie; photographs.
44. WT. Würzburg [inv.] K 1024, in **W. Brashear, Neue Griechische Bruchzahlentabellen**, *Enchoria* 12 (1984), 1–6 and Tafel 1–2; viii AD. Wooden tablet containing tables of 3″ in ff and 11' in af, both with ie, with verifications (or calculations) similar to those in Louvre AF 1196^2 side B and M.P.E.R. 156 (nos. 16 & 26, above); photograph.
45. WT. **SB iii 6219**, first published in G. Plaumann, Antike Schultafeln aus Ägypten, *Amtliche Berichte aus den Königlichen Kunstsammlungen* (Berlin) 34 no. 11 (1913), 210–23, on 222 f; vii AD. Wooden school tablet with tables for 2' and 3″, in ff with ie, and problems in addition and multiplication; photograph.

7.5 Catalogue of tables

7.5(b) *Multiplication and addition tables*

Egyptian, demotic:
46.   P. B.M. [inv.] 10520, Problem 54, in **Parker, *DMP*,** List of multiples of 16 from 1 to 16; photograph.

Coptic:
47.   P. B.M. MS 528 (= no. 7, above). Table, presumably with first page missing, of 7×, 8×, 9×, 10×, 20×, ..., 900×, in third format.
48.   P. Cambridge University Library [inv.] T.-S. Ar. 39:380 (= no. 8, above). Table of 1×, 2×, ..., 9×, 10×, 20×, ..., 90×, 100×, 200×, ..., 1000× in third format; photograph.
49.   G. Monastère de Phoebammon 5 (mural no. 10), 153 (mural no. 186), and possibly 6 (mural no. 12), published by R. Rémondon in **C. Bachatly, *Le monastère de Phoebammon dans la Thébaïde*,** Tome ii: *Graffiti, inscriptions et ostraca,* Cairo: Publications de la Société d'Archéologie Copte, 1965, with review by J. Schwartz in *Chronique d'Égypte* 42 no. 83 (1967), 251–4. Tables of 3× and 4× in first format, and an unidentified table; photograph.
50.   **O. Wadi Sarga 22 and 23** (see no. 11, above). No. 22: tables for 6× and 7× in first format; no. 23: tables for 7× in second format, a palimpsest, with this the younger text.

Greek:
51.   **P. Baden iv 644**, complete publication in F. Bilabel, *Berichtigungsliste* ii part 2, 177–81. Table of 2×, 3×, ..., 9×, 10×, 20×, ..., 90×, 100×, 200×, 300×, 400×, 500× in second format; then 2×, 3×, ..., 9×, 10×, 20×, 30× in third format.
52.   WWT. Leiden Papyrological Institute, in **E. Boswinkel, Schulübungen auf 5 Leidener Wachstäfelchen,** *Proceedings of the XIV International Congress of Papyrologists, Oxford 1974,* London: British Academy, 1975, on 25–28; after 212 AD. Five school tablets, one containing a table of 1 × 40 to 10 × 40 in second format.
53.   P. Lond. iii 737 (p. xxx), published as **M.P.E.R., N.S. xv 150**; iii AD. Table of $n+m$ for $1 \leq n \leq m \leq 9$, then similarly for $10+10, \ldots, 10+90, 20+20, \ldots, 20+90$; photograph.
54.   WT. Louvre [inv.] AF 1196^1, 1196^2, and 1196^3 (= no. 16, above).
        AF 1196^1A: tables for 2×, 3×, ..., 8×, in second format.
        AF 1196^2A: traces of tables, illegible at beginning, for [8×], 9×, 10×, 20×, ..., 50×, in second format.

278   Numbers and fractions   7.5

AF 1196³A: tables of 5000×, 6000×, ..., 9000× in first format.

55. **P. Mich. xv 686** (= no. 22, above). Fragments of a multiplication table up to 10000 × 4 in first format.

56. **M.P.E.R., N.S., xv 151**, the verso of M.P.E.R. i 1; i AD. Addition table in same format as no. 53, above, continuing (with gap for 30+ and 40+) up to 1000 + 1000; photograph.

57. **M.P.E.R., N.S., xv 152**; ii–iii AD. Fragments of tables of multiples and squares. The multiplication table gives 7 × 8, 7 × 9, 7 × 10, 8 × 1, ..., 8 × 10, 9 × 1, ..., 9 × 10, 10 × 1, ..., 10 × 10 in first format, followed immediately by the squares (see no. 68, below); photograph.

58. **M.P.E.R., N.S., xv 153**; ii AD. Fragment of a table of 20 × 500, 20 × 5000, 20 × 6, 20 × 60, ..., 20 × 9000 in third format; photograph.

59. WT. **M.P.E.R., N.S., xv 154**; vii AD. Wooden tablet with tables of 7×, 8×, and 9×, for 1, 2, ..., 9 in second format, and a table of 2′ (no. 25, above); photograph.

60. **M.P.E.R., N.S., xv 157**; ix–x AD. Fragment of tables for 2× and 3× in third format; photograph.

61. **P.S.I. viii 958**, Pack 2307; iv AD. Table headed πολυπλαcιαcμοc, 'multiplication', starting with some unsystematic entries for smaller numbers, then systematically for 30×, 40×, ..., 90×, 100×, 200×, 300×, ..., 1000×, 2000×, 3000×, and 4000× in second format.

62. WWT. **SB iii 6215**, first published in G. Plaumann, Antike Schultafeln aus Ägypten, *Amlichte Berichte aus den Königlichen Kunstsammlungen* (Berlin) 34 no. 11 (1913), 210–23, on 216. Wax tablet with addition table, 8 + 1 to 8 + 9; photograph.

63. WWT. Würzburg [inv.] K1013, in **W. Brashear, Holz- und Wachstafeln der Sammlung Kiseleff,** *Enchoria* 13 (1985), 13–23 and Tafel 1–13; iv–v AD. Parts of five wax tablets containing school exercises including two simple addition tables, second format; photographs.

64. WT. Würzburg [inv.] K1014, see previous entry and no. 43 for publication and details. The multiplications are: side 1B, 60 × 1, 60 × 2, ..., 60 × 10, in first format, then similarly 70×, 80×, 90×, 100×, 200×, and 300×; side 2A, 400×, 500×, 600×, then reverting to 20×, 30×, 40×, 50×; Side 3B (fragment), 700 × 3, ..., 700 × 8, 800 × 8, ..., 900 × 4, 1000 × 4, ..., 1000 × 10, 3000 × 1, ..., 3000 × 7; photographs.

7.5 Catalogue of tables 279

7.5(c) *Tables of squares*

Coptic:
65. G. Monastère de Phoebammon 16 (mural no. 97) (see no. 49, above). Table of squares for $10, 11, \ldots, 29$, and 30; photograph.

Greek:
66. P. *Un livre d'écolier du IIIe siècle avant J.C.*, ed. O. Guéraud & P. Jouget, Cairo: Institut Français d'Archéologie Orientale, 1938 (= P. Cairo J. E. [inv.] 65445), lines 216–34. Table of squares of $1, 2, \ldots, 10, 20, \ldots, 100, 200, \ldots, 800$; complete facsimile, and photograph in Neugebauer, *ESA*.
67. **P. Haun. iii 49**; ii BC? Table of squares for $1, 2, \ldots, 9, 10, 11, 12, \ldots, 19, 20, 30$, and 40; photograph.
68. **M.P.E.R., N.S., xv 152**; ii–iii AD. Multiplication table (see no. 57, above) and two tables of squares, first in the form $1\ 1, 4\ 2, 9\ 3, \ldots, 100\ 10, 400\ 20, \ldots, 6400\ 80, 10000\ 100, 40000\ 200, \ldots, 250000\ 500$, then in the form $1 \times 1\ 1, 2 \times 2\ 4, \ldots, 10000 \times 10000\ 100000000$ $\left(\text{written } \begin{matrix} \alpha & \alpha & \alpha \\ \mu & \dot{\mu} & \mu \\ & \mu & \end{matrix}\right)$; photograph.
69. **P.S.I. vii 763**, Pack 2315; i BC. Fragment of a mathematical lesson, partly in question and answer form. Lines 5 to 11 deal with squares, but not in tabular form:]*ax en α dis duo δ tris tria ennea θ tetrakis tessara ις pentakis pente κε* . . . , up to *dekakis deka ρ*.

PART THREE

LATER DEVELOPMENTS

> Doubtless a vigorous error vigorously pursued has kept the embryos of truth a-burning: the quest for gold being at the same time a questioning of all substances.
> George Eliot, *Middlemarch*, Chapter 48.

The topic of this third part may be paraphrased as ancient and modern reactions to anthyphairesis. In Part One, I proposed that early Greek mathematicians made remarkable investigations and discoveries about ratio, especially anthyphairetic ratios; but when we look to the later Greek commentators to find their reactions to this work, we draw a blank. So, in Chapter 8, as an illustration of what we do find, I take three themes that are discussed in later antiquity, and set them against the early evidence and my interpretation here. I would have liked to have included discussions of several other topics such as analysis and synthesis, the nature of the early *Elements* (*stoicheion*), the role of problems, the deductive method, and so on, and I do not even begin the evaluation of the effects of my proposals on the chronology of the period: but restrictions of space impose, once again, their limits. The purpose of these analyses is to illustrate the quite serious discrepancies between the preoccupations of later commentators and the earlier evidence, such as we now possess.

With the exception of Pappus and Eutocius, the later Greek commentators were not mathematicians. By contrast, many Arab commentators of the eleventh and twelfth centuries were mathematicians in their own right, and I would also have liked to have included a discussion of four known Arab commentators who do refer to anthyphairesis. Unfortunately the time is not yet ripe: more work needs to be done on the manuscripts before one can do more than give a précis of published summaries of the topic. Meanwhile I commend the interested to the discussion and translation in Plooij, *ECR,* and to the translation of Omar Khayyam–Amir-Móez, DDE.

Chapter 9 gives detailed proofs of the main mathematical results underlying the reconstruction, and sketches the history of their development since the seventeenth century. Although it is highly mathematical and is presented in the modern idiom, and its historical content relates only to the period from the seventeenth century up to today, it is far from irrelevant to my historical argument about Greek mathematics. For we can see how anthyphairesis, in this new manifestation of continued fractions, intrigued and inspired many of the most powerful and influential mathematicians of the seventeenth, eighteenth, and nineteenth centuries. Yet if we take a modern textbook of mathematics or a general history of mathematics (our latter-day equivalents of Pappus' *Collection* and Eudemus' *History*, respectively) and search for any hint of the richness of this theory or its historical importance, we are very very unlikely to find anything. A satisfactory explanation would perhaps lead even further afield, into a study of the psychology and sociology of mathematicians, but some awareness of this modern void should take the edge off our surprise about this lack of comment in late antiquity. The final chapter closes with an Epilogue on this theme.

8
LATER INTERPRETATIONS

8.1 Egyptian land measurement as the origin of Greek geometry?

It has been an oft-repeated story since antiquity that the origins of Greek geometry are to be found in Egyptian land measurement; for a discussion of some of the texts, see Heath, *HGM* i, 121–8. The earliest such possible reference comes from Herodotus, *Histories* ii, 109, written in the fifth century BC:

> This king [the semi-mythical Sesostris] moreover (so they said) divided the country among all the Egyptians by giving each an equal square parcel of land, and made this his source of revenue, appointing the payment of a yearly tax. And any man who was robbed by the river of a part of his land would come to Sesostris and declare what had befallen him; then the king would send men to look into it and measure the space by which the land was diminished, so that thereafter it should pay in proportion to (*kata logon*) the tax originally imposed. From this, to my thinking, the Greeks learnt the art of *geōmetria*; the sunclock and the sundial (*polis kai gnōmon*), and the twelve divisions of the day, came to Hellas not from Egypt but from Babylonia [Herodotus–Godley, *H* i, 397–9].

(The "equal square parcel of land" here should not be taken too literally, and the "sunclock and sundial" is more probably the hemispherical bowl and pointer that make up an early sundial; see Goldstein & Bowen, NVEGA, 332 n. 9.) Another version of this story is given in Proclus' *Commentary on the First Book of Euclid's Elements,* written and compiled from various source materials a thousand years later:

> We say, as have most writers of history, that *geōmetria* was first discovered among the Egyptians and originated in the remeasuring of their lands. This was necessary for them because the Nile overflows and obliterates the boundary lines between their properties. It is not surprising that the discovery of this and the other sciences had its origin in necessity, since everything in the world of generation proceeds from imperfection to perfection. Thus they would naturally pass from sense-perception to calculation (*logismos*) and from calculation to reason. Just as among the Phoenicians the necessities of trade and exchange gave the impetus to the accurate study of number, so also among the Egyptians the invention of *geōmetria* came from the cause mentioned. Thales, who had travelled to Egypt, . . . [Proclus–Morrow, *CFBEE*, 64–5],

and there now follows the celebrated summary of early Greek mathematics that is believed to derive from Eudemus. However this introduction shows clear neo-Platonic characteristics and we have no reason to doubt that it was written by Proclus himself.

Let us confront this story with what we know about land measurement in Ptolemaic Egypt, at a period when records become available. In view of what we shall find, it is worth noting immediately that Aristotle proposes a different explanation at *Metaphysics* 981b23, in the context of his critical history of the development of philosophy:

Thus the mathematical sciences originated in the neighbourhood of Egypt, because there the priestly class was allowed leisure,

while Plato, at *Phaedrus* 274c–5b (quoted above in Section 1.2(e)), has Socrates relate the myth that geometry was one of the arts invented by the Egyptian god Theuth. To this Phaedrus replies:

It's easy for you, Socrates, to make up tales from Egypt or anywhere else you fancy.

The style of early Greek mathematics was described in Section 1.2(b), where *Elements* I 35:

Parallelograms which are on the same base and in the same parallels are equal to one another,

was taken as a characteristic example. Proclus says of this proposition:

It may seem a great puzzle to those inexperienced in this science that the parallelograms constructed on the same base [and between the same parallels] should be equal to one another. For when the sides of the areas constructed on the same base can be increased indefinitely ... we may well ask how areas can remain equal when this happens. For if the breadth is the same (since the base is identical) while the side becomes greater, how could the area fail to become greater? This theorem, then, and the following one about triangles belong among what are called the 'paradoxical' theorems in mathematics [Proclus–Morrow, *CFBEE*, 396].

It may seem a great puzzle to us that Proclus should express such surprise over this result and go on to discuss it and the succeeding propositions at such great length, when he is able to regard, for example, the fact that the angle sum of a triangle is equal to two right angles (see *Elements* I 17 and his commentary on it) as reasonably straightforward.

But we can understand something of this surprise when we consider the evidence of Egyptian land surveying, as was described and illustrated in Section 7.1(d). If the nominal area of a field is taken to be the product of the average of the lengths of opposite sides, then there is something to reconcile between the propositions of the mathematicians and the practice of the accountants and tax collectors. It might be objected that this evidence relates to Egyptian, not Greek, practice, and that we have no evidence of land surveys in Greece. But Proclus, born in Lycia, resident in Athens, goes on to tell a story that indicates that it was not unknown to the Greeks:

> The participants in a division of land have sometimes misled their partners in the distribution by misusing the longer boundary line; having acquired a lot with a longer periphery, they later exchanged it for lands with a shorter boundary and so, while getting more than their fellow colonists, have gained a reputation for superior honesty [Proclus–Morrow, *CFBEE,* 403].

Herodotus gives another example in which size was measured by perimeter: three times he called Sardinia "the biggest island in the world" (*Histories* i, 170, v, 106, and vi, 2) when, even in Herodotus' world, Sicily had a greater area though smaller perimeter. (Sardinia: 9187 square miles and 830 miles circumference *v.* Sicily: 9860 square miles and 680 miles circumference; see Rowland, BIW for further literary and geographical details.) Thucydides indicates the obvious way in which such perimeters are estimated by a maritime nation; at the beginning of Book 6 of his *History of the Peloponnesian War* we read:

> The voyage round Sicily takes rather under 8 days in a merchant ship, yet, in spite of the size of the island . . . .

Of course, the mathematicians whose work was later compiled in the *Elements* knew better. But, within a tradition of non-arithmetised geometry and given Plato's attitude to practical mathematics (see Section 2.2 and the end of my S_{35}, above), the issue of land measurement would have held little interest for them. Heron, who worked in the later tradition of arithmetised, metrical geometry that arose as a combination of Greek geometrical and Babylonian arithmetical mathematics, describes correctly how to calculate the precise numerical area of irregular quadrilaterals at *Metrica* I, 14–16.

Did, then, the Greeks discover geometry from Egyptian land measurement? We have so little reliable information about this question that we shall probably never be able to give a sure answer, but I very much doubt the story. Even Plato and Aristotle do not seem to know enough to be able to go beyond myths and generalities. And go back and look again at the texts quoted above: Herodotus only says it if the Greek word *geōmetria* is translated as 'geometry' (as in Heath's discussion, at *HGM* i, 121–8) rather than 'land measurement' (as, for example, in the translation of Godley, in the Loeb Classical Library edition, which I have used here); the same ambiguity is found in the French *géomètre*. In Proclus, the word is surely correctly translated as 'geometry', but in line 2 of O. Bodl. ii 1847 (see Plate 6 and Section 7.1(d)) it clearly must refer to land measurement. So, for example, Herodotus could be contrasting the way that the Greeks learned the measurement of space from Egypt and the measurement of time from Babylon. As to any proposal that the Greeks realised that the basic formula of Egyptian land measurement was wrong, and discovered geometry by trying to correct it, we have no evidence to justify this interpretation, and it runs counter to what evidence we do have.

It may be of interest to describe briefly the medieval European tradition of land surveying. Elaborate surveys of large estates were being made in France in the ninth century, and a survey of unparalleled scope and completeness covering practically the whole of England was made by order of William the Conqueror in 1086: the Domesday Survey. These are usually described as land surveys (see, for example, the *OED*, s.v. Domesday), but the precise measurement of land plays little or no part in them and they are more accurately described as surveys of actual or potential sources of revenue. Take, for example, the Domesday Book, which embodies the result of the Domesday Survey. There are regional variations in the entries; this having been said, a typical entry reads:

The Bishop of Worcester holds Hantone [Hampton Lucy]. There are xii hides. There is land for xxii ploughs. ii are in the demesne, and iiii serfs. And there are xxii villeins and ix bordars with a priest who have xxiiii ploughs. There is a mill worth vi shillings and viii pence, and xv furlongs of meadow in length and one furlong in breadth. In Warwick iii houses worth xvi pence. Wood i league long and another broad. In King Edward's time it was worth iiii pounds, and afterwards the same, now it is worth xx pounds. The same bishop holds and held Stratforde. There are xiv and a half hides ... [translation adapted from *Victoria County History,* Warwickshire i, p. 302].

The land is described as being for so many ploughs; often woods are described as for so many pigs, while in this extract the 'league' used to describe the wood probably had no precise measured length and the wood itself is very unlikely to have been rectangular, or even four-sided. The 'hide' is a unit of assessment, divided into four 'virgates', each of which contained thirty 'acres'; but these were fiscal units, not measures of area, and the acre was the smallest unit used to describe the tax on a holding. There are no maps.

During the thirteenth and early fourteenth centuries, local standards of linear measure were gradually introduced and applied in England, but they were not generally converted into measurements of area; nor could they be, in the absence of any precise way of measuring angles; nor did they need to be in the context of medieval farming, in which valuable arable land was cultivated in narrow strips and other small parcels, mostly rectangular. No further new techniques were applied to surveying during the fourteenth and fifteenth centuries, and there is no trace of the growth of a class of professional surveyors. The earliest English printed treaties on surveying (by Sir Anthony Fitzherbert in 1523 and Richard Benese in 1537) show little advance on thirteenth-century practice, but new methods, particularly of measuring angles, were developed in Germany and the Netherlands, and reached England through books printed in the 1550s. However, two points about these works should be emphasised: first, as some of the writers themselves remark, there is an enormous gap between the theory and practice of surveying; and, second, the increasing accuracy was probably directed towards the new techniques of map-making. Up to the mid-sixteenth century, the result of any kind of survey would be written text, much in the medieval tradition of the Domesday Book; even during the reign of Elizabeth I, most estate surveys were of this form, and very few estate maps survive from before 1550. However, from the 1580s onwards, the results of detailed surveys began to be embodied in maps, without any accompanying written text. There was then an explosion in the use of maps in general, and, with it, a new phase of development began.

The development of the linear measurement of land is described in Jones, *LME*, and a history of map-making and description of all surviving medieval English local maps, in Harvey, *HTM* and Harvey & Skelton, *LMPME*. I would like to thank Paul Harvey for help; this summary is almost entirely based on information provided by him.

8.2 *Neusis*-constructions in Greek geometry

It is widely believed that Greek geometry was restricted to (unmarked) ruler and compass constructions. In fact, our evidence is much more diverse than this, and the repertoire of basic constructions should certainly also include the *neusis*-construction, in which a line is drawn 'verging' or 'inclining' (*neuein*) towards—that is, passing through—a given point and intercepting two given curves or straight lines in a given line segment. In vulgar terms, one might imagine putting a nail at the given point, marking off the given segment on a ruler, and then rotating the ruler against the nail until the intercept between the two given curves is this marked segment. An example will illustrate clearly what is involved.

Our earliest detailed description of a piece of Greek geometry comes at third hand. It is found in Simplicius' commentary on Aristotle's *Physics,* where Simplicius quotes, with his interpolations, a passage from Eudemus' *History of Mathematics* which describes Hippocrates' *Quadrature of Lunes.* (A lune is a plane region bounded by two arcs of circles; Hippocrates described three such regions that can be made equal to rectilinear plane figures, and hence can be squared.) For a discussion of the historical, textual, and mathematical aspects of this passage, see Heath, *HGM* i, 183–200. A purged version of the text is given in Thomas, *SIHGM* i, 234–53, from which the following extract is taken.

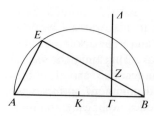

If [the outer circumference of the lune] were less than a semicircle. Hippocrates solved (*kateskeuasen,* constructed) this also, using the following preliminary construction. Let there be a circle with diameter *AB* and centre *K*. Let *ΓΔ* bisect *BK* at right angles; and let the straight line *EZ* be placed between this and the circumference verging (*neuousa*) towards *B* so that the square on it is one-and-a-half (*hēmiolos,* half-whole) times the square on one of the radii. . . .

In other words, draw the line *EZB,* through *B* and such that *EZ* is equal to the specified line.

This particular construction can in fact be accomplished by ruler-and-compass methods, as follows. For any line *EB,* verging towards *B,* the angle *E* in the semicircle will be a right angle, so the triangle *AEB* and *ZΓB* will be similar. Hence $(BZ + EZ) \cdot BZ = AB \cdot B\Gamma = r^2$, where r is the radius of the semicircle. Thus, if so required, we can perform the following ruler-and-compass construction. First construct, separately, a line $E'Z'$ such that $E'Z'^2 = \frac{3}{2}r^2$ (using the techniques of *Elements* II 14), and a line $B'Z'$ such that $B'Z'^2 + B'Z' \cdot E'Z' = r^2$. (This is the elliptic application of areas discussed in Section 5.2(d), and performed in Fig. 5.7(c). Alternatively a construction can be manufactured around the configuration of *Elements* II 6; see Heath, *TBEE* i, 386 ff. for details.) If we then locate *Z* on *ΔΓ* so that $BZ = B'Z'$, we shall have $EZ^2 = \frac{3}{2}r^2$, as required. However, neither Hippocrates nor Simplicius makes any attempt to provide such a ruler-and-compass justification of this *neusis*-construction, even though the text elsewhere gives elaborate descriptions of matters that are far less obvious than this. The *neusis*-construction seems to be invoked as a permitted construction in its own right.

Each *neusis*-construction needs some kind of verification that a solution is possible. For example, here in Hippocrates' *Quadrature,* we need to observe that the line AB will intercept a segment *AΓ* equal to $\frac{3}{2}r^2$, greater than $\sqrt{(\frac{3}{2})}r$, while a line through the intersection of *ΓΔ* and the circle will have a zero intercept. So, by some principle of continuity, an intermediate line with the required intercept will exist. Such an implied condition is analogous to the *diorismos* in the formal division of a Euclidean proposition (see the description in Proclus–Morrow, *CFBEE,* 203 f. quoted in Heath, *TBEE* i, 129 f. and the examples in *Elements* I 22 and VI 28), but no such condition is, to my knowledge, considered in any surviving *neusis*-construction. Nor do I know of any ancient discussion of the underlying principle of continuity, in any context; see, for example, how *Elements* I 1 assumes without comment that the two circles will intersect—but the circles will intersect only if their radii satisfy some implied *diorismos*; and then, here again, something like the principle of continuity is also needed.

Archimedes uses *neusis*-constructions freely in his *Spiral Lines*: the construction appears in Propositions 5, 6, 7, 8, and 9, and some of these propositions play crucial roles later in the book. This time the *neusis*-construction in Propositions 5, 6, and 7 cannot be carried out by ruler-and-compass methods. This has provoked comment and criticism from Pappus, in late antiquity, that Archimedes was using methods inappropriate to the problem (see below), and ancient and modern commentators have shown how Archimedes' material can be reworked by reformulating the later propositions of the treatise which depend on these *neusis*-constructions so that the properties of the spiral can be established differently, by ruler-and-compass methods; see Archimedes–Heath, *WA*, Chapter 5; Heath, *HGM* ii, 556–61; and Knorr, ANCSL for examples, discussions, and references.

Proposition 8 of another work loosely attributed to Archimedes, the *Book of Lemmas*, which survives only in an Arabic translation, describes a geometrical configuration that can immediately be used to give a *neusis*-construction for trisecting an angle. The statement of this proposition is:

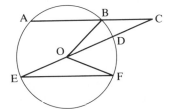

If AB is any chord of a circle whose centre is O, and if AB be produced to C so that BC is equal to the radius; if further CO meet the circle in D and be produced to meet the circle a second time in E, the arc AE will be equal to three-times the arc BD.

The proof is straightforward: if EF is parallel to AB then, since OE = OF = OB = BC, we easily see that the angle DOF is twice BOD, from which the result follows. We can use this configuration to trisect an angle as follows: set up the figure so that AOE is the angle to be trisected and EOD is the diameter produced beyond D. Then use a *neusis*-construction to insert the line CB, verging towards A, between the circle FDB and the extension of line BC; BOD is then one-third of AOE. This construction cannot now be performed by ruler-and-compass methods, though it can be related to construction involving conic sections. For discussion of the *Book of Lemmas* and this kind of construction, see Archimedes–Heath, *WA*, pp. xxii f. and 301–18, and Hogendijk, HTATGIG.

Apollonius, a younger contemporary of Archimedes, wrote a work *On Neuses*, in two books. This is lost, but we have a description by Pappus which seems to indicate that it was an investigation of the techniques that would be needed to carry out different kinds of *neusis*-constructions. See Heath, *HGM* ii, 189–92 for a description.

Nicomedes, a younger contemporary of Apollonius, defined the conchoid, a curve based on a *neusis*-construction, which he used to provide a construction for duplicating the cube; the procedure is described both in Pappus' *Collection* and Eutocius' *Commentary on Archimedes' Sphere and Cylinder* and is excerpted in Thomas, *SIGHM* i, 296–309. The conchoid is the locus of points which, along with a given straight line, cut off a given segment on all lines which verge towards a given point, and it can then be used to determine a particular *neusis*-construction. Nicomedes, as reported by Eutocius, describes in some detail an instrument for drawing the conchoid on the basis of which it is possible to reconstruct, with some confidence, his arrangement of sliding slotted rulers; see Archimedes–Heiberg, *Opera* iii, 99, or Heath, *HGM* i, 239.

In broad terms, we see three phases of development. The early geometers, from Hippocrates to Archimedes, seem to have used the *neusis*-construction freely, as a basic construction of mathematics. This was followed by a period, represented by Apollonius and Nicomedes, in which the scope and implications of the construction were investigated, and its relation with other constructs was explored. Finally, in late antiquity, we find Pappus laying down formal criteria that the use of constructions must satisfy. In this last phase, Pappus does not mince his words in his explicit criticism of Archimedes and Apollonius:

It seems to be a grave error into which geometers fall whenever anyone discovers the solution of a plane problem by means of conics or linear curves, or generally solves it by means of a foreign kind, as is the case, for example, (i) with the problem in the fifth Book of Conics of Apollonius relating to the parabola, and (ii) when Archimedes assumes in his work on the spiral a *neusis* of a solid character with reference to a circle; for it is possible without calling in the aid of anything solid to find the [proof of] the theorem given by the latter [sc. Archimedes], that is, to prove that ... [Archimedes–Heath, *WA*, pp. ciii f. or Heath, *HGM* ii, 68].

I shall not digress here to describe and explore Pappus' description of plane, solid, and linear constructions; nor shall I discuss the status of other kinds of constructions such as Archytas' duplication of the cube; nor shall I describe how tenuous is the evidence for the criticism that Plato is said to have directed against the use of mechanical constructions in geometry; nor shall I describe the other evidence we have for more exotic mechanical devices and constructions in geometry, such as the device attributed by Eutocius to Plato for duplicating the cube, or the mysterious 'horn ruler' invoked in Proposition 4 of Diocles' *On Burning Mirrors*. (On these matters see the references already cited; Niebel, *UBGKA*; Steele, *URZLGM*; and the further references contained therein.) I shall take the well-founded example of the *neusis*-construction as an illustration of the proposal, accepted by most modern commentators, that more general kinds of constructions than ruler-and-compass constructions were admitted freely into early Greek geometry (see Knorr, ANCSL, 86–9 for a discussion), and will now turn to a consideration of Euclid's *Elements*.

Although the *Elements* is set in the middle of the period when *neusis*-constructions were freely used, yet we find there no reference to or use of them. Let me dispose first of one common explanation of this, that the *Elements* provides only an introduction to the basic techniques of mathematics, preparing the way for further treatises that will deal with the more advanced problems where *neusis*-constructions will find their proper place. First, to describe the *Elements* as restricted to basic mathematics seems to me to be a short-sighted judgement based only on the contents of the first few books. The material in Books X, XII, and XIII can scarcely be called straightforward, basic, or easy; see, for example, the discussion in Chapter 5, above. Next, if the aim of the *Elements* were pedagogical, with Euclid the supreme teacher (on this again see Book X, which occupies more than a quarter of the bulk of the *Elements*!), then surely his treatment would anticipate the later enlargement of the geometrical repertoire, and would prepare the reader for the introduction to more general constructions. But the *Elements* does precisely the opposite: in Book I, Propositions 1–3, we find the most restricted possible interpretation of the Postulates, in which the notional, unmentioned 'compass' (see Postulate 3) cannot be used as dividers, to move around freely the endpoints of a line. Far from anticipating a relaxation of the scope of geometrical constructions, this sets the scene for the most severely constrained interpretation of Euclidean compasses. (A possible explanation of these propositions is, I feel, to set up a complete theory of transformation and application of lines founded on the most basic interpretation of the first three postulates, as a model for the later theory of application of rectilinear areas and for the unachieved theory of application of volumes.) Further, the argument that *neusis*-construction belongs only to the more advanced parts of geometry does not stand close scrutiny. See, for example, Aristotle, *Posterior Analytics,* I 10, 76a30 f.:

> By first principles in each genus I mean those the truth of which it is not possible to prove. What is *denoted* by the first [terms] and those derived from them is assumed; but, as regards their *existence,* this must be assumed for the principles but proved for the rest. Thus what a unit is, what the straight [line] is, or what a triangle is [must be assumed]; and the existence of the unit and of magnitude must also be assumed, but the rest must be proved. Now of the premisses used in demonstrative sciences some are peculiar to each science and other common [to all], the latter being common by analogy, for of course they are

actually useful in so far as they are applied to the subject-matter included under the particular science. Instances of first principles peculiar to a science are the assumptions that a line is of such-and-such a character, and similarly for the straight [line]; whereas it is a common principle, for instance, that, if equals be subtracted from equals, the remainders are equal. But it is enough that each of the common principles is true so far as regards the particular genus [subject-matter]; for [in geometry] the effect will be the same even if the common principle be assumed to be true, not of everything, but only of magnitudes, and, in arithmetic, of numbers.

Now the things peculiar to the science, the existence of which must be assumed, are the things with reference to which the science investigates the essential attributes, e.g. arithmetic with reference to units, and geometry with reference to points and lines. With these things it is assumed that they exist and that they are of such-and-such a nature. But, with regard to their essential properties, what is assumed is only the meaning of each term employed: thus arithmetic assumes the answer to the question what is [meant by] 'odd' or 'even', 'a square' or 'a cube', and geometry to the question what is [meant by] 'the *alogon*' or 'deflection' (*keklasthai*) or [the so-called] 'verging' (*neuein*) [to a point]; but that there are such things is proved by means of the common principles and of what has already been demonstrated. Similarly with astronomy. For every demonstrative science has to do with three things, (1) the things which are assumed to exist, namely the genus [subject-matter] in each case, the essential properties of which the science investigates, (2) the common axioms so-called, which are the primary source of demonstration, and (3) the properties with regard to which all that is assumed is the meaning of the respective terms used. There is, however, no reason why some sciences should not omit to speak of one or other of these things [Archimedes–Heath, *WA*, 50 f].

Aristotle's reference to *neusis*-construction as an essential attribute for study is all the more relevant when we consider that he probably learned his mathematics during his nearly twenty-year association with the Academy, up to Plato's death, in contact with Theodorus, Archytas, Theaetetus, and Eudoxus. Thus this passage gives yet further direct evidence of the basic role of *neusis*-constructions in geometry during the pre-Euclidean period.

Consider the constructions invoked by Euclid. First, the scope of the *Elements* is not restricted to ruler-and-compass constructions. While the mean, third, and fourth proportionals for lines are constructed in detail by (implied) ruler-and-compass methods in Book VI 11–13, there are not, and there cannot be, similar constructions to back up the use of the fourth proportional to two circles and a square in XII 2, and to various three-dimensional figures in XII 5, 11, 12, and 18. While these latter fourth proportionals do not have ruler-and-compass constructions, their existence is assumed, without any constraints. (A separate issue, which I have set to one side above and shall continue to ignore here, is how one can talk of three-dimensional figures being 'constructed' in any sense of the word; here the same kinds of difficulties that apply to, for example, Archytas' duplication of a cube are also found throughout the construction of the regular polyhedra in *Elements* XIII.)

We can perhaps discern a later historical stratum: those parts of the *Elements* that invoke these more general kinds of constructions might be associated with the innovations due to Eudoxus. Even if this is indeed the case, our question still applies in a modified form: why do the earlier, pre-Eudoxan parts of the *Elements* make no reference to *neusis*-constructions?

An answer to this can only be very speculative. With this qualification, and viewing the issue from the vantage point of the reconstruction proposed here, we see how the basic geometry that is developed in Books I, III, VI, and XI is applied, in Books II, IV, X, and XIII, in contexts which have very strong anthyphairetic overtones. I have illustrated throughout Chapters 2 to 5 and will go on to describe in Chapter 9 in more detail how quadratic problems and only quadratic problems are amenable to anthyphairetic analysis. The geometrical translation of what I have here loosely called 'quadratic problems' is 'ruler-and-compass constructions'; so, I propose, the bias of the *Elements* towards ruler-and-compass constructions and its exclusion of *neusis*-constructions may be a consequence of its anthyphairetic origins.

As the anthyphairetic preoccupations faded away, as new techniques were introduced, particularly by Eudoxus, as the geometrical language that had been developed was applied to wider and wider contexts, particularly by Archimedes and Apollonius, so would these underlying constructions move into prominence and be of interest in their own right. There would then be a tendency towards an exploration and classification of the underlying geometrical procedures, as in what I called above the second and third phases of development. It is a commonplace in mathematics for the initial developments to be prompted by some quite specific problem which then fades away; that which is left, if successful, then often acquires an autonomous existence as a 'theory' in its own right. The eighteenth- and nineteenth-century investigations of a vibrating string, which provoked the creations of analysis and eventually Cantor's set theory, gives one good such example; many more could be cited without difficulty. Thus the apparent concern for ruler-and-compass constructions in the *Elements* may well be the influence of the language developed around the successful treatment of anthyphairetic phenomena, and the absence of higher degree constructions may reflect the paucity, even today, of any anthyphairetic results about anything other than linear or quadratic phenomena.

8.3 The discovery and role of the phenomenon of incommensurability

8.3(a) *The story*

Part of every literate person's intellectual baggage, along with the second law of thermodynamics and the principles of relativity and indeterminacy, is some version of the story of the discovery of incommensurability by Pythagoras or the Pythagoreans and of the role this discovery played in Greek mathematics. Here, to end this last chapter on Greek mathematics, I shall describe and discuss our evidence on this topic. I shall continue to take pains to maintain the perspective of this book, and so will distinguish between that evidence which comes, or purports to come, from the early period, up to the time of Euclid and Archimedes, and that found only in later sources.

I shall only give a summary treatment, and direct the reader to the account in Knorr *EEE*, Chapter 2, for further details and references. I will not attempt to deal further with the philological issues concerning the words (*a*)*summetros*, (*ar*)*rhētos*, (*a*)*logos*, *diametros*, and *diagōnios*, etc., beyond the comments given earlier and listed in the Index. I start by reviewing the treatment of the subject in authors up to Proclus, in late antiquity. Every reference to anything that may be connected with the topic will, I hope, be included somewhere in this list, and I shall also include occasional examples of the many important witnesses whose surviving writings do not mention the topic.

8.3(b) *The evidence*

(i) Democritus (fl. *c*.460 BC) wrote a treatise in two books called *Peri alogōn grammōn kai nastōn*, 'On unreasonable/unutterable/irrational/ ... lines and solids'. Our source is Diogenes Laertius, *Lives of the Eminent Philosophers*, IX 47 (Diels & Kranz, *FV* ii, Fragment A33, p. 91.20 = Fragment B11p, p. 141.24; Diogenes–Hicks, *LEP* ii, 458–9). Diogenes is a gossipy and uncritical compiler of anecdotes, perhaps of the third century AD, though we know nothing with confidence of his dates and origins. Here he is quoting, perhaps accurately, a catalogue of titles of Democritus' works by Thrasyllus of Alexandria (died 36 AD), who also catalogued and arranged Plato's dialogues. We know of Democritus' ideas only through secondary reports and brief quotations, and none of the ninety treatises in this catalogue, twelve of them described as mathematical, have survived. In particular, we know nothing about the contents of this treatise, and it is not even certain that it had anything to do with the topic of incommensurability. A conspectus of scholarly attitudes to Democritus' mathematics can be found in Guthrie, *HGP* ii, 484–8; for a succinct assessment see the *OCD,* s.v. Democritus: "Little is known (though much is written) about the mathematics of Democritus."

(ii) Plato: Our first unequivocal and explicit reference to incommensurability is in Plato, at *Theaetetus* 147d–148b, where the topic is handled confidently in the context of Theodorus' geometry lesson to the young Theaetetus. This passage gives no indication that the phenomenon itself posed any fundamental conceptual difficulty for mathematicians; rather, it is presented as a source of interesting and fruitful problems. There is a playful metaphorical allusion at *Republic* VII, 534d; it enters the construction of the nuptial number, *Republic* VIII, 546b–d, in a passage which prompts the later descriptions of side and diagonal numbers in the commentaries of Theon of Smyrna and Proclus; it is discussed at *Parmenides* 140b–d; and there is a reference in the possibly spurious *Hippias Major,* 303b–c, to sums of incommensurable lines that is reminiscent of the classification of *Elements* X. Finally, in Plato's last dialogue, at *Laws* VII 817e–820e, there is a celebrated passage that talks of "Ignorance [which] seemed to me worthy not of human beings but of pigs, and I felt ashamed not for myself alone but for all the Greeks". The context is a curriculum similar to that of *Republic* VII, but at a lower level, for the basic education of the citizens of the state. The Athenian Stranger talks not only of the measurability of line with line, surface with surface, and volume with volume, but also:

Again, what of the relations of surface and line to volume, or of surface and line to one another; do not all we Greeks imagine that they are measurable (*metrēta*) in some way or another. ... Then if this is absolutely impossible, though all we Greeks, as I was saying, imagine it possible, are we not bound to blush for them all as we say to them, "Worthy Greeks, this is one of the things of which we said that ignorance is a disgrace and that to know such necessary matters is no great achievement" [820a–b; translation from Thomas, *SIHGM* i, 20–7].

The matter at issue here is not simply the topic of commensurability and incommensurability of homogeneous magnitudes, and it may instead be related to the everyday techniques of measuring and comparing areas that I described in Sections 7.1(d) and 8.1, above.

(iii) One of Aristotle's favourite mathematical illustrations is the incommensurability of the diagonal, which he cites about thirty times, at *Prior Analytics* 41a23–30, 46b26–37, 50a37–8, 65b16–21; *Posterior Analytics* 71b25–6, 89a29–32; *Topics* 106a38–b1, 163a11–13; *Sophistical Refutations* 170a25–6, 176b20; *Physics* 221b23–25, 222a3–7; *De Caelo* 281a6–7, b5–6, b12–14; *De Anima* 430a30–1; *Generation of Animals* 742b27–28; *Metaphysics* 983a12–21, 1012a31–3, 1017a34–5, 1019b24–7, 1024b19–21, 1047b6–12, 1051b20–1, 1053a17–8; *Nicomachean Ethics* 1112a21–3; *Eudeman Ethics* 1226a2–4; and *Rhetoric* 1392a15–8, while in a few other places such as *Metaphysics* 1004b11–2, 1021a3–6, and

1061b1, he introduces (in)commensurability in other contexts. A typical example occurs in *Metaphysics* 983a12–20:

> Yet the acquisition of it [sc. knowledge, *epistēmē*] must in a sense end in something which is the opposite of our original inquiries. For all men begin, as we said, by wondering that things are as they are, as they do about self-moving marionettes, or about the solstices or the incommensurability of the diagonal; for it seems wonderful to all who have not yet seen the reason, that there is a thing which cannot be measured even by the smallest unit. But we must end in the contrary and, according to the proverb, the better state, as is the case in these instances too when men learn the cause; for there is nothing which would surprise a geometer so much as if the diagonal turned out to be measurable (*metrēta*).

(This and the other translations from Aristotle have been taken from the edition of W. D. Ross, and they have all been adapted slightly as will be explained below. I have here also adjusted the translation of the last word, which is the same as that found at Plato's *Laws* VII, 819e–820c, quoted above.) Often Aristotle's reference is very short and stereotyped: *estin he diametros asummetros*, 'the diagonal is incommensurable'. Sometimes he adds bits of extra perplexing information, as at *Metaphysics* 1053a14–20:

> But the measure is not always one in number—sometimes there are several; e.g. the quarter-tones (not to the ear, but as determined by the ratios) are two, and the articulate sounds by which we measure are more than one, and the diagonal and its side are measured by two quantities, and all spatial magnitudes reveal similar varieties of unit . . . .

Here there is variation between the manuscripts that makes the original text unsure; the critical editions should be consulted for details. But one point seems clear from the grammar of the undisputed text of the Greek, and it seems to be supported by the example of the quarter-tones: that the meaning is not that there is one measure for the side, and another for the diagonal. The translation should better read "the diagonal *is* measured by two quantities †and [so is] the side . . . † . . . ," where the obeli indicate the uncertain text.

In the large majority of these examples where Aristotle writes of 'the diagonal', he does not mention 'the side,' *hē pleura*. The only places where these words occur are *Topics* 106a38, 163a13; *Physics* 221b25; *Generation of Animals* 742b28; *Nicomachean Ethics* 1112a23; and the doubtful passage of *Metaphysics* 1053a18, just cited. However, there seems no doubt that 'the side' is always understood, as passages such as *Topics* 163a11–13 make clear:

> For example if he had to show that the diagonal is incommensurable with the side, and were to beg that the side is incommensurable with the diagonal . . . .

More curious is that Aristotle nowhere says explicitly that he is talking of

8.3 The phenomenon of incommensurability   297

a square. I have excised these words 'the side' and 'of the square' from the quoted translations whenever they are not found in the Greek text.

Three examples, all in the *Prior Analytics,* give more information about Aristotle's method of proof. One of these, at 65b16–21, is of doubtful use for further reconstruction since it is an example of an incorrect proof:

> The most obvious case of the irrelevance of an assumption to a conclusion which is false is when a syllogism drawn from middle terms to an impossible conclusion is independent of the hypothesis, as we have explained in the *Topics.* For to put that which is not the cause as the cause, is just this: e.g. if a man, wishing to prove that the diagonal is incommensurate, should try to prove Zeno's theorem that motion is impossible, and so establish a *reductio ad impossibile*: for Zeno's false theorem has no connexion at all with the original assumption.

The other two passages give more details of the argument. At *Prior Analytics* 41a23–30 he writes:

> For all who effect an argument *per impossibile* infer syllogistically what is false, and prove the original conclusion hypothetically when something impossible results from the assumption of its contradictory; e.g. that the diagonal is incommensurate, because odd numbers are equal to evens if it is supposed to be commensurate. One infers syllogistically that odd numbers come out equal to evens, and one proves hypothetically the incommensurability of the diagonal, since a falsehood results through contradicting this:

and at 50a35–8:

> In the latter [sc. arguments which are brought to a conclusion *per impossible*], even if no preliminary agreement has been made, men still accept the reasoning, because the falsity is patent, e.g. the falsity of what follows from the assumption that the diagonal is commensurate, viz. that then odd numbers are equal to evens.

So Aristotle has in mind an indirect argument *per impossibile* (i.e. *reductio ad absurdum*) involving odd and even numbers; but, beyond that, he gives no details.

(iv) Eudemus: The lost *History of Mathematics* by Eudemus is cited several times, either directly or indirectly, by Proclus in his *Commentary on the First Book of Euclid's Elements.* One passage, the so-called Catalogue of Geometers, speaks of Pythagoras and the discovery of incommensurability:

> Following upon these men, Pythagoras transformed mathematical philosophy into a scheme of liberal education, surveying its principles from the highest downwards and investigating its theorems in an immaterial and intellectual manner. He it was who discovered the *alogōn pragmateia* [see below] and the structure of the cosmic figures [65.15–21].

(This translation is adapted from Proclus–Morrow *CFBEE,* 52–3.) The word *alogōn* is frequently emended to *analogōn*, as Proclus–Friedlein, *PEEC*, 65.19 and Morrow's translation (there "the doctrine of proportionals"), but *alogōn* is in fact the reading of all manuscripts; see Wasserstein, THTN, p. 165 n. 3 (on p. 166) and the following reference. The first sentence was shown, by Vogt and Sachs, not to have been in Eudemus' original text, but to have been an addition copied from Iamblichus (see (xiii), below), and the second sentence is almost certainly a neo-Platonic gloss, perhaps by Proclus himself; see Burkert, *LSAP*, 409–12 for a discussion and references. So Eudemus' summary of the beginnings of Greek mathematics may have contained no reference to the discovery of incommensurability.

On the evidence of Proclus' writings, Eudemus' *History* does not seem to have contained any direct reference to incommensurability. For Proclus cites Eudemus by name in his *Commentary* on pp. 125.7, 299.3, 333.6, 352.14, and 419.15, and discusses topics related to incommensurability on pp. 6, 60, 65 (see above), and 74; but Proclus never cites anything *by* Eudemus *on* incommensurability. I know of no other fragment of Eudemus that introduces the topic, but unfortunately Eudemus–Wehrli, *ER,* does not contain any indexes so I cannot be certain.

(v) Euclid: The topic of incommensurability appears in the *Elements* only in Books X and XIII, which were described in Chapter 5, above. The first place that exhibits an incommensurable magnitude is X 10, a proposition that may have been interpolated; see the note in Heath, *TBEE* iii, 32–3. At the end of Book X, we find a proposition with two alternative demonstrations that the diagonal and side of a square are incommensurable; this proposition is unnumbered in the manuscripts but is now generally known as X 117 (Euclid–Stamatis, *EE* iii, 231–4). It is very rarely translated, though a complete French translation can be found in Euclid–Peyrard, *OE,* and a translation of the first demonstration and paraphase of the second are given in Knorr, *EEE,* 23 and 230–1. This proposition is almost certainly an interpolation made after the time of Alexander of Aphrodisias; see (xii) below.

(vi) Archimedes only refers to (in)commensurability once, when describing the principle of the balance, at *Plane Equilibria* I, Propositions 6 and 7; these texts are excerpted in Thomas, *SIHGM* ii, 208–17. Proposition 7 is corrupt and garbled, and has provoked an enormous literature; see Knorr, APEPT for a full discussion and references.

(vii) Apollonius wrote an extension of the theory of *Elements* X. We have reports of this from Proclus, who refers to:

Material that Apollonius has elaborated at considerable length about unordered irrationals (Proclus–Morrow, *CFBEE,* 74),

and from Pappus (see (xv), below). Nothing of this work survives, and I know of no reference to incommensurability anywhere in Apollonius' extant writings.

(viii) Heron of Alexandria (first century AD): The only places in the large amount of heterogeneous material that has been transmitted to us under the name of Heron where the topic of incommensurability is mentioned are passages appended, perhaps during the eleventh century AD, to the *Definitions* (Heron–Heiberg, *Opera* iv, 84–6 and 136–40). This appended material also contains an abbreviated version of Eudemus' Catalogue of Geometers (*Opera* iv, 108; see (iv), above) which does not mention the discovery of incommensurability.

(ix) Plutarch (AD *c*. 50–*c*. 120) often introduces mathematical topics of popular interest in his writings. For example, *Moralia* viii 2 is a dialogue: "What Plato meant by saying that God is always doing geometry", (to which Plutarch adds "if indeed this statement is to be attributed to Plato"!), where we read:

Therefore Plato himself censured Eudoxus and Archytas and Menaechmus for endeavouring to solve the doubling of the cube by instruments and mechanical constructions, thus trying *dicha logou* [see below] to find two mean proportionals, so far as that is allowable [translation from Thomas, *SIHGM* i, 386 ff.].

This passage whose syntax is fragmented, has led editors to propose the emendations: δι' ἀλόγυ, 'by irrational means', as in this translation, or διαλόγου, 'by means of reason', or ἀνάλογον 'in proportion', for the manuscript's δίχα λόγου, 'in separation from reason'; see the critical edition Plutarch–Hubert, *M* iv, 262, for details. The passage has the appearance of a gloss, which disturbed the syntax of the sentence when someone later tried to fit it into the text.

I know of no other passage in Plutarch that could be making any reference to incommensurability.

(x) Nicomachus of Gerasa (AD *c*. 50–150), an early neo-Pythagorean writer of introductory books on arithmetic, geometry, and music theory, which are then often quoted by later neo-Pythagoreans, does not mention incommensurability in any of his surviving writings.

(xi) Theon of Smyrna (fl. AD *c*. 125) does not mention incommensurability in his surviving complete work, *Expositio Rerum Mathematicarum ad Legendum Platonem Utilium*, or, to my knowledge, in any fragments.

(xi) Athenagoras of Athens (second half of the second century AD) reports an aphorism attributed to Lysis, a Pythagorean of *c*. 425 BC: *arrhētos arithmos theos*, 'God is an irrational/ineffable/inexpressible/ ... number'. Athenagoras' own direct or indirect source is almost certainly a neo-Pythagorean forgery which may have been influenced by a Jewish intermediary. (See Diels & Krantz, *FV* i, 421.5; and Thesleff, *PTHP*, 114 and *IPWHP*.) A translation of the complete report is:

> Lysis and Opsimus define as God, the former irrational number, the latter the excess of the greatest of the numbers over the [number] nearest to it; and if the greatest number, according to the Pythagoreans, is the ten, this being the *tetraktys* and containing all the harmonic ratios, and the nine lies close beside it, then God is [a] unit, that is [a] one; for the greatest [number] exceeds the nearest [number] by [a] one, that being the smallest number.

This fragment has no value for matters connected with the discovery and early role of incommensurability, and it may have had nothing to do with mathematics, in the sense of my use of the word in this book.

(xii) Alexander of Aphrodisias (early third century AD): The first explicit proof of the incommensurability of the diagonal and side of a square is found in Alexander's commentary on *Prior Analytics* 41a26 (Alexander–Wallies, *IAP* 260.7–261.28). There is no generally available translation of this long proof, but there is a paraphrase and full discussion in Knorr, *EEE*, 228–31, which should be consulted for further details of the following summary. Alexander's proof is not the same as those at *Elements* X 117, though it includes three citations from the *Elements* that agree, word for word, with our manuscripts; from this we infer that his proof must post-date the composition of those parts of the *Elements*. Also, since Alexander's other dozen citations of the *Elements* also correspond to our text, we infer that his version of the *Elements* did not include X 117; so this proposition was probably interpolated sometime between Alexander and Theon of Alexandria (late fourth century AD), whose recension of the *Elements* has been transmitted to us in all but one of our Euclidean manuscripts. Knorr proposes, very plausibly, that X 117 was interpolated into the *Elements* as a result of the continuing activity of later Aristotelian commentators. For example Philoponus (sixth century AD) also discusses incommensurability in his commentary on the *Posterior Analytics,* on 71a17 (Philoponus–Wallies, *IAP,* 12.4–17.9; see especially 16.22–25) and on 71b25 (*IAP* 26.17–27.11; see especially 26.25–27).

(xiii) Iamblichus (AD *c.* 250–*c.* 325): The first surviving stories of the circumstance and impact of the discovery of incommensurability are found in the neo-Pythagorean commentator Iamblichus, who gives five mutually inconsistent accounts which permute stories of the discovery, the construction of a dodecahedron, the career of Hippasus, and the death at sea of a Pythagorean. I refer the reader, in the first instance, to the accounts in Burkert, *LS,* Chapter 6, and Knorr, *EEE,* Chapter 2, for further details and references.

(xiv) Pappus of Alexandria (fl. AD *c.* 320) opens his *Commentary on Book X of Euclid's Elements,* which survives only in an Arabic translation, with a version of the story. Since Pappus is rare among these later commentarors in being an able mathematician, and since this

8.3   The phenomenon of incommensurability   301

reference occurs in a commentary on Euclid's treatment of incommensurable magnitudes, it is worth quoting the passage in full (Pappus–Thomson & Junge, *CPBXEE*, 63–4, with the editor's emphases, parentheses, and expansions):

Book I of the treatise of Pappus on the rational and irrational continuous quantities, which are discussed in the tenth book of Euclid's treatise on the Elements: translated by Abū 'Uthmān Al-Dimishqī.
§1. The aim of Book X of Euclid's treatise on the Elements is to investigate the commensurable and incommensurable, the rational and irrational continuous quantities. This science (or knowledge) had its origin in the sect (or school) of Pythagoras, but underwent an important development at the hands of the Athenian, Theaetetus, who had a natural aptitude for this as for other branches of mathematics most worthy of admiration. One of the most happily endowed of men, he patiently pursued the investigation of the truth contained in these [branches of] science (or knowledge), as Plato bears witness for him in the book which he called after him, and was in my opinion the chief means of establishing exact distinctions and irrefragable proofs with respect to the above-mentioned quantities. For although later the great Apollonius whose genius for mathematics was of the highest possible order, added some remarkable species of these after much laborious application, it was nevertheless Theaetetus who distinguished the *powers* (i.e. the squares) which are commensurable in length, from those which are incommensurable (i.e. in length), and who divided the more generally known irrational lines according to the different means, assigning the medial line to geometry, the binomial to arithmetic, and the apotome to harmony, as is stated by Eudemus, the Peripatetic. Euclid's object, on the other hand, was the attainment of irrefragable principles, which he established for commensurability and incommensurability in general. For rationals and irrationals he formulated definitions and (specified) differences; determined also many orders of the irrationals; and brought to light, finally, whatever of finitude (or definiteness) is to be found in them. Apollonius explained the species of the ordered irrationals and discovered the science of the so-called unordered, of which he produced an exceedingly large number by exact methods.
§2. Since this treatise (i.e. Book X of Euclid.) has the aforesaid aim and object, it will not be unprofitable for us to consolidate the good which it contains. Indeed the sect (or school) of Pythagoras was so affected by its reverence for these things that a saying became current in it, namely, that he who first disclosed the knowledge of surds or irrationals and spread it abroad among the common herd, perished by drowning: which is most probably a parable by which they sought to express their conviction that firstly, it is better to conceal (or veil) every surd, or irrational, or inconceivable in the universe, and, secondly, that the soul which by error or heedlessness discovers or reveals anything of this nature which is in it or in this world, wanders [thereafter] hither and thither on the sea of non-identity (i.e. lacking all similarity of quality or accident), immersed in the stream of the coming-to-be and the passing-away, where there is no standard of measurement. This was the consideration which Pythagoreans and the Athenian Stranger held to be an incentive to particular care and concern for these things and to imply of

necessity the grossest foolishness in him who imagined these things to be of no account.

§3. Such being the case, he of us who has resolved to banish from his soul such a disgrace as this, will assuredly seek to learn from Plato, the distinguisher of accidents, those things that merit shame, and to grasp those propositions which we have endeavoured to explain, and to examine carefully the wonderful clarity with which Euclid has investigated each of the ideas (or definitions) of this treatise (i.e. Book X.)....

Pappus here tells a version of the story of the discovery and then, in §2, he reports another story that "became current" among the Pythagoreans about the disclosure of the discovery to which he assigns a symbolic meaning, since he says that it is unlikely to be literally true. So Pappus, who also had access to Eudemus' *History* in some form or another, is uncertain of the details and circumstances of the discovery of incommensurability, and of its effect on the wider mathematical community; this also confirms further that Eudemus' *History* did not seem to contain details of the story (see (iv), above). Pappus' story also occurs as an anonymous scholium to *Elements* X (see Euclid–Heiberg, *Opera* v, 417).

Unfortunately Pappus' commentary is of little help in understanding *Elements* X.

(xv) Proclus (AD c. 410–85); see (iv), above, and the passage from his commentary on Plato's *Republic*, quoted in Section 3.6(b), above.

8.3(c) *Discussion of the evidence*

I shall focus my brief discussion of the evidence on two issues that play a prominent role in the received interpretation: did the discovery of incommensurability reveal a fatal flaw in the Pythagorean programme of describing everything by number, and what might have been the details of Aristotle's proof *per impossibile* of the incommensurability of the diagonal? These questions are usually taken as basic to any reconstruction of pre-Euclidean mathematics, and some interpretations of the received answers are often taken as the fundamental facts of the subject.

First, the Pythagorean programme. Our most important early source of information is Aristotle, especially in the *Metaphysics*. For example, we find there several versions of what has become the concise and celebrated epitome of Pythagorean philosophy:

They [sc. the Pythagoreans] reduce all things to numbers [1036b13],

The Pythagoreans, also, believe in one kind of number—the mathematical; only they say it is not separate but sensible substances are formed out of it. For they construct the whole universe out of numbers—only not numbers consisting of abstract units; they suppose the units to have spatial magnitude. But how the first one was constructed so as to have magnitude, they seem unable to say [1080b16–21],

8.3 The phenomenon of incommensurability 303

and

> These thinkers [sc. the Pythagoreans] identify number with real things; at any rate they apply their propositions to bodies as if they consisted of those numbers [1083b17–19].

Aristotle's polemical method, particularly in the *Metaphysics,* is to give a prejudiced summary of his predecessors' philosophies, through which he can point out the flaws in their point of view. See, for a further example, one of his longer descriptions of Pythagorean mathematics:

> Contemporaneously with these philosophers [sc. Leucippus and Democritus] and before them, the so-called Pythagoreans, who were the first to take up mathematics, not only advanced this study, but also having been brought up in it they thought its principles were the principles of all things. Since of these principles numbers are by nature the first, and in numbers they seemed to see many resemblances to the things that exist and come into being—more than in fire and earth and water (such and such a modification of numbers being justice, another being soul and reason, another being opportunity—and similarly almost all other things being numerically expressible); since, again, they saw that the modifications and the ratios of the musical scales were expressible in numbers;—since, then, all other things seemed in their whole nature to be modelled on numbers, and numbers seemed to be the first things in the whole of nature, they supposed the elements of numbers to be the elements of all things, and the whole heaven to be a musical scale and a number. And all the properties of numbers and scales which they could show to agree with the attributes and parts and the whole arrangement of the heavens, they collected and fitted into their scheme; and if there was a gap anywhere, they readily made additions so as to make their whole theory coherent. E.g. as the number ten is thought to be perfect and to comprise the whole nature of numbers, they say that the bodies which move through the heavens are ten, but as the visible bodies are only nine, to meet this they invent a tenth—the 'counter-earth'. We have discussed these matters more exactly elsewhere [985b23–986a12].

This and other passages, and his further objections to Platonic philosophy (see, for example, 987b23–988a16, 1080b16–35, and 1090a16–1091a23) make it quite clear that Aristotle is vehemently opposed to such a numerical explanation of nature. For example:

> Mathematics has come to be identical with philosophy for modern thinkers [sc. Plato and the Academy], though they say that it should be studied for the sake of other things [992a32–b1].

followed, a few lines later, by the complaint that in such a programme:

> The whole study of nature has been annihilated [992b8–9].

I have described in the previous section how Aristotle's favourite mathematical illustration is the incommensurability of the diagonal, which he cites eight times in the *Metaphysics* alone. It is therefore curious that Aristotle nowhere points to the phenomenon of incommensurability as a fatal flaw in the Pythagorean programme, of which he is so unequivocally critical.

Not only Aristotle, but every witness up to the time of Iamblichus, is silent on the issue of the impact of the discovery; the only possible exception is Plato, at *Laws* VII, 817e–820e, in a passage which deals with topics more basic and more general than incommensurability. Finally, we have no explicit evidence about the details of the Pythagorean technique for handling proportions or ratios which is, today, often said to have been shown to be inadequate by the discovery of incommensurability; and to talk of such a theory is to appeal to a speculative reconstruction. Throughout this book, I have tried to respect the overall texture of this early evidence; for example, since our best and earliest sources do not make the topic of incommensurability a subject of special notice, no more have I. It is a fact of mathematical life, as Aristotle in effect points out repeatedly, that the ratio of two lines may not be expressible as a *ratio* of two *arithmoi*; but there is no difficulty in expressing any ratio of lines *using only the arithmoi* (and of completely expressing the ratio of any two expressible lines in a finite number of steps). Only when mathematics becomes fully arithmetised and deductive does the dichotomy between commensurable and incommensurable ratios pose a serious and subtle problem for the foundations of this arithmetised mathematics. I have given here, in this book, an extended illustration of a kind of non-arithmetised mathematics where the split between the tractable and intractable occurs at a different place, just beyond the ratios of expressible lines, and have argued that we do have abundant evidence, in Plato and Euclid, of explorations of the nature of this dichotomy.

Now consider Aristotle's comment on the proof of incommensurability *per impossibile* via odd and even numbers. To widen the context of the discussion I shall start by describing two further proofs which, I shall propose, might also fit Aristotle's remarks. The first begins in the conventional way, so I will model it on the first demonstration of *Elements* X 117 (in the translation of Knorr, *EEE*, 23):

Let it be proposed to us to prove that in square figures the diagonal is incommensurable in length with side.
Let ABCD be a square, of which AC is the diagonal. I say that AC is incommensurable in length with AB. For if possible, let it be commensurable. I say that it will follow that one and the same number is odd and even. Now it is manifest that the square on AC is the double of that on AB. Since AC is commensurable with AB, then AC will have the ratio to AB of one number to another. Let these numbers be [m and n].... And since AC is to AB as [m] is to [n], so also the square on AC is to that on AB as the square of [m] is to that of [n]. The square on AC is double that on AB, so the square of [m] is double that of [n]. [So the odd number of factors of $2n^2$ is equal to the even number of factors of m^2.]

8.3 The phenomenon of incommensurability

The prolixity of an argument often hides the point of the proof, so here, again, is this first proof. The square number m^2 will have an even number of factors, since each factor is repeated twice; but $2n^2$ will have an odd number of factors, since it has the additional factor of 2. Hence, if $m^2 = 2n^2$, an odd number will be equal to an even number.

The second proof is based on the method of diagonals and sides explained in Section 2.2(a). Consider the example of the pentagon. From Fig. 8.1(a), we see that, starting from a smaller pentagon with side s and diagonal d, we construct a larger pentagon with side $S = s + d$ and diagonal $D = s + 2d$. Now suppose that the ratio of the diagonal to side of the smaller pentagon is commensurable, so we can express $s = mb$ and $d = nb$ for some common measure b. Then $d:s = n:m$ will have one of the following forms: either even:even, or even:odd, or odd:even, or odd:odd. Suppose, for example, that $d:s$ is even:odd. Then, for any other pentagon whatsoever with side s' and diagonal d', s' and d' will be commensurable and if $s' = m'b'$ and $d' = n'b'$, then $d':s' = n':m'$ will be either also even:odd or even:even, because, if it were odd:even or odd:odd, then odd numbers would be equal to even numbers. (For example, if $e_1 : o_1 = o_2 : e_2$, then $e_1 \cdot e_2 = o_1 \cdot o_2$.) But, from Fig. 8.1(b), we see that our assumption that $d:s$ is even:odd implies that $D:S = (s + 2d):(s + d)$ is odd:odd; and so that this case cannot occur. Of the four possible cases for $d:s$, we can eliminate the case of even:even by removing factors of two, in the usual way; but then all of the other three cases lead to contradictions, as is illustrated in Figs. 8.1(b), (c), and (d). So, *per impossibile*, the ratio $d:s$ cannot have been commensurable in the first case.

Let us try the same procedure on the diagonal and side of a square. This time we appeal to the construction of Fig. 8.2(a), described by the prescription $S = s + d$, $D = 2s + d$. But, now, when $d:s$ is even:odd, we again have $D:S$ is even:odd, so no contradiction can be deduced by this method for the diagonal of a square.

If we analyse this kind of argument further, we find that the only kind of relations $S = as + bd$, $D = cs + fd$ that will yield contradictions are of

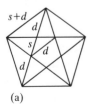
(a)

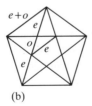
(b)

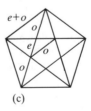
(c)

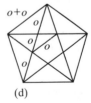
(d)

Fig. 8.1

(a)   (b)   (c)   (d)
$S = s + d$ If $d:s = e:o$ If $d:s = o:e$ If $d:s = o:o$
$D = s + 2d$ then $D:S = o:o$ then $D:S = e:o$ then $D:S = o:e$

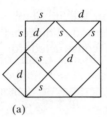 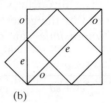

(a)   (b)

Fig. 8.2

$$\begin{array}{ll} \text{(a)} & \text{(b)} \\ S = s + d & \text{If} \quad d:s = e:o \\ D = 2s + d & \text{then} \quad D:S = e:o \end{array}$$

the forms:

$$S = os + od \quad \text{and} \quad S = es + od$$
$$D = os + ed \quad \quad\quad\quad D = os + od.$$

(The argument is summarised in Table 8.1; for example, the case we have just considered, that of the square, is inconclusive because of the exception listed in row 2, column 4, of the second table.) However, the generalised side and diagonal relationships that were explored in Section 3.6 are all of the form $S = as + bd$, $D = nbs + ad$ and so none of these can give rise to a contradiction since, for them, $S = as + \ldots$, $D = \ldots + ad$.

With these two different proofs *per impossibile* in mind, I shall draw my discussion to a close. Our first detailed evidence about Aristotle's proof comes from Alexander of Aphrodisias, and it seems clear from the way he constructs his argument that neither he nor his source had any account of Aristotle's proof before them. It is quite possible that Aristotle's proof did indeed deal with the square, and that a brief description of it had been passed down as something like 'if the diagonal of a square is commensurable with the side then, since the square on the diagonal is twice the square on the side, odd numbers will be equal to even numbers'. But, as my first alternative proof shows, such a summary description is ambiguous, since it could equally well refer either to the numbers themselves, or to the number of their factors; and it is not inconceivable that one proof could be confused with the other during transmission. There is a historical observation to be made about the alternative proof by factorisation, that Euclid does not give a theory of factorisation in the *Elements* and, indeed, the topic of unique factorisation does not become fundamental in number theory before the time of Gauss (see Knorr, PIGNT, for a brief discussion); but this does not, I think, completely rule the alternative proof out of serious consideration as a plausible reconstruction.

Also, it could be significant that Aristotle never says that he is referring

TABLE 8.1. The inconclusive cases in the proof of incommensurability of $d:s$ by the generalised method of diagonal and sides: Proofs *per impossibile* can only be constructed for the cases marked *.

d	s	$es + ed$	$es + od$	$os + ed$	$os + od$
e	o	e	e	o	o
o	e	e	o	e	o
o	o	e	o	o	e

(a) Odd and even sides and diameters:
The parity of $as + bd$ as a & b and s & d are even or odd.

$D = $ \ $S =$	$es + ed$	$es + od$	$os + ed$	$os + od$
$es + ed$	$e:o \rightarrow e:e$ $o:e \rightarrow e:e$ $o:o \rightarrow e:e$	$e:o \rightarrow e:e$	$e:o \rightarrow e:o$ $o:e \rightarrow e:e$	$e:o \rightarrow e:o$ $o:o \rightarrow e:e$
$es + od$	$e:o \rightarrow e:e$ $o:e \rightarrow o:e$	$e:o \rightarrow e:e$ $o:o \rightarrow o:o$	$e:o \rightarrow e:o$ $o:e \rightarrow o:e$ $o:o \rightarrow o:o$	$e:o \rightarrow e:o$
$os + ed$	$o:e \rightarrow e:e$ $o:o \rightarrow o:o$		$o:e \rightarrow e:e$ $o:o \rightarrow o:o$	*
$os + od$	$o:e \rightarrow o:e$ $o:o \rightarrow e:e$	*	$o:e \rightarrow o:e$	$o:o \rightarrow e:e$

(b) The inconclusive cases for $d:s \rightarrow D:S$.

to the diagonal of a square since there is at least one simple proof that does not work for the square, as my second alternative proof illustrates. (Indeed I do not know of any figure other than a pentagon for which this method does work as a proof of incommensurability.) One could then go on to invent explanations for Aristotle's omission, for example that if he had specified the pentagon, then this might have complicated his illustration by introducing Pythagorean overtones which, for some further reason about which we can then speculate further, Aristotle wished to avoid. But our early evidence is so scanty, our late evidence so doubtful, that it is difficult to pin these speculations down to anything more substantial.

Given the tenuous nature of this evidence and the range of its possible interpretations, it seems unwise to base a reconstruction of early Greek mathematics on speculative assumptions about Aristotle's proof of the incommensurability of the diagonal, or the circumstances and effects of the first realisation of the basic fact of incommensurability. And since our earliest and more direct sources do not seem to give any special attention to the discovery and its impact, perhaps it also should play a subordinate role in our reconstruction. There lies behind my alternative reconstruction here the suggestion of excitement over another discovery, that of the power, surprise, and promise of anthyphairetic mathematics, especially the potential in exploring and generalising the periodicity of the anthyphairetic ratios $\sqrt{n}:\sqrt{m}$. But this promise has still to be realised, even today. In the final chapter I shall describe some of the very few further steps that have been carved out in the uphill struggle to anthyphairetic enlightenment.

9
CONTINUED FRACTIONS

> I am sure that, after close examination, you will like the article unless you allow yourself to be put off by the words 'continued fractions', which I used to recoil from too.
>
> (Minkowski to Hilbert, 30 December 1899, translated from Minkowski, *Briefe an David Hilbert,* 118.)

The mathematical techniques developed and deployed so far in the historical reconstructions of this book have not been arithmetised; that is, they have not been expressed in terms of some model, explicit or implicit, of the real numbers and its arithmetic. This chapter will be different. My aim is to prove the main results underlying the reconstruction and to describe some of the more recent investigations into the subject, and I shall give this modern theory in the arithmetised style in which it is now conceived, expressed in the modern algebraic idiom. So the formulae in this chapter, and this chapter only, should be regarded as abstractions of the underlying arithmetic of real numbers.[1]

The mathematico-historical essay that follows makes no pretensions to being more than an introductory sketch to a selection of accessible topics relating to continued fractions and their role in my reconstructions of Greek mathematics. Some parts of the picture have been explored and described in detail by others; in particular, all aspects up to 1731 (the date of Euler's first reference to continued fractions), and the role of continued fractions in number theory up to 1801 (the date of Gauss's *Disquisitiones Arithmeticae*). Beyond that, the subject cries out for adequate historical investigation, but I cannot attempt to fill that need here.[2]

I shall start the historical account at the beginning of the seventeenth

---

[1] The difficulties in describing the real numbers lie not so much in labelling the actual numbers themselves—for some of the techniques described here, like anthyphairetic and astronomical ratios, will provide immediate solutions to this—but, first, in describing how to perform arithmetical operations on these numbers, and, second and much later, in establishing the completeness of the reals. It does not generally seem to be appreciated that Dedekind was the first person to give a satisfactory solution to *both* of these problems; see Section 1.2(e), above, and, for an informal illustration of the problems with arithmetic, my FHYDF.

[2] A history is now in press: Brezinski, *HCF,* which the author describes as containing "more than 2500 references and 1500 quoted mathematicians". I would like to thank Claude Brezinski for giving me a typescript of this book, which has been very useful in several places.

century, and so will omit any discussion of the ascending continued fractions of Fibonacci (see n. 11, below), or the work of Bombelli and Cataldi. My aim is to describe some of the contributions of Euler, Lagrange, and Gauss in the context of the work which more closely preceded and succeeded them.

9.1 The basic theory[3]

9.1(a) *Continued fractions, convergents, and approximation*

Let x_0 be a real number, and set

$$x_0 = n_0 + \varphi_0,$$

where n_0 is the largest integer less than or equal to x_0, written $n_0 = [x_0]$; so φ_0 is the fractional part of x_0, $0 \leq \varphi_0 < 1$. If $\varphi_i \neq 0$ then, similarly, write

$$1/\varphi_0 = x_1 = n_1 + \varphi_1,$$
$$1/\varphi_1 = x_2 = n_2 + \varphi_2,$$
$$\text{etc.};$$

but if some $\varphi_K = 0$, the process will terminate with

$$1/\varphi_{K-1} = x_K = n_K.$$

Now eliminate the φ_is:

$$x_0 = n_0 + \frac{1}{x_1} = n_0 + \cfrac{1}{n_1 + \cfrac{1}{x_2}} = n_0 + \cfrac{1}{n_1 + \cfrac{1}{n_2 + \cfrac{1}{x_3}}} = \text{etc.},$$

to get the continued fraction expansion for x. For typographical convenience, this will hereafter be set out as follows:

$$x_0 = n_0 + \frac{1}{x_1} = n_0 + \frac{1}{n_1 +} \frac{1}{x_2} = n_0 + \frac{1}{n_1 +} \frac{1}{n_2 +} \frac{1}{x_3} = \text{etc.}$$

Note that, as thus described, n_0 is an integer, n_i $(i \geq 1)$ is a positive

---

[3] For a lucid introductory account of the mathematical theory of continued fractions, see Davenport, *HA*; more details can then be sought in the standard references such as Chrystal, *TA*; Hardy & Wright, *ITN*; and Perron, *LK*. However great may be the general neglect of the history of continued fractions, most of the developments of this section have received the attention they deserve. My brief historical notes can be supplemented by, for instance, Dickson, *HTN* ii, Chapters 2 and 7; Diophantus–Heath, *DA*, 277–92; Whiteside, *PMTLSC* (see especially pp. 195–6, 207–13, and 241); and for a magisterial treatment that sets the material on number theory in the widest mathematical and historical context, Weil, *NT* (see the index, s.vv. 'continued fractions', '"Pell's equation"', and 'Ricatti').

integer, and, if the continued fraction terminates with n_K, then $n_K \geq 2$; these conditions describe a simple continued fraction in its standard form. We shall consider these simple continued fractions, together with the similar expressions in which the terms n_i are regarded as indeterminates ('variables') that might take real (i.e. non-integral, as with the x_k above, and even negative) values, and will only make occasional reference to general continued fractions:

$$n_0 + \cfrac{m_1}{n_1 + \cfrac{m_2}{n_2 + \cdots}},$$

which will be set out as $n_0 + \cfrac{m_1}{n_1 +} \cfrac{m_2}{n_2 +} \cdots$.

We immediately relate the continued fraction algorithm to the Euclidean algorithm (sic^4) applied to a and b. Suppose

$$a = n_0 b + c \quad \text{where} \quad c < b,$$
$$b = n_1 c + d \quad \text{where} \quad d < c,$$
$$c = n_2 d + e \quad \text{where} \quad e < d,$$
etc.,

where the process terminates if any remainder is zero. Then

$$a/b = n_0 + c/b, \quad \text{i.e.} \quad a/b = n_0 + \varphi_0,$$
$$b/c = n_1 + d/c, \quad \text{i.e.} \quad 1/\varphi_0 = n_1 + \varphi_1,$$
$$c/d = n_2 + e/d, \quad \text{i.e.} \quad 1/\varphi_1 = n_2 + \varphi_2,$$
etc.,

and the same integers $n_0, n_1, n_2, \ldots$ now describe either the pattern of the Euclidean algorithm applied to a and b or the continued fraction expansion of the real number a/b.[5] So the expressions $n_0 + \cfrac{1}{n_1 +} \cfrac{1}{n_2 +} \cdots$ and $[n_0, n_1, n_2, \ldots]$ now refer to the same thing. I shall, very shortly, say precisely what the three dots mean.

PROPOSITION (see *Elements* X 2 & 3) The continued fraction of a real number x_0 terminates if and only if x_0 is rational.

[4] This chapter is written in the modern idiom! See the note to Section 2.3.
[5] In arithmetised geometry these letters $a, b, c, \ldots$ might, for example, denote ambiguously either lines or the lengths of these lines, and a/b always denotes the quotient of their lengths. There has been a tendency since the seventeenth century to regard $a:b$ and a/b as synonymous; but a/b has no conventional meaning in non-arithmetised geometry, such as has been described in the historical reconstructions in this book.

PROOF Since $x_k = n_k + 1/x_{k+1}$, x_{k+1} will be rational if and only if x_k is rational so, in particular, the algorithm can terminate only if x_0 is rational. Now suppose x_0 is rational, $x_0 = a/b$, where a and b are integers, b positive. Since the remainders of the Euclidean algorithm, applied to a and b, decrease in size (see B_{24} in Section 1.3, above), the algorithm must terminate. QED

Apart from the ambiguity in the last term of a terminating continued fraction, that

$$[n_0, n_1, \ldots, n_K] = [n_0, n_1, \ldots, n_K - 1, 1],$$

the expression as a simple continued fraction is unique.[6] This follows because, if

$$n_0 + \frac{1}{n_1+} \frac{1}{n_2+} \ldots = m_0 + \frac{1}{m_1+} \frac{1}{m_2+} \ldots,$$

then, since the expression $\dfrac{1}{n_1+} \dfrac{1}{n_2+} \ldots$ will always be less than 1 and both n_0 and m_0 are integers, we must have $n_0 = m_0$. Subtracting this term and reciprocating, we get

$$n_1 + \frac{1}{n_2+} \frac{1}{n_3+} \ldots = m_1 + \frac{1}{m_2+} \frac{1}{m_3+} \ldots,$$

and the argument repeats itself. To complete this proof we must show, in the standard style of arithmetised mathematics, that any such non-terminating expression $[0, n_1, n_2, \ldots]$ converges to some real number less than one; this result will be established below.

The convergents (see Sections 2.3(b) and (c)) of the continued fraction $[n_0, n_1, n_2, \ldots]$ are got by truncating and then evaluating the resulting expression as a common fraction in its lowest terms:

$$[n_0, n_1, \ldots, n_k] = p_k/q_k.$$

In fact, any straightforward evaluation of a terminating simple continued fraction will give a common fraction in its lowest terms, or, when the n_i are regarded as indeterminates, a quotient of two polynomials in the n_i with no common factor. This more general point of view, that the n_i are regarded as indeterminates, will often be taken in the following propositions.

---

[6] It is easily seen that this result is false if the n_i can take zero or negative values. For example $[\ldots, x, 0, y, \ldots] = [\ldots, x+y, \ldots]$; $[\ldots, x, -y] = [\ldots, (x-1), 1, (y-1)]$; $[\ldots, x, -y, -z] = [\ldots, x-1, 1, y-2, z-1]$; etc.

9.1 Basic theory

PROPOSITION (see Section 2.3(c) and Table 2.4) For $k \geq 0$

$$p_k = n_k p_{k-1} + p_{k-2}$$
$$q_k = n_k q_{k-1} + q_{k-2},$$

where $p_{-1} = q_{-2} = 1$, $p_{-2} = q_{-1} = 0$.

PROOF I shall adopt the approach of Euler (see, for example, UNAPPS, §19 ff.) and evaluate the p_k and q_k as polynomials in the indeterminates $n_0, n_1, \ldots, n_k$. To emphasise this point of view, write

$$[n_0, n_1, \ldots, n_k] = \frac{P_k(n_0, n_1, \ldots, n_k)}{Q_k(n_0, n_1, \ldots, n_k)},$$

where $P_k(n_0, \ldots, n_k)$ and $Q_k(n_0, \ldots, n_k)$ are the polynomials. For example, we immediately evaluate that

$$\frac{P_0(n_0)}{Q_0(n_0)} = \frac{n_0}{1}, \quad \frac{P_1(n_0, n_1)}{Q_1(n_0, n_1)} = \frac{n_0 n_1 + 1}{n_1},$$

and $\dfrac{P_2(n_0, n_1, n_2)}{Q_2(n_0, n_1, n_2)} = \dfrac{n_0 n_1 n_2 + n_0 + n_2}{n_1 n_2 + 1}.$

The result that

$$Q_k(n_0, n_1, \ldots, n_k) = P_{k-1}(n_1, n_2, \ldots, n_k)$$

now suggests itself; it can be verified by observing that

$$\frac{P_k(n_0, \ldots, n_k)}{Q_k(n_0, \ldots, n_k)} = n_0 + \left(\frac{1}{n_1 +} \cdots \frac{1}{n_k}\right) = n_0 + \frac{Q_{k-1}(n_1, \ldots, n_k)}{P_{k-1}(n_1, \ldots, n_k)}$$
$$= \frac{n_0 P_{k-1}(n_1, \ldots, n_k) + Q_{k-1}(n_1, \ldots, n_k)}{P_{k-1}(n_1, \ldots, n_k)}.$$

(We suppose here and will shortly prove that the P_k and Q_k, as defined by the proposition, do not have any common factor.) We also deduce that

$$P_k(n_0, \ldots, n_k) = n_0 P_{k-1}(n_1, \ldots, n_k) + P_{k-2}(n_2, \ldots, n_k),$$

with $P_{-1} = 1$, $P_{-2} = 0$; and a similar result for the Q_ks with $Q_{-1} = 0$, $Q_{-2} = 1$. This is very close to what we want to prove.

We now deduce that the polynomial $P_k(n_0, \ldots, n_k)$ is got by adding together the following terms: the product $n_0 n_1 \ldots n_k$; every product that arises by omitting any pair of consecutive terms; every product that arises by omitting any two separate pairs of consecutive terms n_i and n_{i+1}, n_j and n_{j+1}; products omitting three separate pairs, etc., where, if $k+1$ is even, the empty product that arises by omitting all the terms has the conventional value 1. Since $P_0(n_0)$ and $P_1(n_0, n_1)$ are of this required form, and P_k satisfies the recurrence relation just established, we get an immediate verification of this description of P_k.

From this it follows that

$$P_k(n_0, n_1, \ldots, n_k) = P_k(n_k, n_{k-1}, \ldots, n_0)$$

and hence we can manipulate the recurrence relation into the form

$$P_k(n_0, n_1, \ldots, n_k) = n_k P_{k-1}(n_0, \ldots, n_{k-1}) + P_{k-2}(n_0, \ldots, n_{k-2}),$$

as required. QED

REMARK The P_k are sometimes called Euler or Gauss brackets. They were introduced by Euler, who used the notation $(n_0, n_1, \ldots, n_k)$ for them; Gauss wrote them as $[n_0, n_1, \ldots, n_k]$ (see Section 2(d), below). They satisfy many identities, first explored by Euler. For example the basic generating relation:

$$P_k(n_0, \ldots, n_k) = n_k P_{k-1}(n_0, \ldots, n_{k-1}) + P_{k-2}(n_0, \ldots, n_{k-2})$$

can be generalised to

$$P_k(n_0, \ldots, n_k) = P_l(n_0, \ldots, n_l) P_{k-l}(n_{l+1}, \ldots, n_k)$$
$$+ P_{l-1}(n_0, \ldots, n_{l-1}) P_{k-l-1}(n_{l+2}, \ldots, n_k).$$

Later the corresponding polynomials for a general continued fraction

$$n_0 + \frac{m_1}{n_1 +} \frac{m_2}{n_2 +} \ldots$$

were seen by Sylvester (1853), Muir (1874), and others to be expressible as special kinds of determinants, called continuants, from which many further identities can be generated. For a historical account see Muir, *TD*.

COROLLARY For $k \geq 2$, $p_{k+1} > p_k$, $q_{k+1} > q_k$.

COROLLARY For $k \geq 0$,

$$x_0 = \frac{x_{k+1} p_k + p_{k-1}}{x_{k+1} q_k + q_{k-1}}.$$

PROOF Apply the proposition to $[n_0, n_1, \ldots, n_k, x_{k+1}] = x_0$. QED

COROLLARY (see Section 2.4(f)) For $k \geq 0$, $p_{k+1} q_k - p_k q_{k+1} = (-1)^k$. Hence p_k and q_k, as calculated by the recurrence relation of the proposition, are coprime, either as numbers or as polynomials.

PROOF

$$p_{k+1} q_k - p_k q_{k+1} = (n_{k+1} p_k + p_{k-1}) q_k - p_k (n_{k+1} q_k + q_{k-1})$$
$$= -(p_k q_{k-1} - p_{k-1} q_k)$$
$$= (-1)^k (p_1 q_0 - p_0 q_1) = (-1)^k.$$

9.1 Basic theory

Hence any common factor of p_k and q_k must also divide 1; so p_k and q_k are coprime. QED

COROLLARY (see S_{21}–B_{24} in Section 1.3)

$$\frac{p_0}{q_0} < \frac{p_2}{q_2} < \frac{p_4}{q_4} < \ldots < x_0 < \ldots < \frac{p_5}{q_5} < \frac{p_3}{q_3} < \frac{p_1}{q_1}$$

where if x_0 is rational and so its continued fraction terminates, then x_0 is equal to the last convergent, $x_0 = [n_0, n_1, \ldots, n_K] = p_K/q_K$; while if x_0 is irrational, then p_k/q_k converges to x_0 as k increases.

PROOF

$$\frac{p_{k+1}}{q_{k+1}} - \frac{p_k}{q_k} = \frac{p_{k+1}q_k - p_k q_{k+1}}{q_k q_{k+1}} = \frac{(-1)^k}{q_k q_{k+1}},$$

and so this difference is positive if k is even, negative if k is odd. Moreover, the numbers q_k increase as k increases, so the difference between consecutive terms will decrease as k increases. Hence we get the chain of inequalities.

If x_0 is rational, its continued fraction will terminate and, by the definition of the convergents, it will be equal to its last convergent. If x_0 is irrational, the q_k will increase without limit, so the difference between consecutive terms, which bracket x_0, will tend to zero. Hence the convergents p_k/q_k will, indeed, converge to x_0 as k increases. QED

REMARK This corollary describes the usual meaning in arithmetised mathematics of the three dots in expressions like

$$x_0 = [n_0, n_1, n_2, \ldots],$$

that either the sequence terminates, in which case it is an arithmetical (or algebraic) identity, or it does not terminate and the numbers $[n_0]$, $[n_0, n_1]$, $[n_0, n_1, n_2]$, ... tend arithmetically to x_0. The chain of inequalities implies that $n_0 \leq x \leq n_0 + 1$, where the left-hand equality can only hold if $x = [n_0]$, the right-hand if $x = [n_0, 1]$.

COROLLARY (see Section 2.2(b)(vii)) If $n_0, n_1, n_2, \ldots$ is any non-terminating sequence of integers with $n_i \geq 1$ for $i \geq 1$, then the continued fraction $[n_0, n_1, n_2, \ldots]$ converges to some irrational number x_0.

PROOF The convergents satisfy the inequalities of the previous corollaries; so p_0/q_0, p_2/q_2, $p_4/q_4, \ldots$, being a bounded increasing sequence, converges to a limit; also p_1/q_1, p_3/q_3, $p_5/q_5, \ldots$, being a bounded decreasing sequence, also converges to a limit; and these two limits must be equal since we can make $p_k/q_k - p_{k+1}/q_{k+1}$ as small as we please. Finally, since the continued fraction expansion is unique, we must have $x_0 = [n_0, n_1, n_2, \ldots]$ and so, since this does not terminate, x_0 is irrational. QED

In fact, consecutive convergents also improve in accuracy, and we can give close estimates of their error and deduce that, in some very precise sense, they give the best approximations to x_0 by rational numbers; see Sections 2.2(b)(iii) and (iv) above. For, since

$$x_0 = \frac{x_{k+1} p_k + p_{k-1}}{x_{k+1} q_k + q_{k-1}},$$

we can either rewrite this to give

$$x_0 - \frac{p_k}{q_k} = \frac{q_{k-1}}{x_{k+1} q_k} \left(\frac{p_{k-1}}{q_{k-1}} - x_0 \right)$$

whence, since $0 < q_{k-1} < x_{k+1} q_k$, we get:

COROLLARY When $k \geq 1$,

$$\left| x_0 - \frac{p_k}{q_k} \right| < \left| x_0 - \frac{p_{k-1}}{q_{k-1}} \right|.$$

Or, alternatively, we can evaluate

$$x_0 - \frac{p_k}{q_k} = \frac{p_{k-1} q_k - p_k q_{k-1}}{q_k (x_{k+1} q_k + q_{k-1})} = \pm \frac{1}{q_k (x_{k+1} q_k + q_{k-1})},$$

and hence, since $n_{k+1} \leq x_{k+1} < n_{k+1} + 1$ and $q_{k-1} < q_k$ for $k \geq 2$, we have the estimates of Sections 2.2(b)(iii) and (iv):

COROLLARY If $k \geq 2$,

$$\frac{1}{(n_{k+1} + 2) q_k^2} < \left| x_0 - \frac{p_k}{q_k} \right| < \frac{1}{q_k q_{k+1}} < \frac{1}{n_{k+1} q_k^2}.$$

Or we can manipulate the formulae to prove that the convergents provide the best approximations:

COROLLARY If $\left| x_0 - \frac{a}{b} \right| < \left| x_0 - \frac{p_k}{q_k} \right|$, where $\frac{p_k}{q_k}$ is a convergent of x_0, then $b > q_k$.

PROOF Suppose, first, that a/b and p_k/q_k lie on the same side of x_0, for example that $p_k/q_k < a/b < x_0 < p_{k+1}/q_{k+1}$. Then

$$0 < \frac{a}{b} - \frac{p_k}{q_k} < \frac{p_{k+1}}{q_{k+1}} - \frac{p_k}{q_k} = \frac{1}{q_k q_{k+1}}.$$

Hence

$$0 < a q_k - n p_k < \frac{b}{q_{k+1}},$$

and since $a q_k - b p_k$ is an integer, $b > q_{k+1}$.

9.1 Basic theory

Now suppose that they lie on opposite sides, for example that $p_k/q_k < x_0 < a/b < p_{k-1}/q_{k-1}$, where the final inequality follows from the previous corollary but one. Then by what we have just proved, $b > q_k$.

QED

HISTORICAL NOTES The convergents to a simple continued fraction, as such, seem to have first been defined in the West since the Renaissance by Daniel Schwenter in his *Geometriae Practicae* of 1618 and *Deliciae Physico-Mathematicae* of 1636, and for a general continued fraction by Wallis, in his *Arithmetica Infinitorum* (1658)—where the expression 'continued fraction' first appeared—though no general algebraic formulae are given before Euler. Approximation methods which turn out to be closely related to the continued fraction process are rather older; see the next section for an early example.

Wallis devoted Chapters 10 and 11 of his *Treatise of Algebra* of 1685 to the problem of the "Reduction of Fractions or Proportions to smaller Terms, as near as may be to the just Value", that is to approximating a given fraction or decimal number by a fraction whose denominator should not be greater than some given number, say 999. The problem, he said, was sent to him in 1663 or 1664; it is due to a certain Dr Davenant and it arose in connection with Metius's approximation of π by $\frac{335}{113}$. Wallis's procedure is tantamount to the computation of the terms, convergents, and intermediate convergents (see the next section) of the continued fraction expansion, but this is concealed by the cumbersome procedure he adopted, which proceeded from observations about and manipulations of the decimal expansion of the given common fraction. Then, in Chapter 11, he demonstrated his procedure on van Ceulen's thirty-five-decimal-place value of π, augmented by a final approximate place:

$$3 \cdot 14159\ 26535\ 89793\ 23846\ 26433\ 83279\ 50288\tfrac{1}{2} \quad [sic]$$

from which he deduced, in effect, that

$\pi = [3, 7, 15, 1, 292, 1, 1, 1, 2, 1, 3, 1, 14, 2, 1, 1, 2, 2, 2, 2, 1,$
$\quad 84, 2, 1, 1, 15, 3, 13, 1, 4, 2, 6, 6, 1 \text{ (should be 99)}, \ldots],$

again with a tabulation of all (!) convergents and intermediate convergents. He did not say why he stopped at n_{33} but, as he explained, he has suppressed much of the numerical calculation, though what remains still fills eight dense pages; we can presume that he noted that the fractional parts of the numbers $q_k \pi$, around which his calculation revolves, had become unreliably close to an integer. Be that as it may, his intuition is almost correct: the significance of the calculation breaks down only with the last term.[7]

---

[7] This error seems to have been noticed first in Lehmer, EALN.

Approximation using the continued fraction process is handled fluently, with descriptions of proofs, by Huygens, in his *Descriptio Automati Planetarii* (written between 1680 and 1687 but only published posthumously in 1703), in connection with the problem of selecting gear ratios for a planetarium.[8] The procedure is then formalised and developed by Euler and Lagrange; the latter called it "une des principales découvertes de ce grand géomètre [sc. Huygens]" in §7 of his *Additions* to Euler's *Algebra*; see Weil, *NT*, 120, for further references and a brief comment.

As illustrations of the understanding and use of continued fractions just before the introduction of Euler's formalism, here are two extracts from Roger Cotes. First, the discovery of the surprising behaviour of e, published first in his article 'Logometrica' and repeated in his posthumous *Harmonia Mensurarum* (1722):

The major term 2·71828 &c. [$= e$] should be divided by the minor 1 ... and once more the minor by the number which is left, and this again by the last remainder, and so continue forward; and the quotients 2, 1, 2, 1, 1, 4, 1, 1, 6, 1, 1, 8, 1, 1, 10, 1, 1, 12, 1, 1, 14, 1, 1, 16, 1, 1 &c. will be produced. Having made these calculations, we must set out two columns of ratios, of which one contains the terms which have a ratio greater than the true one (*rationes vera majores*), and the other contains the terms which have a ratio less than the true one (*rationes vera minores*), beginning the computation with the ratios 1 to 0 and 0 to 1, which are most remote from the true one, and having started there, continuing to deduce the remaining ratios, which approach ever closer to the true one. ... [There follows here a long verbal description of the following table:]

Rationes Vera Majores		*Rationes Vera Minores*	
		0	1
1	0×2	2	0
2	1	2	1×1
3	1×2	6	2
8	3	8	3×1
11	4×1	11	4
76	28	19	7×4
87	32×1	87	32
106	39	106	39×1
193	71×6		
...			

The application of these approximations is widespread, wherefore I have given a somewhat prolix exposition of their invention, by the method which seems to me simplest and easiest. The celebrated men Wallis and Huygens have dealt with the same arguments, slightly differently [*Harmonia Mensurarum*, 7–9].[9]

---

[8] The technique can now be found in any good advanced manual of engineering workshop practice.
[9] I am grateful to Ronald Gowing for sending me material on Cotes; the translation is by him. I have slightly rearranged the order and presentation of Cotes's text: the table precedes its description in the original, and I have introduced a one-line displacement between the

Next, here is an extract from Cotes's correspondence with Newton over the second edition of the *Principia*:[10]

I find that 25 & 21 express the proportion of $\sqrt{\sqrt{2}}$ to 1 as nearly as it is possible for so small numbers to do it, whence it is probable y^t the exact proportion of the diameter of the Hole to y^e diameter of y^e Stream is that of $\sqrt{\sqrt{2}}$ to 1, & then y^e proportion of 44 to 37 will be much nearer the truth than y^t of 25 to 21.

In fact $\sqrt{\sqrt{2}} = [1, 5, 3, 1, 1, 40, \ldots]$, with convergents $\frac{6}{5}, \frac{19}{16}, \frac{25}{21}$, and $\frac{44}{37}$ and, because of the next large term of 40, $\frac{44}{37}$ will be a very good estimate: $\frac{1}{42 \times 37^2} < (\sqrt{\sqrt{2}} - \frac{44}{37}) < \frac{1}{40 \times 37^2} < 1 \cdot 8 \times 10^{-5}$.

Finally, as a postscript to this sketch of the early history of continued fractions and approximations, I note that, rather later, Lagrange wrote a modest little article, EANTF (1799), in which he described different approaches to a similar problem, that of approximating a given common fraction by sums and differences of other particular kinds of fractions:

$$\frac{p}{q} = \frac{m_0}{n_0} \pm \frac{m_1}{n_1} \pm \frac{m_2}{n_2} \pm \ldots,$$

under various different conditions on the ms and ns. Within this general framework, he described expressions like

$$\frac{p}{q} = m_0 \pm \frac{m_1}{n} \pm \frac{m_2}{n^2} \pm \frac{m_3}{n^3} \pm \ldots,$$

which is a generalisation of the sexagesimal or decimal expansion to base n with positive or negative terms; or its more general form

$$\frac{p}{q} = m_0 \pm \frac{m_1}{n_1} \pm \frac{m_2}{n_1 n_2} \pm \frac{m_3}{n_1 n_2 n_3} \pm \ldots,$$

an expansion to changing bases, as with Imperial measures;[11] then

$$\frac{p}{q} = m_0 \pm \frac{1}{n_1} \pm \frac{1}{n_1 n_2} \pm \frac{1}{n_1 n_2 n_3} \pm \ldots,$$

---

two columns of the table to clarify his procedure. Cotes also gives a similar description, with table, in which the intermediate convergents are calculated. Incidentally, I do not believe that Cotes continued his numerical calculation up to the terms "... 16, 1, 1, &c"; this is surely an 'induction', seventeenth-century style, because the sequence given exactly fills one line of text in the original.

[10] Quoted from Westfall, *NR*, 711, where an explanation of the context can be found. The letter is dated 31 March 1711.

[11] These kinds of expansions are, in fact, expressions as ascending continued fractions,

$$m_0 \pm \frac{m_1 \pm \frac{m_2 \pm}{n_2} \ldots}{n_1}$$

They were considered by Fibonacci and other Italian mathematicians up to the sixteenth century.

an expression in sums and differences of unit fractions (precisely the procedure described at B_{114} in Section 7.4(d), above), one example of which he gave being

$$\pi = 3 + \frac{1}{7} - \frac{1}{7 \times 113} - \frac{1}{7 \times 113 \times 4739} + \frac{1}{7 \times 113 \times 4739 \times 47051} + \cdots$$

(contrast note 97 to Chapter 7, above). He then gave a treatment of continued fractions approached via the problem of constructing approximations, similar in spirit but different in its details from the algorithms of Section 2.3. The article is not of great importance mathematically and seems generally to have been forgotten, but parts of it echo and develop themes that have been raised here, particularly in some of my slaveboy's remarks and questions.

9.1(b) *The Parmenides proposition and algorithm*[12]

The historical reconstructions in this book have not been based on the continued fraction algorithms that I have so far been describing,[13] but on the generation of approximations by the *Parmenides* proposition; see Section 2.2, above, where the use of the *Parmenides* proposition to compute approximations and continued fraction expansions is explained. Let us now investigate these propositions and algorithms in more detail. We suppose, for simplicity, that x_0 is positive.[14]

THE *PARMENIDES* PROPOSITION   If $\dfrac{p}{q} < \dfrac{r}{s}$ then $\dfrac{p}{q} < \dfrac{p+r}{q+s} < \dfrac{r}{s}$.

PROOF $\dfrac{p+r}{q+s} - \dfrac{p}{q} = \dfrac{qr-ps}{q(q+s)} > 0$ since $\dfrac{p}{r} < \dfrac{q}{s}$.   QED

THE *PARMENIDES* ALGORITHM   If p/q is an underestimate and r/s is an overestimate for x_0, then $(p+r)/(q+s)$ will be a better approximation than that original estimate which lies on the same side of x_0. Start the algorithm with the formal universal estimates $0/1 < x_0 < 1/0$.

---

[12] This section is inspired by Fletcher, AV, and owes a lot to the meticulous account in Stark, *INT*, Chapter 7. Readers should consult these for more detailed explanations and illustrations.

[13] This, it seems to me, is a fatal flaw in all the other detailed anthyphairetic reconstructions of early Greek mathematics known to me. I repeat, yet again, a theme that runs throughout this book: we have no evidence to indicate that early Greek mathematics was arithmetised, either in conception or in underlying motivation, and we know much which suggests that it was not. Therefore the description so far in this chapter is simply irrelevant to early Greek mathematics. Of course, the later Greek mathematics, for example of Heron, Ptolemy, and Diophantus, is clearly deeply influenced by Babylonian arithmetised techniques, but anthyphairesis does not seem to play any practical or theoretical part during this later period. See Section 9.4, below.

[14] When x_0 is negative, a similar description can be given for the fourth quadrant $X \leq 0$, $Y \geq 0$. Or we can work with $-x_0$.

9.1 Basic theory

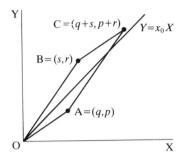

Fig. 9.1. The *Parmenides* algorithm.

I shall develop the properties of the *Parmenides* algorithm using coordinate geometry and properties of the integer lattice, the set of all points having integer coordinates, a description first set out in Smith, NCF. Take the line $Y = x_0 X$ in the positive quadrant of the (X, Y)-plane (see Fig. 9.1); then our aim is to find lines joining the origin to points of the integer lattice whose slopes will closely approximate the slope x_0 of this given line.

If OA, where $A = (q, p)$, is the underestimate in the algorithm, and OB, where $B = (s, r)$, is the overestimate, then the *Parmenides* proposition states that the slope of the diagonal OC, the vector sum of OA and OB, lies between the slopes of OA and OB. In the case illustrated in Fig. 9.1, OC will then be a better estimate than OB. So, at the next step of the algorithm, we repeat the operation with the points A and C. Figure 9.2 illustrates the process applied to $x_0 = \sqrt{3}$, and the annexed table lists the successive approximations; it is identical in content to Table 2.1, save for the mild inconvenience that the slope p/q corresponds to the point (q, p) but the ratio $p:q$. Note also that we distinguish p/q, $2p/2q$, $3p/3q$, etc., since they correspond to different approximating points in the integer lattice; for this reason, I shall generally refer to (q, p) rather than the fraction p/q. Also observe that if (q, p) and (s, r) are the current estimates, then $q \leq s$ and $p \leq r$, or $q \geq s$ and $p \geq r$, where both inequalities are strict after the initial estimates have been superseded, so thereafter any estimate will lie above and to the right of any earlier estimate. I always suppose that this configuration holds.

If A and B are the current approximations, and C the new approximation (as in Fig. 9.1), then call OAB the approximating triangle and OABC the approximating parallelogram. I shall base the analysis of the *Parmenides* algorithm on:

PICK'S THEOREM The area of a simple polygon whose vertices are lattice points is equal to half of the number of points on the perimeter, plus the number of points inside, minus one.

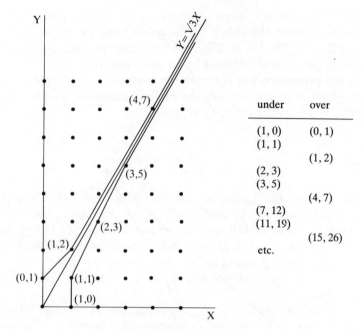

Fig. 9.2. Approximations to $\sqrt{3}$.

A satisfactory discussion and proof of this result would lead us too far from our main interest; for further details see, for example, Coxeter, *IG*, 209–10 or Varberg, *PTR*.

PROPOSITION The approximating triangle and parallelogram contain no points of the integer lattice.

PROOF Each step of the algorithm replaces the approximating triangle OAB by OAC or OBC, and each of these triangles is half of the approximating parallelogram OABC; so the area of the approximating triangle is unchanged. Since it starts with $A = (1, 0)$ and $B = (0, 1)$, its area will always be $\frac{1}{2}$. Since its three vertices are lattice points, Pick's theorem tells us that it can have no lattice points inside. QED

Characteristic sequences of steps are illustrated in Fig. 9.3. Consider Fig. 9.3(a): the estimates $A = (q, p)$ and $B = (s, r)$ give rise to a run of n underestimates $A, A', A'', \ldots, A^{(n-1)}$, followed by a new overestimate $B_1 = (q + ns, p + nr)$. Each of the triangles $OAA', OA'A'', \ldots, OA^{(n-1)}B_1$, being half of its corresponding approximating parallelogram, will contain or meet no lattice point other than its three vertices. Hence neither the triangle OAB_1 nor OB_1nB—which has

9.1 Basic theory

the same area and the same number of perimeter lattice points—will contain or meet any other lattice points than O, A, A', A",..., $A^{(n)}$, B_1, nB, ..., 3B, 2B, B. Figure 9.3(b) illustrates the similar case of a run of n overestimates followed by an underestimate.

Hence the procedure will generate the configuration of Tables 2.2 and 2.3 from which, if we extract only the run-end estimates, we get the more

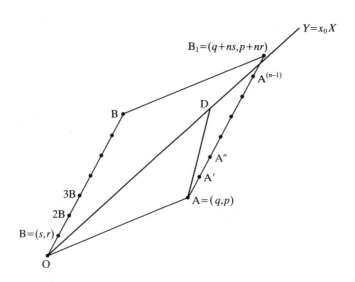

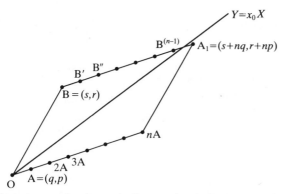

Fig. 9.3. Typical runs of estimates in *Parmenides* algorithm. (The thickness and areas of the parallelograms in all figures except Fig. 9.2 have been greatly exaggerated and distorted to make the drawings possible. The axes have been omitted since they will also have been distorted in direction and scale.)

(a) A run of n underestimates $A, A', \ldots, A^{(n-1)}$, followed by a new overestimate $B_1 = (q + ns, p + nr)$.
(b) A run of n overestimates $B, B', \ldots, B^{(n-1)}$, followed by a new overestimate $A_1 = (nq + s, np + r)$.

schematic configuration of Table 2.4; but this describes precisely the calculation of the convergents of our basic continued fraction algorithm. Hence the two procedures are equivalent. Thus the estimates that occur at the end of the runs of the *Parmenides* algorithm are the convergents of the continued fraction expansion of $x_0 = [n_0, n_1, n_2, \ldots]$, while the other estimates are the so-called intermediate convergents, defined by

$$\frac{p_k^{(m)}}{q_k^{(m)}} = [n_0, n_1, \ldots, n_{k+1}, m] \quad \text{where} \quad 1 \leq m < n_k.$$

The way the current and run-end estimates alternate around the line $Y = x_0 X$ illustrates clearly the way the intermediate convergents and convergents alternate around x_0.

The approximating parallelograms in a correctly drawn figure quickly become very, very long and thin; see, for example, Fig. 9.2. To make drawings possible, gross distortions have had to be introduced into almost all of the figures here. However the proofs ultimately depend on the careful use of words like 'above', 'below', 'inside', and 'outside' in contexts that are not affected by these distortions. This point is examined carefully in Stark, *INT*.

So far we have been considering angular approximations to the line $Y = x_0 X$, i.e. approximations of the form $|x_0 - p/q|$. We now show that the continued fraction process actually locates those points of the integer lattice which lie closest to the line $Y = x_0 X$, although it is in practice more convenient to work with the vertical distance from the point (q, p) to $Y = x_0 X$, given by $|x_0 q - p|$ (see, for example, the line AD in Fig. 9.3(a)); this is the shortest distance multiplied by a constant, $\sec x_0$. Clearly points which are close to the line $Y = x_0 X$ will generate good angular approximations, though not necessarily conversely: consider (nq, np) for large n.

PROPOSITION If $A = (q, p)$ and $B = (s, r)$ are the current under- and overestimates to x_0, with $q < s$ (so A is a convergent) then

$$|x_0 q - p| < |x_0 u - t|$$

for every $F = (u, t)$ different from A and B with $t \leq s$. If, further, B is a convergent, then

$$|x_0 s - r| < |x_0 q - p|.$$

PROOF We suppose that A lies below $Y = x_0 X$ and B lies above. Since $q < s$, A must be a run-end estimate (i.e. a convergent), while B may or may not be a run-end estimate (i.e. it is either an intermediate convergent or a convergent). We therefore have a situation analogous to that shown in Fig. 9.3(b) in which any one of the $B, B', \ldots, B^{(n-1)}$ can now play the role of B. This figure is incorporated in Fig. 9.4, in which

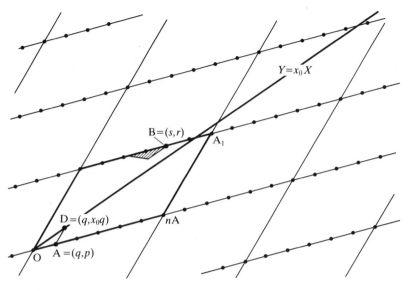

Fig. 9.4

$B^{(iv)}$ is chosen to typify the point B. We have shown, in Fig. 9.3(b), that the only lattice points in or on the parallelogram O_nAA_1B are the marked points O, A, 2A, ..., nA, A_1, $B^{(n-1)}$. If, as in Fig. 9.4, we now tile the plane with translates of the parallelogram, the points of the integer lattice will lie, similarly, on the sides of the translated parallelogram. (We see clearly here an effect of the distortion in this figure. The axes have been omitted from these figures since they, too, will be distorted in direction and scale.)

Suppose first that the point $F = (u, t)$ lies below $Y = x_0X$. Then, since $t \leq s$, it must lie on or below the line OA and before the line $X = s$; and, by hypothesis, it is not equal to A. Since the slope of $Y = x_0X$ is greater than the slope of OA, F must therefore be further away than A from $Y = x_0X$.

Next suppose that F lies above $Y = x_0X$, in which case it must lie on or above the line containing B, and it is not the point B itself. Now the lattice points before B are further from $Y = x_0X$ than is A, because $Y = x_0X$ cannot pass through the small shaded triangle in Fig. 9.4, congruent to OAD, where $D = (q, x_0q)$, the point vertically above A. Hence F must again lie further from $Y = x_0X$ than does A, and again we have

$$|x_0q - p| < |x_0u - t|.$$

Finally, if B is a run-end estimate, then line $Y = x_0X$ will pass through the corresponding small triangle congruent to OAD with vertex at A_1.

Hence
$$|x_0 s - r| < |x_0 q - p|.\qquad\text{QED}$$

PROPOSITION (see Section 2.2(b)(ii)) If $A = (q, p)$ and $B = (s, r)$ are successive convergents to x_0 then
$$\left|x_0 - \frac{p}{q}\right| < \frac{1}{q^2} \quad\text{and}\quad \left|x_0 - \frac{r}{s}\right| < \frac{1}{s^2},$$
and at least one of them satisfies
$$\left|x_0 - \frac{p}{q}\right| < \frac{1}{2q^2} \quad\text{or}\quad \left|x_0 - \frac{r}{s}\right| < \frac{1}{2s^2}.$$

PROOF   See Fig. 9.5, in which we suppose that A is an underestimate, B is an overestimate, and $q < s$. Then the error in each estimate is less than
$$\left|\frac{r}{s} - \frac{p}{q}\right| = \frac{|qr - ps|}{qs} = \frac{2}{qs} \times \text{area of the approximating triangle OAB} = \frac{1}{qs}.$$
So, since $q < s$,
$$\left|x_0 - \frac{p}{q}\right| < \frac{1}{q^2}.$$

Now suppose that both
$$\left|x_0 - \frac{p}{q}\right| \geq \frac{1}{2q^2} \quad\text{and}\quad \left|x_0 - \frac{r}{s}\right| \geq \frac{1}{2s^2};$$
we shall derive a contradiction by analysing Fig. 9.5 more closely, and showing that these inequalities imply that the triangle OAB is too big.

Let C be the point $(q, x_0 q)$, vertically above A, and D be $(s, x_0 s)$, vertically below B; and let AB intersect $Y = x_0 X$ in E. Then, by our assumed inequalities,
$$AC = |x_0 q - p| \geq 1/2q \quad\text{and}\quad BD = |x_0 s - r| \geq 1/2s.$$

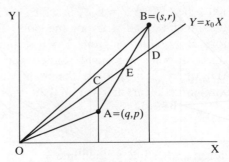

Fig. 9.5

9.1 Basic theory

Since $q < s$, and A and B are both convergents, $AC > BD$; so, since the triangles ACE and BDE are similar, ACE is larger than BDE. Next

$$\text{area OAC} = \tfrac{1}{2}q \cdot AC \geq \tfrac{1}{4} \quad \text{and} \quad \text{area ODB} = \tfrac{1}{2}s \cdot BD \geq \tfrac{1}{4}.$$

Hence

$$\text{area OAB} = \text{area OAC} + \text{area OBD}$$
$$+ (\text{area ACE} - \text{area DBE}) > \tfrac{1}{4} + \tfrac{1}{4} = \tfrac{1}{2},$$

which gives the required contradiction.   QED

There is an important converse to this result:

PROPOSITION If r/s satisfies

$$\left| x_0 - \frac{r}{s} \right| < \frac{1}{2s^2}$$

then r/s is a convergent of x_0.

PROOF Suppose that r/s is in its lowest terms, and that $A = (q, p)$ is the last convergent (i.e. that with largest q) such that $q \leq s$. We will then prove, by contradiction, that $r/s = p/q$. For suppose that $r/s \neq p/q$, i.e. $ps - qr \neq 0$; then, since $ps - qr$ is an integer, $|ps - qr| \geq 1$, so triangle OAB (where $B = (s, r)$) will have area $\geq \tfrac{1}{2}$. But we shall show that the assumption on B does not allow this. As in the previous proposition, this assumption implies that $|sx_0 - r| < 1/2s$, i.e. $BD < 1/2s$ in Figs. 9.6(a) and (b).

We consider the case where $p/q < x_0$, i.e. A lies below $Y = x_0 X$; the case of $p/q > x_0$ is similar. We investigate separately the three situations where $r/s \leq p/q < x_0$; $p/q < r/s \leq x_0$; and $p/q < x_0 < r/s$.

First suppose $r/s \leq p/q < x_0$, Fig. 9.6(a). Then A lies inside or on the edge OB of triangle OBD, so

$$\text{area OAB} < \text{area OBD} = \tfrac{1}{2}r \cdot BD < \tfrac{1}{4},$$

which shows this case cannot occur.

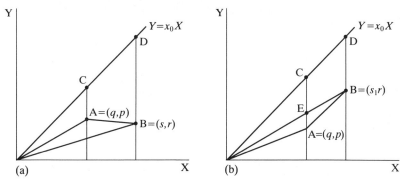

Fig. 9.6

Next suppose $p/q < r/s \leq x_0$, Fig. 9.6(b). Then

$$\text{area OAB} = \text{area OAE} + \text{area AEB}$$
$$= \tfrac{1}{2}q \cdot \text{AE} + \tfrac{1}{2}(s-q) \cdot \text{AE}$$
$$= \tfrac{1}{2}s \cdot \text{AE} < \tfrac{1}{2}s \cdot \text{AC} < \tfrac{1}{2}s \cdot \text{BD} < \tfrac{1}{4},$$

which again cannot occur. (Note that both of these contradictions would follow from the weaker hypothesis that $|x_0 - r/s| < 1/s^2$.)

Finally, suppose that $p/q < r/s < x_0$; here see Fig. 9.5, but in a different interpretation from that in the previous proposition. Let (p', q') be the convergent following (p, q), so that $q \leq s < q'$. Then, by the last proposition but one, $\text{AC} < \text{BD}$. So, since triangles ACE and BDE are similar, ACE is smaller. Then

$$\text{area OAB} = \text{area OAC} + \text{area ODB} - (\text{area BDE} - \text{area ACE})$$
$$< \tfrac{1}{2}q \cdot \text{AC} + \tfrac{1}{2}s \cdot \text{BD} = \tfrac{1}{2}$$

and again we have a contradiction. QED

HISTORICAL NOTES Chuquet described the *Parmenides* algorithm under the name *La règle des nombres moyens* in his *Triparty* of 1484:

This rule serves to find as many numbers intermediate between two neighbouring numbers as one desires. By its means it is possible to find many more numbers and to do more calculations than are found by the rule of three or by one position or by two positions. ... Numerator is added to numerator, and denominator to denominator [Chuquet–Flegg, Hay, & Moss, *NC*, 90].

There was some sixteenth-century exploration of this technique by Juan de Ortega and Buteo (see Dickson, *HTN* ii, 350–1 for references), but modern commentators often seem not to have realised its potential. For example, the *DSB*, s.v. Chuquet, says of it: "... this rule, together with a lot of patience, makes it possible to solve any problem allowing of a rational solution. ... Since he was little concerned about rapid methods of approximation, Chuquet throughout his work used nothing but this one rule ...."

The geometrical description of continued fractions in terms of the integer lattice is sometimes attributed to Klein, but the idea is to be found twenty years earlier in Smith, NCF, and then, shortly afterwards, in Poincaré, GFC. Klein made the following additional observation:

Imagine pegs or needles affixed at all the integral points, and wrap a tightly drawn string about the sets of pegs to the right and then to the left of the ray $[Y = x_0 X]$, then the vertices of the two convex string-polygons which bound our two points sets will be precisely the points (q_k, p_k) whose coordinates are the numerators and denominators of the successive convergents to x_0, the left polygon having the even convergents, the right one the odd. This gives a new and, one may well say, an extremely graphic definition of a continued fraction [Klein, *EMAS* i, 44; he illustrates it by Fig. 9.2, above].

The approach of the *Parmenides* algorithm is closely related to the properties of Farey series, on which see, for example, Dickson, *HNT* i, 155–8, or Hardy & Wright, *ITN*, Chapter 3 (and do not miss the note to §3.1).

Lagrange opened his *Additions* to Euler's *Algebra* with a discussion of approximations of the form $|x_0 q - p|$, and he showed that the best such approximations correspond to the convergents of the continued fraction expansion of x_0. This was extended in Smith, NCF, to the more complicated situation of approximations of the form $|x_0 - p/q|$, whose solution involves the intermediate convergents. This, incidentally, gives a complete solution to Dr Davenant's problem, which was described in the Historical Note to the previous section.

9.1(c) The quadratic theory

Let us start with an example such as delighted eighteenth-century (and, in a different guise, seventeenth-century) mathematicians. This is taken from Euler, UNAPPS (1767) §14, but expressed in the notation that has been established here.

Take $x_0 = \sqrt{54}$; then

$$n_0 = 7 \quad \text{and} \quad x_1 = \frac{1}{\sqrt{54} - 7} = \frac{\sqrt{54} + 7}{5};$$

$$\text{so } n_1 = 2 \quad \text{and} \quad x_2 = \frac{5}{\sqrt{54} - 3} = \frac{5(\sqrt{54} + 3)}{45} = \frac{\sqrt{54} + 3}{9};$$

$$\text{so } n_2 = 1 \quad \text{and} \quad x_3 = \frac{9}{\sqrt{54} - 6} = \frac{9(\sqrt{54} + 6)}{18} = \frac{\sqrt{54} + 6}{2};$$

$$\text{so } n_3 = 6 \quad \text{and} \quad x_4 = \frac{2}{\sqrt{54} - 6} = \frac{2(\sqrt{54} + 6)}{18} = \frac{\sqrt{54} + 6}{9};$$

$$\text{so } n_4 = 1 \quad \text{and} \quad x_5 = \frac{9}{\sqrt{54} - 3} = \frac{9(\sqrt{54} + 3)}{45} = \frac{\sqrt{54} + 3}{5};$$

$$\text{so } n_5 = 2 \quad \text{and} \quad x_6 = \frac{5}{\sqrt{54} - 7} = \frac{5(\sqrt{54} + 7)}{5} = \sqrt{54} + 7;$$

$$\text{so } n_6 = 14 \quad \text{and} \quad x_7 = \frac{1}{\sqrt{54} - 7} = x_1.$$

Hence $\sqrt{54} = [7, \overline{2, 1, 6, 1, 2, 14}]$, where the bar indicates that the term will repeat indefinitely.

Euler frequently gives such examples to illustrate the algorithm. Here, in UNAPPS, he starts with his favourite but less illuminating example $\sqrt{13}$, and then gives $\sqrt{61}$ and $\sqrt{67}$, all in complete detail; then he goes on to $\sqrt{31}$, $\sqrt{46}$, and $\sqrt{54}$ in slightly abbreviated format; then he gives a table of the expansions of $\sqrt{n}$ for $2 \le n \le 120$, n not a square (see Table 3.1, above) and describes their form when $n = m^2 + 1$, $m^2 + 2$, $m^2 + m - 1$, $m^2 + 2m$, $4m^2 + 4$, $9m^2 + 3$, and $9m^2 + 6$; these correspond, of course, to the hypotheses of Section 3.3, above.[15]

We can describe the process in general as follows:

Let $x_0 = \dfrac{a_0 + b_0\sqrt{n}}{c_0}$, where n is not a square (or, equivalently, x_0 is the irrational solution of some quadratic equation with integer coefficients).[16] At each step of the algorithm we write

$$x_k = \frac{a_k + b_k\sqrt{n}}{c_k} = n_k + \frac{(a_k - n_k c_k) + b_k\sqrt{n}}{c_k},$$

where $n_k = [x_k]$. Then

$$x_{k+1} = \frac{c_k}{(a_k - n_k c_k) + b_k\sqrt{n}} = \frac{c_k((a_k - n_k c_k) - b_k\sqrt{n})}{(a_k - n_k c_k)^2 - b_k^2 n},$$

and so we get a quadratic surd of the same kind, in which the new coefficients a_{k+1}, b_{k+1}, and c_{k+1} are simple functions of the a_k, b_k, and c_k. However, it will now be found that these coefficients grow rapidly in size as they pick up new common factors at each step. In the example of $\sqrt{54}$, above, these factors were divided out at each step; in the general case we must be a bit more subtle to ensure that this will happen.

Manipulate the surd into the form $x_0 = \dfrac{a_0 + \sqrt{n}}{c_0}$ where n is not a square and c_0 divides $n - a_0^2$; this can be done by writing

$$\frac{a_0 + b_0\sqrt{n}}{c_0} = \frac{\pm a_0 c_0 + \sqrt{(b_0^2 c_0^2 n)}}{\pm c_0^2},$$

and note that, by fortunate chance, $\sqrt{n} = (0 + \sqrt{n})/1$ is already in this required form. Now look at the step from x_k to x_{k+1}:

$$x_{k+1} = \frac{c_k}{(a_k - n_k c_k) + \sqrt{n}} = \frac{c_k((n_k c_k - a_k) + \sqrt{n})}{n - (n_k c_k - a_k)^2};$$

---

[15] An understanding of the material that follows will be greatly enhanced by exploring some numerical examples of this, and other such, algorithms. A programmable pocket calculating machine gives an advantage undreamt of by previous generations of mathematicians.

[16] All letters except x, y, and z will hereafter denote integers; n and m will denote positive integers, except that n_0, the initial term of the expansion, may sometimes be negative; and $\sqrt{n}$ is always the positive square root.

9.1 Basic theory

if c_k divides $n - (n_k c_k - a_k)^2$, we can then write this as

$$x_{k+1} = \frac{a_{k+1} + \sqrt{n}}{c_{k+1}} \quad \text{where} \quad a_{k+1} = n_k c_k - a_k \quad \text{and} \quad c_{k+1} = \frac{n - a_{k+1}^2}{c_k}.$$

But $c_{k+1} = (n - a_{k+1}^2)/c_k$ divides $n - a_{k+1}^2$ if and only if c_k divides $n - a_{k+1}^2 = n - a_k^2 - c_k(n_k - c_k)$, i.e. if and only if c_k divides $n - a_k^2$, which we have arranged to be so. Hence x_{k+1} will indeed again be a surd of the required form. Let us state this formally:

ALGORITHM FOR QUADRATIC SURDS Express the quadratic surd in the form $x_0 = \dfrac{a_0 + \sqrt{n}}{c_0}$ where n is not a square and c_0 divides $n - a_0^2$. For $k \geq 0$ let

$$n_k = [x_k], \quad \text{the integer part of } x_k,$$
$$a_{k+1} = n_k c_k - a_k, \quad \text{and}$$
$$c_{k+1} = (n - a_{k+1}^2)/c_k.$$

Then x_{k+1} will again be of the required form, and hence $x_0 = [n_0, n_1, n_2, \ldots]$.

There is a remarkable theory of continued fractions of quadratic surds. We start with an easy result:

PROPOSITION If the continued fraction of x_0 is eventually periodic,

$$x_0 = [n_0, n_1, \ldots, n_k, \overline{m_0, m_1, \ldots, m_l}],$$

then x_0 is a quadratic surd.

PROOF Observe first that

$$y_0 = \overline{[m_0, m_0, \ldots, m_l]} = [m_0, m_1, \ldots, m_l, y_0] = \frac{y_0 r_l + r_{l-1}}{y_0 s_l + s_{l-1}}$$

where r_{l-1}/s_{l-1} and r_l/s_l are the last two convergents of the first period of y_0. Hence y_0 satifies the quadratic equation

$$s_l y^2 + (s_{l-1} - r_l) y - r_{l-1} = 0.$$

Since the coefficient of y^2 and the absolute term have opposite signs, there will be one positive and one negative root; so y_0 must be the positive root. Finally

$$x_0 = [n_0, n_1, \ldots, n_k, y_0] = \frac{y_0 p_k + p_{k-1}}{y_0 q_k + q_{k-1}}$$

is again a quadratic surd. QED

The converse of this proposition is also true, but the proof is much more difficult.

PROPOSITION (Lagrange, AMREN (1770)) *The continued fraction of a quadratic surd is eventually periodic.*

PROOF We shall prove that, in the notation above for the algorithm for quadratic surds, there is an index K such that, for all $k \geq K$, $1 \leq a_k < \sqrt{n}$ and $1 \leq c_k < 2\sqrt{n}$. Hence some x_k must eventually recur as x_l with $l > k$, and we shall then have $x_0 = [n_0, n_1, \ldots, \overline{n_k, \ldots, n_{l-1}}]$.

The surd $x_0 = (a_0 + \sqrt{n})/c_0$ is one root of some quadratic equation $Ax^2 + Bx + C = 0$; the other root, called the (algebraic) conjugate of x_0 will then be $x_0' = (a_0 - \sqrt{n})/c_0$. Write, similarly, x_k' for the conjugate of x_k. Now call x_k a reduced quadratic surd if $x_k > 1$ and $-1 < x_k' < 0$. Then, if x_k is reduced:

$$x_k - x_k' > 0, \quad \text{so} \quad \sqrt{n}/c_k > 0, \quad \text{so} \quad c_k > 0,$$
$$x_k + x_k' > 0, \quad \text{so} \quad a_k/c_k > 0, \quad \text{so} \quad a_k > 0,$$
$$x_k' < 0, \quad \text{so} \quad a_k < \sqrt{n}, \quad \text{and}$$
$$x_k > 1, \quad \text{so} \quad c_k < (a_k + \sqrt{n}) < 2\sqrt{n}.$$

These are precisely the inequalities we wish to establish.

Now suppose x_k is reduced. Then $x_{k+1} = 1/(x_k - n_k) > 1$, and $x_{k+1}' = -1/(n_k - x_k')$, where $-1 < x_k' < 0$ and $n_k \geq 1$. Hence $-1 < x_{k+1}' < 0$. Thus if x_k is reduced, x_{k+1} will also be reduced.

It remains to show that some x_k is reduced. Observe first that $x_k \leq 1$ for all $k \geq 1$. Next, we have shown that

$$x_0 = \frac{x_{k+1}p_k + p_{k-1}}{x_{k+1}q_k + q_{k-1}},$$

and so the similar relation holds between x_0' and x_{k+1}'. Hence

$$x_{k+1}' = \frac{x_0'q_{k-1} - p_{k-1}}{x_0 q_k - p_k} = -\frac{q_{k-1}}{q_k}\left(\frac{x_0' - p_{k-1}/q_{k-1}}{x_0' - p_k/q_k}\right).$$

Now both p_k/q_k and p_{k-1}/q_{k-1} tend to x_0 as k increases, and they lie on opposite sides of x_0. Hence the quantity in brackets will tend to 1, and will eventually alternate on each side of 1. Also $0 < q_{k-1} < q_k$. So, eventually, some x_K will satisfy $-1 < x_K < 0$ and the corresponding x_K will be reduced. Hence the proposition is proved. QED

PROPOSITION (Galois, DTFCP (1829)) *A quadratic surd x_0 has a purely periodic continued fraction expansion if and only if it is reduced, and the purely periodic number got by reversing the period is then $-1/x_0'$.*

PROOF Suppose that $x_0 = [\overline{n_0, n_1, \ldots, n_k}]$; observe that for this to be a

9.1 Basic theory

simple continued fraction, we must have $n_0 \geq 1$, so $x_0 > 1$. Now consider the two equations

$$x = [n_0, n_1, \ldots, n_k, x] \quad \text{and} \quad y = [n_k, n_{k-1}, \ldots, n_0, y].$$

We have seen in the previous proposition but one that these are each quadratic equations of which one root is positive, the other root is negative. We now relate their roots.

Manipulate the first equation as follows:

$$x = n_0 + \cfrac{1}{n_1 +} \cdots \cfrac{1}{n_k +} \cfrac{1}{x},$$

so

$$\frac{1}{n_0 - x} = -n_1 - \cfrac{1}{n_2 +} \cdots \cfrac{1}{n_k +} \cfrac{1}{x},$$

so

$$\frac{1}{n_1 +} \frac{1}{n_0 - x} = -n_2 - \cfrac{1}{n_3 +} \cdots \cfrac{1}{n_k +} \cfrac{1}{x},$$

and so, eventually,

$$n_k + \cfrac{1}{n_{k-1} +} \cdots \cfrac{1}{n_1 +} \cfrac{1}{n_0 - x} = \frac{1}{-x},$$

or, writing $y = -1/x$, $y = [n_k, n_{k-1}, \ldots, n_0, y]$. Hence, if the roots of the first equation are x_0 and x_0', then the roots of the second are $-1/x_0$ and $-1/x_0'$. Since $-1/x_0$ is negative, its conjugate $-1/x_0'$ must be positive; it is also greater than 1. Hence $-1 < x_0' < 0$, and x_0 is reduced.

Conversely, suppose that x_0 is reduced; we show that its continued fraction expansion is purely periodic. We know, from Lagrange's theorem, that it is eventually periodic, so that $x_k = x_l$ for some k and l with $k < l$. Now $x_k = n_k + 1/x_{k+1}$, so $x_k' = n_k + 1/x_{k+1}'$. Write y_k for $-1/x_k'$; then, since x_k' and x_{k+1}' lie between -1 and 0, y_k is greater than 1, and we have the relation

$$-\frac{1}{y_k} = n_k - y_{k+1}, \quad \text{i.e.} \quad y_{k+1} = n_k + \frac{1}{y_k}.$$

Hence n_k is the integer part of y_{k+1}. Now if x_k and x_l are equal, then x_k' and x_l' will be equal, hence y_k and y_l will be equal, and hence n_{k-1} and n_{l-1} will be equal. But

$$x_{k-1} = n_{k-1} + 1/x_k \quad \text{and} \quad x_{l-1} = n_{l-1} + 1/x_l,$$

so x_{k-1} and x_{l-1} will be equal. Repeating this argument, we will get eventually that x_0 and x_{l-k} are equal, so the continued fraction is purely periodic. QED

PROPOSITION If n/m is a positive rational number greater than 1 whose square root is irrational, then

$$\sqrt{(n/m)} = [n_0, \overline{n_1, n_2, \ldots, n_2, n_1, 2n_0}],$$

and conversely, if x_0 has an expansion of this form, then it is the irrational square root of a rational number greater than 1. (For numbers between 0 and 1, the expansion starts with a zero term.)

PROOF Consider $y_0 = \sqrt{(n/m)} + n_0$, where $n_0 = [\sqrt{(n/m)}]$; its conjugate $n_0 - \sqrt{(n/m)}$ lies between -1 and 0, and so y_0 is a reduced quadratic surd. Hence, by Galois' theorem, its continued fraction will be purely periodic, of the form

$$\sqrt{(n/m)} + n_0 = [\overline{2n_0, n_1, n_2, \ldots, n_k}],$$

while

$$\frac{1}{\sqrt{(n/m)} - n_0} = [\overline{n_k, n_{k-1}, \ldots, 2n_0}],$$

From this second expansion we get, by reciprocating and adding $2n_0$,

$$\sqrt{(n/m)} + n_0 = [2n_0, \overline{n_k, n_{k-1}, \ldots, 2n_0}],$$

and since the expansion is unqiue, we see that $n_1 = n_k$, $n_2 = n_{k-1}$, etc.

Conversely, suppose that

$$x_0 = [n_0, \overline{n_1, n_2, \ldots, n_2, n_1, 2n_0}].$$

Then $y_0 = x_0 + n_0$ is purely periodic, so its conjugate root y_0' is given by

$$-\frac{1}{y_0'} = [\overline{n_1, n_2, \ldots, n_2, n_1, 2n_0}].$$

so $y_0' = 2n_0 - y_0$. Hence y_0 is the positive root of $(y - y_0)(y - 2n_0 + y_0) = 0$, i.e. of $y^2 - 2n_0 y + y_0 y_0' = 0$. Hence $y = n_0 + \sqrt{z_0}$ for some rational number z_0, and $x_0 = \sqrt{z_0}$. QED

PROPOSITION (see Section 2.4(e)) If

$$\sqrt{n} = [n_0, \overline{n_1, n_2, \ldots, n_2, n_1, 2n_0}]$$

and $p_k/q_k = [n_0, n_1, \ldots, n_2, n_1]$ is the penultimate convergent of the first period of the expansion (so the period contains $k + 1$ terms), then

$$p_k^2 - nq_k^2 = (-1)^{(k+1)}.$$

Similarly, for the penultimate convergent of the second period,

$$p_{2k+1}^2 - nq_{2k+1}^2 = (-1)^{2k+2} = 1,$$

and so on for each succeeding period; so the signs alternate if k is even, and are still all positive if k is odd.

9.1 Basic theory

PROOF We have

$$\sqrt{n} = x_0 = \frac{x_{k+1}p_k + p_{k-1}}{x_{k+1}q_k + q_{k-1}},$$

where

$$x_{k+1} = [2n_0, \overline{n_1, n_2, \ldots, n_2, n_1, 2n_0}] = \sqrt{n} + n_0.$$

Hence

$$\sqrt{n}((\sqrt{n} + n_0)q_k + q_{k-1}) = (\sqrt{n} + n_0)p_k + p_{k-1},$$

that is

$$\sqrt{n}(n_0 q_k + q_{k-1} - p_k) + (nq_k - n_0 p_k - p_{k-1}) = 0.$$

Since $\sqrt{n}$ is irrational, this implies that

$$n_0 q_k + q_{k-1} - p_k = nq_k - n_0 p_k - p_{k-1} = 0.$$

Now substitute for p_{k-1} and q_{k-1} in $p_k q_{k-1} - p_{k-1} q_k = (-1)^{k+1}$. We get

$$p_k^2 - n q_k^2 = (-1)^{k+1}. \qquad \text{QED}$$

PROPOSITION Every solution of $x^2 - ny^2 = \pm 1$ arises as a convergent of $\sqrt{n}$.

PROOF First suppose $p^2 - nq^2 = 1$. Then, since $p > q\sqrt{n} > q$,

$$\frac{p}{q} - \sqrt{n} = \frac{1}{q(p + q\sqrt{n})} < \frac{1}{2pq} < \frac{1}{2q^2}.$$

Hence, by an earlier proposition, p/q is a convergent of $\sqrt{n}$.

Next, if $p^2 - nq^2 = -1$, then $nq^2 - p^2 = 1$. Again, since $q\sqrt{n} > p > q$,

$$\sqrt{n} - \frac{p}{q} = \frac{1}{q(q\sqrt{n} + p)} < \frac{1}{2pq} < \frac{1}{2q^2},$$

and the result again follows. QED

HISTORICAL REMARKS The material of this section first manifested itself in connection with Fermat's challenge of 1657 concerning the so-called Pell's equation, $x^2 - ny^2 = 1$ (see Section 2.4(e), above, for quotations and further details). Euler was the first to publish a continued fraction description of the 'English solution' (by Brouncker & Wallis) in his UNAPPS, written in 1759 but not published until 1767. The first actual proof that Pell's equation can always be solved was by Lagrange, SPA, written in 1768 but not published until 1773; it was very long and, after an initial step in which continued fractions were used to find a subsidiary equation $x^2 - ny^2 = r$ with an infinite number of solutions, it was based on his rediscovery of the 'Indian solution'. (The introduction to this article is quoted in n. 55, below.) By the time this article was published, Lagrange had found another, simpler and clearer, proof, based on the properties of the continued fraction expansion of quadratic surds; the most succinct exposition of the idea of this proof is in §37 of Lagrange's *Additions* to Euler's *Algebra*. The story continues with further contributions by Euler, Lagrange, Legendre, Galois, and others. A masterly account, with its full mathematical and historical context, can be found in Weil, *NT*.

I shall not attempt to summarise or select further episodes from this rich picture, beyond the following comment. An important development in this story occurs with the publication of Euler's UNAPPS, in which we find one of the first published systematic explorations of the properties of the continued fraction expansion of a quadratic surd. Euler's motivating problem is the solution of Pell's equation, but most of the article is taken up in describing the algorithm for evaluating the continued fraction expansion of $\sqrt{n}$, in tabulating many results, and in inferring some properties, exactly parallel to those that underly my exploration of the dimension of squares, in Section 3.3, above. Euler then turns to Pell's equation, and the first cases he considers are again directly analogous to the further developments I proposed in Section 3.4. He explores those cases where when the period of the expansion of $\sqrt{n}$ contains 1, then 2, then 3, and so on up to 8 terms, displaying the solution in each case, and then finishes by inferring the general procedure and tabulating the solutions. The first complete proofs of the behaviour he has uncovered, by Lagrange, follow a completely different path just as, in Section 3.6, the general proofs of the dimension of squares are resolved by other very circuitous means. Euler is rare among mathematicians in having the desire and facilities to lay before his readers his infectious enthusiasm for the initial steps of exploration and conjecture that precede much of deductive mathematics; if all we possessed on number theory up to the early nineteenth century was Gauss's *Disquisitiones Arithmeticae* (see Section 9.2(d), below), then it would be very difficult to reconstruct this early heuristic phase.

For Greek mathematics, we know very little of the developments that led up to the *Elements* and any reconstruction must therefore of necessity be very speculative. This being admitted, I have tried to show how one can find hints of a similar substrate of problems, explorations, and conjectures that eventually give rise to some of the most formal and arid parts of the *Elements,* for example Books II and X. And I was again surprised to discover subsequently how the analogous explorations by eighteenth-century mathematicians followed a very similar path; and then to see the extent to which the greater part of these explorations and manipulations have already been set aside and forgotten by later generations of mathematicians.

9.1(d) Analytic properties[17]

As so far described, most of the developments can be loosely categorised as arithmetical and number theoretical. In addition, alongside, and even in advance of, these discoveries, there was a developing analytic theory of continued fractions, starting with Brouncker's discovery of 1654 that

$$\frac{\pi}{4} = \frac{1}{1+} \frac{1}{2+} \frac{3^2}{2+} \frac{5^2}{2+} \cdots,$$

an illustration of the promise in passing between expressions as infinite products, continued fractions, and infinite series. Another such hint, perhaps the first suggestion of interesting analytic behaviour involving simple continued fractions and in itself as surprising as the expansion of $\sqrt{n}$, was the expansion

$$e = [2, 1, 2, 1, 1, 4, 1, 1, 6, 1, \ldots]$$

first uncovered by Cotes in 1714. These results in arithmetic and analysis excited and inspired Euler and Lambert no less than the number-theoretic explorations; in fact Euler's first references to continued fractions are found in his correspondence in connection with the Riccati differential equation and the expansion of e.[18] This is not the place to attempt a survey of these and later analytic developments,[19] and I shall restrict myself here to two very brief observations about the analytic theory during the eighteenth century.

First, it is an immediate consequence of a proof of the expansion of e that it is irrational (see Section 2.2(v)(vii), above), and indeed this is how it and associated results were first used by Lambert and Euler. Second, as an illustration of this developing analytic theory, let me describe Chapter 18, 'De Fractionibus Continuis', of Euler's *Introductio in Analysin Infinitorum*, his influential introductory text on the techniques of the newly developing subject of analysis.

The chapter opens with a description of the convergents of both simple and general continued fractions, and then passes immediately to transformations between the convergents of continued fractions and the partial

---

[17] Here I use the words 'analysis' and 'analytic' in their modern sense, as connected with limiting processes, infinite operations, the theory of functions, the use of the calculus, etc. Seventeenth-century use of 'analysis' was different, and started from speculations about the lost methods used by the Greeks in their mathematical discoveries.

[18] See Euler, FCD, or Euler–Wyman & Wyman, ECF, for further details. The reference to the correspondence with Goldbach can be found in the note in the *Opera Omnia* to §28, or in Weil, *NT*, 183 or 230.

[19] At this point, historical work on the topic starts to become seriously deficient. The broad lines of development can be inferred from Perron, *LK*; and the later part of the sketch in Brezinski, *LHCF*, expanded in his *HCF*, is biased towards analytic properties. I know of no other general historical surveys of the subject.

sums of infinite series. These are then applied to several examples:

$$\log 2 = 1 - \frac{1}{2} + \frac{1}{3} - \frac{1}{4} + \ldots = \frac{1}{1+} \frac{1}{1+} \frac{4}{1+} \frac{9}{1+} \frac{16}{1+} \ldots ;$$

$$\frac{\pi}{4} = 1 - \frac{1}{3} + \frac{1}{5} - \frac{1}{7} + \ldots = \frac{1}{1+} \frac{1}{2+} \frac{9}{2+} \frac{25}{2+} \frac{49}{2+} \ldots ;$$

$$\frac{\pi}{n} \cot \frac{m\pi}{n} = \frac{1}{m} - \frac{1}{n-m} + \frac{1}{n+m} - \frac{1}{2n-m} + \frac{1}{2n+m} - \ldots$$

$$= \frac{1}{m+} \frac{m^2}{n-2m+} \frac{(n-m)^2}{2m+} \frac{(n+m)^2}{n-m+} \frac{(2n-m)^2}{2m+} \frac{(2n+m)^2}{n-2m+} \ldots ;$$

$$\frac{1}{e} = 1 - \frac{1}{1!} + \frac{1}{2!} - \frac{1}{3!} + \ldots, \quad \text{whence} \quad \frac{1}{e-1} = \frac{1}{1+} \frac{1}{2+} \frac{2}{3+} \frac{3}{4+} \frac{4}{} \ldots ;$$

and more. Only at this stage does Euler turn to periodic simple continued fractions and quadratic surds, which he introduces with a view to making some remarks about convergence and approximation; for elsewhere in this chapter, as throughout his works, he spends little time on such justifications of his manipulations, though he operated with a sure intuition for what was permitted or not. The chapter finishes with some comments on the conversion of decimal fractions into continued fractions, which he illustrates on the expansion of

$$\frac{e-1}{2} = [1, 6, 10, 14, 18, 22, \ldots],$$

$$\pi = [3, 7, 15, 1, 292, 1, 1, \ldots],$$

and an astronomical illustration, the ratio of the day to the solar year, $[365, 4, 7, 1, 6, 1, 2, 2, 4, \ldots]$, and the Julian and Gregorian calendars.

Such was the basic introduction that Euler set out. It is attractive, accessible, and directed more to describing manipulations than offering rigorous proofs. It is really more arithmetic than analysis, for he considers only numbers that arise in analytic contexts, like e and π, rather than the functions of analysis like

$$\cotan x = [x^{-1}, -3x^{-1}, 5x^{-1}, -7x^{-1}, \ldots],$$

such as appear abundantly in his articles, and which were also used by Lambert. It is an inspiration to further investigation.

I refer those readers who are rightly dissatisfied with this *hors d'oeuvre* to the comprehensive descriptions of the theory that has subsequently developed, for example in Jones & Thron, *CF*; Henrici, *ACCA* ii; Perron, *LK*; or Wall, *ATCF*; and I shall give an example of a more ambitious analytic manipulation in Section 9.2(c) below.

9.1(e) *Lagrange and the solution of equations*

The solution of a general polynomial equation with integer coefficients is a pervasive theme in mathematics. One approach—solution in radicals—reaches a climax in the work of Galois, and Lagrange played no small part in this development with his long survey article RRAE. This successful resolution, albeit negative, of this much studied problem has subsequently overshadowed many of the previous developments and rendered them obsolete or irrelevant. So, for example, the already obsolete theory of the construction of equations, initiated by Descartes, has now disappeared almost completely from the view of even all but the most specialised histories of mathematics (for discussion, see Bos, AMRDMT); and Lagrange's other approach, which I shall now briefly describe, has now also almost vanished, even though it is a very serviceable technique for computing the roots of equations.

Lagrange's work is contained in a series of memoirs that start with his REN (1769), followed a year later by AMREN, almost twice as long as the original memoir, then followed by further memoirs, all of them published by the Berlin Academy; then later all of this material was brought together into a full-length treatise, *TREN* (1798), with a second edition in 1808. His treatment ranges widely, but its focus is the following observation. Take the equation

$$a_k x^k + a_{k-1} x^{k-1} + \ldots + a_1 x + a_0 = 0$$

and suppose, for simplicity, that it has only one positive root x_0, which we can further suppose is greater than 1 (otherwise replace x by $1/x$). Then $n_0 \leq x_0 < n_0 + 1$ for some n_0. So, writing $x_0 = n_0 + 1/x_1$, we see that x_1 will be the unique positive root, in fact greater than 1, of a new equation

$$b_k x^k + b_{k-1} x^{k-1} + \ldots + b_1 x + b_0 = 0,$$

where

$$b_k = a_k n_0^k + a_{k-1} n_0^{k-1} + \ldots + a_0,$$
$$b_{k-1} = k a_k n_0^{k-1} + (k-1) a_{k-1} n_0^{k-2} + \ldots + a_1,$$

etc.

This equation, again, will have only one positive root x_1, again greater than 1, so we can again express this as $x_1 = n_1 + 1/x_2$, etc. Thus we evaluate the solution of the original equation as a continued fraction, $x_0 = [n_0, n_1, n_2, \ldots]$. Lagrange dealt with many other aspects of this procedure, and also described procedures for evaluating the approximations n_0, for computing coefficients of the new equation, for separating roots that lie close together, for finding complex roots, and so on.

The first illustration that he gives is of the time-honoured cubic

TABLE 9.1. Evaluation of $\sqrt[3]{2}$ = [1, 3, 1, 5, 1, 1, ...] by Lagrange's algorithm.

k	Equation	n_k
0	$x^3 - 2 = 0$	1
1	$x^3 - 3x^2 - 3x - 1 = 0$	3
2	$10x^3 - 6x^2 - 6x - x = 0$	1
3	$3x^3 - 12x^2 - 24x - 10 = 0$	5
4	$55x^3 - 81x^2 - 33x - 3 = 0$	1
5	$62x^3 + 30x^2 - 84x - 55 = 0$	1
etc.		

equation $x^3 - 2x - 5 = 0$ which, since Newton's time, had been the standard test for methods of solving equations (see Whiteside, PMTLSC, 206). He gets $x = [2, 10, 1, 1, 2, 1, 3, 1, 12, ...]$ from which he evaluates the convergents and gives estimates of their errors. But here, in a deliberate reference to what I have earlier called the problem of the dimension of cubes (see Section 4.3, above), I shall illustrate the process on the equation $x^3 - 2 = 0$.

We start by writing out the formulae explicitly for $k = 3$. The equation

$$f(x) = a_3 x^3 + a_2 x^2 + a_1 x + a_0 = 0,$$

in which we may suppose $a_3 > 0$, becomes, under the transformation $x \to n + 1/x$, the equation

$$-(a_3 n^3 + a_2 n^2 + a_1 n + a_0) x^3$$
$$- (3a_3 n^3 + 2a_2 n + a_1) x^2 - (3a_3 n + a_2) x - a_3 = 0.$$

Since we have supposed that $f(x) = 0$ has only one positive root which satisfies $n < x_0 < n + 1$, and since $a_3 > 0$, we must have $f(n) < 0$; thus again $b_3 = -f(n) > 0$. We then iterate the procedure.

Consider now $x^3 - 2 = 0$. Then $1 < x_0 < 2$ so $n_0 = 1$; $b_3 = 1$, $b_2 = -3$, $b_1 = -3$, $b_0 = -1$; then the positive root x_1 of $x^3 - 3x^2 - 3x - 1 = 0$ satisfies $3 < x_1 < 4$, so $n_1 = 3$, etc. The first six steps are set out in Table 9.1; we get $\sqrt[3]{2} = [1, 3, 1, 5, 1, 1, ...]$. The arithmetical operations can be considerably refined and economised to make this a practical algorithm for hand or machine calculation.[20]

Lagrange's treatise contains a rich study of general properties of polynomial equations, dealing, for example, with positive and negative, and real and complex, roots, but all directed towards describing and

---

[20] For a succinct modern description of this and other continued fraction algorithms, see the exercises to Knuth, ACP ii, §4.5.3; this is exercise 20. A detailed practical implementation is given in Rosen & Shallit, CFA. Also see n. 32, below.

analysing his algorithm for expressing the solutions as continued fractions—a procedure that he clearly considered as important and valuable. (Some of this work is taken up and extended by Legendre, Fourier, Vincent, and others; see, for example, Serret, *CAS*, Chapters 6 & 7.) It also contains the first proof of the periodicity of the expansion of a quadratic surd, at AMREN, §§32–43 or *TREN,* §§52–59. Hence, he observes, the method will identify all linear and quadratic rational irreducible factors of the polynomial, since the corresponding roots, being rational numbers or quadratic surds, will be revealed by his procedure.[21] He then concludes:

> It is to be hoped that one may also find some characteristic that may serve to reveal the rational divisors of third, fourth,... degrees, which the given equations may have. This seems to me to be a topic very worthy of the attention of mathematicians (AMREN §49, or *TREN* §64).[22]

As I shall describe in Section 9.3(b), below, the search for some such criterion, based on some generalisation of continued fractions, is still active, but has not really yielded the results that Lagrange had hoped.

9.2 Gauss and continued fractions

9.2(a) *Introduction*

At the end of the eighteenth century, a chorus of mathematicians sang forth the wonders of continued fractions. If we restrict our attention to material published more or less in the previous fifty years, we see, in the front rank: Euler, with his *Introductio in Analysin Infinitorum* (1748) and *Anleitung zur Algebra* (1770), as well as a bundle of articles on the topic; Lambert, who, in several articles, discussed the expansion of functions as continued fractions, including details of convergence, and then proved the irrationality of numbers connected with exponential and trigonometric functions; Lagrange, with his *Additions, De l'analyse indéterminée* (1773) to the French translation of Euler's *Algebra,* his *Traité de la résolution des équations numériques de tous les degrés* (1769–1798), and a series of memoirs that build on and establish rigorously results uncovered

---

[21] It must be added that this information will not be effectively computable, since we have no idea how far the computation must be extended for the periodicity to reveal itself. Lagrange was well aware of this problem, and described mathematical techniques for identifying periodicity, based on the idea of generating functions. For comments on this problem of recognising periodic behaviour, see Ferguson, RRQN, and Fowler, BTEE2, 34.

[22] "Il serait à souhaiter que l'on pût trouver aussi quelque caractère qui pût servir à faire reconnaitre les diviseurs commensurables du troisième, quatrième,... degré, lorsqu'il y en a dans l'équation proposée; c'est du moins une recherche qui me parait très-digne d'occuper les Géomètres."

by Euler; and Legendre, with *Recherches d'Analyse Indéterminée* (1785) and *Essai sur la Théorie des Nombres* (1798) which are, in great part, based on continued fraction techniques. Around them are a host of supporting figures.

Gauss (1777–1855) was heir to this material, and one can be sure that he assimilated it. For example, continued fractions play a significant part in his published work in analytic contexts, in connection with hypergeometric series and numerical integration, and we know from his unpublished notes that he investigated the probability theory of continued fractions over a twelve-year period, at least. But his enormously influential book on number theory, the *Disquisitiones Arithmeticae*, makes almost no reference to continued fractions. I shall now look in turn at Gauss's treatment of each of these topics.

9.2(b) *Continued fractions and the hypergeometric series*

Many elementary functions of analysis are particular cases of the hypergeometric series

$$F(\alpha, \beta, \gamma, x) = 1 + \frac{\alpha\beta}{\gamma}x + \frac{\alpha(\alpha+1)\beta(\beta+1)}{2!\,\gamma(\gamma+1)}x^2 + \dots,$$

which, for α, β, and γ different from $0, -1, -2, -3, \dots$, is an infinite series which converges for $|x| < 1$. The proposal to generalise the geometric series goes back to Wallis' *Arithmetica Infinitorum*, while the study of this particular generalisation was initiated by Euler. Gauss's mastery of the topic was complete.[23]

The hypergeometric satisfies many identities. Here we shall start from

$$F(\alpha, \beta, \gamma, x)$$
$$= F(\alpha, \beta+1, \gamma+1, x) - \frac{\alpha(\gamma - \beta)}{\gamma(\gamma+1)} F(\alpha+1, \beta+1, \gamma+2, x)x,$$

which we write as

$$\frac{F(\alpha, \beta+1, \gamma+1, x)}{F(\alpha, \beta, \gamma, x)} = \left(1 - \frac{\alpha(\gamma - \beta)}{\gamma(\gamma+1)} \frac{F(\alpha+1, \beta+1, \gamma+2, x)}{F(\alpha, \beta+1, \gamma+1, x)} x\right)^{-1},$$

whence, by exchanging α and β, and then replacing β by $\beta + 1$ and γ by

---

[23] For details of the early and rich theory of the hypergeometric function, from Wallis to Gauss, see Dutka, EHHF. The manipulations that follow are described and explained at greater length in Wall, *ATCF*, Chapter 18.

9.2 Gauss and continued fractions

$\gamma + 1$, we get

$$\frac{F(\alpha+1, \beta+1, \gamma+2, x)}{F(\alpha, \beta+1, \gamma+1, x)} = \left(1 - \frac{(\beta+1)(\gamma-\alpha+1)}{(\gamma+1)(\gamma+2)} \frac{F(\alpha+1, \beta+2, \gamma+3, x)}{F(\alpha+1, \beta+1, \gamma+2, x)} x\right)^{-1}.$$

Now apply these identities turn and turn about to give

$$\frac{F(\alpha, \beta+1, \gamma+1, x)}{F(\alpha, \beta, \gamma, x)} = \frac{1}{1-} \frac{\frac{\alpha(\gamma-\beta)}{\gamma(\gamma+1)}x}{1-} \frac{\frac{(\beta+1)(\gamma-\alpha+1)}{(\gamma+1)(\gamma+2)}x}{1-} \frac{\frac{(\alpha+1)(\gamma-\beta+1)}{(\gamma+2)(\gamma+3)}x}{1-} \frac{\frac{(\beta+2)(\gamma-\alpha+2)}{(\gamma+3)(\gamma+4)}x}{1-} \ldots.$$

This is the formal derivation of Gauss's continued fraction, to be found in his DGCSI; it is a development of a procedure of Lambert's, who described an algorithm for manipulating two power series analogous to the Euclidean algorithm. While the hypergeometric functions converge only for $|x|<1$, the continued fraction converges for most complex values of x (specifically, when x is not real with $1 \leq x < \infty$, or not equal to an isolated set of values), and so it provides an analytic continuation of $F(\alpha, \beta+1, \gamma+1, x)/F(\alpha, \beta, \gamma, x)$ outside the unit circle, thus possibly providing a meaning for the series in regions where it is divergent. (This convergence theory was not considered by Gauss; it was later investigated by Riemann, in SQSI, a posthumously published paper completed by Schwarz, and by Heine; then the question was finally settled in Thomae, KGQ.)

Put $\alpha = 1$, $\beta = 0$, $\gamma = 1$, and write $-x$ for x. Then, since

$$F(1, 1, 2, -x) = 1 - \frac{x}{2} + \frac{x^3}{3} - \ldots = \frac{\log(1+x)}{x}$$

and $F(1, 0, 1, -x) = 1$, we get, after a little adjustment,

$$\log(1+x) = \frac{x}{1+} \frac{1^2 x}{2+} \frac{1^2 x}{3+} \frac{2^2 x}{4+} \frac{2^2 x}{5+} \ldots,$$

provided x is not real and ≤ -1. Or, put $\alpha = \frac{1}{2}$, $\beta = 0$, $\gamma = \frac{1}{2}$, and write x^2 for x. Then since

$$F(\tfrac{1}{2}, 1, \tfrac{3}{2}, x^2) = \frac{1}{2x} \log \frac{1+x}{1-x}$$

and $F(\frac{1}{2}, 0, \frac{1}{2}, x^2) = 1$, we get, again after adjustment,

$$\log\frac{1+x}{1-x} = \frac{2x}{1-}\frac{1^2 x^2}{3-}\frac{2^2 x^2}{5-}\frac{3^2 x^2}{7-}\cdots$$

For another illustration, write x/α for x,

$$F\left(\alpha, \beta, \gamma, \frac{x}{\alpha}\right) = 1 + \frac{\beta}{\gamma}x + \frac{\beta(\beta+1)}{\gamma(\gamma+1)}\frac{x^2}{2!} \cdot \frac{\alpha+1}{\alpha}$$
$$+ \frac{\beta(\beta+1)(\beta+2)}{\gamma(\gamma+1)(\gamma+2)}\frac{x^3}{3!} \cdot \frac{(\alpha+1)(\alpha+2)}{\alpha^2} + \cdots,$$

and now let α tend to infinity, to give the closely related function

$$\Phi(\beta, \gamma, x) = 1 + \frac{\beta}{\gamma}x + \frac{\beta(\beta+1)}{\gamma(\gamma+1)}\frac{x^2}{2!} + \cdots.$$

In particular $\Phi(1, 1, x) = e^x$. So do the same to Gauss's continued fractions. We first take the expansion for $F(\alpha, \beta+1, \gamma+1, x/\alpha)/F(\alpha, \beta, \gamma, x/\alpha)$ and let α tend to infinity to get

$$\frac{\Phi(\beta+1, \gamma+1, x)}{\Phi(\beta, \gamma, x)} = \frac{1}{1-}$$

$$\frac{\frac{(\gamma-\beta)}{\gamma(\gamma+1)}x}{1+} \frac{\frac{(\beta+1)}{(\gamma+1)(\gamma+2)}x}{1-} \frac{\frac{(\gamma-\beta+1)}{(\gamma+2)(\gamma+3)}x}{1+} \frac{\frac{(\beta+2)}{(\gamma+3)(\gamma+4)}x}{1-}\cdots,$$

then set $\beta = 0$, write γ for $\gamma + 1$, and adjust to give

$$\Phi(1, \gamma, x) = \frac{1}{1-}\frac{x}{\gamma+}\frac{x}{\gamma+1-}\frac{\gamma x}{\gamma+2+}\frac{2x}{\gamma+3-}\frac{(\gamma+1)x}{\gamma+4+}\frac{3x}{\gamma+5-}\cdots.$$

Now set $\gamma = 1$ to give

$$\Phi(1, 1, x) = e^x = \frac{1}{1-}\frac{x}{1+}\frac{x}{2-}\frac{x}{3+}\frac{x}{2-}\frac{x}{5+}\frac{x}{2-}*\cdots.$$

If, finally, we set $x = 1$ and adjust the expression to eliminate the negative terms and zero terms, exactly as Lagrange tells us in his AMREN §§50–67 or *TREN* §§65–76,[24] we get

$$e = [2, 1, 2, 1, 1, 4, 1, 1, 6, 1, \ldots].$$

These manipulations are offered here as illustrations of Gauss's fascination with the hypergeometric function and mastery of the idiom of continued fractions, as displayed in his memoir DGCSI of 1813.

---

[24] These kinds of identities are described in n. 6, above.

9.2 Gauss and continued fractions

In the memoir MNIVAI (1816) on numerical integration. Gauss used continued fractions to initiate what has developed into an important and far-reaching technique. He took the expansion of $\log(1+x/1-x)$ from his earlier memoir DGCSI and rewrote it, by replacing x by x^{-1}, as

$$\log\frac{x+1}{x-1} = \frac{1}{x} + \frac{1}{3x^3} + \frac{1}{5x^5} + \ldots = \frac{1}{x-}\;\frac{\frac{1\cdot 1}{1\cdot 3}}{x-}\;\frac{\frac{2\cdot 2}{3\cdot 5}}{x-}\;\frac{\frac{3\cdot 3}{5\cdot 7}}{x-}\;\frac{\frac{4\cdot 4}{7\cdot 9}}{x-}\ldots$$

The kth convergent of this continued fraction can then be expressed as a rational function, a quotient of a polynomial of degree $k-1$ by a polynomial of degree k; and if this rational function is developed as a power series, its terms will agree with the original power series up to degree $2n$, the best that one can hope for in general. (This kind of approximation is now called a Padé approximant, and rapidly growing interest in this topic, together with the stimulus provided by automatic computation, has given rise to a recent revival of interest in the analytic theory of continued fractions.) Gauss then calculated the poles and residues of this convergent, which he used as the nodes and weights for what is now known as the Gaussian method of numerical integration. The procedure was later developed by Jacobi and Christoffel, and eventually led to the Runge–Kutta method of integrating differential equations; and it is connected with the general theory of orthogonal polynomials.

Two entries in Gauss's celebrated *Tagebuch* bear further witness to his interest in the manipulations of infinite series into continued fractions, and in divergent series. The sixth note, dated 23 May 1796 (and so written less than two months after the notebook was started, and just after his nineteenth birthday) recorded:

The transformation of the series $1 - 2 + 8 - 64 + \ldots$ into the continued fraction

$$\frac{1}{1+}\;\frac{2}{1+}\;\frac{2}{1+}\;\frac{8}{1+}\;\frac{12}{1+}\;\frac{32}{1+}\;\frac{56}{1+}\;\frac{128}{1+}\ldots,$$

$$1 - 1 + 1\cdot 3 - 1\cdot 3\cdot 7 + 1\cdot 3\cdot 7\cdot 15 - \ldots = \frac{1}{1+}\;\frac{1}{1+}\;\frac{2}{1+}\;\frac{6}{1+}\;\frac{12}{1+}\;\frac{28}{1+}\ldots,$$

and others,

while, on 16 February 1797, he returned to the subject:

An amplification of the penultimate proposition on page 1, namely

$$1 - a + a^3 - a^6 + a^{10}\ldots = \frac{1}{1+}\;\frac{a}{1+}\;\frac{a^2-a}{1+}\;\frac{a^3}{1+}\;\frac{a^4-a^2}{1+}\;\frac{a^5}{1+}\;\text{etc.}$$

From this all series where the exponents from a series of the second order are easily transformed.

For discussion of these entries, I refer the reader to the commentary by Schlesinger in Gauss, *Werke* x_1, 490–3 and 513–4.

9.2(c) *Continued fractions and probability theory*

Gauss initiated the probability theory of continued fractions, though he published nothing on the topic. An entry in his *Tagebuch* for 25 October 1800 records: *Problema e calculo probabilitatis circa fractiones continuas olim frustra tentatum solvimus* (We are solving problems in the calculus of probabilities about continued fractions that once we attempted in vain). A work-book for 5 February 1799 records an earlier attempt at a solution, in which he calculated and tried to explain some functions $P(n, x)$ which will be described below in his words. His calculation finishes with *Tam complicatae evadunt, ut nulla spes superesse videatur* (They come out so complicated that no hope appears to be left); see *Werke* x, 552–6 for the text and commentary. On 30 January 1812, he described the problem in a letter to Laplace:

...I do recall however an intriguing problem which I worked on 12 years ago, but which I did not then succeed in resolving to my satisfaction. Perhaps you might take the trouble of spending some moments with it: in that case I am sure that you will find a more complete solution. Here it is. Let M be an unknown quantity, between 0 and 1, for which all values are either equally probable, or are distributed according to a given law. Suppose it converted into a continued fraction

$$M = \cfrac{1}{a' + \cfrac{1}{a'' + \text{etc.}}}$$

What is the probability, if we terminate the expansion at a finite term $a^{(n)}$, that the remaining fraction

$$\cfrac{1}{a^{(n+1)} + \cfrac{1}{a^{(n+2)} + \text{etc.}}}$$

should lie between 0 and x? I denote this by $P(n, x)$. When all the values of M are equally probable, so that

$$P(0, x) = x,$$

then $P(1, x)$ is a transcendental function depending on the function

$$1 + \frac{1}{2} + \frac{1}{3} + \ldots + \frac{1}{x}$$

which Euler called inexplicable and on which I have just given some researches in a memoir presented to our scientific society which will soon be published.[25] But

---

[25] This is Gauss, DGCSI; Euler describes this 'inexplicable function' in, for example, his *ICD*, Chapter 16. Just as the gamma function is defined by extending $n! = 1.2.3.\ldots.n$ to non-integral arguments, so this inexplicable function is the extension of $1 + 2^{-1} + 3^{-1} + \ldots + n^{-1}$ to non-integral arguments.

for larger values of n, the exact value of $P(n, x)$ seems intractable. However I have found by very simple arguments that for infinite n we have

$$P(n, x) = \log(1+x)/\log 2,$$

but my attempts to evaluate

$$P(n, x) - \log(1+x)/\log 2$$

when n is larger but not infinite have been unfruitful [Gauss, *Werke* x_1, 371–4].[26]

He never published his "*raisonnemens tres simples*". The first proof of the limiting distribution $\log(1+x)/\log 2 = \log_2(1+x)$ was given by Kuz'min in 1928, while the final question about the asymptotic behaviour was only resolved in 1974, by Wirsing. The classical exposition is in Khinchin, *CF*; and a convenient account of recent developments can be found in Knuth, *ACP* ii, §4.5.3, where it is interesting to compare the treatments in the first edition of 1969 (which concludes "The world's most famous algorithm [sc. Euclid's algorithm] deserves a complete analysis!") and the second edition of 1981 ("Fortunately it is now possible to supply rigorous proofs, based on careful analysis by several mathematicians"). To say more here would take much space, so I shall restrict myself to a few illustrations.

First, an illustration of Gauss's result: if we take a real number x at random[27]—and it must be emphasised that most of the numbers we have considered so far, such as rational numbers, quadratic surds, e, tanh 1, etc. do not behave like these randomly chosen numbers—and look at the terms of its continued fraction, then we should expect that:

about $\log_2 \tfrac{4}{3} = 41.50\%$ of the n_k should be 1,
about $\log_2 \tfrac{9}{8} = 16.99\%$ of the n_k should be 2,
about $\log_2 \tfrac{16}{15} = 9.31\%$ of the n_k should be 3,
etc.

So we should expect more than two-thirds of the terms of the expansions of randomly chosen numbers to be 1, 2, or 3. Next, Lévy proved in 1929 that we should expect that the convergents should satisfy

$$\sqrt[k]{q_{k-1}} \to \exp(\pi^2/12 \log_e 2) = 3.27582\ldots.$$

Khinchin proved, in 1935, a similar result about the geometric means of the n_k, that

$$\sqrt[k]{(n_1 n_2 \ldots n_k)} \to c$$

---

[26] Gauss's original, in French, is also quoted in my *REGM*, 826–7.
[27] In modern terms, the following results will hold for all x not belonging to some set of Lebesgue measure zero. On the other hand, it must be emphasised that no example is known of any number, arising in any other context, which displays any of the following kinds of typical behaviour. See Section 9.2(e), below.

where

$$c = \prod_{r=1}^{\infty} \left[1 + \frac{1}{r(r+2)}\right]^{\log_2 r} = 2\cdot 68555\ldots,$$

while the arithmetic mean should grow like $\log_2 k$. Finally, the work of Dixon, Heilbron, Knuth, Mendes France, and Wirsing, since around 1970, enables us to give some kind of answer to my slaveboy's reflection at B_{24} of Section 1.3, above: we should expect the number of terms in the terminating expansion of m/n, averaged over all m less than and prime to n, for large n, to be approximately equal to $(12/\pi^2)\log_2 n + 1.46708\ldots,$ and further closer estimates can be given for the error.[28]

9.2(d) *Gauss's number theory*

Gauss had a complete mastery of the techniques of continued fractions and a profound insight into their properties; the descriptions so far illustrate this clearly. And continued fractions played a very significant role in the number theory of Gauss's predecessors; we only need to look at Legendre's *Essai sur la Théorie des Nombres* of 1798 to see this illustrated at length. Yet there is only *one* explicit reference to continued fractions in Gauss's *Disquisitiones Arithmeticae* of 1801, and this found in a historical passage and contains a confusing typographical error! Three other passages introduce material closely related to continued fractions, without explicating the connections. Here are the details.

In §27, Gauss describes how to solve $ax = by \pm 1$ in integers by the Euclidean algorithm, and he defines the Euler brackets.[29] But he says nothing of continued fractions or convergents, and only identifies the process as "the known algorithm for finding the greatest common divisor of two numbers". A footnote gives two further identities for these brackets, without explanation of context or proof. Then, in §28, which I shall now quote in full with my annotations, we find the only explicit mention of continued fractions.

Euler was the first[a] to give the general solution for indeterminate equations of this type (*Comment. Petrop.* T. VII *p.* 46[b]). The method he used consisted in substituting other unknowns for x, y and it is a method that is well known today. Lagrange treated the problem a little differently. As he noted, it is clear from the theory of continued fractions that if the fraction b/a is converted into the

---

[28] Simple upper bounds for the number of steps can be given, based on the fact that the longest expansions occur with the quotient of successive Fibonacci numbers, $[1, 1, \ldots, 1] = F_k/F_{k-1}$.
[29] See Section 9.1(a), above: $[n_0, \ldots, n_k] = P_k(n_0, \ldots, n_k)/P_{k-1}(n_1, \ldots, n_k)$. Euler uses the notation $(n_0, \ldots, n_k)$ and Gauss uses the notation $[n_0, \ldots, n_k]$ for my $P_k(n_0, \ldots, n_k)$.

9.2 Gauss and continued fractions

continued fraction

$$\cfrac{1}{\alpha + \cfrac{1}{\beta + \cfrac{1}{\gamma + \text{etc} + \cfrac{1}{\mu + \cfrac{x}{n}}}}}$$

and if the last part, x/n,[c] is deleted and the result reconverted into a common fraction, x/y then $ax = by \pm 1$, provided a is prime relative to b. For the rest, the same algorithm is derived from the two methods. The investigations of Lagrange appear in *Hist. de l'Ac. de Berlin Année* 1767 p. 175,[d] and with others in *Supplementis versioni gallicae Algebrae Eulerianae adiectis*[e] [Gauss–Clarke, *DA*, 28].

Notes:

[a] This is corrected in Gauss's *Additamenta* as follows: "The solution of the indeterminate equation $ax = by \pm 1$ was not first accomplished by the illustrious Euler (as stated in this section) but a geometer of the seventeenth century, Bachet de Meziriac, the celebrated editor and commentator of Diophantus. It was the illustrious Lagrange who restored this honour to him (*Add. à l'Algèbre d'Euler* p. 525 [§45], where at the same time he indicates the nature of the method). Bachet published his discovery in the second edition of the book *Problêmes plaisans et délectables qui se font par les nombres* (1624). In the first edition (Lyon, 1612) which was the only one I saw, it was not included, although it was mentioned" [Gauss–Clarke, *DA*, 461].

[b] I have given all of the citations in the form used by Gauss, and will identify them in notes; abbreviated titles can be found in the bibliography. This is 'Solutio problematis arithmetici de inveniendo numero, qui per datos numeros divisus, relinquat data residua', *Opera* I ii, 18–31.

[c] This final x/n should, of course, be $1/n$; and this error is not corrected in the two pages of errata bound in the end of the volume, nor in the version reprinted in *Werke* i. This is the only explicit continued fraction to be found in the *Disquisitiones Arithmeticae*, though the topic is implicitly contained in his occasional use of Euler brackets, as I shall explain below.

[d] 'Sur la solution des problèmes indéterminés du second degrée', *Oeuvres* ii, 377–535.

[e] I.e. Lagrange's *Additions* to Euler's *Algebra*, §42–45.

Section 5, §§153–265, of the *Disquisitiones* is devoted to solving equations of the second degree in two variables. A hint of continued fractions appears in §199 which starts "In practice, more suitable formulae can be developed". These formulae are expressed in terms of the Euler brackets, which were introduced in §27 and developed further in §189, and which Gauss identifies retrospectively in a footnote to the passage to be quoted next.

In §202, we find another historical review. Again I shall quote in full, with my annotations:

A particular case of the problem of solving the equation $t^2 - Du^2 = 1$ had already been treated in the last century. That extremely shrewd geometer Fermat proposed the problem to the English analysts, and Wallis called Brouncker the discoverer of the solution which he reported in *Alg. Cap. 98, Opp.* T. II *p.* 418 sqq;[a] Ozanam claims that it was Fermat;[b] and Euler, who treated of it in *Comm. Petr.* VI *p.* 175, *Comm. nov.* XI *p.* 28,[c] *Algebra P.* II *p.* 226, *Opusc. An.* I *p.* 310[d] claims that Pell was the discoverer, and for that reason the problem is called *Pellian* by some authors. All these solutions are essentially the same as the one we would have obtained if in §198 we had used the reduced form with $a = 1$; but no one before Lagrange was able to show that the prescribed operation necessarily comes to an *end*, that is that the problem is *really soluble*.[e] Consult *Mélanges de la Soc. de Turin T.* IV *p.* 19;[f] and for a more elegant presentation, *Hist. de l'Ac. de Berlin*, 1767, *p.* 237.[g] There is also an investigation of this question in the *supplementis ad Euleri Algebram*[h] which we have frequently commended. But our method (starting from totally different principles and not being restricted to the case $m = 1$) gives many ways of arriving at a solution because in §198 we can begin from any reduced form $(a, b, -a')$ [Gauss–Clarke, *DA*, 185].

Notes:

[a] This is Chapter 98, 'A method of Approaches for Numerical Questions; occasioned by a problem of Mons. Fermat' of the Latin translation of Wallis' *Treatise of Algebra* ... (1685), which appeared in his *Opera Omnia Mathematica* ii (1693).

[b] J. Ozanam, *Nouveaux éléments d'algèbre* (1702), described a solution of Pell's equation identical with Brouncker's, illustrated it with the cases $n = 23$ and 19, and attributed it to Fermat. Weil, *NT*, 93, observes: "Did he know more than we do, or had he merely misread the *Commercium Epistolicum*?"

[c] Gauss inserts a footnote that reads: "In this commentary the algorithm we considered in §27 is used with similar notation. We neglected to note it at that time". The algorithm in question is the basic algorithm for computing the Euler brackets: $P_{k+1}(n_0, \ldots, n_k, n_{k+1}) = n_{k+1} P_k(n_0, \ldots, n_k) + P_{k-1}(n_0, \ldots, n_{k-1})$, $P_{-1} = 1$, $P_0(n_0) = n_0$.

[d] These articles by Euler are, respectively, 'De solutione problematum Diophanteorum per numeros integros', *Opera* I ii, 6–17; UNAPPS; *VAA* II ii, Chapter 7'; and 'Nova subsida pro resolutione formulae $axx + 1 = yy$', *Opera* I iv, 76–90.

[e] Gauss's italics. He inserts a footnote: "What Wallis, *loc. cit.* pp. 427–8, proposes for this purpose carries no weight. The paralogism consists in the fact that on p. 428, line 4, he presupposes that, given a quantity p, integers a, z, can be found such that z/a is less than p and that the difference is less than an *assigned* number. This is true when the assigned difference is a *given quantity* but not when, as in the present case, it depends on a and z and this is variable". Also see n. 55, below.

[f] Lagrange, SPA.

[g] Lagrange, 'Sur la solution des problèmes indéterminées du second degrée', *Oeuvres* ii, 377–535.

[h] Lagrange, *Additions* to Euler's *Algebra*, §85–87.

What has happened?

The majority of the significant discoveries and developments in the theory of continued fractions in number theory, in the seventeenth and eighteenth centuries, have centred on the solution of Pell's equation $x^2 - ny^2 = 1$. But Pell's equation is only one example of a very general class of problems of finding for what integers p the equation $ax^2 + bxy + cy^2 = p$ can be solved in integers. This more general approach focuses attention on the so-called binary quadratic forms

$$F(x, y) = ax^2 + bxy + cy^2,$$

and, as number theory developed, so more and more general examples of binary quadratic forms came to be considered: Fermat only dealt with some specific examples $x^2 + ay^2$; Euler extended this systematically to the class $ax^2 + by^2$; and Lagrange, in his *Recherches d'Arithmétique* (1775) with the supplement of 1777, embraced the set of all binary forms. Lagrange is still inspired by his use of continued fractions in Pell's equation, which he called "la clef de tous les autres problèmes de ce genre",[30] and Legendre, in his *ETN* (Part 1, §XIII), made their role even more prominent. But Gauss's intuition seems to have been that little further significant progress was possible using this technique; and, with his eyes on the most general theory, he seems not to want to distract his reader into any byways, especially since he may have come to believe that however charming these byways may be, further progress along them will be extremely difficult if not impossible. He reformulated the proofs of his number theory so as to omit any use of or intuitions about continued fractions, and only included the briefest mention of them in those passages which sketch the history of the subject and give algorithms for finding solutions.

For tiros, I recommend the basic introduction to binary quadratic forms in Davenport, *HA*, Chapter 6; and for a combined historical and mathematical treatment, I again direct everybody to Weil, *NT*. The final appendix of this latter book ('A proof of Lagrange's on Indefinite Binary Quadratic Forms', pp. 350–9) and its back-references give a detailed description of the evolution of the continued fraction process into the new theory of binary quadratic forms that rendered it obsolete.

9.2(e) *Gauss's legacy in number theory*

We see number theory today through Gauss's eyes. For example, unique factorisation into products of primes is, today, *the* fundamental theorem of arithmetic (i.e. *arithmētikē*!), and it is generally referred to as such; but this theorem and attitude is not to be found explicitly before the *Disquisitiones Arithmeticae*, and it may be a historical error to look for it in

---

[30] In the introduction to his SPA (quoted in n. 54, below), his first article on the topic.

Greek mathematics, behind what is preserved in the *Elements*.[31] So, as Legendre's *Essai sur la théorie des nombres* went through revisions, emerging as a two-volume treatise *Théorie des Nombres* in its third edition of 1830, its influence declined, and slowly Gauss's point of view came to dominate the subject. We see here, once again, the same evolutionary process that operated earlier with Euclid's *Elements* and Ptolemy's *Syntaxis*, in which new definitive treatments supersede, dominate, and can even annihilate their predecessors. Continued fractions moved from being a central source of insight and problems to being merely a collection of special techniques to be applied in a few restricted contexts; and, for the most part, that is how they are perceived today. Moreover, during the eighteenth and early nineteenth centuries, number theory had provided a central, coherent, and accessible nucleus to a developing theory of continued fractions; when this base was dismantled, the other topics in the theory were marginalised. But many natural problems connected with the process remain unresolved, and are still studied ardently by mathematicians.

Let me illustrate one such problem, what I called, in Section 4.3, the problem of the 'dimension of cubes': Formulate and prove any hypothesis about the continued fraction expansion of $\sqrt[3]{n}$. A natural first step is to compute some terms of some simple example, and Lagrange's algorithm (see Section 9.1(e) above) provides the means:[32]

$$\sqrt[3]{2} = [1, 3, 1, 5, 1, 1, 4, 1, 1, 8, 1, 14, 1, 10, 2, 1, 4, 12, 2, \ldots].$$

But, faced with such irregular behaviour, we cannot at first formulate anything other than very coarse questions like: Are these terms bounded, or can limits be given their growth? Even then, it is only very recently that any kind of answer to such basic questions has been found, and the

---

[31] For a discussion, see Knorr, PIGNT.

[32] In fact there is no real difficulty in extending the procedure for computing a quadratic surd to the case of a cubic surd. Here is an algorithm which covers three indices and expresses the $(k + 1)$th coefficient in terms of the kth and $(k - 1)$th: If $x_k = (a_k \xi^2 + b_k \xi + c_k)/d_k$ where $\xi = \sqrt[3]{n}$ and n is not a cube, and $x_k = n_k + x_{k+1}^{-1}$, then, for $k \geq 1$,

$$a_{k+1} = n_k a_k + a_{k-1}$$
$$b_{k+1} = n_k b_k + b_{k-1}$$
$$c_{k+1} = n_k^2 d_k - 2n_k c_k - n_{k-1} d_{k-1} + c_{k-1}$$
$$d_{k+1} = -n_k^3 d_k + 3n_k^2 c_k - 3n_k(c_{k-1} - n_{k-1} d_{k-1}) + d_{k-1}.$$

When the algorithm is started with $a_0 = c_0 = 0$, $b_0 = d_0 = 1$, i.e. $x_0 = \sqrt[3]{n}$, so $a_1 = 1$, $b_1 = n_0$, $c_0 = n_0^2$, $d_0 = n - n_0^3$, then $a_k = q_{k-1}$, $b_k = p_{k-1}$, and $b_k^3 - na_k^3 = (-1)^k d_k$. Such three index algorithms are well known for quadratic surds—see e.g. Chrystal, *TA* ii, 454–5—and their derivation would have been immediately accessible to Euler, but I know of no published description of them for cubic or higher order irrationalities. The higher order versions are most conveniently expressed in terms of determinantal identities, as Bernard Teissier has explained to me.

bounds that have been found are themselves also very coarse: $n_k \leq a^{b^k}$ (see Davenport & Roth, RAAN, and Baker, RACTAN). These proofs depend on Roth's theorem, a deep result on Diophantine approximation, which means that, in general, the values of a and b cannot be effectively computed. Recently an elementary proof of such a result for some cube roots, which does not use much beyond the techniques described in Section 1, has been given in Wolfskill, GBPQCN. But let me emphasise: we still do not know any example of an algebraic irrational of degree higher than two whose terms either are bounded, or are not bounded.

Lagrange's optimistic hope about recognising algebraic numbers, quoted at the end of Section 9.1(e), suggests another approach: Can the classical continued fraction algorithm be extended so as to identify other algebraic numbers than the rationals and quadratic surds? This runs against the spirit of Gauss's work in number theory, and so I shall defer comment on it to Section 9.3(b).

Another approach to the expansion of $\sqrt[3]{2}$ is suggested by the later developments in Gauss's unpublished work on probability theory: Does this expansion conform to the expected behaviour of a typical number? I know no evidence that Gauss computed any terms of the expansion of $\sqrt[3]{2}$, but I would be very surprised if he, and Lagrange and Euler before him, did not do so;[33] but it would be very difficult, using only hand calculations, to get enough statistical evidence to be suggestive. However, since the development of automatic computers, the numerical exploration of continued fractions has become a minor industry. The first such investigation, a weekend's calculation on one of the first electronic computers, is reported in von Neumann & Tuckerman, CFE (1955); their one-page article is inconclusive ("We do not know whether this deviation [from the expected distribution] is significant"). A more recent example of the genre, Lang & Trotter, CFSAN, is longer and more comprehensive (they list one thousand places of the expansions of $\sqrt[3]{2}$, $\sqrt[3]{3}$, $\sqrt[3]{4}$, $\sqrt[3]{5}$, and $\sqrt[3]{7}$, and other numbers, together with statistical and theoretical analyses), but is scarcely more conclusive in proportion to its length and electronic effort. A systematic search by Delone & Faddeev of all cubic forms $x^3 + ax + b = 0$, where $|a|$ and $|b| \leq 9$, yielded nothing of interest (though some examples, like $\sqrt[3]{2}$ and $\sqrt[3]{5}$, seem to contain more large terms than one might expect). Then, in 1964, Brillhart hit on one spectacular example that had just evaded the earlier search, that the expansion of the real root $x^3 - 8x - 10 = 0$ has some extraordinarily large terms: $n_{121} = 16\,467\,250$, $n_{33} = 1\,501\,790$, $n_{161} = 325\,927$; and five other n_k with $k \leq 139$ are greater than $20\,000$. Moreover the discriminant of $x^3 - 8x - 10 = 0$ is -4×163, and -163 is a very significant discriminant

---

[33] See how, when there is some systematic behaviour to be found, Euler will often describe it. For example in his FCD, §§21–22, he calculates or infers that $\sqrt{e} = [1, 1, 1, 1, 5, 1, 1, 9, 1, 1, 13, \ldots]$ and $(\sqrt[3]{e} - 1)/2 = [0, 5, 18, 30, 42, 54, \ldots]$.

for *quadratic* forms. A beautifully lucid explanation of why these results are so surprising and how they are connected is given in Churchhouse & Muir, CFANMI; and a more sophisticated mathematical exploration can be found in Stark, EECFFB.[34] This one extraordinary example enables us to perceive dimly some of the subtle and complicated features that may eventually be involved if continued fraction expansions of higher degree irrationals are ever to be understood.

9.3 Two recent developments

9.3(a) *Continued fraction arithmetic*

I have already described some of the difficulties that seem to lie in the way of an algorithm for continued fraction arithmetic; see Sections 4.2 (especially the discussion of B_{32-34}) and 9.5(b). The first approach to any such algorithm seems to be found in Hurwitz, KEZE (1891) and KTAR (1896). We find there the basic rules:[35]

$$2[n_0, 2n_1, n_2, x_3] = [2n_0, n_1, 2n, \tfrac{1}{2}x_3],$$

and

$$2[n_0, 2n_1 + 1, x_2] = [2n_0, n_1, 1, 1, \tfrac{1}{2}(x_2 - 1)],$$

to which, to make the calculation explicit, we can add

$$\tfrac{1}{2}[2n_0, x_1] = [n_0, 2x_1],$$

and

$$\tfrac{1}{2}[2n_0 + 1, x_1] = [n_0, 1, 1, \tfrac{1}{2}(x_1 - 1)].$$

With these, and also the relations described in n. 6, above, we can now double or halve any simple continued fraction. For example, take one of

---

[34] The discriminant of the binary quadratic form $f(x, y) = ax^2 + bxy + cy^2$ is the polynomial $D = b^2 - 4ac$. If $D = 0$, then f is a perfect square or, equivalently, $f(x, 1) = 0$ has coincident roots. If $D < 0$ and $a > 0$, then $f(x, y) > 0$ for all $(x, y) \neq (0, 0)$; it is called positive definite. The theory of binary quadratic forms divides the forms into equivalence classes under a very natural transformation of coordinates by unimodular transformations and the class number is the number of equivalence classes. It has been long known that, for negative discriminants, the class number is finite and is equal to one for $D = -1, -4, -7, -8, -11, -19, -43, -67$, and -163; also that class number one implies that the algebraic numbers in $\mathbb{Q}(\sqrt{D})$ have unique factorisation; but it was only 1967 that Stark gave a generally acceptable proof that -163 is the largest negative class one discriminant.

The discriminant of a binary cubic form $f(x, y, z) = ax^3 + bx^2y + cxy^2 + dy^3$ is $D = 18abcd + b^2c^2 - 27a^2d^2 - 4ac^3 - 4b^3d$; hence the discriminant for Brillhart's cubic is -4×163; but -163 has no special significance for the discriminants of cubic forms. Not unrelated to the eventual explanation is also the fact that $\exp(\pi\sqrt{163})$ is very close to an integer, $\exp(\pi\sqrt{163}) = 262537412640768743.999999999999250\ldots$. I hope that these tantalising titbits encourage the interested reader to consult, at least, Churchhouse & Muir, CFANMI.

[35] I know of no earlier occurrence of these rules, though they would have been quite accessible to Euler and Lagrange, and Euler several times juxtaposed the expansions of e and $\tfrac{1}{2}(e - 1)$; see, for example, Section 9.1(d), above.

9.3 Recent developments

Euler's favourite examples:

$$\tfrac{1}{2}(e-1) = [0, 1, 6, 10, 14, \ldots].$$

This gives

$$\begin{aligned}
e - 1 &= 2[0, 1, 6, 10, \ldots] \\
&= [0, 0, 1, 1, \tfrac{1}{2}[5, 10, 14, \ldots]] \\
&= [1, 1, 2, 1, 1, \tfrac{1}{2}[9, 14, \ldots]] \\
&= [1, 1, 2, 1, 1, 4, 1, \ldots, 1, 2t, 1, \ldots]
\end{aligned}$$

and so the expansion of e. Hurwitz also investigated these kinds of numbers x_0, which have the property that there is a set of polynomials $p_1, p_2, \ldots, p_m$ such that, after some initial acyclic terms, their expansion becomes

$$x_0 = [\ldots, p_1(1), p_2(1), \ldots, p_m(1), p_1(2), p_2(2), \ldots, p_m(2), p_1(3) \ldots].$$

For example, for e, $m = 3$ and $p_1(t) = p_3(t) = 1$, $p_2(t) = 2t$; while for $\tfrac{1}{2}(e-1)$, $m = 1$ and $p_1(t) = 4t + 2$; and for quadratic surds, the polynomials are all constant. He showed that if x_0 is such a number, and $y_0 = (a + bx_0)/(c + dx_0)$ where the integers a, b, c, and d satisfy $ad - bc \neq 0$, then y_0 will also have the same property, though usually for a different value of m.[36] This, together with the fact that e, e^2, and $\sqrt[t]{e}$ are of this kind, explains in part the expansions of many exponential and hyperbolic numbers. Numbers which display this kind of regular behaviour are now known as Hurwitz numbers.

The problem of arithmetic was again raised in Whittaker, TCF (1915):

> The great impediment to the use of continued-fractions in Theory of Functions and Differential Equations is the want of algorithms for adding, multiplying, and differentiating them. The object of the present paper is to supply in some measure this deficiency. I think it would be a mistake to propose the problem in the form: *Given two continued-fractions, to find a third continued-fraction which is equal to their sum or product* ... for I doubt if the problems so formulated possess any simple solutions. But Sylvester showed long ago [in 1853] that any continued-fraction may be regarded as the quotient of two determinants [see the remark in Section 9.1(a), above, on continuants]; and if we regard continued-fractions in this light, advancing boldly from their theory to the theory of determinants, and aiming to express the products or sums of derivates of continued-fractions *in the form of determinants*, the situation becomes much more promising ....

---

[36] If $ad - bc = 0$, then the transformation degenerates in $y_0 =$ constant. There is an older result, due to Serret, that the tails of the expansions of two numbers x_0 and y_0 are eventually equal (i.e. $x_{K+j} = y_{L+j}$ for $j = 0, 1, 2, \ldots$) if and only if there exist integers a, b, c, and d such that $ad - bc = \pm 1$ and $y_0 = (a + bx_0)/(c + dx_0)$; see the third edition of Serret, *CAS* i, 34–7. (This textbook contains an influential account of continued fractions which then forms the basis of later textbooks, for example, Chrystal, *TA*. Serret was also the editor of Lagrange's collected works.)

Then, in the body of the article, Whittaker discusses only differentiation of a certain type of function expressed as a continued fraction, and makes no attempt to consider continued fraction arithmetic.

General algorithms for evaluating sums and products are given in Hall, SPCF (1947), and Rayney, CFFA (1973), but the description of Khinchin, CF, quoted above in Section 4.2, applies to these: they are "exceedingly complicated and unworkable in computational practice". Yet a simple algorithm exists! It was found by R. W. Gosper in the 1970s and has not yet been properly published, though enough details are given in Knuth, APC ii, §4.5.3, Exercise 15, with its answer, to reconstruct the procedure.[37] Here it is:

Consider the problem of evaluating the continued fraction expansion of the real number

$$x = \frac{a + bx + cy + dxy}{a' + b'x + c'y + d'xy},$$

given the continued fraction expansions of x and y; it will become clear that the algorithm itself can be described purely in terms of integer arithmetic and manipulations of the expansions $x_0 = [n_0, n_1, n_2, \ldots]$ and $y_0 = [m_1, m_2, m_3, \ldots]$ of x and y.[38] By choosing appropriate initial values of $a, b, \ldots, d'$, we then get procedures for evaluating the sum $\frac{0 + 1x + 1y + 0xy}{1 + 0x + 0y + 0xy}$, the difference, the product, etc., of two continued fractions. I shall describe the algorithm in terms of inputting the terms of x and y, and outputting the terms of z. For convenience, suppose that everything is positive, though there is no real difficulty in adapting the procedure to negative values.

First consider the simpler case of $z = \dfrac{a + bx}{a' + b'x}$. Inputting a term of the expansion of x corresponds to supplying the information that $x = n + 1/x'$ where $1 \leqslant x'$, and so

$$z = \frac{a + b\left(n + \dfrac{1}{x'}\right)}{a' + b'\left(n + \dfrac{1}{x'}\right)} = \frac{b + (a + nb)x'}{b' + (a' + nb')x'}$$

This step has the effect of replacing the matrix of coefficients

---

[37] I would like to thank Bill Gosper warmly for sending me unpublished material on continued fractions and allowing me to describe his algorithm here. Also many thanks to Mike Paterson for explaining much, here and elsewhere, about continued fractions and algorithms.

[38] I shall drop the suffixes on x_n & x_{n+1} and refer to the basic step of the continued fraction algorithm as $x = n + 1/x'$.

9.3 Recent developments

$\begin{bmatrix} a & b \\ a' & b' \end{bmatrix}$ by $\begin{bmatrix} b & a+nb \\ b' & a'+nb' \end{bmatrix}$, which contains more information about z.
This step can be repeated indefinitely; it generalises the procedure of calculating the convergent, when $z = x = \dfrac{0 + 1x}{1 + 0x}$, and it can be conveniently set out as in the following example of the calculation of $z = \dfrac{9 + 4\sqrt{2}}{1 + 2\sqrt{2}}$. (Recall that $\sqrt{2} = [1, 2, 2, 2, \ldots]$.)

	x		1	2	2	...
Matrix of	9	4	13	30	73	
coefficients	1	2	3	8	19	

(For example, at the last step of an input of 2 has changed the coefficients $\begin{bmatrix} 13 & 30 \\ 3 & 8 \end{bmatrix}$ into $\begin{bmatrix} 30 & 73 \\ 8 & 19 \end{bmatrix}$; here $73 = 2 \times 30 + 13$ and $19 = 2 \times 8 + 3$.)

Next consider the output of z. We know that $0 \leq x \leq \infty$; hence z will lie between a/a' and b/b'.[39] Hence if a/a' and b/b' have the same integer part p, we know that $z = p + 1/z'$ with $z' > 1$, and so the next term of the expansion of z will be p. This gives

$$p + \frac{1}{z'} = \frac{a + bx}{a' + b'x}, \quad \text{so} \quad z' = \frac{a' + b'x}{(a - pa') + (b - pb')x}.$$

Hence the effect of outputting the term p will be to replace the matrix $\begin{bmatrix} a & b \\ a' & b' \end{bmatrix}$ by $\begin{bmatrix} a' & b' \\ a - pa' & b - pb' \end{bmatrix}$. Look again at the example of $\dfrac{9 + 4\sqrt{2}}{1 + 2\sqrt{2}}$; sufficient input has been made above to determine that the initial term of the output is 3. We can now incorporate the operation of extracting this output and continue the calculation further, as in Table 9.2 where the input is written horizontally and the output vertically. In this example, with the last displayed output, the matrix of coefficients reverts to an earlier value of $\begin{bmatrix} 6 & 16 \\ 2 & 3 \end{bmatrix}$, so an input of 2,2 will then again generate an output of 4,1. Hence the output is periodic, with

$$\frac{9 + 4\sqrt{2}}{1 + 2\sqrt{2}} = [3, 1, \overline{4, 1}].$$

Now consider the case of

$$z = \frac{a + bx + cy + dxy}{a' + b'x + c'y + d'xy}.$$

---

[39] In fact $1 \leq x \leq \infty$, so z will actually lie between $(a + b)/(a' + b')$ and b/b', but the weaker condition is more convenient for hand calculation. These statements are slightly more complicated if any of the a, b, a', and b' can take negative values.

TABLE 9.2

				Input x					
O				1	2	2	2	2	...
u			9	4	13	30	73		
t			1	2	3	8	19		
p		3				6	16	38	92
u		1				2	3	8	19
t		4						6	16
		1						2	3
y		⋮							

Here we can write the coefficients in $2 \times 2 \times 2$ array, $\begin{bmatrix} a_{a'} & b_{b'} \\ c_{c'} & d_{d'} \end{bmatrix}$ in which we think of the primed coefficients of the denominator as lying below the coefficients of the numerator. (It is sometimes convenient, in hand calculations, to use two colours to distinguish alternate layers.) We then can immediately verify the following operations:

Input $x = n + \dfrac{1}{x'}$, and $\begin{bmatrix} a_{a'} & b_{b'} \\ c_{c'} & d_{d'} \end{bmatrix}$ becomes $\begin{bmatrix} b_{b'} & a + nb_{a' + nb'} \\ d_{d'} & c + nd_{c' + nd'} \end{bmatrix}$.

Input $y = m + \dfrac{1}{y'}$, and $\begin{bmatrix} a_{a'} & b_{b'} \\ c_{c'} & d_{d'} \end{bmatrix}$

becomes $\begin{bmatrix} c & c' & d & d' \\ a + mc_{a' + mc'} & & b + md_{b' + md'} & \end{bmatrix}$.

Output $z = p + \dfrac{1}{z'}$, and $\begin{bmatrix} a_{a'} & b_{b'} \\ c_{c'} & d_{d'} \end{bmatrix}$ becomes $\begin{bmatrix} a'_{a - pa'} & b'_{b - pb'} \\ c'_{c - pc'} & d'_{d - pd'} \end{bmatrix}$.

Again, if the coefficients are all positive, the value of z will lie within the smallest interval containing a/a', b/b', c/c', and d/d'. When sufficient input has been made to determine the value of p, it can be outputted. Observe that the previous one-variable calculation is, in effect, a vertical slice through the plane $y = 0$ of the two-variable calculation.

Finally, if x terminates, the most convenient way of incorporating this is, after the final input, to replace the matrix of coefficients $\begin{bmatrix} a_{a'} & b_{b'} \\ c_{c'} & d_{d'} \end{bmatrix}$ by $\begin{bmatrix} b_{b'} & b_{b'} \\ d_{d'} & d_{d'} \end{bmatrix}$; and similarly for terminating y.

Table 9.3 sets out the first few steps of the calculation of $\sqrt{2} \times \sqrt{3}$; alternate layers of the calculation are set in roman and italic type. The

TABLE 9.3. $[1,2,2,2,2,2,2,2,2,\ldots] \times [1,1,2,1,2,1,2,1,2,1,\ldots] = [2,2,4,2,4,2,4,\ldots]$. Alternate layers are shown in roman and italic type; outputs are encircled.

				Input x					
		1	2	2	2	2	2	2	etc.
	0_1	0_1							
	0_0	1_0							
I 1		0_2	0 5						
n 1		3_0	7 0						
p 2		3_2	7 5						
u		6_2	14 5						
		15_6 3	35_{15} 5						
			②						
t 1		21_8 5	49_{20} 9_2	48_{23} 2	55 6				
y 2		22_{13}	55_{23} $_9$	132_{59} $_{14}$	141 37				
			②						
1				82_{16}	196 43				
2				223_{46} $_{39}$	533_{123} $_{41}$	292_{121} 50	707 283_{141}		
1				305_{62} $_{57}$	④ 729_{166} $_{65}$	394_{187} 20	954 439 76		
							②		
2						495 90	1161_{293}	2817_{676} $_{113}$	6795_{1645} 215
1						682_{110}	1600_{369}	3882_{848} $_{490}$	9364_{2065} $_{1104}$
									④
etc.									

periodic inputs provoke an output of $[2, 2, 4, 2, 4, \ldots]$, but the apparent periodicity of this output will not show up, this time, in the recurrence of a matrix of coefficients, and I cannot conceive of any way of proving, within the restriction of integer arithmetic, that the output will indeed be periodic.

The algorithm can easily be extended to accept a general continued fraction

$$x_0 = n_0 + \frac{m_1}{n_1 +} \frac{m_2}{n_2 +} \ldots$$

as input, and Gosper has used it in this form to compute and analyse 204 103 terms of π, by converting a Ramanujan expansion for $\pi/4$, similar to Brouncker's expansion but converging much more quickly, into a simple continued fraction; see Gosper, TSCFP; this calculation has now been modified by him to exploit fast Fourier transform integer arithmetic and extended to 17 million terms. On this evidence, π appears to exhibit the behaviour of a typical number, in the sense of Section 9.2(b), above.

9.3(b) *Higher dimensional algorithms*

We can describe the classical Euclidean algorithm as follows: Given (x_0, y_0)—two positive integers, or two positive real numbers, or two lines, or two areas, etc.—then, at each step, if the smaller term is non-zero, subtract it from the larger; if it is zero, terminate. We have seen that this simple algorithm possesses a host of remarkable properties, which we can enumerate as follows:

(i) When x_0 and y_0 are integers, it will terminate with the non-zero term equal to the greatest common factor or measure of x_0 and y_0; this is *Elements* VII 1.

(ii) It will terminate if and only if x_0/y_0 is rational. In this case, the last non-zero term will be the greatest common measure of x_0 and y_0, and the last convergent will be equal to x_0/y_0, expressed as a common fraction in its lowest terms; see *Elements* X 2, 3, and 5. We can generalise this by saying that the algorithm will terminate if and only if there are integers n_0 and m_0 such that $n_0 x_0 - m_0 y_0 = 0$, a condition now expressed today by saying that x_0 and y_0 are linearly dependent over the integers; and the algorithm can be extended to evaluate the least such (n_0, m_0), from which all others can be calculated.

(iii) The procedure will generate best possible approximations. There are various ways of describing this precisely. For example, in the geometrical description of Section 9.1(b), above, we can say that the algorithm will locate those lattice points (p, q) that lie closest to the ray through 0 and (x_0, y_0).

(iv) If x_0 and y_0 are not linearly dependent over the integers, but are quadratically dependent, so there are integers a_0, b_0, and c_0 such that $a_0 x_0^2 + b_0 x_0 y_0 + c_0 y_0^2 = 0$, then the algorithm will become periodic.

How far can we generalise some or all of these properties to apply to higher-dimensional situations $(x_0, y_0, \ldots, z_0)$?[40] For example, there is no difficulty in generalising (i), and Euclid already presents one such generalisation at *Elements* VII 3: Given three integers (x_0, y_0, z_0), apply the classical two-dimensional algorithm to (x_0, y_0), which will then terminate with the non-zero term w_0. Then apply the algorithm to (w_0, z_0); this will terminate with the highest common factor of x_0, y_0, and z_0. In this manner, it is clear that Euclid now knows that this algorithm will apply in any dimension $(x_0, y_0, \ldots, z_0)$, and he uses the general form in VII 33. Then he applies the algorithm to three mutually commensurable magnitudes in *Elements* X 4 and this time he notes, in a porism, that the procedure is general.

It is easy to see that, in three or more dimensions, (ii), generalised as 'termination if and only if linearly dependent', and (iii) are incompatible. Look at the problem geometrically, as described in Section 9.1(b), above. In the two-dimensional case, if $n_0 x_0 - m_0 y_0 = 0$, then the line $Y = (x_0/y_0)X$ will pass through the lattice point (n_0, m_0), and a terminating algorithm will reveal this best possible approximation. But in three or more dimensions, we can have examples like $(1, \sqrt{2}, 1 + \sqrt{2})$ which are linearly dependent but which do not lie on any ray through the origin and a lattice point.[41] Hence any algorithm that reveals this linear dependence by terminating will not go on to generate an infinite sequence of improving rational approximations by lattice points.

Before going further, let me describe two more algorithms. First, Brun's algorithm (see Brun, GK and AETQN, and, for a nice elementary exposition and proof, Ferguson, RRQN[42]): Given $(x_0, y_0, \ldots, z_0)$, at each step of the algorithm, if every term is non-zero, subtract the second largest term from the largest; if any term is zero, terminate. The algorithm will satisfy (i) for any set of integers; and it will satisfy (ii) for triples, but may fail to satisfy it for quadruples or more of

---

[40] I would like to thank and acknowledge the help I have had in understanding these higher dimensional algorithms from letters and unpublished work by George Bergman, Helaman Ferguson, and Rodney Forcade.
[41] We might write $1:\sqrt{2}:(1+\sqrt{2}) \neq p:q:r$ for any integers p, q, and r, in a notation that suddenly stretches all the discussions of ratio in this book. Also note that simultaneous approximation has entered briefly into the discussion of Callippus' calendar in Section 2.4(b) and could very easily have been introduced by my Eudoxus in Section 4.4(b); and it is built into the well-tempered diatonic scale which is based on the approximation of $4:7:12$ for $\log \frac{5}{4} : \log \frac{3}{2} : \log 2$, which could have entered the discussion with my Archytas in Section 4.5(b).
[42] Ferguson comments, in his RRQN: "It seems to be easier to reprove Brun's algorithm independently than to find and read the original!".

real numbers. (Ponder that: if x_0 is the solution of a quadratic equation, and we apply the algorithm to $(1, x_0, x_0^2)$, then it will terminate; if x_0 is a solution of a cubic equation and we apply the algorithm to $(1, x_0, x_0^2, x_0^3)$, it may or may not terminate. For a specific example, the algorithm will not terminate if x_0 is the largest root of $x^3 - 3x^2 + 1 = 0$.) The algorithm will only generate best approximations in special circumstances, and I do not know when it will exhibit periodic behaviour.

Contrast Brun's algorithm with the following: Given $(x_0, y_0, \ldots, z_0)$, at each step of the algorithm, if every term is non-zero, subtract the smallest non-zero term from the largest; if any term is zero, terminate. Again it is easy to see that this will satisfy (i), but, beyond that, its behaviour is completely different from Brun's algorithm. For example, applied to $(1, \sqrt{2}, 1 + \sqrt{2})$, it is easy to see that it will not terminate. I do not know its further properties.

So far I have been describing Euclidean type algorithms based on subtraction. There are also continued fraction type algorithms which involve division, generalisations of the basic step $x_k = [x_k] + 1/x_{k+1}$, and which may also possibly involve renormalisation (such as replacing $(x_0, y_0, \ldots, z_0, w_0)$ by $(x_0/w_0, y_0/w_0, \ldots, z_0/w_0)$, thus seeming to deal with one less dimension). The Jacobi–Perron algorithm is an example of such a division algorithm. In its normalised form it is: Given $(x_0, y_0, \ldots, z_0)$, if x_0 is not an integer, replace each term by its fractional part, shift one place to the left, insert a 1 in the final place, and divide by $x_0 - [x_0]$. Thus

$$(x_k, y_k, \ldots, z_k) \to \left(\frac{y_k - [y_k]}{x_k - [x_k]}, \ldots, \frac{z_k - [z_k]}{x_k - [x_k]}, \frac{1}{x_k - [x_k]} \right).$$

If we perform this curious ritual on the singleton (x_0), we get $x_{k+1} = (x_k - [x_k])^{-1}$, the classical continued fraction algorithm.[43] But why should such a complicated generalisation have provoked so much investigation and counterproposal, as indeed the Jacobi–Perron algorithm has?

The periodicity of the Euclidean algorithm for quadratic surds, property (iv), is a most remarkable mathematical phenomenon. It lies at the heart of my proposed reconstructions of Greek mathematics. It was clearly a property that fascinated Euler and Lagrange, as I have tried to illustrate in Section 9.1 above. That Gauss may have thoroughly understood this mathematical fact and its implications, but yet been unable to extend this kind of behaviour to any other algebraic numbers, may have been one of the reasons why he purged continued fractions from his number theory in favour of the more general and promising theory of quadratic forms, and turned instead towards the probability

---

[43] For a more sympathetic geometrical description of the Jacobi–Perron algorithm as a generalisation of the *Parmenides* algorithm, and related comments, see Pisot, SM.

9.3 Recent developments

theory of continued fractions. Lagrange posed his problem about higher-degree algebraic numbers, quoted in Section 9.1(e) above, from firmly within the context of periodic continued fractions. Seventy years later, in 1839, Hermite posed exactly the same kind of question, again making explicit reference to the periodicity of the continued fraction expansion of quadratic surds, to Jacobi, who responded with his algorithm, an unnormalised three-dimensional version of that described above, and with examples of cases where it was and was not periodic. Jacobi's algorithm was published only posthumously in 1869 in an article, ATKA, completed by Heine from Jacobi's notes, and it was later generalised and studied intensively by Perron, who proved that it will converge, though it may not generate the best possible approximations, except for very special cases in low dimensions. And the fundamental issue of its periodicity still remains unresolved, despite much effort.[44]

Now we come to the final irony. By setting condition (iv) aside, so turning away from the fascination of the periodicity of the classical algorithm, and by concentrating on the subtractive formulation of the Euclidean algorithm, rather than division and a continued fraction algorithm, a new and successful kind of generalisation to all dimensions has been found by Helaman Ferguson & Rodney Forcade; see their GEA and MEA.[45] These algorithms are far too complicated to describe here; but they are more geometrical, illuminating, and already seem to be

---

[44] For references and details of the history and theory of the Jacobi–Perron algorithm, see Berstein, *JPA*. Higher dimensional algorithms seem to inspire a purple prose; Berstein refers to the attempts to generalise this periodic theory in terms of "Dantesque (sic) despair" of "abandoned hope", relieved only by the "great master" Jacobi and the nonagenarian Perron "still working feverishly on the periodicity question of the Jacobi algorithm", which however, is "still waiting for the master-mind to decipher it completely" (see the historical sketch, pp. 1–10). One can also detect the excitement in the warm but sober letters of Minkowski to Hilbert: "I am nearly finished with the continued fractions for two real numbers; this final version of this examination will, I think, be pretty instructive. It is very similar to the Jacobi algorithm; however my algorithm has a few more frills (*Chicanen*) which allow for all sorts of questions to be asked which would be arbitrary for Jacobi's algorithm. Hermite has now scrutinised that part of my book which I have sent him. He wrote very delightedly to the translator about it: Je crois voir la terre promise, etc." (Minkowski to Hilbert, 20 August, 1894). Alas, on 10 February 1896, he was to write: "The complete presentation of my investigations on continued fractions has reached almost a hundred printed pages but the all-satisfying conclusion is still missing: the vaguely conceived characteristic criterion for cubic irrational numbers." (Both extracts translated from Minkowski, *BDH*, 62 & 77.) Also see the next note.

[45] In an unpublished historical survey of higher dimensional algorithms Ferguson & Forcade call periodicity a "fluorescent red herring", and a "track laid by Lagrange down which many more trains were to roar impressively if fruitlessly". Of the Jacobi–Perron algorithm, they observe "there is undoubtedly a hoary counterexample of degree seven involving five digit coefficients". Minkowski's algorithm (see the previous note) does in some sense answer Lagrange and Hermite's question and also sets strict limitations on the scope of periodicity; but it is not the iterative kind of algorithm considered here. A modern account of Minkowski's algorithm and a report of extensive computer explorations using the Jacobi–Perron algorithm can be found in Ferguson, *ACGLSE*.

more successful than the earlier attempts at periodic algorithms have been.

9.4 Epilogue

At first sight, the single most serious objection to the reconstructions of early Greek mathematics developed here must be that if the exploration of ratio theories in general and anthyphairesis in particular did play a role such as I have proposed, then why do our sources, both early and late, not refer to these topics? Here, then, are some reactions to this issue.

First, our sources are not silent. Some of them may be explicitly referring to these topics, if only we could understand them; in other cases, we get a comprehensive and straightforward explanation of otherwise perplexing features when the material is explained from this point of view. For example, two perplexing fragments of Archytas concern *logistikē* and *logismos*; and Plato may tell us a great deal more about this 'ratio theory', in his many further references, if only we interpret him correctly.[46] Aristotle, in possibly our most explicit piece of information—perhaps even our only surviving explicit piece of information—about any early definition of ratio, tells us of *antanairesis*.[47] Euclid never gives us motivation for any of the mathematics he presents, but the only mathematically meaningful explanation of the basic distinction between the *rhētē* and *alogoi* lines of *Elements* X and XIII that I know is the one based on anthyphairesis described above in Chapter 5; and I believe that further relevant material can be found in *Elements* VII. The idiosyncratic features of Archimedes' *Measurement of a Circle* fit more naturally in an anthyphairetic context than an arithmetised one, as I have tried to illustrate in Sections 2.2(b) and 2.4(d). Nicomachus, one of our earliest commentators on *arithmētikē*, writing in the second century AD, gives a discussion of what he called *antaphairesis* and an elaboration of the classification into multiple, epimoric, and epimeric ratios.[48] The later Arab commentators, some of whom are beginning to be revealed as considerable mathematicians in their own right, do several times invoke anthyphairesis in their commentaries on *Elements* V.[49] The systematic way of calculating gear ratios such as are found in the Antikythera mechanism uses anthyphairesis.[50] And suggestive material on astronomi-

---

[46] See the index in the Appendix to Chapter 4, and the proposals in that chapter.
[47] Aristotle, *Topics* VIII 3, 158b29–35, quoted in Section 1.2(d), above.
[48] See Nicomachus–D'Ooge, *IA*, I xiii and xvii seqq., and Section 4.5, above, especially A_{80}–B_{83}.
[49] The Arabic commentaries on ratio are summarised and discussed in Plooij, *ECR*. Also see Omar Khayyam–Amir-Móez, DDE.
[50] See Price, *GG*, especially p. 58: "In this [now lost] planetarium Archimedes would have used, perhaps for the first time, sets of gears arranged to mesh in parallel planes, and he would have been led to the rather elegant number manipulation which is necessary to get a set of correct ratios for turning the various planetary markers". This is, of course, the same kind of investigation that we later find in Huygens, DAP; see the Historical Notes to Section 9.1(a), above.

cal ratios is found in the fifth and fourth centuries BC, and on musical ratios throughout early and late antiquity.[51]

Nevertheless, the situation is curious. If anthyphairesis, for example, did provide an impetus to the development of mathematics, as in my reconstruction, then why does nobody tell us this more clearly? I shall approach this more refined version of the question from several different directions. I consider first the question in its historical context of early Greek mathematics; then consider the reactions of the later commentators; then introduce and develop an analogy from more recent times.

So, first, consider our early sources. My study here has been restricted to only a part of the evidence deriving: from Plato and his friends, who met in their suburb of Athens and communicated amongst themselves by conversations and letters; from Archimedes, whose isolation from the mathematicians of Alexandria is testified by the prefaces to his works; and from the elusive figure of Euclid. We know the names and, sometimes, a few details of the work of many other mathematicians, working in other places, or at other times, or on other topics; and we also know that my small, selected group of mathematicians were not working exclusively or even primarily on mathematics, or exclusively on the topics in mathematics that I have been considering. So we should not think of the mathematicians of the fourth and third centuries BC as a homogeneous group, in close and constant contact; and the specialism of my study here should not be thought of as reflecting any exclusive specialism of its restricted cast of characters. And in this Greek world of intense intellectual, artistic, and political exploration, the excitement of the explosion of ideas across such a vast front might leave little time for, and less interest in, a clear and correct historical account by the participants.

Now turn to the later commentators, from whom the main structure of the received interpretation derives. If they had got main details of their story right, if they still had reliable sources and had been able to use them with insight, then there might not be the fundamental problems about understanding early Greek mathematics that still face us today. But we cannot have that confidence, since a long process of refinement and critical analysis of their writings has only led to deeper and deeper problems and controversies.[52] If we did have a consistent and convincing knowledge of early Greek mathematics, there would be only the details to resolve and historians would have moved on to other fields. But this is manifestly not the case; it is still not possible to arrive at any general consensus even over some of the most fundamental issues.

The proposals set out here are only the beginning of a longer argument, since they have so far not taken into account the reactions of

---

[51] See the references in Section 4.4 and 5.
[52] The discussion, in Section 3.6(b) and Chapter 8, above, gives an indication of some of the difficulties with this material.

these later commentators whose possible misunderstandings can, I believe, be retrospectively understood and, to some extent, explained. To introduce my brief discussion here of this, which will be based on an analogy with more recent times, I first propose an experiment to my reader. I have tried, in this chapter, to indicate how the study of continued fractions, our version of anthyphairetic ratio theory, runs broad and deep through our mathematical heritage since the Renaissance, how some of the most influential and celebrated mathematicians have drawn inspiration from, and made contributions to, the theory, and how its properties lie behind some well-known and widely studied parts of mathematics today. But take down the fattest general history of mathematics from your shelf and look up 'continued fractions', 'Euclidean algorithm', or any other such synonym in the index; go to your nearest specialised mathematics library and search there; ask your neighbourhood mathematician for information about the subject; and so on. See if any of these sources give any hint of any of these developments. You will, I anticipate, be unlikely to find much beyond a general discussion of Pell's equation. Why do our commentators and compilers today make so little reference to this important theme in the modern history of the subject? There have been, since the seventeenth century, increasingly coherent communities of professional mathematicians and historians, drawn together through increasingly elaborate networks of personal and professional publications. Do we have any good reason to expect that the much more isolated ancient commentators and compilers could have done better?

It is a commonplace of the contemporary mathematical community that mathematical research fills out a broad spectrum. At one end are the informal and ephemeral activities of lectures, letters, and conversations, where underlying motivation is discussed, promising problems are formulated with suggestions for solution, blind alleys are described, and on, and where communication is carried out quickly and efficiently in what is often an inchoate, fragmentary, ambiguous, improvised language. The other end of the spectrum is dominated by the published tradition. Here much of the accompanying matter of the informal tradition is suppressed, and only successful investigations are presented, however irrelevant they may be to the original motivating problems. Few unsuccessful investigations, however relevant, get published because, quite simply, what can one say beyond that one has not succeeded?[53] The published works are

---

[53] Occasionally, however, an unsuccessful investigation may succeed in diverting mathematical activity away from the problem in hand. Poincaré's proof of the impossibility of one kind of solution of the three-body problem diverted attention away from the 2300-year-old mathematical study of the movement of sun and moon and towards topology; and I have proposed in Section 9.2(d), above, that Gauss's explorations of continued fractions in number theory, of which to my knowledge no trace survives, may have led him to develop further and promote the theory of quadratic forms.

set out in a very formal style, often involving a highly developed symbolic language that may be accessible only to those other mathematicians who specialise in the immediate topic. It is then only possible to get behind this published presentation and uncover with certainty the unexpected deeper motivations and the unpublished record of failed investigations when we can back up this published record with the conversations, letters, working papers, and rubbish baskets that belong to the informal tradition. Moreover, published mathematics can remain unread or unassimilated, even when it comes from the pen of a master. There are few who have the time, breadth of interest, and ability to work through the eighty-four volumes of Euler's collected works, or the insight to understand Riemann's single volume, or the patience and persistence to read *Elements* X.

Such is the situation today and I see little reason to expect anything different at the dawn of this kind of mathematics; surely no one can believe that there was not some rich prehistory to what we find in Euclid's *Elements*. But nothing of what was then written has survived in its original form: still less can we, or could the later commentators, know of the conversations in the garden of Academe and elsewhere. In the case of early Greek mathematics, the important informal tradition has been completely lost.

With the *Elements* we find ourselves facing a further barrier. No mathematician I can conceive talks like this:

If a straight line be bisected and a straight line be added to it in a straight line, the rectangle contained by the whole with the added straight line and the straight line together with the square on the half is equal to the square on the straight line made up of the half and the added straight line [*Elements* II 6],

though, unfortunately, many write thus, confident thereby of their place in an ancient tradition. The convoluted language is necessary here because formal Euclidean style does not allow any reference to any specific figure in its enunciation.[54] But many of the explorations and algorithms I have described in Part One simply cannot be expressed within the constraints of this formal Euclidean style; see S_{27}, in Section 1.3. Moreover, while these algorithms and explorations play an essential role in the understanding of anthyphairetic ratio theory, they have no part in the associated deductive theory. Observe, for example, how in Chapters 2 to 5 I have described and used in a fundamental way, but have not proved, any of the surprising and important behaviour that we

---

[54] The formal style of the *Elements,* in particular the structure of a Euclidean proposition, apparently as strict as a piece of music in the strictest of classical styles and as often broken in practice, is described in Heath, *TBEE* i, Chapter 9, 114–51, and Mueller, *PMDSEE,* Chapter 1, 1–57. Here, again, our main source is Proclus, *Commentary on the First Book of Euclid's Elements,* 203–5.

observe in the *Parmenides* algorithm: that its run-lengths do in fact generate the anthyphairesis, that its behaviour will in fact be periodic when applied to $\sqrt{n}:\sqrt{m}$, that it will in fact always generate solutions to Pell's equation, and so on.[55] Here again, it is not easy to see how to give a deductive proof of these results without using the machinery of symbolic algebra and complete induction, let alone to see how this, also, could be expressed in formal Euclidean style. This then becomes the kind of material that belongs to the mathematician's craft, the everyday skills that are learnt orally, which are an essential part of his trade, but which often die along with the community that is investigating a particular corner of mathematics.

There is a further characteristic of continued fractions that makes them all the more prone to periods of fascination followed by neglect. On the one hand, the route via continued fractions is one of the quickest into the mathematical way of thinking. The *Parmenides* algorithm applied to $\sqrt{n}:\sqrt{m}$—or any of its equivalent manifestations—leads directly into subtle, surprising, and useful mathematics whose exploration will initiate and train a gifted novice. Further exploration will reveal rich further developments, and promise of continuing spectacular progress. But continued fractions are also the quickest route to a second mathematical phenomenon: the unsolved problem and, worse, the unfruitful unsolved problem. Mathematics can be a gratuitous activity and mathematicians are often wilful, as Plato complained at *Republic* VII, 528b–c. When the scent of an unsolved problem goes stale, when a new and promising point of view is revealed elsewhere, they will move off in another direction. The general ignorance of and little progress made in understanding the continued fraction expansion of higher-degree algebraic numbers, or

---

[55] This is exactly the kind of criticism that Lagrange levelled against earlier treatments of Pell's equation and continued fractions. See, for example, the introduction to his SPA: "Given any non-square positive integer find a square integer such that the product of these two numbers, increased by one, will be a square number. This is one of the problems that Mr. Fermat proposed, as a kind of challenge, to English mathematicians, in particular to Mr. Wallis who was the only one, to my knowledge, who resolved it or at least who published its solution. (See Chapter 98 of his *Algebra* and letters 17 & 19 of his *Commercium Epistolicum*.) But his method only consists of a kind of groping (*tâtonnement*) by which he arrives at the solution only in an uncertain manner, without knowing that he will actually ever get there. However, one should above all show that the solution of the problem is always possible, whatever the given number, a proposition that is generally regarded as true but which has to my knowledge not yet been firmly and rigorously proved. It is true that Mr. Wallis claims to have proved it, but by an argument that mathematicians find very unsatisfactory and which to me seems to beg the question (see Chapter 19 of his *Algebra* [and, in Section 9.2(d), above, Gauss's note *e* to the *Disquisitiones Arithmeticae*, §202]). Thus the problem has not yet been satisfactorily resolved in a way that leaves nothing to be desired. This is what determined me to make it the object of my study, the more so since the solution of this problem is like a key to all other problems of this type." (The original French is quoted in my BTEE2, 205 n. 18.) In fact many of the explorations before Lagrange were in the nature of what I have called 'heuristics' but which Lagrange characterised, here and elsewhere, by the very Lagrangian word '*tâtonnement*'.

higher-dimensional continued fractions, since the time of Gauss, despite the concentrated effort of a few maverick mathematicians, give us clear illustrations of stale problems and mathematicians' reactions to them. I have proposed here that analogous problems could also have arisen in the Greek context and they would, by the same token, have gone stale there too. In a process that is found many times throughout mathematics, the original problems themselves then disappear but the language that has developed in the pursuit of their solution lives on, transformed into the status of a theory. Many of the great theories of modern mathematics arose in this way, and I suggest that the same may have happened with early Greek mathematics, with geometry and *arithmētikē*.

So far in this final section I have considered only anthyphairetic ratio theory; but similar comments apply with even more force to the astonomical and musical ratio theories whose reconstruction I sketched in Chapter 4. Astronomical ratio theory, as I described it in Section 4.4, is so subtle that it scarcely gets beyond the stage of a definition and initial exploration before it peters out; though, as I suggested at E_{57}, it may not have disappeared but survived, transformed, as the proportion theory of *Elements* V. And the bridge between musical ratio theory and the other ratio theories—namely, arithmetic with ratios—is so well hidden in the mathematical jungle that my proposed investigation could never have got beyond a few, faltering steps.

In addition to these general characteristics of the mathematical context, there is a further specific historical development that would compound the possibility of a later misunderstanding of any early Greek work on ratio theories. I have here throughout insisted on the non-arithmetised character of early Greek mathematics. We have abundant evidence of an earlier, profoundly arithmetical, Babylonian mathematics, but our historical record so far shows no explicit trace of any influence of Babylonian techniques on *early* Greek mathematics or astronomy.[56] Then, in the second century BC, with Hipparchus and Hypsicles, we find the first Greek usage of sexagesimal fractions; and thereafter Greek mathematics and science, as seen in the works of Ptolemy, Heron, Diophantus, and the later commentators, becomes a powerful mixture of the arithmetical and geometrical points of view. Now the attitudes and problems associated with different kinds of non-arithmetised and arithmetised mathematics can sometimes be very different, even incompatible; and this is particularly so for the case of ratio theories. Arithmetised ratio theory makes a sharp distinction between commensurable (our 'rational') and incommensurable (our 'irrational') ratios, the former being amenable to basic mathematical techniques, the latter requiring a much more

---

[56] This is a typical example of a fundamental issue on which there is no general consensus. For a brief summary of the different positions, see Berggren, HGM, 397–8.

subtle treatment; then when the ratios are conceived as sexagesimal or decimal numbers, there are other kinds of distinctions, such as whether the ratios terminate or not, or are periodic or not. To be sure, the dichotomy between commensurable and incommensurable also manifests itself in other ratio theories, for example as terminating or non-terminating anthyphairetic ratios, but this distinction is much less striking than the difference between anthyphairetic ratios of the form $\sqrt{n}:\sqrt{m}$ (Euclid's *rhētos*) and the rest (Euclid's *alogos*), a difference which lies at the heart of my reconstruction here. Reciprocally, the difference between $\sqrt{2}:1$ and $\sqrt[3]{2}:1$ is much less striking within an arithmetised ratio theory than an anthyphairetic theory. Or, for astronomical ratios, commensurability will manifest itself as a periodicity, with or without coincidences, but it is difficult, without very careful analysis, to comprehend the general commensurable astronomical ratio; here the understandable astronomical ratios are practically confined to multiple, epimoric, and multiple epimoric ratios.[57] Now arithmetised mathematics leads to a kind of algebraic reasoning that is absent from non-arithmetised mathematics, since the arithmetised operations which are then generalised and abstracted to give this algebra are either completely absent, or have a very special character, specific to each context.[58] So it may become difficult to appreciate the motivation, strengths, and weaknesses of one approach from within the context of another. For example, the ratio of the diagonal to side of a square poses a serious problem of understanding to arithmetised ratio theory; in the non-arithmetised anthyphairetic ratio theory, once an appropriate figure has been drawn, it becomes a basic and vivid example which marks an entrance into the theory. *Elements* V, Definition 5, posed a conundrum in arithmetised mathematics that was not really understood before Dedekind, but it corresponds to the first insight into the nature of astronomical ratios. Algebraic manipulations are abstractions from arithmetic; but the problem of describing the basic arithmetical operations within anthyphairetic ratio theory has only recently been solved, and I know of no such algorithms for astronomical ratio theory. Now extend these specific examples to a more global point of view: if, as I have proposed, early Greek mathematics was non-arithmetised, then we may not be able to understand or translate its motivations and some of its methods into our arithmetised point of view. So there would be a tendency for it to be misunderstood from the second century BC onwards.

So while I can understand a sceptical reader remaining worried by the lack of any clear and explicit anthyphairetic explanation of early Greek mathematics in the last two thousand and more years, I cannot agree that

---

[57] Witness the complications of the descriptions in E_{67} and B_{68} of Section 4.4(b).
[58] See Section 4.5(b), A_{86-88}.

this absence is fatal to my reconstruction. Most of what has been said on the topic in this period has been speculation, much of it has not been written by mathematicians, and most of it comes from within alien and incompatible mathematical traditions. However much it may tell us about the intellectual climate of the times at which it was written and of its author's ideas, most of it may be simply irrelevant to early Greek mathematics. I hope it is clear, too, that I am aware that however plausible and appealing my proposals here might be, they too may be irrelevant or wrong, like so many before them; though while I can see the possibility of quite drastic revisions to their details, I am not yet aware of any cogent arguments for rejecting them outright. And I do not think that one should confuse the exploration of the documented stages of some intellectual development, such as we have for some episodes in mathematics since the seventeenth century, with the kind of plausible rational reconstructions that have been attempted here. But, given the state of our evidence, no other approach to early Greek mathematics is possible.

I have dwelt very little on the historical and biographical part of early Greek mathematics here. Our sources give us very little reliable biographical information, even about matters as basic as chronology, and most of the details have to be reconstructed. In a subject with the logical coherence of mathematics, there is a tendency for the mathematical reconstruction to precede and determine the biographical reconstruction; for example, details about the discovery of incommensurability tend to affect reconstructions of pre-Socratic philosophy and mathematics, while the reconstructions of proportion theory have had an important part in the biographical reconstructions of the fifth and fourth centuries BC.[59] The drastic revision of the reconstruction of pre-Euclidean mathematics proposed here will provoke some revision of the historical picture, but I did not wish to complicate the issue by introducing these matters at this stage. However, I cannot resist finishing with a historical analogy. It is a convenient tendency, sometimes not without truth, to personify trends and locate movements in individuals. In this spirit, it is tempting to speculate that, just as I have suggested that Gauss rewrote number theory without continued fractions and thus may have prompted the decline in their study, so Eudoxus may have rewritten anthyphairetic and astronomical ratio theories as proportion theory with the same effect; and that both worked with a deep understanding of what they were omitting.

---

[59] See, for example, Knorr, *EEE,* Chapter 2, and de Santillana, EPSC.

BIBLIOGRAPHY

The general principles of the referencing system by acronym have been described in the Introduction. To these can be added: authors are listed alphabetically by first upper case letter of family name when abbreviations have been expanded. (Thus V. De Falco under D; G. E. M. de Ste. Croix under S, between Saffrey and Sandys; and D. de Solla Price under P.) Items that have also appeared conveniently in collected works are also cross-referenced to these works, which are identified by a keyword (*Works, Werke, Opera,* etc.), not an acronym; and collected works precede any references to items they contain. In general, the main reference is to the most convenient modern edition or reprint, and the information about original publication may be restricted to a parenthetic date immediately after the title. Items within an author heading are arranged chronologically by first publication; single authors come before joint authors. Occasional parenthetic comments give references to reviews or other information.

This bibliography is restricted to works that have been cited in the text. For more comprehensive bibliographies see, for the history of mathematics in general: Dauben, *HMAP*; for ancient mathematics, in particular: Knorr, *EEE* 353–68, Procissi, *BMGA*, and Thomas, *SIHGM* i (repr. 1980) pp. xvii–lii; on the history of continued fractions: the forthcoming Brezinski, *HCF*; on scientific biography: the *DSB*; on Plato: Cherniss, P50–57, Brisson, P58–75, Brisson & Ioannidi, P75–80; and for editions of classical texts and general topics in classical antiquity: the Introduction to Liddell, Scott, & Jones, *GEL,* and the *OCD.*

Alexander of Aphrodisias–Wallies, *IAP = In Aristotelis Analyticorum Priorum Librum 1 Commentarium,* ed. M. Wallies, Commentaria in Aristotelem Graeca 2_1, Berlin: Hayduck, 1883.

Alexander of Aphrodisias–Wallies, *IT = In Aristotelis Topicorum Libros Octo Commentaria,* ed. M. Wallies, Commentaria in Aristotelem Graeca 2_2, Berlin: Hayduck, 1891.

Archimedes–Heath, *WA = The Works of Archimedes Edited in Modern Notation* (1897) with supplement *The Method of Archimedes* (1912), tr. T. L. Heath, Cambridge: Cambridge University Press, 1912, repr. New York: Dover, n.d.

Archimedes–Heiberg, *Opera = Opera Omnia cum Commentariis Eutocii,* ed. J. L. Heiberg, 2nd edn. 3 vols., Leipzig: Teubner, 1910–15, repr. Stuttgart: Teubner, 1972.

Aristarchus–Heath, *AS = Aristarchus of Samos, The Ancient Copernicus,* ed. & tr. T. L. Heath, Oxford: Clarendon Press, 1913, repr. New York: Dover, 1981.

Aristotle, Individual works with the Greek text and parallel English translation in the Loeb Classical Library, Cambridge, Massachusetts & London: Harvard University Press & Heinemann.

Aristotle, Critical editions of the Greek text of individual works in the Scriptorum Classicorum Bibliotheca Oxoniensis (= The Oxford Classical Texts), Oxford: Clarendon Press.

Aristotle–Barnes, *CW* = *The Complete Works of Aristotle*, ed. J. Barnes, 2 vols., Bollingen Series 71_2, Princeton: Princeton University Press, 1984. (A revised version of Aristotle–Ross, *WA*.)

Aristotle–Bekker, *Opera* = *Aristotelis Opera*, ed. I. Bekker, 5 vols., Berlin: Königliche Akademie der Wissenschaften, 1831–70, often reprinted.

Aristotle–Ross, *WA* = *The Works of Aristotle Translated into English*, ed. W. D. Ross, various translators, 12 vols., Oxford: Clarendon Press, 1908–52. (Also see Aristotle–Barnes, *CW*.)

Baker, A., RACTAN = Rational approximation to $\sqrt[3]{2}$ and other algebraic numbers, *Quarterly Journal of Mathematics* (2nd series) 15 (1964), 375–83.

Barker, A., MPSA = Music and perception: A study in Aristoxenus, *Journal of Hellenic Studies* 98 (1978), 9–16.

Barker, A., MAESC = Methods and aims in the Euclidean *Sectio Canonis*, *Journal of Hellenic Studies* 101 (1981), 1–16.

Barker, A., *GMW* = *Greek Musical Writings*, 2 vols., Cambridge: Cambridge University Press, vol. 1 1984, vol. 2 in preparation.

Becker, O., ESI = Eudoxus-Studien I. Eine voreudoxische Proportionenlehre und ihre Spuren bei Aristotles und Euklid, *Quellen und Studien zur Geschichte der Mathematik, Astronomie und Physik*, Abteilung B: *Studien* 2 (1933), 311–33.

Berggren, J. L., HGM = History of Greek mathematics, A survey of recent research, *Historia Mathematica* 11 (1984), 394–410.

Berstein, L., *JPA* = *The Jacobi–Perron Algorithm, Its Theory and Application*, Lecture Notes in Mathematics 207, Berlin, Heidelberg, New York: Springer, 1971.

Bilabel, F., *SGUA* = *Sammelbuch Griechischer Urkunden aus Ägypten*, ed. F. Bilabel & others, 13 vols. to 1979, Berlin & Leipzig: de Gruyter. (Cited by papyrologists under the abbreviation SB.)

Blanchard, A., *SAPDG* = *Sigles et Abreviations dans les Papyrus Documentaires Grecs: Recherches de Paleographie*, London: Bulletin of the Institute of Classical Studies Supplement 30, 1974.

Boethius–Friedlein, *IM* = *De Institutione Arithmetica & De Institutione Musica*, ed. G. Friedlein, Leipzig: Teubner, 1867.

Bonitz, H., *IA* = *Index Aristotelicus*, Berlin: Reimer, 1831, often reprinted. (= vol. 5 of Aristotle–Bekker, *Opera*.)

Bos, H. J. M., AMRDMT = Arguments on motivation in the rise and decline of a mathematical theory; the "Construction of Equations", 1637—ca.1750, *Archive for History of Exact Sciences* 30 (1984), 331–80.

Bowen, A. C., FEPHS = The foundations of early Pythagorean harmonic science: Archytas, fragment 1, *Ancient Philosophy* 2 (1982), 79–104.

Brandwood, L., *WIP* = *A Word Index to Plato*, Leeds: W. S. Maney, 1976.

Brashear, W., CTAL = Corrections à des tablettes arithmétiques du Louvre, *Revue des Études Grecques* 97 (1984), 214–17.

Brashear, W., MS = The myrias-symbol in CPR vii 8, *Zeitschrift für Papyrologie und Epigraphik* 60 (1985), 239–42.

Brezinski, C., LHCF = The long history of continued fractions and Padé approximants, pp. 1–27 in *Padé Approximation and its Applications, Amsterdam 1980,* ed. M. G. de Bruin & H. van Rossum, Lecture Notes in Mathematics 888, Berlin, Heidelberg, New York: Springer, 1981.

Brezinski, C., *HCF = History of Continued Fractions and Padé Approximants,* Berlin, Heidelberg, New York: Springer, in press.

Brisson, L., P58–75 = Platon 1958–1975, *Lustrum* 20 (1977), 5–304.

Brisson, L. & Ioannidi, H., P75–80 = Platon 1975–1980, *Lustrum* 25 (1983), 31–320.

Brown, M., PDSA = Plato disapproves of the slave-boy's answer. *The Review of Metaphysics* 20 (1967), 57–93, repr. in Brown, *PM,* 198–242.

Brown, M., *PM = Plato's Meno,* tr. W. K. C. Guthrie, with essays, ed. M. Brown, Indianapolis: Bobbs-Merrill, 1971.

Brun, V., GK = En generalisation av kjedebroken I & II, *Norske Videnkapsselskapets Skifter I, Matematisk-Naturvidenskapelig Klasse* 6 (1919), 1–29 & 6 (1920) 1–24.

Brun, V., AETQN = Algorithms euclidiens pour trois et quatre nombres, *Treizième congres des mathematiciens scandinaves, Helsinki* (1958), 46–64.

Bulmer-Thomas [= Thomas], I., PA = Plato's Astronomy, *Classical Quarterly* 34 (1984), 107–12.

Burkert, W., *LSAP = Lore and Science in Ancient Pythagoreanism,* tr. E. L. Minar of *Weisheit und Wissenschaft: Studien zu Pythagoras, Philolaos und Platon* (1962), Cambridge, Massachusetts: Harvard University Press, 1972.

Burnyeat, M. F., MSPD = The material and sources of Plato's dream, *Phronesis* 15 (1970), 101–22.

Burnyeat, M. F., PSTM = The philosophical sense of Theaetetus' mathematics, *Isis* 69 (1978), 489–513. (Also see Knorr, MPP.)

Butler, A. J., *ACE = The Arab Conquest of Egypt and the Last Thirty Years of the Roman Dominion* (1902), 2nd edn. ed. P. M. Fraser, Oxford: Clarendon Press, 1978.

Cameron, A., LDAA = The last days of the Academy in Athens, *Proceedings of the Cambridge Philological Society* 95 (New Series 15) (1969), 7–29.

Cavallo, G., LSSE = Libri scritture scribe a Ercolano, *Cronache Ercolanesi* 13, Supplemento 1, 1983.

Černý, J. *PBAE = Paper and Books in Ancient Egypt* (An inaugural lecture, May, 1947), London: University College & H. K. Lewis, 1952, repr. Chicago: Ares, 1977.

Chace, A. B., *RMP = The Rhind Mathematical Papyrus,* 2 vols., Oberlin, Ohio: Mathematical Association of America, 1927–29, abridged repr. in 1 vol., Classics in Mathematics Education 8, Reston, Virginia: The National Council for Teachers of Mathematics, 1979.

Cherniss, H., *REA = The Riddle of the Early Academy,* Berkeley: University of California Press, 1945, repr. with index by L. Tarán, New York & London: Garland, 1980.

Cherniss, H., P50–57 = Plato 1950–1957, *Lustrum* 4 (1959), 5–308 & 5 (1960), 321–656.

Cherniss, H., *Papers* = *Selected Papers*, ed. L. Tarán, Leiden: Brill, 1977.
Cherniss, H., PM = Plato as mathematician, *The Review of Metaphysics* 4 (1951), 395–425, repr. in *Papers*, 222–52.
Christoffel, E. B., OA = Observatio arithmetica, *Annali di Mathematica* (Series 2) 6 (1875), 148–52.
Chrystal, G., *TA* = *A Textbook of Algebra* (1886), 7th edn. 2 vols., repr. New York: Chelsea, 1964.
Chuquet–Flegg, Hay, & Moss, *NC* = *Nicholas Chuquet, Renaissance Mathematician, A Study with Extensive Translation of Chuquet's Mathematical Manuscript Completed in 1484 [= Le Triparty en la Science des Nombres]*, ed. G. Flegg, C. Hay, & B. Moss, Dordrecht: Reidel, 1985.
Churchhouse, R. F. & Muir, S. T. E., CFANMI = Continued fractions, algebraic numbers, and modular invariants, *Journal of the Institute of Applied Mathematics* 5 (1969), 318–28.
Clagett, M., *AMA* = *Archimedes in the Middle Ages*, 5 vols. in 10 parts, Philadelphia: American Philosophical Society, 1964–84.
Cockle, W. E. H., RCP = Restoring and conserving papyri, *Bulletin of the Institute of Classical Studies* 30 (1983), 147–65.
Coldstream, J. N., *GG* = *Geometric Greece*, London: Benn, 1977.
Connelly, R., FS = Flexing surfaces, pp. 79–89 in *The Mathematical Gardner*, ed. D. A. Klarner, Boston: Prindle, Webster, & Schmidt and Belmont, California: Wadsworth, 1981.
Cotes, R. L., L = Logometria, *Philosophical Transactions of the Royal Society* 29 (1714), 5–47.
Cotes, R., *HM* = *Harmonium Mensurarum, sive Analysis et Synthesis per Rationum et Angulorum Mensuras Promotae: Accedent alia Opuscula Mathematica per Rogerum Cotesium*, London, 1722.
Coulton, J., TUGTD = Towards understanding Greek temple design: General considerations, *Annual of the British School at Athens* 70 (1975), 59–99; corrigenda, ibid. 71 (1976), 149–50.
Coxeter, H. S. M., *IG* = *Introduction to Geometry*, New York: Wiley, 1961.
Crawford, D. J. *KEVPD* = *Kerkeosiris: An Egyptian Village in the Ptolemaic Period*, Cambridge: Cambridge University Press, 1971.
Crawford, D. S., MT = A mathematical table [= P. Michael. 62], *Aegyptus* 33 (1953), 222–40.
Curchin, L. & Fischler, R., HANTDEMR = Hero of Alexandria's numerical treatment of division in extreme and mean ratio and its implications, *Phoenix* 35 (1981), 129–33.
Curchin, L. & Herz-Fischler, R., DQDPR = De quand date le premier rapprochement entre la suite de Fibonacci et la division en extrême et moyenne raison?, *Centaurus* 28 (1985), 129–38.
Dauben, J. W., *HMAP* = *The History of Mathematics from Antiquity to the Present, A Selective Bibliography*, New York & London: Garland, 1985.
Davenport, H., *HA* = *The Higher Arithmetic, An Introduction to the Theory of Numbers* (1952), 5th edn., Cambridge: Cambridge University Press, 1982.
Davenport, H. & Roth, K. F., RAAN = Rational approximations to algebraic numbers, *Mathematika* 2 (1955), 160–67.
Davy, H., SOEPFRH = Some observations and experiments on the papyri found

in the ruins of Herculaneum, *Philosophical Transactions of the Royal Society*, 111 (1821), 191–208.

Dedekind, R., *ETN* = *Essay on the Theory of Numbers: I. Continuity and Irrational Numbers [Stetigkeit und die Irrationale Zahlen (1872)], II. The Nature and Meaning of Numbers [Was sind und was sollen die Zahlen (1888)]*, tr. W. W. Beman, Chicago: Open Court, 1901, repr. New York: Dover, 1963.

De Falco, V., *EDL* = *L'Epicureo Demetrio Lacone*, Naples: Achille Cimmarua, 1923.

Delone, B. N. & Faddeev, D. K., *TITD* = *The Theory of Irrationalities of the Third Degree*, tr. E. Lehmer & S. A. Walker of *Teoriya Irratsional'nostei Tretéi Stepeni* (1940), Translations of Mathematical Monographs 10, Providence: American Mathematical Society, 1964.

Dicks, D. R., *GFH* = *The Geographical Fragments of Hipparchus*, London: University of London Press, 1960.

Dickson, L. E., *HTN* = *History of the Theory of Numbers*, 3 vols., Washington, D.C.: Carnegie Insitute 1919–23, repr. New York: Chelsea, 1971.

Diels, H. & Kranz, W., *FV* = *Die Fragmente der Vorsokratiker*, 6th edn., Zurich: Weidmann, 1951–2. (Also see Freeman, *APSP*.)

Diogenes Laertius–Hicks, *LEP* = *Lives of the Eminent Philosophers*, tr. R. D. Hicks, Loeb Classical Library, Cambridge, Massachusetts & London: Harvard University Press & Heinemann, 1925, rev. 1938.

Diophantus–Heath, *DA* = *Diophantus of Alexandria: A Study in the History of Greek Algebra*, ed. & tr. T. L. Heath, 2nd edn., Cambridge: Cambridge University Press, 1910.

Diophantus–Sesiano, *BFSDA* = *Books IV to VII of Diophantus' Arithmetica in the Arabic Translation Attributed to Quṣṭā ibn Lūqā*, ed. & tr. J. Sesiano, Sources in the History of Mathematics and Physical Sciences 3, Berlin, Heidelburg, New York: Springer, 1982.

Diophantus–Tannery, *Opera* = *Opera Omnia cum Graecia Commentariis*, ed. P. Tannery, 2 vols., Leipzig: Teubner, 1893–5.

DSB = *Dictionary of Scientific Biography*, ed. C. C. Gillispie, 15 vols., New York: Charles Schribner's Sons, 1970–8.

Dutka, J., *EHHF* = The early history of the hypergeometric function, *Archive for History of Exact Sciences* 31 (1984), 15–34.

Edwards, H. M., *FLT* = *Fermat's Last Theorem*, Berlin, Heidelberg, New York: Springer, 1979.

Erichsen, W., *DG* = *Demotisches Glossar*, Copenhagen: Munksgaard, 1954.

Euclid. For an English translation, see Heath, *TBEE*.

Euclid–Heiberg, *Opera* = *Opera Omnia*, ed. J. L. Heiberg, H. Menge, & M. Curtze, 9 vols., Leipzig: Teubner, 1883–1916.

Euclid–Peyrard, *OE* = *Les Oeuvres d'Euclide* (1804), traduites littéralement par F. Peyrard, Nouveau Tirage ed. J. Itard, Paris: Blanchard, 1966.

Euclid–Stamatis, *EE* = *Euclides Elementa*, ed. J. L. Heiberg, rev. E. S. Stamatis, 5 vols. in 6 parts, Leipzig: Teubner, 1969–77.

Eudemus–Wehrli, *ER* = *Eudemos von Rhodos, Die Schule des Aristoteles: Texte und Kommentar*, ed. F. Wehrili, Basel: Schwabe, 1955.

Eudoxus–Laserre, *FEK* = *Die Fragmente des Eudoxos von Knidos*, ed. F.

Laserre, Berlin: de Gruyter, 1966. (See the review by G. J. Toomer, *Gnomon* 40 (1968), 334–7.)
Euler, L., *Opera* = *Leonhardi Euleri Opera Omnia*, various editors, Series I (*Opera Mathematica*) 29 vols., Series II (*Opera Mechanica et Asronomica*) 31 vols., Series III (*Opera Physica, Miscellanea*) 12 vols., Series IVA (*Epistolae*) in preparation, Series IVB (Unpublished Manuscripts) in preparation, Schweizerische Naturforschende Gesellschaft, Leipzig: Teubner, 1911– .
Euler, L., FCD = De fractionibus continuis dissertatio, *Commentarii academiae scientiarum Petropolitanae* 9 (1737), 98–137, repr. in *Opera* I xiv, 187–215.
Euler, L., IAI = *Introductio in Analysin Infinitorum* (1748), repr. as *Opera* I viii & ix.
Euler, L., ICD = *Institutiones Calculi Differentialis* (1755), repr. as Opera I x.
Euler, L., UNAPPS = De usu novi algorithmi in problemate Pelliano solvendo, *Novi commentarii academiae scientiarum Petropolitanae* 11 (1765), 28–66, repr. in *Opera* I iii, 73–111.
Euler, L., VAA = *Vollständige Anleitung zur Algebra,* (1770), repr. with Lagrange, *AEAE* as *Opera* I i.
Euler–Wyman & Wyman, ECF = An essay on continued fractions = Euler, FCD, tr. M. F. & B. F. Wyman, *Mathematical Systems Theory,* 18 (1895), 295–328.
Ferguson, H. R. P., ACGLSE = Arithmetic criteria, GL(n, Z[1/n!]) and some experiments, *1977 Annual Seminar of the Canadian Mathematical Congress, Queen's Papers* 48 (1978), 398–433.
Ferguson, H. R. P., RRQN = Recognize rational and quadratic numbers with Euclid's and Brun's algorithms, *Mathematics Magazine,* to appear.
Ferguson, H. R. P. & Forcade, R., GEA = Generalization of the Euclidean algorithm for real numbers to all dimensions higher than two, *Bulletin of the American Mathematical Society* (New Series) 1 (1979), 912–14.
Ferguson, H. R. P. & Forcade, R., MEA = Multidimensional Euclidean algorithms, *Journal für die reine und angewandte Mathematik* 334 (1982), 171–81.
Fischler [= Herz-Fischler], R., RE = A remark on Euclid II, 11, *Historia Mathematica* 6 (1979), 418–22.
Fletcher, T. J., AV = Approximating by vectors, *Mathematics Teaching* 63 (1973), 4–9 & 64 (1973), 42–4.
Fowler, D. H., REGM = Ratio in early Greek mathematics, *Bulletin of the American Mathematical Society* (New Series) 1 (1979), 807–46.
Fowler, D. H., BTEE = Book II of Euclid's *Elements* and a pre-Eudoxan theory of ratio, *Archive for History of Exact Sciences* 22 (1980), 5–36.
Fowler, D. H., ACPPCM = Archimedes' *Cattle Problem* and the pocket calculating machine, Coventry: University of Warwick, Mathematics Institute Preprint, 1980, supplemented 1981.
Fowler, D. H., AREP = Anthyphairetic ratio and Eudoxan proportion, *Archive for History of Exact Sciences* 24 (1981), 69–72.
Fowler, D. H., GGS = A generalisation of the golden section, *Fibonacci Quarterly* 20 (1982), 146–58.
Fowler, D. H., BTEE2 = Book II of Euclid's *Elements* and a pre-Eudoxan theory of ratio, Part 2: Sides and diameters, *Archive for History of Exact Sciences* 26 (1982), 193–209.

Fowler, D. H., IEE = Investigating Euclid's *Elements*, *British Journal for the Philosophy of Science* 34 (1983), 57–70. (A review of Mueller, *PMDSEE*.)
Fowler, D. H., NFA = A note on fractions of the artaba, *Zeitschrift für Papyrologie und Epigraphik* 52 (1983), 273–4.
Fowler, D. H., TP = Tables of parts, *Zeitschrift für Papyrologie und Epigraphik* 53 (1983), 263–4.
Fowler, D. H., EROE = Eratosthenes' ratio for the obliquity of the ecliptic, with a reply by Dennis Rawlins, *Isis* 74 (1983), 556–62. (But also see the review of Thomas, *SIHGM*, by O. Neugebauer, *American Journal of Philology* 64 (1943), 452–7, on 453.)
Fowler, D. H., CQ = A review of Taisbak, *CQ*, *Mathematical Intelligencer* 5 (1983), 69–72.
Fowler, D. H., FHYDF = 400 years of decimal fractions, *Mathematics Teaching* 110 (1985), 20–1 & 111 (1985), 30–1.
Fowler, D. H. & Turner, E. G., HP = Hibeh Papyrus i 27: An early example of Greek arithmetical notation, *Historia Mathematica* 10 (1983), 344–59.
Fraser, P. M., *PA = Ptolemaic Alexandria*, 3 vols., Oxford: Clarendon Press, 1972.
Freeman, *APSP = Ancilla to the Pre-Socratic Philosophers*, Oxford: Blackwell, 1962. (English translations of the 'B fragments' of Diels & Kranz, *FV*.)
von Fritz, K., DIHM = The discovery of incommensurability by Hippasus of Metapontum, *Annals of Mathematics* 46 (1945), 242–64.
Furlani, G., SIBA = Sull'incendio della biblioteca di Alessandria, *Aegyptus* 5 (1924), 205–12.
Furlani, G., GFIBA = Giovanni il Filopono et l'incendio della biblioteca di Alessandria, *Bulletin de la Sociéta Archéologique d'Alexandria* 21 (New Series 6) (1925), 58–77.
Galen–Kühn, *Opera* = Galen, *Opera Quae Existant*, ed. C. G. Kühn, 20 vols., Leipzig: Officina Libraria Car. Cnoblochii, 1821–33.
Galois, E., *Oeuvres = Oeuvres mathématiques d'Évariste Galois*, ed. É. Picard, Société Mathématique de France, Paris: Gauthier-Villars, 1897.
Galois, E., DTFCP = Démonstration d'un théorème sur les fractions continues périodiques, *Annales de Mathématiques* 19 (1828–29), repr. in *Oeuvres* 1–8.
Gardiner, A., *EG = Egyptian Grammar*, 3rd edn., London and Oxford: Oxford University Press for the Griffith Institute, Ashmolean Museum, 1957.
Gauss, C. F., *Werke = Werke*, 12 vols., Königlichen Gesellschaft der Wissenschaften zu Göttingen, Hildesheim: Olms, 1863–1933.
Gauss, C. F., *Tagebuch = Nachbildung und Abdruck des Tagebuchs (Notizenjournals) mit Erlanterungen*, in *Werke* x_1, 483–574.
Gauss, C. G., *DA = Disquisitiones Arithmeticae* (1801), repr. as *Werke* i.
Gauss, C. F., DGCSI = Disquisitiones generales circa seriem infinitam..., Pars prior, *Commentationes societatis regiae scientiarum Gottingensis recentiores* 2 (1813), repr. in *Werke* iii, 125–62.
Gauss, C. F., MNIVAI = Methodus nova integralium valores per approximationem inveniendi, *Commentationes societatis regiae scientiarum Gottingensis recentiores* 3 (1816), repr. in *Werke* iii, 165–96.
Gauss–Clarke, *DA = Disquisitiones Arithmeticae*, tr. A. A. Clarke, New Haven: Yale University Press, 1966.

Geffcken, J., *ZGA = Zwei Griechische Apologeten* [*Aristides & Anathagoras*], Leipzig: Teubner, 1907.

Gerstinger, H. & Vogel, K., *MNWPER = Mitteilungen aus der Nationalbibliothek in Wien: Papyrus Erzherzog Rainer,* Neue Series i [= M.P.E.R., N.S. i], Vienna: Osterreichischen Staatsdruckerei, 1932.

Gigante, M., *CPE = Catalogo dei Papiri Ercolanesi,* Naples: Bibliopolis, 1979.

Glucker, J., *ALA = Antiochus and the Late Academy,* Hypomnemata 56, Göttingen: Vandenhoeck & Ruprecht, 1978.

Goldstein B. R. NMC = A note on the Metonic cycle, *Isis* 57 (1966), 115–6.

Goldstein, B. R., OEAGA = The obliquity of the ecliptic in ancient Greek astronomy, *Archives Internationales d'Histoire des Sciences* 33 (1983), 3–14.

Goldstein, B. R. & Bowen, A. C., NVEGA = A new view of early Greek astronomy, *Isis* 74 (1983), 330–40.

Gombrich, E. H., *SI = Symbolic Images, Studies in the Art of the Renaissance,* London: Phaidon, 1972.

Goody, J. *DSM = The Domestication of the Savage Mind,* Cambridge: Cambridge University Press, 1977.

Gosper, R. W., TSCFP = Table of the simple continued fraction of π and the derived decimal approximation; see the review by J. W. Wrench, *Mathematics of Computation* 31 (1977), 1044.

Grenfell, B. P., Hunt, A. S., & Hogarth, D. G., *FTP = Fayûm Towns and their Papyri* [= P. Fay.], London: Egypt Exploration Fund, 1900.

Grenfell, B. P., Hunt, A. S., & others, *HP = The Hibeh Papyri* [= P. Hib.], 2 vols., London: Egypt Exploration Society, 1906–55.

Grenfell, B. P., Hunt, A. S., & others, *OP = The Oxyrhynchus Papyri* [= P. Oxy.], 53 vols. to 1986, London: Egypt Exploration Society, 1898– .

Griffith, F. Ll., *HPKG = Hieratic Papyri from Kahun and Gurob,* [= P. Kahun], 2 vols., London: Quaritch, 1897–8.

Guarducci, M., *EG = Epigrafia Greca,* 4 vols., Rome: Instituto Poligrafico, 1967–78.

Guthrie, W. K. C., *HGP = A History of Greek Philosophy,* 6 vols., Cambridge: Cambridge University Press, 1962–81.

Hall, F. W., *CCT = A Companion to Classical Texts,* Oxford: Clarendon Press, 1913.

Hall, M., SPCF = On the sum and product of continued fractions, *Mathematische Annalen* 48 (1947), 966–93.

Hardy, G. H. & Wright, E. M., *ITN = Introduction to the Theory of Numbers,* Oxford: Clarendon Press, 1938, 5th edn., 1979.

Harrauer, H. & Sijpesteijn, P. J., *NTAU = Neue Texte aus dem Antiken Unterricht* [= M.P.E.R., N.S. xv], 2 vols., Vienna: Hollinek, 1985.

Harvey, F. D., TKE = Two kinds of equality. *Classica ed Mediaevalia* 26 (1965), 101–46; corrigenda, ibid. 27 (1966), 99–100.

Harvey, P. D. A., *HTM = The History of Topographical Maps: Symbols, Pictures, and Surveys,* London: Thames and Hudson, 1980.

Harvey, P. D. A. & Skelton, R. A., *LMPME = Local Maps and Plans from Medieval England,* Oxford: Clarendon Press, 1986.

Havelock, E. A., *LRG = The Literate Revolution in Greece and its Cultural Consequences,* Princeton: Princeton University Press, 1982.

Heath, T. L., *HGM = A History of Greek Mathematics*, 2 vols., Oxford: Clarendon Press, 1921, repr. New York: Dover, 1981.
Heath, T. L., *TBEE = The Thirteen Books of Euclid's Elements*, 2nd edn. 3 vols., Cambridge: Cambridge University Press, 1926, repr. New York: Dover, 1956.
Heath, T. L., *MA = Mathematics in Aristotle*, Oxford: Clarendon Press, 1949, repr. New York: Garland, 1980.
Heath, T. L., also see Aristarchus-Heath, *AS* & Diophantus-Heath, *DA*.
Heiberg, J. L., PEE = Ein palimpsest der Elemente Euklidis, *Philologus* 44 (1885), 353–66.
Heiberg, J. L., QPTM = Quelques papyrus traitant de mathématiques, *Oversigt over det Kgl. Danske Videnskabernes Selskabs Forhandlinger* [= *Proceedings of the Danish Academy of Sciences*] 2 (1900), 147–71.
Heiberg, J. L., ENA = Eine neue Archimedeschandschrift, *Hermes* 62 (1907), 235–303.
Heiberg, J. L., PE = Paralipomena zu Euklid, *Hermes* 39 (1930), 46–74, 161–201, & 321–56.
Heiberg, J. L. Also see Archimedes-Heiberg, *Opera*; Euclid-Heiberg, *Opera*; and Heron-Heiberg, Schöne, & others, *Opera*.
Henrici, P., *ACCA* ii = *Applied and Computational Complex Analysis* ii, New York: Wiley, 1977.
Herodotus-Godley, *H = Herodotus*, tr. A. D. Godley, 4 vols., Loeb Classical Library, Cambridge, Massachusetts: Harvard University Press & London: Heinemann, 1920–4.
Heron-Bruins, *CC = Codex Constantinopolitanus Palatii Veteris No. 1*, ed. E. M. Bruins, 3 vols., (i, Photographic reproduction of the manuscript; ii, Greek transcription; iii, Translation & commentary), Leiden: Brill, 1964. (A re-edition of our principal manuscript of Heron's mathematical writings and our sole source of his *Metrica*.)
Heron-Heiberg, Schöne, & others, *Opera = Opera Quae Supersunt Omnia*, 5 vols.: i, *Pneumatica et Automata*, ed. G. Schmidt; ii *Mechanica et Catoprica*, ed. L. Nix & W. Schmidt; iii, *Rationes Dimetiendi* [= *Metrica*] & *Commentatio Dioptrica*, ed. H. Schöne; iv, *Definitiones*, ed. J. L. Heiberg; v., *Stereometrica & De Mensuris*, ed. J. L. Heiberg, Leipzig: Teubner, 1895–1914, repr. 1976. (Also see Heron-Bruins, *CC*.)
Herz-Fischler [= Fischler], R., *MHDEMR = A Mathematical History of Division in Extreme and Mean Ratio*, Waterloo: Laurier, in press.
Hogendijk, J., HTATGIG = How trisections of the angle were transmitted from Greek to Islamic geometry, *Historia Mathematica* 8 (1981), 417–38.
Horsfall, N., SB = Stesichorus at Bovillae?, *Journal of Hellenic Studies* 99 (179), 26–48, with addendum: *Tabulae Iliacae* in the Collection Froehner, Paris, ibid. 103 (1983), 144–7.
Huffman, C. A., AAF1 = The authenticity of Archytas Fr. 1, *Classical Quarterly* 35 (1985), 344–8.
Hunt, A. S. & Edgar, C. C., *SP* i = *Select Papyri* i, *Non-Literary Papyri*, Loeb Classical Library, Cambridge, Massachusetts & London: Harvard University Press & Heinemann, 1932.

Hurwitz, A., *Werke* = *Mathematische Werke*, 2 vols., Eidgenössischen Technischen Hochschule in Zürich, Basel: Birkhäuser, 1932-3.

Hurwitz, A., KEZE = Über die Kettenbruch-Entwicklung der Zahl *e*, *Schriften der Physikalisch-Ökonomischen Gessellschaft zu Königsberg* 32 (1981), 59-62, repr. in *Werke* ii, 129-133.

Hurwitz, A., KTAR = Über die Kettenbrüche, deren Teilnenner arithmetische Reihen bilden, *Vierteljahrsschrift der Naturforschenden Gessellschaft in Zürich* 41 (1986), 34-64, repr. in *Werke* ii, 276-302.

Huygens, C., *Ouevres* = *Oeuvres Complètes*, various editors, 22 vols., The Hague: Société Hollandaise des Sciences, 1888-1950.

Huygens, C., DAP = Descriptio automati planetarii, repr. in *Oeuvres* xxii, 579-652.

Hypsicles-De Falco, Krause, & Neugebauer, AG = Hypsikles, Die Aufgangszeiten der Gestirne, ed. & tr. V. De Falco, M. Krause, & O. Neugebauer, *Abhandlungen der Akademie der Wissenschaften in Göttingen, Philologisch-Historische Klasse* 63 (1966).

Itard, J., *LAE* = *Les Livres Arithmétiques d'Euclide*, Histoire de la pensée 10, Paris: Hermann, 1962.

Jacobi, C. G. J., *Werke* = *Gesammelte Werke*, ed. C. W. Borchardt & K. Weierstrass, 8 vols., Berlin, 1881-91, repr. New York: Chelsea, 1969.

Jacobi, C. G. J., ATKA = Allgemeine Theorie der Kettenbruchähnlichen Algorithmen, in welchen jede Zahl, aus drei Vorhergehenden gebildet wird (aus den hinterlassenen Papieren C. G. J. Jacobi's mitgetheit durch E. Heine), *Journal für die reine und angewandte Mathematik* 69 (1868), 29-64, repr. in *Werke* vi, 385-426.

James, T. G. H., *EEEES* = *Excavating in Egypt: The Egypt Exploration Society 1882-1982*, ed. T. G. H. James, London: British Museum Publications, 1982.

Jan, K. von, *MSG* = *Musici Scriptores Graeci*, ed. K. von Jan, Leipzig: Teubner, 1895, repr. Hildesheim: Olms, 1962.

Johnston, A., *TGV* = *Trademarks on Greek Vases*, Warminster: Arts & Phillips, 1979.

Jones, A., LME = Land measurement in England 1150-1350, *Agricultural History Review* 27 (1979), 10-18.

Jones, W. B. & Thron, W. J., *CF* = *Continued Fractions: Analytic Theory and Applications*, Encyclopedia of Mathematics and its Applications 11, Reading, Massachusetts: Addison-Wesley, 1980.

Kapsomenos, S. G., OPRT = The Orphic papyrus roll of Thessalonica, *Bulletin of the American Society of Papyrologists* 2 (1964-5), 3-31.

Keaney, J. J., ETTHP = The early tradition of Theophrastus' *Historia Plantarum*, *Hermes* 96 (1968), 293-98.

Kenyon, F. G., Bell, H. I., & others, *GPBM* = *Greek Papyri in the British Museum* [= P. Lond.], 7 vols. to 1974, London: British Museum 1893- .

Khinchin, A. Ya., *CF* = *Continued Fractions*, tr. Scripta Technica Inc. of the 3rd Russian edn. (1936) of *Cepnye Drobi* (1961), Chicago: University of Chicago Press, 1964.

Klein, F., *EMAS* i = *Elementary Mathematics from an Advanced Standpoint*, vol.

1: *Arithmetic, Algebra, Analysis* (1908), 3rd edn., tr. E. A. Hedrick & C. A. Noble, repr. New York: Dover, 1945.

Klein, J., *GMTOA* = *Greek Mathematical Thought and the Origin of Algebra*, tr. E. Brann of Die griechische Logistik und die Entstehung der Algebra, *Quellen und Studien zu Geschichte der Mathematik, Astronomie und Physik*, Section B: *Studien*, 3_1 (1934), 18–105 & 3_2 (1936), 122–235, with Appendix containing François Viète's *Introduction to the Analytic Art*, tr. J. W. Smith of *Isagoge* (1591), Cambridge, Massachusetts: MIT Press, 1968.

Knorr, W. R., *EEE* = *The Evolution of the Euclidean Elements: A Study of the Theory of Incommensurable Magnitudes and Its Significance for Early Greek Geometry*, Dordrecht: Reidel, 1975.

Knorr, W. R., PIGNT = Problems in the interpretation of Greek number theory: Euclid and the 'Fundamental Theorem of Arithmetic', *Studies in the History and Philosophy of Science* 7 (1976), 353–68.

Knorr, W. R., AMC = Archimedes and the measurement of the circle: A new interpretation, *Archive for History of Exact Sciences* 15 (1976), 115–40.

Knorr, W. R., AE = Archimedes and the *Elements*: Proposal for a revised chronological ordering of the Archimedean corpus, *Archive for History of Exact Sciences* 19 (1978), 211–90.

Knorr, W. R., ANCSL = Archimedes' neusis-construction in spiral lines, *Centaurus* 22 (1978), 77–98.

Knorr, W. R., MPP = Methodology, philology, and philosophy, with a reply by M. Burnyeat, *Isis* 70 (1979), 565–70.

Knorr, W. R., APEPT = Archimedes and the pre-Euclidean proportion theory, *Archives Internationales d'Histoire des Sciences* 28 (1980), 183–244.

Knorr, W. R., TFAEG = Techniques of fractions in ancient Egypt and Greece *Historia Mathematica* 9 (1982), 133–71.

Knorr, W. R., CM = "La Croix des Mathématiciens": The Euclidean theory of irrational lines, *Bulletin of the American Mathematical Society* (New Series) 9 (1983), 41–69.

Knorr, W. R., ETB = Euclid's tenth book: An analytic survey, *Historia Scientiarum* 29 (1985), 17–35.

Knuth, D. E., *ACP* ii = *The Art of Computer Programming* ii, *Seminumerical Algorithms*, 2nd edn., Reading, Massachusetts: Addison-Wesley, 1981.

Lagrange, J. L., *Oeuvres* = *Oeuvres des Lagrange*, ed. J.-A. Serret, 14 vols., Le Ministre de l'Instruction Publique, Paris: Gauthier–Villars, 1867–92.

Lagrange, J. L., SPA = Solution d'un problème d'arithmétique, *Miscellanea Taurinensia* 4 (1766–9), repr. in *Oeuvres* i, 671–731.

Lagrange, J. L., REN = Sur la résolution des équations numériques, *Mémoires de l'Académie royale des Sciences et Belles-Lettres de Berlin* 23 (1769), repr. in *Oeuvres* ii, 539–78.

Lagrange, J. L., AMREN = Additions au mémoire sur la résolution des équations numériques, *Memoires de l'Académie royale des Sciences et Belles-Lettres de Berlin* 24 (1770), repr. in *Oeuvres* ii, 581–652.

Lagrange, J. L., RRAE = Réflexions sur la résolution algébrique des équations, *Nouveaux Mémoires de l'Académie royale des Sciences et Belles-Lettres de Berlin*, 1770 & 1771, repr. in *Oeuvres* iii, 205–421.

Lagrange, J. L., *AEAE* = *Additions aux Éléments d'Algèbre d'Euler: Analyse*

Indéterminée, i.e. Appendix to the French translation (1774), by Jean III Bernoulli, of Euler, *VAA*; repr. in Lagrange, *Oeuvres* vii, 5–180 & Euler, *Opera* I i, 499–651.

Lagrange, J. L., *TREN* = *Traité de la résolution des équations numérique des tous les degrés*, 1st edn. 1798, revised & augmented 2nd edn. 1808, repr. in *Oeuvres* viii, 11–370.

Lagrange, J. L., EANTF = Essai d'analyse numérique sur la transformation des fractions, *Journal de l'École Polytechnique* 2 (prairial an vi [= 28 May–18 June, 1799]), repr. in *Oeuvres* vii, 291–313.

Landau, E., *FA* = *Foundations of Analysis*, tr. F. Steinhardt of *Grundlagen der Analysis* (1930), New York: Chelsea, 1951.

Lang, M., NNGV = Numerical notation on Greek vases, *Hesperia* 25 (1956), 1–24.

Lang, S. & Trotter, H., CFSAN = Continued fractions for some algebraic numbers, *Journal für die reine und angewandte Mathematik* 255 (1972), 112–34.

Legendre, A. M., *ETN* & *TN* = *Essai sur la théorie des nombres*, 1st edn. 1798; supplemented 2nd edn. 1808; further supplemented 3rd edn. in 2 vols.: *Théorie des nombres*, 1830, repr. Paris: Blanchard, 1955.

Lehmer, D. H., EALN = Euclid's algorithm for large numbers, *American Mathematical Monthly* 45 (1938), 227–33.

Lemerle, P., *PHB* = *Le Premier Humanisme Byzantin*, Paris: Presses Universitaires, 1971.

Lewis, N., *PCA* = *Papyrus in Classical Antiquity*, Oxford: Clarendon Press, 1974.

Liddell, H. G., Scott, R., & Jones, H. S., *GEL* = *A Greek–English Lexicon*, 9th edn. (1940) with *Supplement* (1968), Oxford: Clarendon Press, often reprinted.

Lucas, A., *AEMI* = *Ancient Egyptian Material and Industries* (1926), 4th edn. rev. & enl. by J. R. Harris, London: Arnold, 1962.

Lynch, J. P., *AS* = *Aristotle's School: A Study of a Greek Educational Institution*, Berkeley: University of California Press, 1972.

McNamee, K., *AGLPO* = *Abbreviations in Greek Literary Papyri and Ostraca*, Bulletin of the American Society of Papyrologists Supplement 3, Chicago: Scholars Press, 1981.

Mattha, G., *DO* = *Demotic Ostraca from the Collections at Oxford, Paris, Berlin, Vienna, and Cairo*, Publications de la Societé Fouad I de Papyrologie, Cairo: Institute Française d'Archéologie Orientale, 1945.

Mau, J. & Müller, W., MOBS = Mathematische ostraka aus der Berliner Sammlung, *Archiv für Papyrusforschung* 17 (1982), 1–10.

Meyerhof, M., JGPAAM = Joannes Grammatikos (Philoponos) von Alexandrien und die Arabische Medizin, *Mitteilungen des Deutschen Instituts für Ägyptische Altertumskunde in Kairo* 2 (1932), 1–21.

Milne, H. J. M. & Skeat, T. C., *SCCS* = *Scribes and Correctors of the Codex Sinaiticus*, London: British Museum, 1938.

Minkowski, H., *Abhandlungen* = *Gesammelte Abhandlungen von Hermann Minkowski*, ed. A. Speiser, H. Weyl, & D. Hilbert, 2 vols., Leipzig: Teubner, 1911.

Minkowski, H., TK = Zur Theorie der Kettenbruche, *Abhandlungen* i, 278–92; French translation by L. Laugel, Généralisation de la théorie des fractions continues, *Annales de l'École Normale Supérieur*, (3e série) 13 (1896), 41–60.

Minkowski, H., KAZ = Ein Kriterium für die algebraischen Zahlen, *Nachrichten der K. Gesellschaft der Wissenschaften zu Göttingen, mathematische-physikalische Klasse* (1900), 64–88, repr. in *Abhandlungen* i, 293–315.

Minkowski, H., UP = Über Periodische Approximationen Algebraischen Zahlen, *Acta Mathematica* 26 (1902), 333–51, repr. in *Abhandlungen* i, 357–71.

Minkowski, H., *BDH = Briefe an David Hilbert*, ed. L. Rüdenberg & H. Zassenhaus, Berlin, Heidelberg, New York: Springer, 1973.

Mueller, I., *PMDSEE = Philosophy of Mathematics and Deductive Structure in Euclid's Elements*, Cambridge, Massachusetts: MIT Press, 1981.

Muir, T., *EQSCF = The Expression of a Quadratic Surd as a Continued Fraction*, Glasgow: Maclehose, 1874.

Muir, T., *TD = The Theory of Determinants, in the Historical Order of Development*, 4 vols., London: St Martin's Press, 1906–23, repr. in 2 vols., New York: Dover, 1960; and *Contributions to the Theory of Determinants 1900–1920*, London: Blackie, 1930.

Nelson, H. L., SACP = A Solution to Archimedes' Cattle Problem, *Journal of Recreational Mathematics* 13 (1980–1), 162–76.

Neuenschwander, E., EVBEE = Die ersten vier Bücher der Elemente Euklidis, *Archive for History of Exact Sciences* 9 (1973), 325–80.

Neuenschwander, E., SBEE = Die stereometrischen Bücher der Elemente Euklidis, *Archive for History of Exact Sciences* 14 (1975), 91–125.

Neugebauer, O., *ESA = The Exact Sciences in Antiquity*, 2nd edn., Providence: Brown University Press, 1957, repr. New York: Dover, 1969.

Neugebauer, O., APO = Astronomical papyri and ostraca: Biographical notes, *Proceedings of the American Philosophical Society* 106 (1962), 383–91.

Neugebauer, O., *HAMA = A History of Ancient Mathematical Astronomy*, 3 vols., Berlin, Heidelberg, New York: Springer, 1975.

Neugebauer, O. & Parker, R. A., *EAT = Egyptian Astronomical Texts*, 3 vols., Providence: Brown University Press, 1960–9.

von Neumann, J. & Tuckerman, B., CFE = Continued fraction expansion of $\sqrt[3]{2}$, *Mathematics of Computation* 9 (1955), 23–4.

Nicomachus–D'Ooge, *IA = Introduction to Arithmetic*, tr. M. L. D'Ooge, New York: Macmillan, 1926.

Nicomachus–Hoche, *IA = Introductionis Arithmeticae Libri II*, ed. R. Hoch, Leipzig: Teubner, 1866.

Niebel, E., *UBGKA = Untersuchungen über die Bedeutung der geometrischen Konstruktion in der Antike*, Kantstudien 76, Cologne: Kölner Universitäts-Verlag, 1959.

OCD = The Oxford Classical Dictionary, 2nd edn., ed. N. G. L. Hammond & H. H. Scullard, Oxford: Clarendon Press, 1970.

Odom, G., EP = Elementary Problem E 3007, *American Mathematical Monthly* 90 (1983), 482.

OED = The Oxford English Dictionary, Oxford: Clarendon Press, 1933, often reprinted; and *Supplement*, 4 vols., ed. R. W. Burchfield, Oxford: Clarendon Press, 1972–86.

Omar Khayyam–Amir-Móez, DDE = Discussion of Difficulties in Euclid, by Omar ibn Abraham al-Khayyami (Omar Khayyam), tr. A. R. Amir-Móez, *Scripta Mathematica* 24 (1959), 275–303.

Pack, R. A., *GLLTGRE* = *The Greek and Latin Literary Texts from Greco-Roman Egypt*, 2nd edn., Ann Arbor: University of Michigan Press, 1965.
Pappus–Thomson & Junge, *CPBXEE* = *The Commentary of Pappus on Book X of Euclid's Elements*, ed. & tr. W. Thomson & G. Junge, Harvard Semitic Series 8, Cambridge, Massachusetts: Harvard University Press, 1930, repr. New York: Johnson Reprint Company, 1968.
Parker, R. A., *CAE* = *The Calendars of Ancient Egypt*, Studies in Ancient Oriental Civilization 26, Chicago: University of Chicago Press, 1950.
Parker, R. A., *DMPF* = A demotic mathematical papyrus fragment, *Journal of Near Eastern Studies* 18 (1959), 275–9.
Parker, R. A., *DMP* = *Demotic Mathematical Papyri*, Providence: Brown University Press, 1972.
Parker, R. A., ME = A mathematical exercise—P. dem. Heidelberg 663, *Journal of Egyptian Archaeology* 61 (1975), 189–96.
Perron, O., *LK* = *Die Lehre von den Kettenbrüchen*, 3rd edn. 2 vols., Stuttgart: Teubner, 1954.
Pestman, P. W., *GDTZA* = *Greek and Demotic Texts from the Zenon Archive* [= P.L. Bat. xx], Leiden: Brill, 1980.
Pestman, P. W., *GZA* = *A Guide to the Zenon Archive* [= P. L. Bat. xxi], 2 vols., Leiden: Brill, 1981.
Pfeiffer, R., *HCS* = *History of Classical Scholarship: From the Beginnings to the End of the Hellenistic Age*, Oxford: Clarendon Press, 1968.
Philoponus–Wallies, *IAP* = *In Aristotelis Analytica Posteriora Commentaria*, ed. M. Wallies, Commentaria in Aristotes Graeca 5_3, Berlin: Reimer, 1909.
Pintaudi, R. & Sijpesteijn, P. J., *TLC* = *Tavolette Lignee e Cerate della Biblioteca Apostolica Vaticana*, with Appendices, *Quatre Cahiers Scolaires* (P. Cauderlier) and *A Tablet from the Pierpont Library* (R. S. Bagnall), Florence: Gonnelli, in press.
Pisot, C., SM = Souvenirs mathématiques, *L'Enseignement Mathématique* 26 (1980), 185–92.
Plato, Individual works with the Greek text and parallel English translation in the Loeb Classical Library, Cambridge, Massachusetts & London: Harvard University Press & Heinemann.
Plato, Critical editions of the Greek text of individual works in the Scriptorum Classicorum Bibliotheca Oxoniensis (= The Oxford Classical Texts), Oxford: Clarendon Press.
Plato, *CD* = *The Collected Dialogues of Plato including the Letters*, ed. E. Hamilton & H. Cairns, Bollingen Series 71, Princeton: Princeton University Press, 1961.
Plato–Allen, *PP* = *Plato's Parmenides, Translation and Analysis*, ed. & tr. R. E. Allen, Oxford: Blackwell, 1983.
Plato–Bluck, *PM* = *Plato's Meno*, ed. R. S. Bluck, Cambridge: Cambridge University Press, 1961. (Critical edition of Greek text, with extensive commentary.)
Plato–Guthrie, *PM* = *Protagoras and Meno*, tr. W. K. C. Guthrie, Harmondsworth: Penguin, 1956. (Also repr. in Plato, *CD* and Brown, *PM*.)
Plooij, E. B., *ECR* = *Euclid's Conception of Ratio and his Definition of*

Proportional Magnitudes as Criticised by Arabian Commentators, Rotterdam: van Hengel, 1950.
Plutarch–Hubert, *M* = *Moralia* iv; ed. C. Hubert, Leipzig: Teubner, 1938, repr. 1971.
Poincaré, H., *Oeuvres* = *Oeuvres de Henri Poincaré*, 11 vols., L'Académie des Sciences, Paris: Gauthier-Villars, 1916–54.
Poincaré, H., GCF = Sur une généralisation des fractions continues, *Comptes rendus de l'Académie des Sciences* 99 (1884), 1014–16, repr. in *Oeuvres* v, 185–8.
Popper, K., *OSE* i = *The Open Society and its Enemies* i, *The Spell of Plato*, London: Routledge & Kegan Paul, 1945, 5th edn. 1966.
Price, D. de Solla, *GG* = *Gears from the Greeks: The Antikythera Mechanism—A Calendar Computer from ca. 80 B.C.*, Philadelphia: Transactions of the American Philosophical Society (New Series) 64 Part 7 (1974), repr. New York: Science History Publications, 1975.
Procissi, A., BMGA = Bibliografia della mathematica Greca antica, *Bollettino di Storia della Scienze Matematiche* 1 (1981), 7–149.
Proclus–Festugière, *CR* = *Commentaire sur la République*, tr. A. J. Festugière, 3 vols., Paris: Vrin, 1970.
Proculus–Friedlein, *PEEC* = *In Primum Euclidis Elementorum Commentarii*, ed. G. Friedlein, Leipzig: Teubner, 1873.
Proclus–Kroll, *IR* = *In Platonis Rem Publicam Commentarii*, ed. W. Kroll, 2 vols., Leipzig: Teubner, 1899–1901.
Proclus–Morrow, *CFBEE* = *A Commentary on the First Book of Euclid's Elements*, tr. G. R. Morrow, Princeton: Princeton University Press, 1970. (All references to this book will use the pagination of Proclus-Friedlein, *PEEC*.)
Ptolemy–Düring, *HKP* = *Die Harmonielehre des Klaudios Ptolemaios*, ed. I. Düring, Göteborg: Elander, 1930.
Ptolemy–Toomer, *PA* = *Ptolemy's Almagest*, tr. G. J. Toomer, London: Duckworth, 1984.
Rawlins, D., EGU = Eratosthenes' geodesy unravelled: Was there a high-accuracy Hellenistic astronomy?, *Isis* 73 (1982), 259–62. (Also see Fowler, EROE.)
Rayney, G. N., CFFA = On continued fractions and finite automata, *Mathematische Annalen* 206 (1973), 265–83.
Reid, C., *H* = *Hilbert*, Berlin, Heidelberg, New York: Springer, 1970.
Reynolds, L. D. & Wilson, N. G., *SS* = *Scribes and Scholars: A Guide to the Transmission of Greek and Latin Literature*, 2nd edn., Oxford: Clarendon Press, 1974.
Riddell, R. C., EMES = Eudoxan mathematics and the Eudoxan spheres, *Archive for History of Exact Sciences* 20 (1979), 1–19.
Riemann, B., *Werke* = *Gesammelte Mathematische Werke und Wissenschaftlicher Nachlass*, 2nd edn. (1892), ed. R. Dedekind & H. Weber, with *Nachträge* (1902), ed. M. Noether & W. Wirtinger, repr. New York: Dover, 1953.
Riemann, B., SQSI = Sullo svolgimento del quoziente di due serie ipergeometriche in frazione continus infinia (edited from notes of 1863 by H. A. Schwarz), in *Werke*, 424–30.

Riginos, A. S., *PACLWP* = *Platonica: The Anecdotes Concerning the Life and Writings of Plato*, Leiden: Brill, 1976.
Roberts, C. H., & Skeat, T. C., *BC* = *The Birth of the Codex*, Oxford: Oxford University Press, for the British Academy, 1983.
Rosen, D. & Shallit, J., CFA = A continued fraction algorithm for approximating all real polynomial roots, *Mathematics Magazine* 51 (1978), 112–16.
Rowland, R. J., BIW = The biggest island in the world, *The Classical World* 68 (1975), 438–9.
Sadie, S., *NGDMM* = *The New Grove Dictionary of Music and Musicians*, ed. S. Sadie, 20 vols., London: Macmillan, 1980.
Saffrey, H. D., AME = *ΑΓΕΩΜΕΤΡΗΤΟΣ ΜΗΔΕΙΣ ΕΙΣΙΤΩ*, une inscription légendaire, *Revue des Études Grecques* 81 (1968), 67–87.
Ste. Croix, G. E. M. de, GRA = Greek and Roman accounting, in *Studies in the History of Accounting*, ed. A. C. Littleton & B. S. Yamey, London: Sweet & Maxwell, 1956.
Ste. Croix, G. E. M. de, *CSAGW* = *The Class Struggle in the Ancient Greek World*, London: Duckworth, 1981.
Sandys, J. E., *HCS* = *History of Classical Scholarship* (1903), 3rd edn. 3 vols., Cambridge: Cambridge University Press, 1921.
Santillana, G. de, EPSC = Eudoxus and Plato: A study in chronology, *Isis* 32 (1940/1947), 248–62.
Schwenter, D., *GP* = *Geometriae Practicae Novae et Auctae Tractatus*, Nurenberg, 1627.
Schwenter, D., *DPM* = *Deliciae Physico-Mathematicae*, Nurenberg, 1636.
Sedley, D., EA = The end of the Academy, *Phronesis* (1981), 67–75.
Series, C., GMN = The geometry of Markoff numbers, *The Mathematical Intelligencer* 7 (1983), 20–9.
Serret, J.-A., *CAS* = *Cours d'Algèbre Supérieure*, 3rd edn. 3 vols., Paris: Gauthier-Villars, 1866.
Sethe, K. H., *ZZAA* = *Von Zahlen und Zahlworten bei Alten Ägyptern*, Strassburg: Trübner, 1916.
Sharples, R. W., NETDP = Nemesius of Emesa and some theories of divine providence, *Vigilae Christianae* 37 (1983), 141–56.
Sijpesteijn, P., WTMC = A wooden tablet in the Moen Collection, *Chronique d'Égypte* III (New Series 56) (1981), 99–101.
Smith, H. J. S., *Papers* = *The Collected Mathematical Papers of Henry John Stephen Smith*, ed. J. W. L. Glaisher, 2 vols., repr. New York: Chelsea, 1965.
Smith, H. J. S., NCF = Note on continued fractions, *Messenger of Mathematics* (Series 2) 6 (1876), 1–14, repr. in *Papers* ii, 135–47.
Smyly, J. G., EAGL = The employment of the alphabet in Greek logistic, pp. 515–30 in *Mélanges Nicole, Recueil des Memoirs de Philologie*, Geneva: Kundig, 1905.
Smyly, J. G., NTS = Notes on Theon of Smyrna, *Hermathena* 14 (1907), 261–79.
Stark, H. M., *INT* = *An Introduction to Number Theory*, Chicago: Markham, 1970, repr. Cambridge, Massachusetts: MIT Press, 1978.
Stark, H. M., EECFFB = An explanation of some exotic continued fractions found by Brillhart, pp. 21–35 in *Computers in Number Theory, Proceedings of*

the *Science Research Council Atlas Symposium 2, Oxford 1969,* ed. A. O. L. Atkin & B. J. Birch, London & New York: Academic Press, 1971.

Steele, A. D., URZLGM = Über die Rolle von Zirkel und Lineal in der griechischen Mathematik, *Quellen und Studien zur Geschichte der Mathematik, Astronimie und Physik,* Abteilung B: *Studien* 3 (1936), 287–369.

Steele, D. A., MRCP = A mathematical reappraisal of the Corpus Platonicum, *Scripta Mathematica* 17 (1951), 173–89.

Taisbak, C. M., *DL = Division and Logos, A Theory of Equivalent Couples and Sets of Integers,* Odense: Odense University Press, 1971.

Taisbak, C. M., *CQ = Coloured Quadrangles: A Guide to the Tenth Book of Euclid's Elements,* Opuscula Graecolatina 24, Copenhagen: Museum Tusculanum Press, 1982.

Taisbak, C. M., EET = Eleven Eighty-Thirds. Ptolemy's reference to Eratosthenes in *Almagest* I.12, *Centaurus* 27 (1984), 165–7.

Tait, J. G., Préaux, C., & others, *GOBLO = Greek Ostraca in the Bodleian Library at Oxford and Various Other Collections* [= O. Bodl. or O. Tait], 3 vols., London: Egypt Exploration Society, 1930–64.

Tannery, P., *Memoires = Memoires Scientifiques,* 17 vols., Paris: Gauthier-Villars, 1912–50.

Tannery, P. SSCG = Sur la symbole de soustration chez les Grecs, *Bibliotheca Mathematica* 5 (1904–5), 342–9, repr. in *Memoires* xiii, 208–12.

Theon of Smyrna–Hiller, *ERMLPU = Expositio Rerum Mathematicarum ad Legendum Platonem Utilium,* ed. E. Hillier, Leipzig: Teubner, 1878.

Theon of Smyrna–Lawlor, *MUUP = Mathematics Useful for Understanding Plato,* tr. from the Greek/French edn. of J. Dupuis (1892) by R. & D. Lawlor, Secret Doctrine Reference Series, San Diego: Wizards Bookshelf, 1979. (See the review by R. D. McKirahan, *Historia Mathematica* 9 (1982), 100–4.)

Thesleff, H., *IPWHP = An Introduction to the Pythagorean Writings of the Hellenistic Period,* Åbo: Academie, 1961.

Thesleff, H., *PTHP = The Pythagorean Texts of the Hellenistic Period,* Åbo: Academie, 1965.

Thomae, L. W., KGQ = Über die Kettenbrüchentwicklung der Gauss'chen Quotienten..., *Journal für die reine und angewandte Mathematik* 67 (1867), 299–309.

Thomas [= Bulmer-Thomas], I., *SIHGM = Selections Illustrating the History of Greek Mathematics,* 2 vols., Loeb Classical Library, Cambridge, Massachusetts & London: Harvard University Press & Heinemann, vol. 1, 1939, supplemented repr. 1980; vol. 2, 1941. (See the review by O. Neugebauer, *American Journal of Philology* 64 (1943), 452–7.)

Tod, M. N., *AGNS = Ancient Greek Numerical Systems: Six Studies,* Chicago: Ares, 1979. (Six articles on the epigraphical evidence for numerical systems, originally published in the *Annual of the British School at Athens* vols. 18, 28, 37, 45, & 49 and the *Journal of Hellenic Studies* 33.)

Tod, M. N. GNN = The Greek numeral notation, *Annual of the British School of Athens,* 18 (1911/12) 98–132, repr. in Tod, *AGNS,* 1–35.

Tod, M. N., LLGI = Letter-labels in Greek inscriptions, *Annual of the British School at Athens,* 49 (1954), 1–8, repr. in Tod, *AGNS,* 98–105.

Bibliography 389

Toomer, G. J., LGMWAT = Lost Greek mathematical works in the Arabic tradition, *Mathematical Intelligencer* 6 (1984), 32–8.

Turner, E. G., ABFFCBC = *Athenian Books in the Fifth and Fourth Centuries B.C.* (An inaugural lecture, May 1951), London: University College & H. K. Lewis, 1952, 2nd edn. 1977.

Turner, E. G., FODMS = Four obols a day men at Saqqara, pp. 573–7 & Plate X in *Hommages à Claire Préaux*, ed. J. Bingen, G. Cambier, & G. Nachtergael, Brussels: Editions de l'Université de Bruxelles, 1975.

Turner, E. G., GMAW = *Greek Manuscripts of the Ancient World*, Oxford: Clarendon Press, 1971.

Turner, E. G., PW = *The Papyrologist at Work*, Durham; North Carolina: Greek, Roman and Byzantine Monographs 6, 1973.

Turner, E. G., CCOS = A commander-in-chief's order from Saqqâra, *Journal of Egyptian Archaeology* 60 (1974), 239–42.

Turner, E. G., GP = *Greek Papyri: An Introduction*, 2nd edn., Oxford: Clarendon Press, 1980.

Turner, E. G., Fowler, D. H., Koenen, L., & Youtie, L. C., EEI = Euclid, Elements I, Definitions 1–10 (P. Mich. iii 143), *Yale Classical Studies* 28 (1985), 13–24.

Turner, E. G. & Neugebauer, O., GDNM = Gymnasium debts and new moons [= P. Ryl. iv 589], *Bulletin of the John Rylands Library* 32 (1949), 80–96.

Turner, E. G., Tsantsanoglou, K., & Parássoglou, G. M., ODP = On the Derveni papyrus, *Gnomon* 54 (1982), 855–6.

Varberg, D. E., PTR = Pick's theorem revisited, *American Mathematical Monthly* 92 (1985), 584–7.

Vogel, K., BGL = Beiträge zur griechischen Logistik, *Sitzungsberichte der Bayerischen Akademie der Wissenschaften, Mathematisch-Naturwissenschaftliche Abteilung* (1936), 357–472.

Vogel, K., BIZB = Buchstabenrechnung und indische Ziffern in Byzanz, pp. 660–4 in *Akten des XI. Internationalen Byzantinistenkongresses, München 1958*, Munich: Beck, 1960.

van der Waerden, B. L., SA = *Science Awakening*, tr. A. Dresden of *Ontwakende Wetenschap* (1950), Groningen: Noordhoff, 1954.

van der Waerden, B. L., GAC = Greek astronomical calendars: I, The parapegma of Euctemon; II, Callippos and his calendar; III, The calendar of Dionysios; IV, The Parapegma of the Egyptians and their "Perpetual Tables", *Archive for History of Exact Sciences* 29 (1984), 101–14, 115–24, 125–30, & 32 (1985), 95–104.

Wall, H. S., ATCF = *Analytic Theory of Continued Fractions*, New York: Van Nostrand, 1948, repr. New York: Chelsea, 1973.

Wallis, J., AI = *Arithmetica Infinitorum, sive nova methodus inquirendi in curvilineorum quadraturam aliaque difficiliora matheseos problemata*, Oxford, 1656.

Wallis, J., CE = *Commercium Epistolicum de quaestionbus quibusdam mathematicis nuper habitum inter nobilissimos viros...*, Oxford, 1658. (Correspondence concerning Pell's equation between Brouncker, Digby, Fermat, Frenicle de Bessy, van Schooten, & Wallis.)

Wallis, J., *TA* = *A Treatise of Algebra, both Historical and Practical, shewing the Original, Progress, and Advancement thereof, from time to time, and by what Steps it hath attained to the Heighth at which now it is. . . .* , London, 1685.

Wasserstein, A., THTN = Theaetetus and the history of the theory of numbers, *Classical Quarterly* 8 (1958), 165–79.

Waterhouse, W. C., DRS = The discovery of the regular solids, *Archive for History of Exact Sciences* 9 (1972), 212–21.

Watson, J. S., *LRP* = *The Life of Richard Porson, M.A.*, London: Longman, 1861.

Weil, A., *NT* = *Number Theory, An Approach through History from Hammurapi to Legendre*, Boston: Birkhäuser, 1984.

Weitzmann, K. *ABI* = *Ancient Book Illumination*, Cambridge, Massachusetts: Harvard University Press, 1959.

Westfall, R. S., *NR* = *Never at Rest, A Biography of Isaac Newton*, Cambridge: Cambridge University Press, 1980, repr. with additions 1983.

Whiteside, D. T., PMTLSC = Patterns of mathematical thought in the later seventeenth century, *Archive for History of Exact Sciences* 1 (1961), 179–388.

Whittaker, E. T., TCF = On the theory of continued-fractions, *Proceedings of the Royal Society of Edinburgh* 36 (1915–16), 243–55.

Whycherly, R. E., PAPS = Peripatos: The Athenian philosophical scene, *Greece and Rome* (2nd Series) 8 (1961), 152–63 & 9 (1962), 2–21.

Wilson, N. G., TBS = Three Byzantine scribes, *Greek, Roman and Byzantine Studies* 14 (1973), 223–8.

Wilson, N. G., MP = Miscellanea palaeographica, *Greek, Roman and Byzantine Studies* 22 (1981), 395–404.

Wilson, N. G., *SB* = *Scholars of Byzantium*, London: Duckworth, 1983.

Winter, J. G., Robbins, F. E., & others, *PUMC* = *Papyri from the University of Michigan Collection* [= P. Mich.], 15 vols to 1982, Ann Arbor: University of Michigan Press, 1931– .

Wolfskill, J., GBPQCN = A growth bound on the partial quotients of cubic numbers, *Journal für die reine und angewandt Mathematik* 346 (1984), 129–40.

Youtie, H. C., *TCDP* = *The Textual Criticism of Documentary Papyri: Prolegomena*, 2nd edn., London: Bulletin of the Institute of Classical Studies Supplement 33, 1974.

Zeeman, E. C., GG = Gears from the Greeks, *Proceedings of the Royal Institution*, 58, in press.

Zeuthen, H. G., CLAEE = Sur la constitution des livres arithmétiques des Éléments d'Euclide et leur rapport à la question de l'irrationalité, *Oversigt over det Kgl. Danske Videnskabernes Selskabs Forhandlinger* [= *Proceedings of the Danish Academy of Sciences*] 5 (1910), 395–435.

INDEX OF CITED PASSAGES

This index covers only passages from ancient authors, and does not include incidental references. Papyri (Egyptian and Greek) are all listed together under 'Papyri', and minuscule manuscripts under 'Manuscripts'. Plates fall between pages 202 and 203.

Alexander of Aphrodisias
  In Analytica Priora 260-1: 300
  In Topica 545: 32, 63
Archimedes
  Book of Lemmas: 153, 289
  Cattle Problem: 60–2, 66, 116, 225
  Measurement of a Circle: 36, 54–7, 65–6, 150, 240–6, 364
  Plane Equilibria I 6 & 7: 298
  Sandreckoner: 225
  Sphere and Cylinder I: 36, 42, 203
  Spiral Lines: 289
Archytas
  Fragments B1: 145–6; B3 & 4: 156, 202, 364
Aristarchus
  On the Sizes and Distances of the Sun and Moon: 53–4, 65, 246–8
Aristotle
  Prior Analytics 41a23–30: 297; 50a35–8: 297; 65a16–21: 297
  Posterior Analytics 74a17–25: 203; 76a30–b16: 291–2; 77b12–13: 201
  Topics 158a31–159a2 (*see also* Euclid, *Elements* VI 1): 17–18, 30, 32, 364; 163a11–13: 296; 163b24–9: 239
  Nicomachean Ethics 1129a32–b1: 201; 1131a30–b16: 149
  Metaphysics 981b23: 284; 983a12–20: 296; 985b23–6a12: 303; 992a32–b1: 202, 303; 992b8–9: 303; 1036b13: 302; 1053a14–20: 296; 1080b16–21: 302; 1083b17–19: 303
  [*De Audibilibus*] 803b27–4a9: 146
  [*Problems*] 921a7–31: 146
  See also: (*a*)*logos* and (*ar*)*rhētos* in Aristotle: 193; *logistikē* and *logsmos* in Aristotle: 156; incommensurability of the diagonal in Aristotle: 295–6
Autolycus
  On the Moving Sphere and *On Risings and Settings*: 154, 217

Boethius
  De Institutione Musica: 149

Democritus
  Fragments A33 = B11p: 168, 294–5
Diogenes Laertus
  Lives of the Eminent Philosophers III 37: 204; IV 1: 199

Eutocius
  Commentary on Archimedes' *Measurement of a Circle*: 241–5;
  Commentary on Archimedes' *Sphere and Cylinder* II: 7, 290
Euclid
  Elements: 202–220
    Book I: 13, 29; Definitions: 209, 214, 221; Postulates: 11, 153, 154; Common Notions: 11, 103; I 1: 11; I 1–3: 13, 29, 209, 291; I 9–10: 209; I 17: 12, 154, 284; I 20: 98; I 34: 63–4; I 35: 10, 29, 284; I 39–41: 212–14; I 43–5: 13, 72, 172; I 46: 118; I 47: 21, 72, 191
    Book II: 29, 67–105, 121, 160, 166, 190–1; Definitions: 24, 68; II 4: 69–71; II 5: 209–12; II 6: 103, 367; II 8: 103; 'II 8a': 21, 34, 73, 104; II 9–10: 72, 103–4; II 11: 72, 87, 182–3; II 12–13: 72, 98, 100, 104; II 14: 13, 24, 72
    Book IV: 121, 158–61; IV 10–11: 87, 158–61
    Book V: 21, 30, 33, 133, 142, 364; Definitions: 16, 20, 31, 32, 139, 140, 142, 149, 152, 227, 370; V 4: 140; V 10: 32; V 12: 44; V 18: 44, 139; V 19: 103, 140
    Book VI: 29; Definitions: 21, 30, 87, 141, 143; VI 1 (The *Topics* Proposition): 18–20, 30, 32, 129, 141, 172; VI 11–13: 292; V 16: 19, VI 22: 139; VI 23: 20, 132, 141–3

Euclid (cont.)
  Elements: (cont.)
    Book VII: 33, 114, 138–9, 142;
      Definitions: 15, 17, 142; VII 1: 31, 360;
      VII 2: 31; VII 4: 17; VII 11: 103;
      VII 19: 19, 43; VII 22: 149; VII 28:
      142; VII 33: 361
    Book VIII: 142; VIII 2: 151; VIII 4: 141;
      VIII 5: 20, 132, 141–3, 148; VIII 7:
      148–9; VIII 8: 149; VIII 11–12: 141;
      VIII 19: 29; VIII 21: 29; VIII 26–7: 190
    Book IX: 142; IX 12: 29
    Book X: 21, 30, 121, 136, 166–92, 291,
      298, 302, 364, 367; Definitions: 167;
      X 2: 31, 50, 360; X 3–8: 51, 360; X 4:
      361; X 9: 190; X 10: 29, 298; X 30: 29;
      X 33: 142; X 55: 29; X 56: 29; X 117:
      298, 300, 304
    Book XI: 29, 142; Definitions: 11, XI 25:
      191; XI 28: 64; XI 38: 64
    Book XII: 29, 142, 291; XII 2: 12, 292;
      XII 5: 292; XII 9: 142; XII 11–2: 292;
      XII 18: 292
    Book XIII: 87, 121, 143, 158, 160, 191,
      291, 298, 364; XIII 1–5: 67, 87, 90–2,
      159–61; XIII 6: 162, 183, 191; XIII 8:
      35, 87; XIII 9: 87, 159; XIII 10: 208;
      XIII 11: 87, 139–40, 160–1, 186, 191,
      221; XIII 12: 35, 258; XIII 12–15: 158;
      XIII 16: 87, 191, 208; XIII 17: 12, 87,
      191; XIII 18: 191
  Sectio Canonis: 144–53, 221

Heraclitus
  Fragments 1 & 50: 108
Herodotus
  Histories ii 109: 283
Heron
  Definitions: 110, 299
  Metrica: 64, 265; I 14–16: 285; I 17: 224;
    I 25: 37, 56; I 37: 245; II 7: 231–2
  Stereometrica: 257–8
Hippocrates, see Simplicius, In Physica
Hypsicles
  'Book XIV of the Elements': 87
  On Rising-times: 218, 222

Lysis
  Doubtful Writings 4, 168, 299–300

Manuscripts
  Bodleian D'Orville 301: 217, 218, 219,
    Plate 1
  Bodleian E. D. Clarke 39: 218
  British Library Add. gr. 17211: 214
  Cambridge University Library
    Add. 1879.23: 243
  Leiden B. P. G. 78: 217
  Leningrad gr. 219: 217
  Paris gr. 2389: 217, 264
  Vatican gr. 190: 217, 219; 204: 217, 222,
    246; 218: 218, 220; 303: 226; 1594: 217,
    218, 264

Nicomachus
  Introduction to Arithmetic I 13: 33, 364

Pappus
  Collection: 87; II 17–21: 225; IV 26: 290;
    IV 36: 290; VII 8: 44, 64
  Commentary on Book X of Euclid's
    Elements 1 & 2: 300–2
Papyri
  Achmîm Mathematical Papyrus: 224, 237,
    265
  Corpus dei Papyri Filosofici: 204, 209
  Demotic Mathematical Papyri: 227, 259–63
  M.P.E.R., N.S. i 1: 227, 231, 254–9, 265,
    Plate 8; xv 151: 255
  O. Bodl. ii 1847: 230–4, 265, 285, Plate 6
  O. Tait, see O. Bodl.
  Pack no. 1537 (Timotheus, Persae):
    207; 2323 (Euclid): 208
  P. Cair. Zen iii: 59355: 32
  P. Carlsberg 9: 127–8
  P. Eleph. 1: 207
  P. Fay. 9: 212–4, Plate 3
  P. Herc. 1061: 209, 213, Plate 1; 1151: 222
  P. Hib. i 27: 207, 229–30, Plate 5
  P. Lond. ii 265 (p. 257): 222, 227, 248–54,
    259, 265, Plate 7; vii 1994: 32; vii 1995:
    32
  P. Louvre MND552K: 228
  P. Mich. iii 143: 214; iii 145: 248; iii 146:
    205, 235, 237
  P. Oxy. i 29: 205, 209–12, 213, Plate 2;
    x 1231: 222; xii 1446: 228; xlii 3000: 222
  P. Par. 1: 127
  P. Rainer: see M.P.E.R., N.S.
  P. Ryl. iii 540: 222; iv 589: 127–8
  P. Vindob.: see M.P.E.R., N.S.
  Rhind Mathematical Papyrus: 234, 236–7
  Also see the catalogue of tables: 271–9
Philoponus
  In Analytica Posteriora 12–17 & 26–7: 300
Plato
  Charmides 165e–66a: 109, 110, 111
  Gorgias 451a–c: 22, 30, 32, 109, 111; 508a:
    202
  Hippias Major 285c: 157; 303b–c: 295
  Laws VII 817e–22c: 107, 295, 304

Index of cited passages

Meno 81e–5d: 3–7, 31, 67, 118–9, 238; 82c: 10, 34, 138; 82c–d: 13–4; 83c: 15; 83e: 14; 84d–e: 104; 86e–7b: 67
Parmenides 140b–d: 295; 154b–d: 42, 64
Phaedrus 274c–5b: 23, 30, 284
Philebus 56d–7a: 109–10
Protagoras 318e: 157; 321c: 194; 356a–7b: 109
Republic VI 495c–6a: 28; 509b: 118
Republic VII 521b: 107; 521c–531c: 8, 107–54; 522c: 111; 522c–6c: 113–7; 525c–6a: 111–2, 139; 526c: 28; 526d–27c: 117; 527a: 12; 528a–d: 117–21, 368; 529c–30c: 121–30; 530b: 121; 530d–1c: 130–1; 531c–d: 108; 534d: 295; 536d–41b: 107
Republic VIII 546b–d: 295
Statesman 259e: 109
Theaetetus 144b: 25; 147a–8b: 295; 154d: 157; 201e–2c: 168
Timaeus 36a–b: 227; 37c–9e: 122
See also: *logistikē* and *logismos* in Plato: 154–5; (*a*)*logos* and (*ar*)*rhētos* in Plato: 192–3

Plutarch
   Moralia viii 2: 299
Proclus
   Commentary on ... Euclid's Elements
      38–40: 110; 64–5: 283, 297; 74: 298; 203–5: 367; 396: 284; 403: 285
   In Rem Publicam ii 24–9: 63, 96, 100–4
Ptolemy
   Almagest I 10: 87; I 12: 51; VI 2: 128
   Harmonica: 147

Simplicuis
   In Physica 60–8: 7, 287–8

Theon of Smyrna
   Expositio ... Utilium: 299; 42–5: 58–9, 63, 96, 100; 126–7: 226
Thucydides
   History of the Peloponnesian War VI: 285
Tzetzes
   Chiliades VIII 974–7: 201

INDEX OF NAMES

This index does not cover the following: the Preface and other front material; the Bibliography; the names of editors and translators of texts, unless there is some special reference to them; names that are only mentioned incidentally; and terms such as Farey series, the *Parminedes* proposition, Pythagoras' theorem, etc., which can be found in the General Index.

Alexandria 214–16, 365
Alexander of Aphrodisias 32, 63, 298, 300, 306
Allen, R. E. 64
Allman, G. J. 63
Anatolius 110
Anaxagoras 157
Apéry, R. 42
Apollonius 218, 225–6, 289–90, 293, 298, 301
Archimedes 7, 36–7, 41, 42, 54–7, 60–2, 64, 65–6, 116, 138, 153, 203, 218, 225, 240–6, 289–90, 293, 298, 364–5
Archytas 7, 25, 28, 118, 131–8, 145–7, 156, 157, 202, 292–3, 299, 364
Arethas 218
Aristarchus 53–4, 65, 217, 226, 246–8
Aristides Quintilianus 146
Aristotle 7, 11, 17–18, 20, 32, 33, 200–2, 203, 239, 291–2, 295–7, 302–8, 364
Aristoxenus 146
Athenagoras 299–300
Athens 23, 197–200, 203, 365
Autolycus 154, 217

Bachet de Méziriac, C. G. 349
Bagnall, R. S. 271
Baker, A. 41, 353
Barker, A. 143–53, 200, 201
Barrow, I. 20
Becker, O. 63
Berggren, J. L. 105, 369
Bergman, G. 361
Berstein, L. 363
Blanchard, A. 231–2, 255, 265
Bluck, R. S. 29
Boethius 149
Bombelli, R. 310
Bonitz, H. 156
Bos, H. J. M. 339

Bowen, A. C. 126, 143, 145, 283
Brandwood, L. 154
Brashear, W. 224, 236, 271
Brezinski, C. 309, 337
Brillhart, J. 353
Brouncker, W. 66, 99, 335, 337, 350, 360
Brown, M. 29, 64, 119
Bruins, E. M. 63
Brun, V. 361–2
Bulmer–Thomas (=Thomas), I. 126, 203, 241, and *passim*
Burkert, W. 104, 111, 143, 149, 251, 298, 300
Buteo, J. 328
Butler, A. J. 215–16

Callippus 52–3, 65, 126–7, 361
Cameron, A. 197–8
Cantor, G. 9, 293
Cassels, J. W. S. 105
Cataldi, P. A. 310
Cauderlier, P. 271
Cavallo, G. 208
Černý, J. 205
Chance, A. B. 237
Cherniss, H. 106–7, 197–9, 202
Christoffel, E. B. 129, 345
Chrystal, G. 310, 352, 355
Chuquet, N. 44, 328
Churchouse, R. F. 120, 354
Clagett, M. 66, 243, 246
Cockle, W. E. H. 205, 206, 208, 254, 271
Coldstream, J. N. 204, 223
Commandinus 246
Coner, A. 243, 246
Connelly, R. 29
Cotes, R. 318–19, 337
Coulton, J. 223
Coxeter, H. S. M. 322
Crawford, D. J. 230

Index of names

Crawford, D. S. 236
Curchhin, L. 105

Darius 52
Davenport, H. 310, 351, 353
Davey, H. 208
Dedekind, R. 9, 24, 29, 370
De Falco, V. 209
Delone, B. N. 353
Demetrius Lacon 209
Democritus 126, 157, 168, 194, 294–5, 303
Derveni 206
Descartes, R. 339
Dicks, D. R. 65
Dickson, L. E. 310, 328, 329
Diocles 290
Diogenes Laertius 204, 294
Diophantus 9, 57, 226, 231–2, 264, 265, 310, 369
Dorandi, T. 208, 209
Dutka, J. 342

Edgar, C. C. 207
Edmonds, A. 213
Edwards, H. M. 66
Elephantine Island 207–8
Eliot, G. (=Evans, M. A.) 281
Empedocles 194
Epicharmus 157
Epicurus 101, 209
Erathostenes 51, 65, 222
Erichsen, W. 259
Euclid 7, 25, 202–4, and *passim*
Euctemon 126
Eudemus 7, 106, 282, 283, 287, 297–8, 299, 301–2
Eudoxus 16–17, 21, 25, 28, 29, 106, 118, 122–30, 133, 207, 229, 392–3, 299, 361, 371
Euler, L. 42, 48, 59, 65, 99, 309–10, 313–38, 341, 346, 348–51, 353, 354–5, 362, 367
Eutecmon 52
Eutocius 7, 240–6, 281, 290

Faddeev, D. K. 353
Fayûm 212
Ferguson, H. R. P. 341, 361–4
Fermat, P. de 57–9, 64, 66, 99, 335, 350, 368
Fibonacci 310, 319
Fischler (=Herz–Fischler), R. 105
Fletcher, T. J. 320
Forcade, R. 361–4
Fourier, J. 341

Fowler, H. N. 109
Fraser, P. M. 203, 215
Frenicle de Bessy 66
von Fritz, K. 63
Furlani, G. 215

Galois, E. 99, 332–3, 339
Gardiner, A. 259
Gauss, C. F. 306, 309–10, 336, 341–54, 362, 366, 368, 371
Gelfond, A. O. 41, 64
Geminus 110, 127
Gerstinger, H. 254, 265
Gigante, M. 208
Glucker, J. 197–200
Goldstein, B. R. 52, 65, 126, 271, 283
Goody, J. 30
Gosper, R. W. 356–60
Gowing, R. 318
Grenfell, B. P. 209–13, 229
Guarducci, M. 222
Guthrie, W. K. C. 3, 15, 29, 294

Hall, F. W. 203
Hall, M. 356
Hardy, G. H. 65, 310, 329
Harrauer, H. 254
Harvey, F. D. 202
Harvey, P. D. A. 287
Havelock, E. A. 30
Heath, T. L. *passim*
Heiberg, J. L. 209–14, 219, 242–6
Heine, E. 343, 363
Heller, S. 63
Henrici, P. 338
Heraclitus 108, 194
Hermite, C. 363
Herodotus 283–6
Heron 9, 37, 64, 110, 219, 224, 226–7, 231–2, 245, 254, 257–8, 265, 285, 299, 369
Herz–Fischler (=Fischler), R. 105
Hilbert, D. 64, 309, 363
Hipparchus 129, 222, 369
Hippasus 300
Hippias 157
Hippocrates of Chios 7, 118, 287–8
Hippocrates of Cos (physician) 157
Hogarth, D. G. 212
Hogendijk, J. 289
Horsfall, N. 105
Huffman, C. A. 145
Hunt, A. S. 207, 209–13, 229
Hurwitz, A. 354–5
Huygens, C. 318, 364
Hypatia 198
Hypsicles 87, 218, 222, 369

Index of names

Iamblichus 8, 300, 304
Itard, J. 169, 203

Jacobi, C. G. J. 345, 363
Johnston, A. 222
Jones, A(lexander) 220
Jones, A(ndrew) 287
Jones, W. B. 338
Julian the Apostate 200–2
Junge, G. 181, 191
Justinian 197–8

Kapsomenos, D. G. 206
Keaney, J. J. 223
Khinchin, A. Ya. 115, 347, 356
Klein, F. 328–9
Klein, J. 29, 30, 110–12
Knorr, W. R. 30, 33, 63, 64, 66, 73, 149, 168–70, 173, 181, 183, 188, 203, 236, 265, 289, 291, 294, 298, 300, 304, 306, 352, 371
Knuth, D. E. 29, 340, 347, 356
Koenen, L. 205, 214, 221, 271
Kroll, W. 102
Kuz'min, R. O. 347

Lagrange, J. L. 99, 310, 318, 319–20, 329, 332–6, 339–41, 349–51, 353, 354, 362–3, 368
Lambert, J. H. 41, 337, 341, 343
Landau, E. 223
Lang, M. 222
Lang, S. 353
Laplace, P. S. 346–7
Legendre, A.-M. 99, 341, 342, 348, 351, 352
Lehmer, D. H. 317
Lemerle, P. 216
Leo the Philosopher 218, 242
Leodamus 106
Leucippus 194
Lévy, P. 347
Lewis, N. 205
Lindemann, C. L. F. 41
Loria, G. 168
Lucas, A. 205
Lynch, J. P. 197–200
Lysis 168, 194, 299–300

McNamee, K. 222, 231, 265
Mau, J. 208
Meray, C. 9
Meton 52, 65, 126
Meyerhoff, M. 215
Milne, H. J. M. 224
Minkowski, H. 309, 363

Morrow, G. R. 297–8
Mountford, J. F. 143
Mueller, I. 29, 105, 114, 141, 169, 170, 188, 191, 367
Muir, S. T. E. 120, 354
Muir, T. 76, 314
Müller, W. 208
Mudoch, J. 203

Nelson, H. L. 225
Neuenschwander, E. 29
Neugebauer, O. 65, 105, 126–30, 217, 223, 229, 239, 259, 267, 268
Neumann, J. von 120, 353
Newton, I. 319, 340
Nicomachus 33, 299, 364
Nicomedes 290
Niebel, E. 290

Odom, G. 105
Ohm, M. 105
Omar Khayyam 1, 281, 364
Ortega, J. de 328
Oxyrhynchus 209, 214
Ozanam, J. 350

Pack, R. 204, 210
Pappus 44, 64, 87, 181, 191, 218, 220, 225, 268, 281–2, 289–90, 298, 300–2
Paràssoglou, G. M. 206
Parker, R. A. 126–8, 259–63
Parmenides 194
Paterson, M. 356
Pattie, T. S. 249, 271
Pausanias 199–200
Pell, J. 59, 350
Perron, O. 310, 337, 338, 363
Peukestas 207
Peyrard, F. 217, 219
Pfeiffer, R. 204
Phillip of Opus 106, 204
Philodemus 106
Philolaus 194
Philoponus 201, 215, 300
Pintandi, R. 271
Pisot, C. 362
Plato 8, 23, 28, 30, 106–8, 197–202, 284, 299 364, and *passim*
Plooij, E. B. 281
Plutarch 198, 299
Poincaré, H. 328, 366
Popper, K. 163
Porson, R. 205–6
Préaux, C. 230
Price, D. de Solla 223, 364

Index of names

Proclus 59, 63, 96, 100–4, 106, 110, 198, 218, 234, 283–5, 295, 297–8, 302, 367
Protagoras 194
Ptolemy (astronomer) 65, 87, 128–9, 217–18, 222, 369
Ptolemy (king) 127
Pythagoras and Pythagoreans 7–8, 100–1, 130–1, 157, 294, 297–8, 301–4, 307

Rawlins, D. 65
Rayney, G. N. 356
Reid, C. 64
Reynolds, L. D. 214, 216
Riddell, R. 126, 130
Riemann, B. 64, 343, 367
Riginos, A. S. 200
Robbins, F. E. 214
Roberts, C. H. 206
Rosen, D. 340
Ross, W. D. 296
Roth, K. F. 353
Rowland, R. J. 285

Sachs, E. 298
Sadie, S. 143
Saffrey, H. D. 200
de Ste. Croix, G. E. M. 202, 223, 233, 238
Sandys, J. E. 205
Santillana, G. de 371
Saqqara 205, 207, 222
Schlesinger, J. 345
Schwarz, H. A. 343
Schwenter, D. 317
Sedley, D. 198
Series, C. 129
Serret, J.-A. 341, 355
Shallit, J. 340
Sharples, R. W. 223
Shorey, P. 108
Siegel, C. L. 64
Sijpesteijn, P. J. 236, 271
Simplicius 7, 198, 287–8
Skeat, T. C. 206, 224
Skelton, R. A. 287
Skemp, J. B. 109
slaveboy *passim*
  and the accountant 268–70
  and Archytas 131–8
  and Eudoxus 122–6
  and Socrates 3–7, 25–8, 47, 114
Smith, H. J. S. 129, 328–9
Smyly, J. G. 223, 226, 238
Socrates 3, 10, 25, 67, 132, 138, 168, 284, and *passim*
Solmsen, F. 198
Speusippus 197

Stark, H. M. 320, 354
Steele, A. D. 290
Steele, D. A. 64
Stephanus 218
Stevin, S. 168
Sylvester, J. J. 314, 355
Synesius 198

Taisbak, C. 65, 114, 158, 168–70, 175, 180, 189
Tait, J. G. 230
Tait, W. G. 259
Tannery, P. 105, 232, 265
Teissier, B. 352
Theaetetus 21, 25, 28, 106, 118, 292, 295, 301
Theodorus 157, 292, 295
Theon of Alexandria 71, 217–19, 268, 300
Theon of Smyrna 58, 63, 66, 96, 100–3, 198, 226, 295, 299
Thesleff, H. 299
Thucydides 285
Theuth 23, 284
Thomae, L. W. 343
Thomas (=Bulmer–Thomas), I. 126, 203, 241, and *passim*
Thomson, W. 181, 191
Thrasyllus 294
Thron, W. J. 338
Tod, M. N. 222, 223
Toomer, G. J. 65, 204
Trotter, H. 353
Tsantsanoglou, K. 206
Tuckerman, B. 120, 353
Turner, E. G. 30, 63, 127, 204–14, 222, 229, 233, 255, 271
Tzetzes 201

Varberg, D. E. 322
Vincent, A. J. H. 341
Vogel, K. 221, 254, 265
Vogt, H. 298
van der Waerden, B. L. 63, 114, 126, 138, 168–70

Wall, H. S. 338, 342
Wallis, J. 66, 99, 246, 317, 318, 335, 350, 368
Wasserstein, A. 298
Watson, J. S. 205
Weierstrass, K. 9
Weil, A. 42, 310, 318, 336, 337, 350, 351
Weitzmann, K. 255
Westfall, R. S. 319
Whiteside, D. T. 310, 340

Whittaker, E. T. 355
Whycherly, R. E. 200
William of Moerbeke 66
Wilson, N. G. 214, 216, 218–19, 221, 243
Winnington-Ingram, R. P. 143
Winter, J. G. 214
Wirsing, E. 347–8
Wolfskill, J. 353
Wright, E. M. 65, 310, 329

Xenocrates 197

Youtie, H. C. 195
Youtie, L. C. 205, 214, 221

Zeeman, E. C. 95, 129
Zeuthen, H. G. 63, 105

GENERAL INDEX

abbreviations 231–2, 255, 265–6
Academy 197–202
  and Aristotle 202, 292
Alexandrian Library 214–16
algebra 9, 24, 68, 93, 171, 181
algorism v. algorithm 65; *see also* Euclidean algorithm
alogos 115, 158–94, 161, 167–8, 191, 192–4, 292, 294–302, 364; *see also logos*
analogon 16, 297–8, 299
antanairesis 32, 63, 364
antaphairesis 33, 364
anthuphairesis 31, 63
anthyphairesis, *see* ratio (anthyphairetic), Euclidean algorithm
Antikythera mechanism 223, 364
apotome, *see* incommensurables
application of areas 177–80
approximation 36–41, 45–57, 59–62, 310–29, 360–2
Arabic astronomy and mathematics 9, 129, 281
area 9, 11
Aristotle's inscription 200
arithmeō 118–19
arithmetic 8–9, 114, 134, 138–9, 268, 354–60
arithmētikē 14–16, 22–3, 29, 108–17, 134–7, 369
  v. arithmetic 114, 134–6
  practical v. theoretical 110
arithmetised mathematics 8–10, 24, 33, 43, 54, 191–2, 195, 226, 246, 248, 285, 304, 309–11, 320
arithmos 14–16, 134, 234–5
arrhētos, *see rhētos*
artaba 233, 248–9
asummetros 294–308
astronomy
  Academic 121–30
  Egyptian and early Greek 126–30
  later 129
  see also calendars, parapegma
auxē 118–19

Babylonian mathematics 9, 195, 222, 369
binomial, *see* incommensurables
Brun's algorithm 361–2

calendars
  cyclical 52–3, 65, 126–30
  see also parapegma
circumference and diameter 9, 35–41, 54–7, 240–6
codex 206
compounding, *see* ratio
congruence 11, 29
construction of equations 339
continued fractions 9, 31, 45, 63, 65, 281–2, 309–71
  arithmetic 115–16, 354–60
  convergents 48–51, 312
  general 311
  higher dimensional 360–4
  probability theory 120, 346–8
  simple 311
  see also ratio (anthyphairetic), Euclidean algorithm
convergents 44, 48–51, 312

deductive mathematics 121, 281, 367–8
Demotic papyri 259–63
dia tōn arithmōn v. *grammōn* 100–5
diagonal and side, *see* side and diagonal
diagōnios 64, 294
diagramma 33
dialogues, *see* slaveboy
diametros 63–4, 294–302
diaphoron schoinismou 233
diastēma 132, 147–8
dimension of cubes 28, 118–21, 352–4
dimension of squares 28, 35, 67–105, 118–19
diorismos 180, 288–9
division 16, 234–8, 253–4, 267–8, 271–6
duplication of cube 118, 290, 299

Egyptian mathematics 9, 259–63, 266–7, 283–6
elegant mathematics 101–4
epimorion, *see* ratio (epimoric)
equivalence relations 20, 114
errors 151, 229, 233, 243–6
Euclidean algorithm 31, 65, 311–12, 361–4
Euler brackets 314, 349

400 General index

even and odd 32, 116, 292, 297, 304–6
evidence 7–8, 195–308, 364–71
exhaustion 28
expressibles, *see* ratio

factorisation 306, 351–2
Farey series 31, 44, 329
financial accounts 109–10, 114–15
foundations crisis 17, 294, 302–4
fractions 112–13, 221–82
   and arithmetic 8–9, 29, 114, 138
   common 20, 195, 226–7
   decimal 29
   fractional quantities 8–9, 226
   parts, or unit fractions 41, 226–38, 268
   see also arithmetic, arithmetised mathematics, ratio

Galois theory 339
Gauss brackets, *see* Euler brackets
geometrical algebra 105, 190
geōmetria 136, 283–6
geometry
   non-arithmetised 10–14
   plane and solid 117–21
gnomon 68, 73, 76–86
golden section 105; *see also* ratio (extreme and mean)
greater and less 52, 113, 116, 134–5

Hurwitz numbers 355
hypergeometric series 342–4

incommensurables 17, 41, and *passim*
   apotome 162, 182
   bimedial 185
   binomial 115, 182
   discovery of 112, 294–308
   major/minor 161, 185
   medial 173
   rational/irrational 167–8
   rhētos and *alogos* 167–8
inscriptions 205, 222
isos 10–11, 13, 29

Jacobi–Perron algorithm 362

land measurement 230–4, 283–7
lattice points 321, 361
less and greater, *see* greater and less
logismos, *see logistikē*

logistikē 2, 22–3, 108–17, 154–7, 364
   practical v. theoretical 110
logos 16, 20, 58, 108–9, 147–8; *see also* ratio

means 65, 132
meros 15, 17, 226–8, 268
Metonic cycle 52–3
metrētikē 109–10, 295–6
minuscule script 205, 216–20
morion, *see meros*
music theory 143–53
myriads 225, 266

neusis-constructions 287–93
non-Euclidean geometry 12, 136–8, 153–4, 284
numbers 8–10
   Greek 14–16, 221–6, 234–5, 238
   irrational 41–2, 64, 167
   large 163, 225–6
   place-value 238
   real 9, 29, 31, 309
   repetition 15, 29, 31, 112
   sexagesimal 195, 222, 267, 268

obliquity of ecliptic 51–2, 65
odd and even, *see* even and odd
ostraca 204

Padé approximant 345
palimpsest 214, 243–4
palindromes 75–6, 105
paper, introduction of 216
papyrus 205–6, 219
   referencing system 63, 204
parapegma 126–30, 207, 229–30
parchment 216
Parmenides proposition and algorithm 42–8, 64, 320–9
parts, *see* fractions, *meros*
Pell's equation 57–60, 66, 119, 334–6, 351, 366, 368
periodic phenomena 34–5, 59, 128–9, 146, 308, 329–36, 341, 362–4, 368
pi 41, 268, 317, 337–8, 360; *see also* circumference and diameter
Pick's theorem 321–2
Plato
   aristocratic atitude 28, 202
   and astronomy 121–2
   curriculum in *Laws* VII 107, 295, 304
   curriculum in *Republic* VII 107–54
   and mathematics 106–8
   and music theory 130–1

General index

Plato's inscription 197–202
probability theory 346–8
problems 121, 131, 281, 290, 295
proportion 16–20, 31, 202
   Eudoxan 16–17
   see also ratio
Ptolemaic astronomy 9, 129, 217–18, 222
Pythagoras' theorem 21, 72–3, 104
Pythagorean science 7–8, 100–1, 130–1, 294, 297–8, 300, 301–4, 307

quadratic forms 351, 354
quadratic problems 191–2, 293, 329–36

ratio 16, 20, and *passim*
   accountant's 28, 268–9
   anthyphairetic 25–7, 31–66, 191–2, 364–71
      and *passim*
   astronomical 28, 122–30
   commensurable 41, 124–5, 132, 369–70
   comparison of 31–2, 47
   compounding 21, 27–8, 114–6, 131–4, 138–43, 153
   in the *Elements* 16–21, 141, 191–2, 364
   epimoric 132–3, 146–8, 251, 364
   expressible 167–8
   extreme and mean 86–95, 105, 161–2
   and fractions 9, 112–13, 138–43, 195, 240–2, 247–8
   medial 173
   musical 28, 132–4, 152–3
   musical anthyphairetic 135
   musical astronomical 135
   noem 87–95
   v. proportion 16–21, 63
   synthesis 86–95, 281
   theory, see *logistikē*
   in the *Topics* 17–19, 32, 364
   see also alogos, analogos, logos, proportion, rhētos
rational/irrational 167–8, 369–70; *see also* numbers
rhētos 161, 167–8, 191, 192–4, 294, 299, 364
rule-and-compass constructions 287, 292–3

side and diagonal 33–5, 58–9, 305–6
   generalised 95–105
   numbers v. lines 58–9, 96, 100–2
slaveboy
   and the accountant (A_{97}-B_{115}) 268–70
   and Archytas (A_{70}-A_{96}) 131–8
   and Eudoxus (B_{44}-E_{69}) 122–6
   in the *Meno* 3–7
   and Socrates (S_1-S_{43}) 25–8, 47, 114
suntithēmi 142–3; *see also* ratio (compounding)
synodic month 53, 126
synthesis 86–95, 143

tables 234–40, 270–9
tōn m to ń 235–6, 260, 264–6
Topics proposition 18–19, 30, 141, 172
trisection 289

writing 23, 30, 202–20

Zenon archive 32, 63